Nanofabrication

Nanolithography techniques and their applications

Nanofabrication

Nanolithography techniques and their applications

Edited by
José María De Teresa
Instituto de Nanociencia y Materiales de Aragón (INMA), CSIC-Universidad de Zaragoza, Zaragoza 50009, Spain

IOP Publishing, Bristol, UK

Multimedia content is available for this book from http://iopscience.iop.org/book/978-0-7503-2608-7.

ISBN 978-0-7503-2608-7 (ebook)
ISBN 978-0-7503-2606-3 (print)
ISBN 978-0-7503-2609-4 (myPrint)
ISBN 978-0-7503-2607-0 (mobi)

DOI 10.1088/978-0-7503-2608-7

Version: 20201201

IOP ebooks

British Library Cataloguing-in-Publication Data: A catalogue record for this book is available from the British Library.

Published by IOP Publishing, wholly owned by The Institute of Physics, London

IOP Publishing, Temple Circus, Temple Way, Bristol, BS1 6HG, UK

US Office: IOP Publishing, Inc., 190 North Independence Mall West, Suite 601, Philadelphia, PA 19106, USA

This book is dedicated to so many school teachers, university professors, colleagues, collaborators, students, friends and family members who have taught me not only the secrets of science, and in particular of nanolithography, but also the secrets of a full life. And especially to those who like spending their time around me: my wife Anabel, my children Sofía and Isaac, my parents Cándida and José María and my sister Helena, given that the long hours devoted to this book have been mostly taken from our time for common recreation.

Contents

Preface

Nowadays, researchers are frequently requested to contribute to journals and books, and, unfortunately, we have to decline most of these invitations due to the lack of time to take additional tasks. But, in this case, when IOP approached me to make a proposal on a book about nanolithography, I realized that this time it could be different. It should be different. For more than 15 years, I have been teaching nanotechnology and nanolithography in postgraduate courses and summer schools. And I have always been faced by the same question: can I select a single book on nanolithography to allow the students to go deeper in the topic? A Master or PhD student following such courses has generally many tasks to do and cannot afford to spend much time on bibliography. However, the situation is such that most books on nanolithography go very deep into some particular technique but do not show the global picture. This can be very frustrating for the student, who does not wish to know all the details of a particular technique but understand the fundamentals of the main nanolithography techniques and know their main applications. Some review articles have addressed this global picture, but these are normally limited to a few tens of pages at most, thus providing the flavour of the topic but without providing a solid view of it. With this background in mind, I proposed to IOP that I should edit such a *missing book*.

The book covers in great detail the following nanolithography techniques: optical lithography (chapter 2), electron beam lithography (chapter 3), focused electron beam induced deposition (chapter 4), focused ion beam and focused ion beam induced deposition (chapter 5), scanning probe lithography (chapter 6), stamp lithography (chapter 7), stencil lithography (chapter 8) and ice lithography (chapter 9). From the beginning, I proposed that all the co-authors should address in their chapters not only the techniques but also the main applications. In my opinion, it is not pretentious to state that nanolithography currently infiltrates most of the existing technology by enabling the digital technology. Hiding behind the scenes, nanolithography allows building the integrated circuits that rule (overrule?) our daily life. Apart from this classical application of nanolithography, the reader will discover other applications in a material's nanopatterning, device prototyping, nano-optics and plasmonics, nano-superconductivity, nanomagnetism, electrical nanocontacting, nanosensors, energy storage and conversion, as well as in new exciting fields like biology, medicine, robotics and neuroscience. The last three chapters of the book are in fact devoted to two specific applications, which are thoroughly discussed: magnetic nanostructures (chapters 10 and 11) and two-dimensional materials (chapter 12). Chapter 3 also addresses in great detail the application of electron beam lithography to two-dimensional materials.

In summary, I hope that the book will provide the reader with the key fundaments of the main nanolithography techniques and their applications, and, at the same time, it will open the reader's mind towards recent developments and emergent applications in the field. Last, but not least, each chapter provides a student's corner, with some questions and exercises as well as further hints to go deeper in the topics. In an appendix at the end of the book, the reader will find the responses to the questions and exercises.

Acknowledgements

First, I want to acknowledge the enthusiastic collaboration of all the book's co-authors, who made this project happen and helped to put together an immense amount of knowledge in the field of nanolithography with the hope that the book will be useful to students and colleagues. By choosing contributors who lead their respective research fields, each chapter is in its own right a precious piece of work. Their collection is intended to constitute the reference book on nanolithography for the upcoming generations of Master and PhD students as well as for researchers seeking a comprehensive view of the field. Thank you, my dear colleagues and friends for your involvement!

Secondly, I want to express my gratitude to the colleagues from the Spanish network on Nanolithography (NANOLITO) for sharing their knowledge on the topic for the last 12 years in the course of summer schools and workshops.

Finally, I would like to highlight the managers, colleagues, technicians and students working at the Laboratory of Advanced Microscopies (University of Zaragoza, Spain). Thanks to their hard work, the lab is currently a reference place for nanolithography and also this book will see the light.

Editor biography

José Maria De Teresa

Physicist José María De Teresa is research professor at the Institute of Nanoscience and Materials of Aragon (INMA, CSIC-University of Zaragoza, Spain). After obtaining a PhD degree from University of Zaragoza in 1997 under supervision of Professor del Moral and Professor Ibarra, he carried out postdoctoral stays at IFW (Dresden, Germany) and CNRS (Paris, France), where he worked under the guidance of Professor Albert Fert, Nobel prize winner in Physics in 2007. During this postdoctoral period, he learnt to use lithography techniques, which he later implemented in his current institute, where he leads the group of Nanofabrication and Advanced Microscopies (NANOMIDAS). He coordinates the Spanish network on Nanolithography (NANOLITO) and the FIB-SEM area in the Spanish National facility for Advanced Microscopies. From 1st January 2021, he will chair the Condensed Matter Division (CMD) Board in the European Physical Society (EPS). His main research interests are nanofabrication with focused electron and ion beams and its application to nanoelectronics, magnetic materials, nano-superconductors and new materials. He has published more than 200 research articles and given about 100 invited conferences.

List of contributors

Edoardo Albisetti
Dipartimento di Fisica, Politecnico di Milano, Milan, Italy

Mario Amado
University of Salamanca, Salamanca, Spain

Riccardo Bertacco
Dipartimento di Fisica, Politecnico di Milano, Milan, Italy

Vito Clericò
University of Salamanca, Salamanca, Spain

Rosa Córdoba
Institute of Molecular Science (ICMol), University of Valencia, Paterna, Spain

José María De Teresa
Instituto de Nanociencia y Materiales de Aragón (INMA), CSIC-Universidad de Zaragoza, Zaragoza, Spain

Enrique Diez
University of Salamanca, Salamanca, Spain

Claire Donnelly
Cavendish Laboratory, University of Cambridge, Cambridge, UK

Andrei Fedorov
Georgia Institute of Technology, Atlanta, GA, USA

Amalio Fernández-Pacheco
University of Glasgow, Glasgow, UK

Ricardo Garcia
Instituto de Ciencia de Materiales de Madrid, CSIC, Madrid, Spain

Anpan Han
Department of Mechanical Engineering, Section of Manufacturing Engineering, Technical University of Denmark, Lyngby, Denmark

Jaime Hernández
Madrid Institute for Advanced Studies in Nanoscience (IMDEA Nanoscience), Madrid, Spain

Songkil Kim
Pusan National University, Busan, South Korea

César Magén
Instituto de Nanociencia y Materiales de Aragón (INMA), CSIC-Universidad de Zaragoza, Zaragoza, Spain

José Ignacio Martín
University of Oviedo and Nanomaterials & Nanotechnology Research Center (CINN), Asturias, Spain

Pablo Orús
Instituto de Nanociencia y Materiales de Aragón (INMA), CSIC-Universidad de Zaragoza, Zaragoza, Spain

Javier Pablo-Navarro
Institute of Ion Beam Physics and Materials Research, Helmholtz-Zentrum Dresden-Rossendorf, Dresden, Germany

Lucas Pérez
Universidad Complutense de Madrid and Madrid Institute for Advanced Studies in Nanoscience (IMDEA Nanoscience), Madrid, Spain

Francesc Perez-Murano
Institute of Microelectronics of Barcelona (IMB-CNM, CSIC), Bellaterra, Spain

Daniela Petti
Dipartimento di Fisica, Politecnico di Milano, Milan, Italy

Min Qiu
Westlake University, Hangzhou, Zhejiang Province, People's Republic of China

Elisa Riedo
Tandon School of Engineering, New York University, New York, USA

Isabel Rodríguez
Madrid Institute for Advanced Studies in Nanoscience (IMDEA Nanoscience), Madrid, Spain

Soraya Sangiao
Instituto de Nanociencia y Materiales de Aragón (INMA), CSIC-Universidad de Zaragoza, Zaragoza, Spain

Dédalo Sanz-Hernández
Unité Mixte de Physique, CNRS, Thales, Université Paris-Saclay, Palaiseau, France

Oscar Vázquez Mena
UCSD—University of California San Diego, La Jolla, CA, USA

Guillermo Villanueva
EPFL—Ecole Polytechnique Fédérale de Lausanne, Lausanne, Switzerland

Ding Zhao
Westlake University, Hangzhou, Zhejiang Province, People's Republic of China

Chapter 1

Introduction to nanolithography techniques and their applications

José María De Teresa

Nanolithography encompasses the set of top-down fabrication techniques that allow patterning materials and building devices with nanoscale resolution. Nanolithography is supplemented by thin-film-deposition techniques as well as self-assembly and self-organization to provide an immense variety of nanofabrication strategies. In this chapter, the key concepts underlying nanolithography will be described and the main nanolithography techniques will be introduced and compared in terms of the achievable resolution, throughput, pattern complexity, process complexity and cost. The main applications of nanolithography will be subsequently reviewed, with an emphasis on the overwhelming role of nanolithography in the semiconductor industry, in nanotechnology and in emergent applications.

1.1 Key concepts in nanolithography

1.1.1 Top-down versus self-assembly and self-organization processes

Nature has generally opted for self-assembly and self-organization processes to build complex structures, with the fabrication of proteins from amino acids and peptide bonds as one illustrative example. This is generally known as the **bottom-up approach**. On the other hand, when faced with the problem of building artificial structures, our species (and other species not far from us in the evolutionary chain) relies either on combining building blocks that are smaller than the targeted structure or on extracting it from a larger structure, which is known as the **top-down approach**. One example of the former case is the use of bricks to build a house and one example of the latter case is a Greek sculpture modeled from a marble block. The bottom-up versus top-down dichotomy also comes up when processes need to be designed in order to produce small-sized materials and build small devices.

doi:10.1088/978-0-7503-2608-7ch1

In figure 1.1 an example of the usefulness of the self-assembly and self-organization bottom-up strategies to synthesize Ag nanowires is shown. First, two different polymers (PMMA and PS) are mixed in solution and spin coated on a substrate. Due to specific interactions amongst these polymers and the substrate (which will be reviewed in detail in chapter 2), in certain conditions the two types of polymers tend to form the intertwined patterned displayed in figure 1.1(a). As can be identified in the zoom-in image of that pattern, a local order between both polymers exists, forming nanowire-like domains with periodicity in the range of 50 nm. This pattern becomes very useful if, subsequently, Ag is evaporated at room temperature, which, after landing on the PMMA-PS layer, tends to cluster on top of the PS polymer due to the lower diffusion mobility of Ag on PS compared to PMMA. Thus, Ag follows the underlying nanowire-like structure of the polymer scaffold that extends undisrupted along various microns. However, the formed Ag nanowires present some unwanted features, as a closer look at figure 1.1(b) reveals. A significant nanowire roughness is observed, as well as nanowire interruption at certain points and deviation from straight shape, all of which are commonly sources of deterioration of the physical properties of the nanowires. In addition, the long-range order of the array of nanowires fabricated following this strategy is not

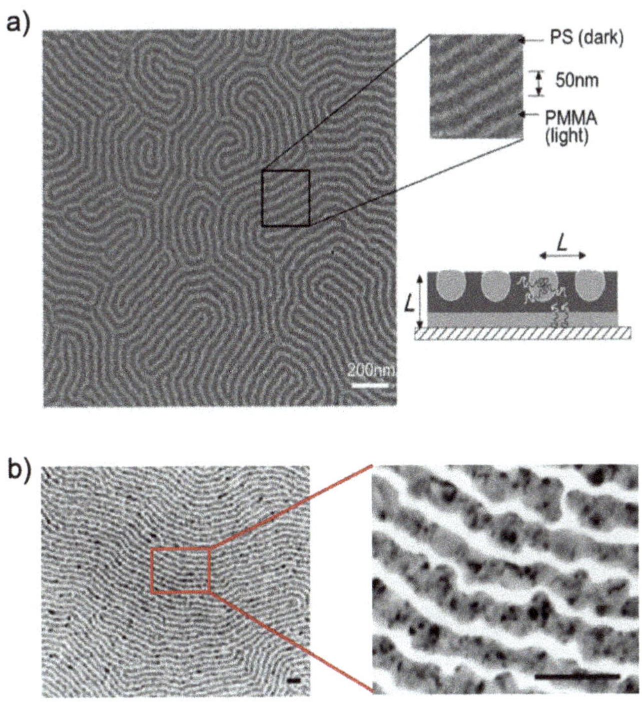

Figure 1.1. (a) Spin-coated film of PMMA-PS block copolymers, with 50 nm periodicity. (b) Scanning electron microscopy micrograph after evaporating Ag on the film shown in (a). Due to diffusion effects, the Ag atoms tend to cluster on top of the PS polymer, forming self-assembled Ag nanowires. Reprinted and adapted by permission from Springer [1]. Copyright (2001).

achieved, which is necessary in certain applications such as those in the semiconductor industry.

Top-down approaches, which represent most of the content of the present book, aim to develop strategies and processes leading to the fabrication of artificial structures with fine control of all dimensions (length, width, height) and other features such as roughness, edge shape, materials inter-diffusion, etc. One example of the accuracy achievable with currently existing top-down lithography techniques is shown in figure 1.2. In that example, the technique called focused electron beam induced deposition (FEBID) is used to decompose the $Co_2(CO)_8$ precursor material by means of a scanning electron microscope (SEM), forming arrays of cobalt nanowires with fine control of their periodicity (down to 30 nm). As can be observed in figure 1.2, due to the underlying physical mechanisms governing the growth in FEBID (to be explained in chapter 4), neighbour cobalt nanowires are not isolated one from another. However, a post-growth Ar^+ milling process allows the nanowire isolation, as shown in the corresponding publication [2].

In general, top-down approaches make use of the synergic use of thin-film deposition techniques (sputtering, molecular beam epitaxy, pulsed laser deposition,

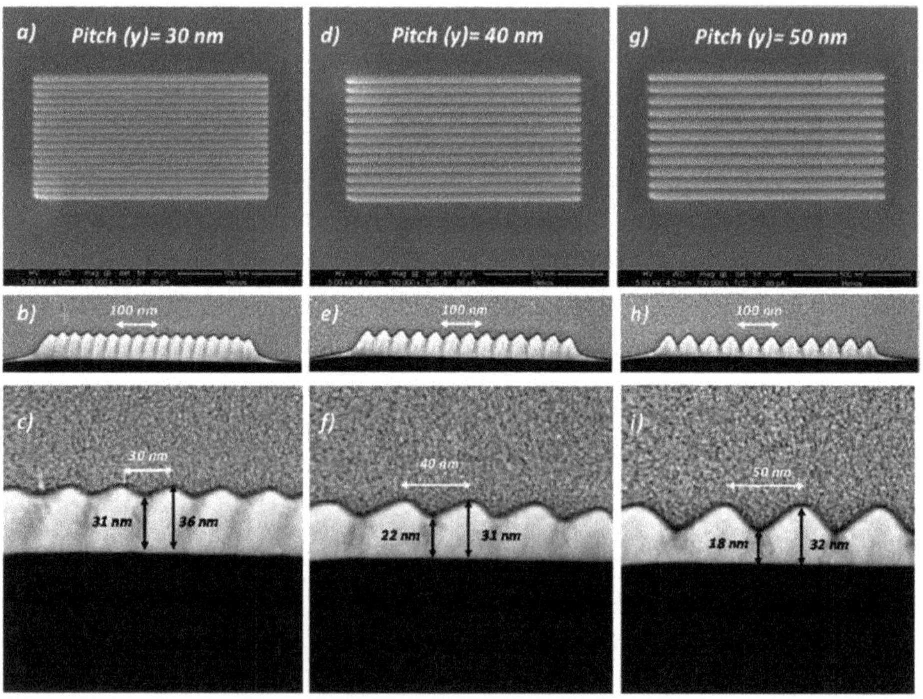

Figure 1.2. Arrays of cobalt nanowires grown by FEBID with fine control of the nanowire inter-distance: 30, 40, and 50 nm. The top-down SEM micrographs of the arrays are shown in (a), (d), and (g), whereas cross-sectional transmission electron microscopy (TEM) images of the arrays are shown in low magnification in (b), (e), and (h) and in high magnification in (c), (f), and (i). Reprinted with permission from [2]. Copyright (2014) American Chemistry Society.

chemical vapour deposition, thermal evaporation, electrodeposition, etc) and nanolithography techniques. Two approaches are commonly followed. In the first approach, a thin film (or multi-layered film) is grown on a substrate, with accurate control of the layer's thickness. Depending on the homogeneity of the growth technique and the growth equipment used, the area of the substrate can reach large values in the range of tens of cm^2 or beyond. In a second step, a lithography technique is applied to pattern the film to micrometric or nanometric lateral dimensions, removing material on particular locations (to be described in chapters 2 and 3). In the second approach, the nanolithography step is first performed on the bare substrate, which is followed by the growth of the materials. This is the case of the so-called lift-off technique (to be described in chapter 3). A combination of both approaches in multi-step processes is commonly needed in order to build complex devices. Various examples of this will be described throughout the present book.

Moreover, smart combinations of bottom-up and top-down approaches have been developed. In the first example, an array of metal dots (Au, Cu, Ni) is created by means of top-down photolithography. These metallic dots act as catalyst centres for the growth of Si nanowires by means of vapour-liquid-solid (VLS) technique. Without the use of well-ordered patterned metal dots as catalysts (for instance using randomly-distributed metallic nanoparticles), the Si nanowires would grow without preferred direction or determined spatial periodicity. The well-ordered arrays of vertical Si nanowires shown in figure 1.3 are well suited for application in photovoltaics [3]. In the second example, in a first step, electron beam lithography (EBL) plus immersion in a chemical solution are used to define long NH_2-terminated channels. When a drop of a solution containing dispersed GaP nanowires is poured onto the substrate, they preferentially assemble on the NH_2-terminated surfaces, leading to a well-ordered array of GaP nanowires [4].

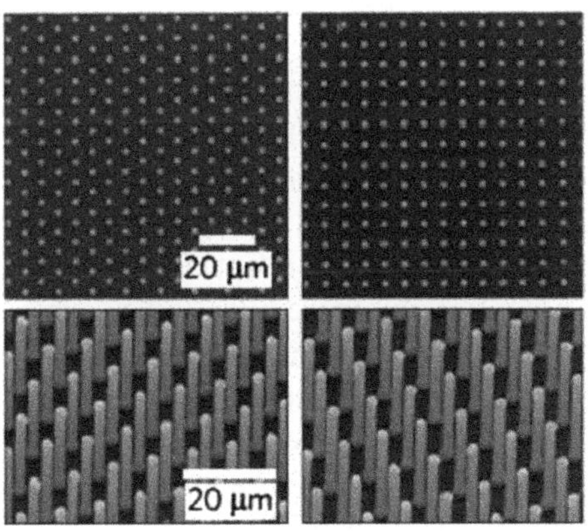

Figure 1.3. Triangular (left) and square (right) arrays of metal dots patterned by photolithography (top) and the corresponding arrays of Si nanowires after VLS growth (bottom). Reprinted and adapted by permission from Springer [3]. Copyright (2010).

1.1.2 Resolution, cost, throughput and complexity

One of the most critical parameters in nanolithography is the maximum attainable **resolution**. However, the resolution will change whether the targeted pattern is dense or the features are isolated. In general, any nanolithography technique will suffer from proximity effects and will require specific strategies in the case of dense patterns. Conventionally, the resolution of a nanolithography technique is defined as the minimum distance achievable between two individual isolated or resolved features. Nevertheless, the resolution is sometimes considered to be the minimum lateral dimension of an isolated feature fabricated with the corresponding nanolithography technique. We inform the reader that throughout this book both definitions will be indiscriminately used. Also, we would like to make the reader aware that the phrase *higher resolution* or *better resolution* is commonly used to express that *smaller features* can be fabricated. In the next chapters, the specific limitations of each technique to achieve a better resolution will be discussed in detail. However, in some cases the potential of a lithography technique cannot be fully deployed in the laboratory due to practical limitations. Obviously, one of these limitations can be the equipment or the process **cost**. For example, optical lithography is nowadays capable of the fabrication of structures in the range of 10 nm. However, the required equipment, a stepper, is too expensive to be afforded by a standard research laboratory. As a consequence, the use of conventional equipment for optical lithography in a university lab will typically lead to a resolution of 0.5 μm at best.

The concept of **throughput**, defined as the capacity of a technique for mass production, is borrowed from the industrial environment, where the *cost per unit* is crucial. If the nanolithography technique allows the production of a large amount of identical patterns in a short period of time, the technique's throughput will be high. The high throughput of a given process is beneficial for its industrial implementation. From the perspective of throughput, the ideal nanolithography technique should be able to achieve the simultaneous patterning of as many as features as possible. In the next chapters, the reader will get detailed examples of techniques such as optical lithography or nanoimprint lithography, which are well suited for the fabrication of many features in parallel. However, other techniques such as electron beam lithography or scanning probe lithography produce features in a sequential mode, which leads to limited throughput. Whereas throughput is not an important factor for the development of proof-of-concept devices and prototyping, it becomes very important for the potential commercialization of new technologies.

The **complexity** of a lithography technique is generally associated with the number and difficulty of the involved steps. In this book, we use this term very broadly, encompassing aspects such as the number of processing steps required, the degree of reproducibility of the process, the required human resources or user skills, etc. All these aspects need to be considered before choosing the nanolithography technique to use. It is also important to consider the capability of the technique to fabricate complex patterns, which vary from one to another. In particular, not all nanolithography techniques have the capability to create three-dimensional (3D) structures.

1.2 Nanolithography techniques using resists

1.2.1 Optical lithography (OL)

In conventional optical lithography (OL), a photoresist (a resist whose solubility in a developer will change after photon irradiation) is first patterned with the same design of an optical mask. This involves three steps, as shown in figure 1.4: (1) resist spin coating, (2) resist exposure and (3) resist development. The next two steps consist in (4) transferring the pattern from the resist to the sample of interest by means of an etching process of the unprotected part of the sample and (5) removing the resist, which leads to achieving the goal of a clean patterned sample. Instead of being grown before the lithography process starts, the sample of interest can be grown at an intermediate step (after the resist development), in a process called lift-off (see chapter 3 for details). Lift-off processes are in general advantageous given that the etching step is no longer required. However, lift-off cannot be performed if, for example, the material of interest needs to be grown at high temperatures that cause the resist to deteriorate.

OL will be described in detail in chapter 2. Here, we will limit the discussion to a few important points:

- Photoresists are of utmost importance, given that the pattern is first transferred to them and they should endure either the etching process or the sample growth and lift-off. Thanks to the availability of positive and negative

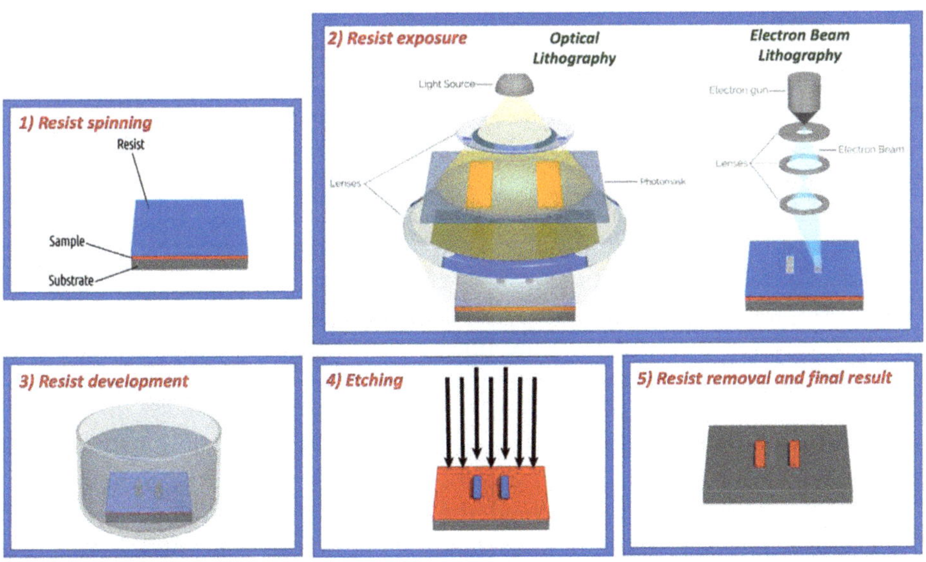

Figure 1.4. Various steps commonly performed in optical and electron beam lithography techniques: resist coating (1), resist exposure (2), resist development (3), sample etching (4) and resist removal with the final result (5).

resists, which, respectively, increase or decrease their solubility in the developer under photon exposure, the overall photolithography process can be designed to produce either the same pattern present in the mask or the supplementary one.

- The wavelength of the photon source will determine to a great extent the maximum resolution of the process. The technology has seen the evolution of light sources from NUV (near ultraviolet) to DUV (deep ultraviolet) to EUV (extreme ultraviolet). Typical resolution with laboratory mask aligners is in the range of 1 μm, whereas steppers target resolution in the 10 nm range.
- OL is well suited for large-area patterning given the availability of appropriate photoresists, light sources, lenses and masks to expose a large area of the resist in a very short period of time (typically a few seconds). In addition, a high throughput is achievable in OL with industrial equipment called a stepper. A stepper is very expensive and fine equipment with the capability to handle the exposure of hundreds of wafers per hour, as required in the industry of integrated circuits [5].
- Several variants of OL exist and the most important ones will be addressed in chapter 2. Such related techniques exploit for example the use of dynamical masks (instead of physical masks) [6], the properties of two-photon non-linear absorption [7], coherent light interference [8], etc.

1.2.2 Electron beam lithography (EBL)

Similarly to OL, in the case of EBL the process relies on the change of solubility of a resist after exposure. However, in EBL one needs to use an electron-sensitive resist instead of a photoresist. EBL is typically performed inside an SEM or in equipment designed specifically for this task. In both cases, a finely-focused electron beam is scanned sequentially along the sample to expose the resist. The rest of the steps are common to those already shown for OL (see figure 1.4).

EBL will be described in detail in chapter 3. Here, we will limit the discussion to a few important points:

- The EBL process is intrinsically slow in comparison with OL given that the electron beam needs to dwell a finite time on each exposure point. It is not rare to see exposure times of several hours for a single sample. Some developments for multi-electron-beam exposure are underway [9].
- In return for the long exposure time, a very high resolution (sub-100 nm) is achieved routinely by EBL, reaching the realm of 10 nm with some optimization and appropriate equipment [10].
- As occurs in OL, positive and negative resists are available in EBL as well as the capability for performing lift-off processes that save the etching step.
- EBL is a technique well suited for laboratory research and prototyping, considering that throughput is not an important issue and a very good resolution is obtained.

1.2.3 Nanoimprint lithography (NIL)

Although nanoimprint lithography (NIL) also uses a resist, resist exposure is generally not needed, as shown in figure 1.5. Instead, a mould with the targeted pattern is imprinted on the resist. By means of a high imprinting pressure and a high working temperature, the resist adopts the shape of the mould. After sample etching and resist removal, the sample is revealed with a pattern that is supplementary to that imprinted in the mould [11]. Some variants of NIL include microcontact printing, which relies on inks and self-assembly processes without the need of high pressure and high temperature [12], and step-and-flash imprint lithography, where the resist is hardened during the imprint process at room temperature by means of UV light exposure [13].

NIL and related techniques (also known as stamp techniques) will be described in detail in chapter 7. Here, we will limit the discussion to a few important points:

- The mould fabrication and its quality is key in order to achieve high-resolution patterning on large working areas. The ultimate resolution of the technique will be largely determined by the features in the mould. As the mould is fabricated with another nanolithography technique, it can host features of only a few nanometers [14].
- The NIL process is capable of high throughput. In a single imprint step, a very large sample area can be patterned if the appropriate mould and mechanical system is used.
- In general, some residual resist remains on the sample's surface after the imprint step, which calls for an additional step to eliminate it. Thus, the etching step becomes double: a breakthrough etch and a transfer etch.

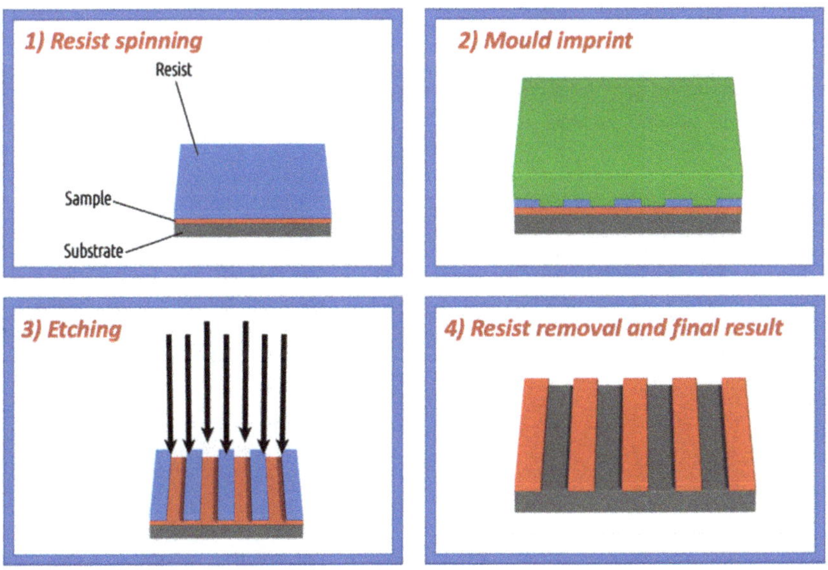

Figure 1.5. Various steps commonly performed in nanoimprint lithography: resist coating (1), mould imprint (2), sample etching (3) and resist removal with the final result (4).

- Stamp techniques are well suited for industrial applications given their relative simplicity and their capability for high throughput, good resolution and large-area patterning. However, the patterns cannot be in general complex. Despite this limiting factor, its use is broad in a few niche applications and some domains where hybrid fabrication processes are required, such as in biotechnology or to create smart surfaces.

1.3 Direct nanolithography techniques

1.3.1 Focused ion beam (FIB)

The capability of impinging accelerated ions to sputter the atoms at the surface of materials is very convenient for subtractive lithography. Thanks to the development in the early 1980s of the technology of the production of finely-focused ion beams in ion microscopes, this became a nanolithography technique with very high lateral resolution, in the sub-10 nm range [15]. FIB commercial sources were overwhelmingly based on Ga^+ given its stability and easy of use. Scanning a FIB on the surface of a material produces material removal with the desired pattern and with great nanoscale precision, as sketched in figure 1.6. The conceptual design of the FIB microscope leverages from the SEM, integrating similar parts: the source, the extraction and acceleration, the optics, the scan coils, the sample stage, the electron detectors, etc. Interestingly, combined FIB-SEM equipment was soon commercialized, providing the user with all the imaging, nanofabrication and analysis capabilities of both technologies in a single platform. For this reason, the FIB technology has become very popular to carry out certain tasks such as cross-sectional imaging, TEM lamella preparation, material nanopatterning and circuit edit.

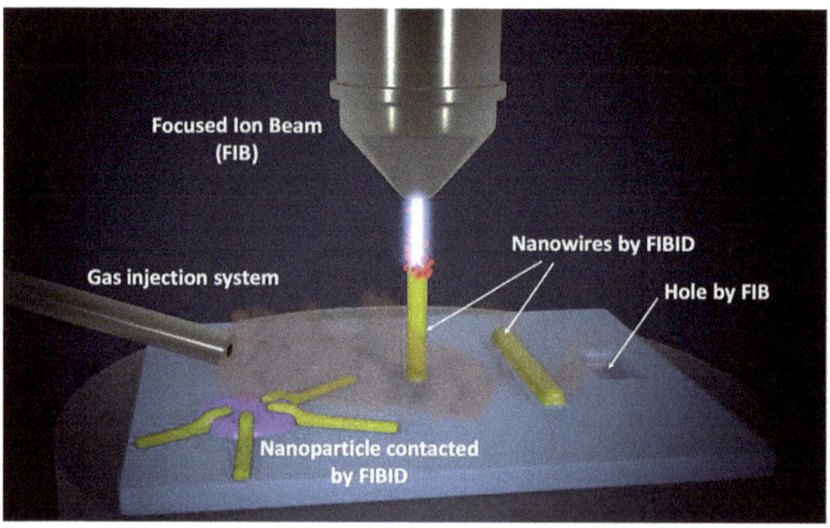

Figure 1.6. Sketch of the FIB equipment and some of its applications: creation of local milling (hole) and growth of 2D and 3D nanostructures with the assistance of a gas injection system.

The FIB technique will be described in detail in chapter 5. Here, we will limit the discussion to a few important points:

- FIB is capable of direct removal of material without the cumbersome use of resists. Being a direct nanolithography technique, the number of processing steps is minimized compared to OL, EBL and NIL.
- Being a sequential nanolithography technique, FIB is intrinsically slow, with much lower throughput than OL or NIL techniques.
- The Ga^+-based liquid metal ion source has become the most widespread type of source in FIB equipment. However, in recent years, new developments in sources such as those based on gas field ion sources (He^+, Ne^+), plasma sources (Xe^+) and metal-alloy sources represent a further step in terms of resolution or throughput.
- As the ion–matter interaction is stronger than the electron–matter one, harmful effects can be produced on the remaining material, changing its physical and chemical properties.
- Niche but key applications have been found for FIB technology in the semiconductor industry, in nanotechnology and in materials science.

1.3.2 Focused electron/ion beam induced deposition (FEBID and FIBID)

Soon after the first commercialization of FIB equipment, a related technique was developed and coined focused ion beam induced deposition (FIBID). FIBID requires a gas injection system (GIS) to produce, out of a precursor material delivered in gas form, a local deposit by precursor dissociation triggered by the FIB irradiation [16], as sketched in figure 1.6. The main advantage of this technique is the selective growth of a material on the area of interest in a single step. Given the high resolution of the FIB technique, the deposits can be grown with high lateral resolution [2]. Although the main property of interest in FIBID is commonly a high electrical conduction [17], other functional properties such as ferromagnetism [18], superconductivity [19] or good optical [20] and mechanical performance [21] can be obtained. It was soon realized that FIBID is suitable for the growth of 3D structures [22], with superb resolution recently obtained thanks to He^+-FIBID [23].

The sister technique of FIBID uses a focused electron beam instead, being known as FEBID (focused electron beam induced deposition). FEBID is carried out inside a SEM with the assistance of a GIS. FEBID targets the nanoscale growth of functional materials, similarly to FIBID, but with the advantage of the much lower damage created on the substrate given the lower linear momentum of electrons compared to ions. In return, the growth rate and the metal content in the deposits are generally lower for FEBID compared to FIBID [24].

The FIBID technique will be described in chapter 5 in the context of FIB technology, whereas FEBID will be described separately in chapter 4. Here, we will limit the discussion to a few important points common to both techniques:

- FIBID and FEBID, using, respectively, a focused ion or electron beam to dissociate a precursor gas, are direct additive nanolithography techniques allowing the growth of specific materials on targeted locations without the

need of resists. They can be used on substrates not appropriate for resist-based processing such as cantilever tips [25].
- The purity of the materials deposited by FIBID and FEBID can be an issue and intense research exists on purification methods during or after growth [26].
- Both, FIBID and FEBID, are appropriate for the growth of 3D structures with very high lateral resolution.
- Being sequential techniques, FIBID and FEBID are intrinsically slow and are not appropriate for large-area nanopatterning. Recent developments using cryogenic temperatures to condense the precursor material improve the growth speed a few orders of magnitude, the techniques being known as Cryo-FEBID [27] and Cryo-FIBID [28].
- Applications of FIBID and FEBID are generally connected to (a) circuit edit and mask repair in the semiconductor industry, (b) protection for cross-sectional FIB-SEM imaging and TEM lamella preparation, and (c) prototyping and nanodevice fabrication in nanotechnology fields.

1.3.3 Scanning probe lithography (SPL)

The tip of a scanning tunneling microscope (STM) or an atomic force microscope (AFM) provides not only imaging capabilities down to the atomic resolution but also nanofabrication capabilities with such great resolution. A paradigmatic example is the fabrication of a quantum corral by manipulation of individual atoms on a surface by means of a STM [29]. Although the speed of such fabrication has recently improved [30], it is a process with great difficulties in being scaled up and integrated with the semiconductor industry. The use of AFM for scanning probe lithography (SPL) also suffers from throughput issues, but it is more suitable than STM for this task given the less stringent requirements of the technique: no ultra-high vacuum conditions, conductive surface or very-fine control of the tip-to-sample distance are required. SPL based on AFM can be carried out through various mechanisms, offering a wide range of possibilities [31]. Thus, the AFM tip can produce local changes in the composition, height or physical/chemical properties of surfaces via thermal effects, mechanical effects, deposition, chemical effects, etc. [32]. The principle of the technique is sketched in figure 1.7(a) and the example of fabrication of Si nanowires based on oxidation SPL is shown in figure 1.7(b) [33]. A derived technique that has become very popular is dip-pen nanolithography, where the tip deposits specific inks on targeted locations with great resolution [34].

SPL techniques will be described in detail in chapter 6 and one application using thermal SPL will be discussed in detail in chapter 10. Here, a few important points can be highlighted:

- The main advantage of SPL techniques is the great resolution achievable, which can attain atomic resolution by STM in very specific experiments [29], and sub-20 nm resolution in a more routine way by AFM [35].

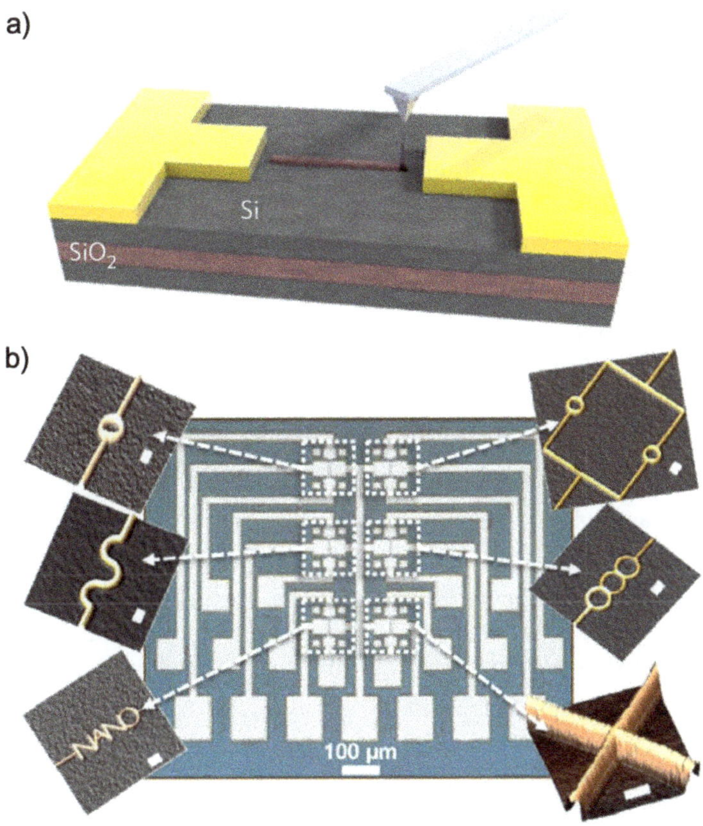

Figure 1.7. (a) Sketch of an atomic-force-microscope-based process called oxidation nanolithography. Reprinted and adapted by permission from Springer [31]. Copyright (2014). (b) Si nanowires with different shapes fabricated with sub-20 nm resolution (scale bar is 100 nm except in the central image) by this scanning probe lithography technique. Reproduced from [33]. Copyright IOP Publishing. Reproduced with permission. All rights reserved.

- SPL techniques are very slow by nature, but combined with other large-area lithography techniques can be of great interest for prototyping or niche applications [36].
- SPL techniques based on AFM are flexible with respect to the type of substrate, environment conditions, driving force of the lithography (thermal, mechanical, oxidation, deposition, etc), the absence of physical masks, etc. Thus, a wide range of SPL techniques exists nowadays, with thermal SPL, oxidation SPL and dip-pen being the most common ones [37].
- Intense effort is currently being deployed towards the fabrication of active customized probes, which can be used for SPL applications. These efforts include downsizing of self-sensed and self-actuated probes and upscaling of cantilever arrays amongst others [38].

1.4 Other nanolithography techniques and hybrid approaches

In recent years, the field of nanolithography has been enriched by new processes or techniques that have expanded the horizons for nanofabrication. A few of them can be mentioned: stencil lithography [39], nanosphere/colloidal lithography [40], roll-to-roll lithography [41], Talbot lithography [42], soft x-ray lithography using resists [43] or precursors [44], proton-beam lithography [45], plasmonic lithography [46], ice lithography [47], fountain-pen nanolithography [48], crack lithography [49], grain-boundary lithography [50], DNA lithography [51], self-assembled-monolayer lithography [52], beam-pen lithography [53], scanning near-field lithography [54], nanoskiving [55], etc. Given the limited space and the focus of the present chapter, only stencil and nanosphere/colloidal lithography will be introduced in detail in this section. New developments in coming years in the field of nanolithography could promote the wider use of some of the above mentioned techniques.

1.4.1 Stencil lithography

In this technique, a thin plate with designed apertures, called a stencil mask, is placed in proximity to the substrate and used in combination with material evaporation, as shown in figure 1.8(a). The evaporated material is stopped on the top surface of the stencil mask, except in the apertures. The material is thus grown

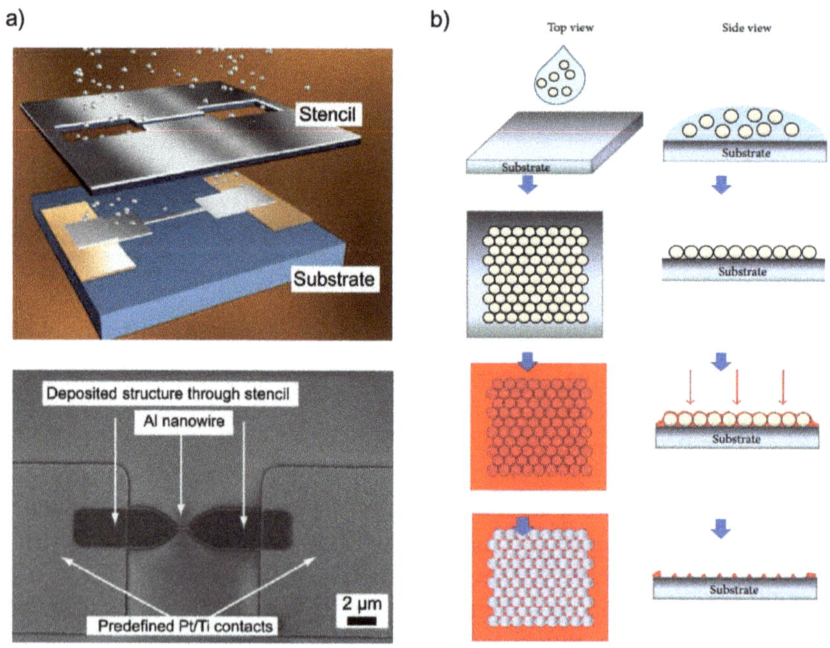

Figure 1.8. Sketches of two nanolithography techniques that allow the deposition of material only through small apertures: (a) stencil lithography. Reprinted with permission from [56]. Copyright (2008) American Chemical Society. (b) Nanosphere/colloidal lithography. Reprinted under the Creative Commons Attribution License from [40].

with specific patterns determined by those of the stencil mask [56]. With appropriate stencil masks, nanopatterned materials can be grown at the wafer scale in a single shot. Resolution down to 20 nm has been obtained and it can be used on unconventional substrates like cantilevers [39], but the technique bears some difficulties related to shadow deposition below the stencil mask and lifetime issues due to the deterioration of the stencil mask caused by material deposition at the edges of the holes, eventually blocking them (clogging effect). Developments such as the use of dynamic stencil masks have enlarged the range of applications of the technique [57], which will be discussed in detail in chapter 8.

1.4.2 Nanosphere/colloidal lithography

Under this name we include a set of techniques making use of a solution of nanoparticles, which form a very thin layer on the substrate of interest, and are used as a sacrificial layer during the nanopatterning process. Due to the interaction amongst the nanoparticles, they can self-organize forming a thin layer that leaves holes in between them, as sketched in figure 1.8(b). The technique was originally termed natural lithography [58]. Provided the monodisperse nature of the colloidal particles and their hydrophilic character, they form a colloidal crystal with ordered holes through which the material of interest will penetrate and be deposited on the substrate [40]. For example, polystyrene latex nanospheres can be used. The material deposited on top of the nanoparticles will disappear after the sample is immersed in an adequate solvent and sonicated. The process is similar to a lift-off process described in section 1.1.1. Advantages of this technique are the capability for large-area patterning, simplicity, good resolution and the potential to be combined with other lithography techniques. On the other hand, the technique presents difficulties with respect to the limited shapes available for the patterned functional material, the long-range order of the nanopatterns and the presence of point defects.

1.4.3 Hybrid approaches

In order to cover all the possibilities, the term *hybrid* can mean in this context:

(a) The use of two or more of the top-down nanolithography techniques to complete the patterning of a material or to fabricate a nanodevice;
(b) The combination of a top-down nanolithography technique with a bottom-up nanofabrication process to achieve a certain nanopatterning process or nanodevice or the unconventional use of a top-down nanolithography or a bottom-up nanofabrication technique with thin-film or other growth techniques with the aim of nanopatterning and building nanodevices.

Just to give the reader the flavor of such hybrid strategies, we consider in the following a few selected examples within each category:

(a) If two lithography techniques complement each other in terms of resolution or patterning area, it makes sense to combine them in a single instrument. This approach has been used to fabricate a single-electron transistor, as shown in figure 1.9 [36], integrating direct laser writing for large-area

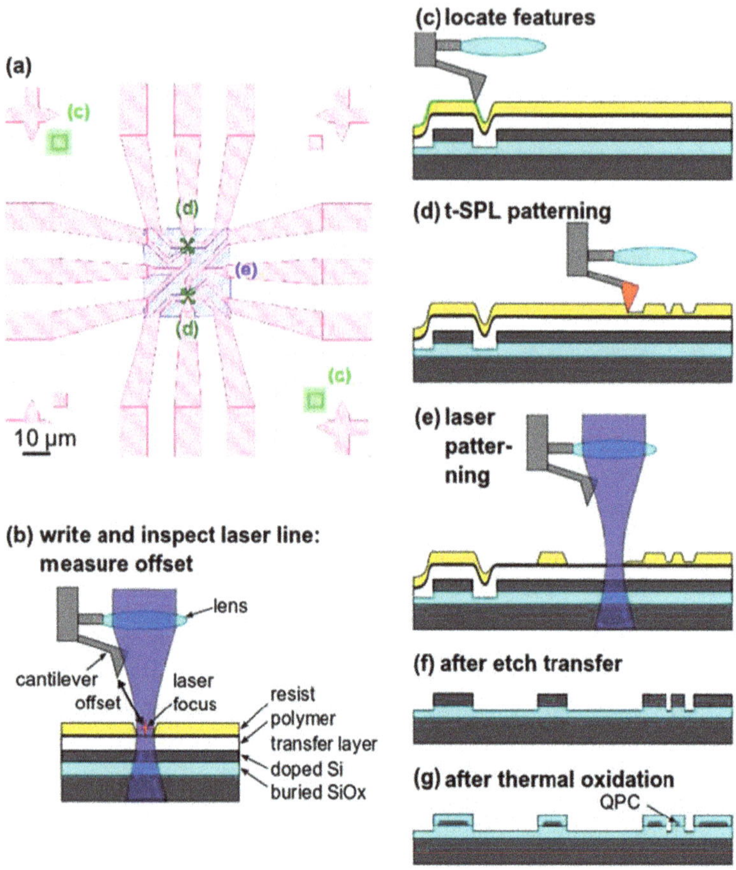

Figure 1.9. Example of a hybrid lithography technique combining two classical lithography techniques: fabrication of a single-electron transistor based on a quantum point contact (QPC) combining direct laser writing and thermal SPL (t-SPL). Reprinted from [36]. Copyright IOP Publishing. Reproduced with permission. All rights reserved.

exposure and thermal scanning probe lithography for high-resolution patterning. Another example that can be found in literature is the combination of electron beam lithography and scanning probe lithography using a single calixarene resist layer [59], in a process that is coined *mix-and-match*. This concept normally entails the consecutive use of a large-area lithography technique to pattern non-critical dimensions and a high-resolution lithography technique to pattern the smallest dimensions, with appropriate protocols to align both lithography steps. *Mix-and-match* has also been used to perform the consecutive resist exposure of chemically-amplified resists by DUV optical lithography and electron beam lithography [60].

(b) As will be shown in detail in chapter 2, directed block-copolymer lithography combines top-down lithography and bottom-up self-organization of

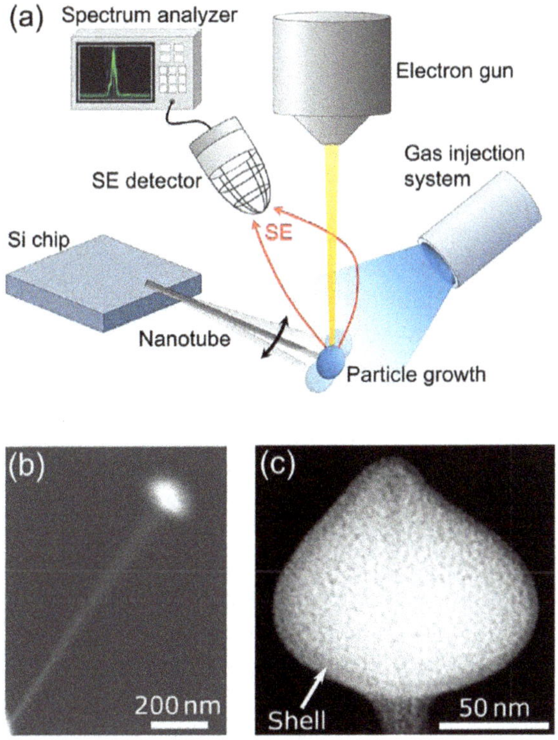

Figure 1.10. Example of a hybrid unconventional lithography processing for demonstrating mass nanosensing. (a) Growth of carbon nanotubes by chemical vapour deposition, which become attached to the substrate by Van-der-Waals forces and in some cases overhang the substrate. The resonance frequency of the carbon nanotubes is measured in a SEM through the generated secondary-electron generation and a vector network analyser. This resonance frequency becomes modified by the change in mass through FEBID deposition at the carbon nanotube's tip. The technique shows 13 zg mass resolution; (b) SEM micrograph of the Pt-C nanoparticle (total mass of 1.33 fg) deposited by FEBID; (c) TEM image of the nanoparticle shown in (b). Reprinted from [63]. Copyright (2019) American Chemical Society.

two polymers in order to produce high-resolution nanopatterning in large areas [61]. Normally, the block-copolymer self-organization is randomly oriented and lacks long-range order (see figure 1.1), but the previous top-down patterning provides the platform for a directed block-copolymer lithography. In another example to be described in chapter 5, FIB irradiation of a Si substrate gives rise to the preferential growth of semiconductor Ge material on the irradiated areas, which can be used to fabricate ordered arrays of semiconducting dots [62]. In another example of this category that is shown in figure 1.10, the FEBID technique is used to demonstrate the unsurpassed capabilities of carbon nanotubes for room-temperature mass detection, in the zg range (1 zg = 10^{-21} g) [63].

Further examples of the broad topic of hybrid lithography can be found in the following references by Saavedra *et al* [64], by Kim *et al* [65] and by Lu *et al* [66].

1.5 Comparison of nanolithography techniques

In figure 1.11, four selected lithography techniques are compared in terms of five key indicators: resolution, throughput, cost of the equipment, process complexity and the capability to fabricate complex patterns. In the following, we discuss these indicators:

- *Resolution*: it refers to the smallest features achievable with the technique, and good resolution refers to sub-10 nm patterning. In the semiconductor industry, optical lithography using steppers (industrial optical lithography) shows capability for fabrication of integrated circuits down to sub-10 nm resolution [5]. Electron beam lithography is also capable of sub-10 nm patterning using appropriate SEM equipment, resist and processing [10, 67]. Nanoimprint lithography with suitable moulds and resists also leads to sub-10 nm patterning [68]. Given the intrinsically small tip diameter, scanning probe lithography has so far shown the best resolution in nanopatterning [31]. Other techniques with sub-10 nm patterning capability are FEBID [69] and focused ion beam milling [70].
- *Throughput*: it refers to the amount of sample area patterned per unit of time. García *et al* mention that high-volume production requires throughput values of 10^{12} μm^2 h^{-1}, which means the processing of 100 wafers per hour [31]. Existing equipment with such throughput can be found in industrial optical lithography, where steppers can handle such a high number of wafers per hour. Here, we use a less stringent definition to consider a nanolithography

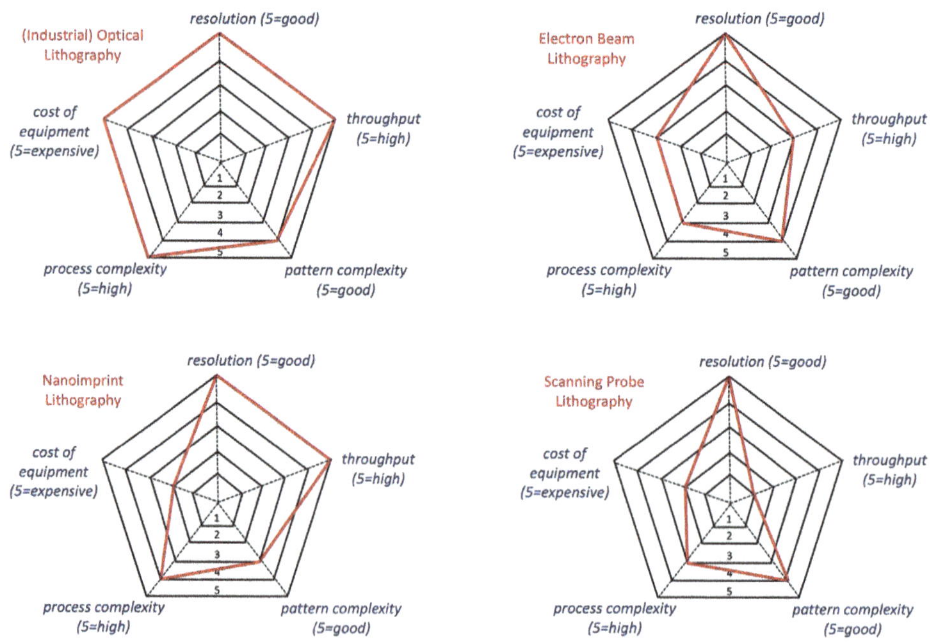

Figure 1.11. Comparison of the main lithography techniques in terms of resolution, throughput, cost of the equipment, process complexity and the capability to fabricate complex patterns.

technique as a high throughput one. We consider high-throughput techniques those capable of processing a full wafer in a single lithography shot, such as optical lithography and nanoimprint lithography. Electron beam lithography and scanning probe lithography can improve the throughput by using multiple-beam [9] and multiple-tip approaches [37], respectively, but are still far from industrial optical lithography. Other nanolithography techniques with high throughput are stencil lithography and nanosphere/colloidal lithography.

- *Pattern complexity*: this concept refers to the capability of the technique to generate complex patterns, ideally, in 2D and 3D. Optical lithography using masks are good to generate arbitrary patterns in 2D but not in 3D. However, two-photon-absorption lithography is well suited for 3D patterns. Standard electron beam lithography can generate arbitrary patterns in 2D but shows greater difficulty to achieve 3D patterning [71]. Nanoimprint lithography is commonly used to generate relatively simple 2D patterns. The fact that the patterning should be first imprinted on the mould is an added difficulty. However, recent developments in the field are promising towards 3D nanopatterning [72]. Scanning probe lithography is well suited to generate arbitrary patterns in 2D but not so much for 3D despite the fact that thermal scanning probe lithography has shown capability for 3D nanopatterning [73]. Regarding the rest of nanolithography techniques, FEBID and FIBID have shown great capability to generate nanopatterns in 2D and 3D [23, 74].
- *Process complexity*: within this concept, we evaluate the amount of steps required to perform the processing and the difficulty of these steps. Processing in industrial optical lithography is very sophisticated. The industry of integrated circuits has refined the involved lithography processes for several decades, collecting high-end knowledge on key aspects [75]. This has led to working devices in sub-10 nm nodes, which seemed unthinkable a few years ago. A huge complexity gap exists between an industrial optical lithography process carried out using a state-of-the-art stepper and a standard optical lithography process carried out using a mask aligner. In sharp contrast, electron beam lithography and scanning probe lithography can attain similar resolution to industrial optical lithography with much simpler processes, but with immensely lower throughput. In the case of nanoimprint lithography, the imprint process is relatively simple, but some extra complexity arises from the mould fabrication and the resist leftovers. On the other hand, given the absence of resists, the simplicity of focused ion beam milling for nanopatterning should be singled out.
- *Cost*: here, one should mainly consider the lithography equipment cost. By far, a state-of-the-art stepper used in industrial optical lithography is the most expensive lithography equipment one can think of, and its cost can reach 100 million euros. This is why only a few companies in the world can afford such investment in order to fabricate the latest devices in the microelectronics industry. The cost of electron beam lithography and nanoimprint lithography equipment is much lower, typically in the range of 1 million euros or less, but

the equipment price will strongly vary with the handled wafer size. On the other hand, an atomic force microscope with nanolithography capabilities will be surely cheaper than that. Stencil lithography and nanosphere/colloidal lithography excel with respect to the equipment cost, which is certainly low compared to the other techniques.

Definitively, an appropriate lithography technique exists for each particular need, and the user should consider all the key aspects discussed in this section in order to find the most suited one.

1.6 Applications of nanolithography techniques

1.6.1 Classical applications

The fabrication of integrated circuits is the paradigm of the need of lithography techniques in modern technology. Since the monolithic idea was conceived and developed by Jack Kilby and Robert Noyce in the late 1950s, integrating on the same semiconductor block all the circuit elements needed for an electronic device [76], the development of appropriate fabrication techniques was pursued. Optical lithography was found to be the right technique for this fabrication, given its high throughput. In addition, it could cope with the increasing resolution demands in each new generation of electronic devices, as will be shown in chapter 2. In fact, optical lithography is considered to be one of the cornerstones of electronics. Modern electronics would not be possible without the extraordinary properties of semiconductor materials and an enabling technology such as optical lithography. Thanks to optical lithography, Moore's law, which states that in each new generation of integrated circuits the number of transistors doubles, would simply have not been possible [5]. Taking into account that the total footprint of a single transistor today is around 100 nm × 100 nm, or equivalently 10^{10} transistors per cm^2, it is difficult to think that significant further steps will come from size reduction. Nevertheless, More-than-Moore and Beyond-Moore approaches are being intensively investigated as potential replacement to the existing paradigms, with nanolithography techniques remaining as a cornerstone of all new technologies [77]. On the other hand, nanolithography is not only important in the fabrication of the microelectronic components themselves, but also in other related processes. One can cite for example circuit edit and mask repair processes, where focused ion beam, focused ion beam induced deposition, focused electron beam induced etching and focused electron beam induced deposition are employed [78–80].

The reader will find in the book many other classical applications of nanolithography beyond those in microelectronics. Let us cite here a few of them, but we invite the reader to go through the details in the different chapters:

- Optical elements and devices: display panels, optical masks, diffraction gratings, micromirrors, micro-lenses such as Fresnel lenses, optical waveguides, photonic crystals, plasmonic elements, etc.
- Magnetic elements and devices: for information storage, such as memory elements and reading heads, for logics and communication devices.

- Structural elements and functional surfaces: cantilevers for MEMS and probe tips, micro-implants, nano-tweezers, surfaces with specific hydrophobocity/hydrophilicity properties, etc.
- Superconducting devices: arrays of Josephson junctions for quantum computing, high-frequency electronics, single-photon detectors, x-ray detectors, etc.
- Elements and devices for energy storage or conversion: solar cells, batteries, energy-harvesting devices, etc.
- Enabling nanofabrication for advanced imaging: TEM lamellae preparation, 3D FIB-SEM imaging, atom probe tomography, etc.

It should not be forgotten that nanolithography is commonly used in laboratory research for the patterning of materials and the fabrication of proof-of-concept devices. Although most of the investigations will not eventually lead to a commercialized application, many resources and nanolithography equipment are devoted to it. In fact, it is well known that the properties of materials change as their dimensions become comparable to the relevant physical lengths such as the electron mean free path, the magnetic interactions, the superconducting coherence length, etc. As a consequence, research activities involving the study of the change in the physical properties and the discovery of new phenomena at the nanoscale are ubiquitous and call for patterning the materials of interest. Also, exploration of new devices with potential for commercialization requires as a first step to fabricate a small batch of them. Such prototyping or proof-of-concept fabrication needs lithography infrastructure. For a small number of units in the prototyping phase, the cost of fabrication is generally not the main issue and various nanolithography techniques can be generally used.

1.6.2 Interdisciplinary applications

A growing interest of nanolithography exists for interdisciplinary applications, mainly within the fields of biology and medicine. Nanolithography enters the game in two ways. First, it can be used to fabricate the electronic devices that provide a digital readout of a certain measurement or allows applying a certain external action. Secondly, given the micrometric or nanometric size of objects such as bacteria, virus, proteins, DNA, etc, artificial objects of similar size are relevant to interact with those entities. This has prompted several applications in the following topics:

- bionsensing and medical diagnosis [81],
- DNA and protein sequencing [82],
- microfluidic and lab-on-a-chip devices [83, 84],
- medical treatments (drug delivery, hyperthermia, mechanical treatment, etc) [85],
- medical imaging techniques [86],
- neuroscience [87],
- scaffolds for tissue growth and cell differentiation [88, 89].

1.6.3 Emergent and future applications

One can think of the following emergent and future applications:

- The development of flexible electronics has expanded the use of nanolithography into new emergent applications in biomedicine (biopsy diagnosis, implants for optical stimulation, cardiac patches, brain mapping), flexible LEDs, flexible touch screens, and strain and temperature sensors [90]. More futuristic, but approaching fast to reality, flexible electronic components are being developed to accomplish the interfacing between humans and machines, such as in the artificial retina [91, 92]. Cyborgs will surely rely on lithography to some extent.
- New challenges in chemical warfare, contaminant substances, disease diagnosis and drug detection are fostering development under the umbrella of the electronic nose concept, where a small device must be able to detect the targeted substances and trigger a fast response. Nanofabrication is crucial towards device miniaturization, the fabrication of the electronic components and their integration. The electronic nose must ideally perform the sensing, the signal processing and the signal transmission [93].
- With the rising importance of robots in our daily life, nanolithography will play a key role in the fabrication of smaller and faster sensors and actuators as well as suitable electronics for the advanced control and communication of the robot's components [94].
- The change of paradigm from classical computation to quantum computation to perform certain computational tasks will bring about new fabrication challenges to build the qubits as well as the optimized electronics. The scenario is still wide open, but in most of the proposed approaches lithography is one of the key enabling technologies [95–97]. Other alternative computation schemes such as neuromorphic computing also rely on nanolithography [98].
- With the upcoming implantation of autonomous cars, nanolithography will be a key player in the fabrication of advanced sensors and efficient electronics for data treatment, decision making and communication. Important in business terms and technological demands, the Internet of Things will also bring new challenges for sensing and microelectronic fabrication [99].

Other emergent fields could be further mentioned, but I leave the last word to the reader's imagination. Nevertheless, I can share my futuristic thoughts. Artificial intelligence (AI) is expected to change our lives in the next decades. Recent developments have unveiled its potential impact in better designs and fabrication concepts of integrated circuits beyond conventional approaches. Who knows what can happen if AI uses all its power to improve the integrated circuits on which the AI runs itself…will the process converge or diverge?

1.7 End-of-chapter summary

In this introductory chapter, we have provided a broad view on the field of nanofabrication based on nanolithography techniques, with an emphasis on a few aspects that are hereafter summarized:

- Key concepts in nanolithography:
  - (a) Whereas bottom-up approaches rely on self-assembly and self-organization processes that merge smaller building blocks to achieve larger structures, top-down approaches rely on creating small structures out of larger ones.
  - (b) In general, top-down nanolithography is combined with thin-film growth techniques, providing capabilities for fine control of dimensions down to the nanoscale, high throughput and creation of complex structures.
  - (c) The combination of bottom-up and top-down strategies offers additional opportunities for advanced nanofabrication.
  - (d) The choice of an appropriate nanolithography technique for each particular case should consider aspects such as the resolution, cost, throughput and complexity.
- Types of nanolithography techniques:
  - (a) Optical-based lithography includes: UV lithography in contact, proximity and projection modes, extreme UV lithography, direct laser writing, interference lithography, two-photon-absorption lithography, plasmonic lithography, soft x-ray lithography, Talbot lithography, beam-pen lithography, scanning near-field lithography, etc.
  - (b) Electron and ion beam based lithography comprises: electron beam lithography, focused ion beam, focused electron and beam induced deposition (including precursor in gas, liquid or solid phase), ice lithography, etc.
  - (c) Stamp lithography encompasses: nanoimprint lithography, microcontact printing, step-and-flash lithography, self-assembled-monolayer lithography, roll-to-roll lithography, etc.
  - (d) Scanning probe lithography covers: oxidation nanolithography, thermal scanning probe lithography, nanoscratching lithography, dip-pen lithography, fountain-pen lithography, etc.
  - (e) Lithography based on small physical apertures includes: stencil lithography, nanosphere/colloidal lithography, grain-boundary lithography, etc.
  - (f) Other lithography techniques not categorized under the previous epigraphs and mentioned in the present chapter are crack lithography and DNA lithography.
  - (g) In many applications, two or more of these techniques need to be used, which has been coined in section 1.4.3 with the term hybrid lithography approach. Commercial equipment exists in some cases that integrates two or more of these lithography techniques.
- Applications of nanolithography:
  - (a) The classical application of nanolithography is the fabrication of integrated circuits and other electronic devices, in which optical lithography is widespread. In addition, various types of nanolithography

techniques are used in research activities targeting the patterning of materials and realization of prototype and proof-of-concept devices.

(b) In recent years, the penetration of nanolithography in interdisciplinary applications has rocketed, especially in biology and medicine.

(c) A bright future is waiting for nanolithography in emergent applications such as new computation schemes, biomedicine, robotics, the Internet of Things, the autonomous car and human–machine interfacing.

1.8 Students' corner

- An introductory video to the field of nanolithography techniques and their applications can be viewed at https://iopscience.iop.org/book/978-0-7503-2608-7.
- Draw similar diagrams to those shown in figure 1.11 for the following lithography techniques: FIB, FEBID, stencil lithography.
- Brainstorming: think about one application not mentioned in this chapter in which you could apply nanolithography techniques. Try to make it real.

Acknowledgements

I am extremely grateful to all colleagues, collaborators and students whom with I have learnt so many things on nanolithography and their applications along my scientific career. Without so many enlightening discussions, this chapter would not exist. It is with great pleasure that I acknowledge the decisive role played by the participants of NANOLITO, the Spanish network on nanolithography, in sharing their knowledge on this thrilling topic in the last decade. I acknowledge financial support from the Spanish Ministry of Economy and Competitiveness through projects MAT2018-102627-T and MAT2017-82970-C2, from CSIC through project PIE202060E187 and from the Aragon Regional Government (Construyendo Europa desde Aragón) through Project E13_20R, with European Social Fund funding.

References

[1] Lopes W A and Jaeger H M 2001 Hierarchical self-assembly of metal nanostructures on diblock copolymer scaffolds *Nature* **414** 735–8

[2] De Teresa J M and Córdoba R 2014 Arrays of densely-packed isolated nanowires by focused beam induced deposition plus Ar^+ milling *ACS Nano* **8** 3788–95

[3] Kelzenberg M D, Boettcher S W, Petykiewicz J A, Turner-Evans D B, Putnam M C and Warren E L *et al* 2010 Enhanced absorption and carrier collection in Si wire arrays for photovoltaic applications *Nat. Mater.* **9** 239–44

[4] Huang Y, Duan X, Wei Q and Lieber C M 2001 Directed assembly of one-dimensional nanostructures into functional networks *Science* **291** 630–3

[5] Hutcheson G D 2018 Moore's law, lithography, and how optics drive the semiconductor industry *Proc. SPIE* **1058303** 501

[6] Van Kessel P F, Hornbeck L J, Meier R E and Douglass M R 1998 A MEMS-based projection display *Proc. IEEE* **86** 1687–704

[7] LaFratta C N, Fourkas J T, Baldacchini T and Farrer R A 2007 Multiphoton fabrication *Angew Chem. Int. Ed.* **46** 6238–58

[8] Lai N D, Liang W P, Lin J H, Hsu C C and Lin C H 2005 Fabrication of two- and three-dimensional periodic structures by multi-exposure of two-beam interference technique *Opt. Express* **13** 9605

[9] Ichimura T, Ren Y and Kruit P 2014 A large current scanning electron microscope with MEMS-based multi-beam optics *Microelectron. Eng.* **113** 109–13

[10] Hu W, Sarveswaran K, Lieberman M and Bernstein G H 2004 Sub-10 nm electron beam lithography using cold development of poly(methylmethacrylate) *J. Vac. Sci. Technol.* B **22** 1711–6

[11] Chou S Y, Krauss P R and Renstrom P J 1995 Imprint of sub-25 nm vias and trenches in polymers *Appl. Phys. Lett.* **67** 3114–6

[12] Kaufmann T and Ravoo B J 2010 Stamps, inks and substrates: polymers in microcontact printing *Polym. Chem.* **1** 371–87

[13] Bailey T C, Resnick D J, Mancini D, Nordquist K J, Dauksher W J and Ainley E *et al* 2002 Template fabrication schemes for step and flash imprint lithography *Microelectron. Eng.* **61–62** 461–7

[14] Chou S Y, Krauss P R, Zhang W, Guo L and Zhuang L 1997 Sub-10 nm imprint lithography and applications *J. Vac. Sci. Technol.* B **15** 2897–904

[15] Kim C S, Ahn S H and Jang D Y 2012 Review: developments in micro/nanoscale fabrication by focused ion beams *Vacuum* **86** 1014–35

[16] Utke I, Hoffmann P and Melngailis J 2008 Gas-assisted focused electron beam and ion beam processing and fabrication *J. Vac. Sci. Technol.* B **26** 1197–276

[17] Fernández-Pacheco A, De Teresa J M, Córdoba R and Ibarra M R 2009 Metal-insulator transition in Pt-C nanowires grown by focused-ion-beam- induced deposition *Phys. Rev.* B **79** 174204

[18] Sanz-Martín C, Magén C and De Teresa J M 2019 High volume-per-dose and low resistivity of cobalt nanowires grown by Ga^+ focused ion beam induced deposition *Nanomaterials* **9** 1715

[19] Córdoba R, Orús P, Jelić Ž L, Sesé J, Ibarra M R and Guillamón I *et al* 2019 Long-range vortex transfer in superconducting nanowires *Sci. Rep.* **9** 12386

[20] Esposito M, Tasco V, Cuscunà M, Todisco F, Benedetti A and Tarantini I *et al* 2015 Nanoscale 3D chiral plasmonic helices with circular dichroism at visible frequencies *ACS Photonics* **2** 105–14

[21] Córdoba R, Lorenzoni M, Pablo-Navarro J, Magén C, Pérez-Murano F and De Teresa J M 2017 Suspended tungsten-based nanowires with enhanced mechanical properties grown by focused ion beam induced deposition *Nanotechnology* **28** 445301

[22] Matsui S, Kaito T, Fujita J I, Komuro M, Kanda K and Haruyama Y 2000 Three-dimensional nanostructure fabrication by focused-ion-beam chemical vapor deposition *J. Vac. Sci. Technol.* B **18** 3181–4

[23] Córdoba R, Mailly D, Rezaev R O, Smirnova E I, Schmidt O G and Fomin V M *et al* 2019 Three-dimensional superconducting nanohelices grown by He^+-focused-ion-beam direct writing *Nano Lett.* **19** 8597–604

[24] De Teresa J M, Cárdoba R, Fernández-Pacheco A, Montero O, Strichovanec P and Ibarra M R 2009 Origin of the difference in the resistivity of as-grown focused-ion- and focused-electron-beam-induced Pt nanodeposits *J. Nanomater.* **2009** 936863

[25] Castagné M, Benfedda M, Lahimer S, Falgayrettes P and Fillard J P 1999 Near field optical behaviour of C supertips *Ultramicroscopy* **76** 187–94

[26] Pablo-Navarro J, Magén C and de Teresa J M 2018 Purified and crystalline three-dimensional electron-beam-induced deposits: the successful case of cobalt for high-performance magnetic nanowires *ACS Appl. Nano Mater.* **1** 38–46

[27] Bresin M, Thiel B L, Toth M, Dunn K A and Toth M 2011 Focused electron beam-induced deposition at cryogenic temperatures *J. Mater. Res.* **26** 357–64

[28] Córdoba R, Strohauer S, Torres T E and De Teresa J M 2019 Ultra-fast direct growth of metallic micro- and nano-structures by focused ion beam irradiation *Sci. Rep.* **9** 14076

[29] Crommie M F, Lutz C P and Eigler D M 1993 Confinement of electrons to quantum corrals on a metal surface *Science* **262** 218–20

[30] Kalff F E, Rebergen M P, Fahrenfort E, Girovsky J, Toskovic R and Lado J L *et al* 2016 A kilobyte rewritable atomic memory *Nat. Nanotechnol.* **11** 926–9

[31] Garcia R, Knoll A W and Riedo E 2014 Advanced scanning probe lithography *Nat. Nanotechnol.* **9** 577–87

[32] Howell S T, Grushina A, Holzner F and Brugger J 2020 Thermal scanning probe lithography—a review *Microsyst. Nanoeng* **6** 21

[33] Martínez R V, Martínez J and Garcia R 2010 Silicon nanowire circuits fabricated by AFM oxidation nanolithography *Nanotechnology* **21** 245301

[34] Piner R D, Zhu J, Xu F and Hong S 1999 'Dip-Pen' nanolithography *Science* **283** 661–4

[35] Dago A I, Sangiao S, Fernández-Pacheco R, De Teresa J M and Garcia R 2018 Chemical and structural analysis of sub-20 nm graphene patterns generated by scanning probe lithography *Carbon* **129** 281–5

[36] Rawlings C, Ryu Y K, Rüegg M, Lassaline N, Schwemmer C and Duerig U *et al* 2018 Fast turnaround fabrication of silicon point-contact quantum-dot transistors using combined thermal scanning probe lithography and laser writing *Nanotechnology* **29** 505302

[37] Hu H, Kim H J and Somnath S 2017 Tip-based nanofabrication for scalable manufacturing *Micromachines* **8** 90

[38] Rangelow I W, Ivanov T, Ahmad A, Kaestner M, Lenk C and Bozchalooi I S *et al* 2017 Review article: active scanning probes: a versatile toolkit for fast imaging and emerging nanofabrication *J. Vac. Sci. Technol.* B **35** 06G101

[39] Vazquez-Mena O, Gross L, Xie S, Villanueva L G and Brugger J 2015 Resistless nanofabrication by stencil lithography: a review *Microelectron. Eng.* **132** 236–54

[40] Colson P, Henrist C and Cloots R 2013 Nanosphere lithography: a powerful method for the controlled manufacturing of nanomaterials *J. Nanomater.* **2013** 948510

[41] Kooy N, Mohamed K, Pin L T and Guan O S 2014 A review of roll-to-roll nanoimprint lithography *Nanoscale Res. Lett.* **9** 320

[42] Li W, Urbanski L and Marconi M C 2015 Invited article: progress in coherent lithography using table-top extreme ultraviolet lasers *Rev. Sci. Instrum.* **86** 121301

[43] Chao W, Harteneck B D, Liddle J A, Anderson E H and Attwood D T 2005 Soft X-ray microscopy at a spatial resolution better than 15 nm *Nature* **435** 1210–3

[44] Späth A 2019 Additive nano-lithography with focused soft X-rays: basics, challenges, and opportunities *Micromachines* **10** 834

[45] Van Kan J A, Malar P and Wang Y H 2014 Resist materials for proton beam writing: a review *Appl. Surf. Sci.* **310** 100–11
[46] Hong F and Blaikie R 2019 Plasmonic lithography: recent progress *Adv. Opt. Mater.* **7** 1801653
[47] Han A, Kuan A, Golovchenko J and Branton D 2012 Nanopatterning on nonplanar and fragile substrates with ice resists *Nano Lett.* **12** 1018–21
[48] Kim K H, Moldovan N and Espinosa H D 2005 A nanofountain probe with Sub-100 nm molecular writing resolution *Small* **1** 632–5
[49] Kim M, Kim D J, Ha D and Kim T 2016 Cracking-assisted fabrication of nanoscale patterns for micro/nanotechnological applications *Nanoscale* **8** 9461–79
[50] Guo C F, Sun T, Liu Q, Suo Z and Ren Z 2014 Highly stretchable and transparent nanomesh electrodes made by grain boundary lithography *Nat. Commun.* **5** 3121
[51] Keren K, Krueger M, Gilad R, Ben-Yoseph G, Sivan U and Braun E 2002 Sequence-specific molecular lithography on single DNA molecules *Science* **297** 72–5
[52] Smith R K, Lewis P A and Weiss P S 2004 Patterning self-assembled monolayers *Prog. Surf. Sci.* **75** 1–68
[53] Huo F, Zheng G, Liao X, Giam L R, Chai J and Chen X *et al* 2010 Beam pen lithography *Nat. Nanotechnol.* **5** 637–40
[54] Sun S and Leggett G J 2004 Matching the resolution of electron beam lithography by scanning near-field photolithography *Nano Lett.* **4** 1381–4
[55] Xu Q, Rioux R M, Dickey M D and Whitesides G M 2008 Nanoskiving: a new method to produce arrays of nanostructures *Acc. Chem. Res.* **41** 1566–77
[56] Vazquez-Mena O, Villanueva G, Savu V, Sidler K, Van Den Boogaart M A F and Brugger J 2008 Metallic nanowires by full wafer stencil lithography *Nano Lett.* **8** 3675–82
[57] Du K, Ding J, Liu Y, Wathuthanthri I and Choi C H 2017 Stencil lithography for scalable micro- and nanomanufacturing *Micromachines* **8** 131
[58] Deckman H W and Dunsmuir J H 1982 Natural lithography *Appl. Phys. Lett.* **41** 377–9
[59] Kaestner M, Hofer M and Rangelow I W 2013 Nanolithography by scanning probes on calixarene molecular glass resist using mix-and-match lithography *J. Micro/Nanolithogr. MEMS MOEMS* **12** 031111
[60] Tedesco S 1998 Resist processes for hybrid (electron-beam/deep ultraviolet) lithography *J. Vac. Sci. Technol.* **B 16** 3676
[61] Park S M, Liang X, Harteneck B D, Pick T E, Hiroshiba N and Wu Y *et al* 2011 Sub-10 nm nanofabrication via nanoimprint directed self-assembly of block copolymers *ACS Nano* **5** 8523–31
[62] Portavoce A, Kammler M, Hull R, Reuter M C and Ross F M 2006 Mechanism of the nanoscale localization of Ge quantum dot nucleation on focused ion beam templated Si(001) surfaces *Nanotechnology* **17** 4451–5
[63] Gruber G, Urgell C, Tavernarakis A, Stavrinadis A, Tepsic S and Magén C *et al* 2019 Mass sensing for the advanced fabrication of nanomechanical resonators *Nano Lett.* **19** 6987–92
[64] Saavedra H M, Mullen T J, Zhang P, Dewey D C, Claridge S A and Weiss P S 2010 Hybrid strategies in nanolithography *Rep. Prog. Phys.* **73** 036501
[65] Kim J U, Lee S and Kim T-I 2016 Recent advances in unconventional lithography for challenging 3D hierarchical structures and their applications *J. Nanomater.* **2016** 7602395
[66] Lu W and Lieber C M 2007 Nanoelectronics from the bottom up *Nat. Mater* **6** 841–50

[67] Duan H, Hu H, Kumar K, Shen Z and Yang J K W 2011 Direct and reliable patterning of plasmonic nanostructures with sub-10-nm gaps *ACS Nano* **5** 7593–600
[68] Austin M D, Zhang W, Ge H X, Wasserman D, Lyon S A and Chou S Y 2005 6 nm half-pitch lines and 0.04 mu m(2) static random access memory patterns by nanoimprint lithography *Nanotechnology* **16** 1058–61
[69] van Oven J C, Berwald F, Berggren K K, Kruit P and Hagen C W 2011 Electron-beam-induced deposition of 3-nm-half-pitch patterns on bulk Si *J. Vac. Sci. Technol.* B **29** 06F305
[70] Gierak J, Bourhis E, Faini G, Patriarche G, Madouri A and Jede R *et al* 2009 Exploration of the ultimate patterning potential achievable with focused ion beams *Ultramicroscopy* **109** 457–62
[71] Mortelmans T, Kazazis D, Guzenko V A, Padeste C, Braun T and Stahlberg H *et al* 2020 Grayscale e-beam lithography: effects of a delayed development for well-controlled 3D patterning *Microelectron. Eng.* **225** 11272
[72] Schift H 2015 Nanoimprint lithography: 2D or not 2D? A review *Appl. Phys.* A **121** 415–35
[73] Knoll A W, Pires D, Coulembier O, Dubois P, Hedrick J L and Frommer J *et al* 2010 Probe-based 3-D nanolithography using self-amplified depolymerization polymers *Adv. Mater.* **22** 3361–5
[74] Winkler R, Fowlkes J D, Rack P D and Plank H 2019 3D nanoprinting via focused electron beams *J. Appl. Phys.* **125** 210901
[75] Lin B J 2015 Making lithography work for the 7-nm node and beyond in overlay accuracy, resolution, defect, and cost *Microelectron. Eng.* **143** 91–101
[76] Kilby J S 1976 Invention of the integrated circuit *IEEE Trans. Electron Devices* **23** 648–54
[77] Cavin R K, Lugli P and Zhirnov V V 2012 Science and engineering beyond Moore's law *Proc. IEEE* **100** 1720–49
[78] Mohiuddin T 2014 Focused ion beam (FIB) circuit edit *Electron Device Fail. Anal.* **16** 20–3
[79] Drezner Y, Fishman D, Greenzweig Y and Raveh A 2011 Characterization of damage induced by FIB etch and tungsten deposition in high aspect ratio vias *J. Vac. Sci. Technol.* B **29** 011026
[80] Bret T, Jonckheere R, Van den Heuvel D, Baur C, Waiblinger M and Baralia G 2012 Closing the gap for EUV mask repair *Proc. SPIE* **8322** 83220C
[81] Grodzinski P, Silver M and Molnar L K 2006 Nanotechnology for cancer diagnostics: promises and challenges *Expert Rev. Mol. Diagn.* **6** 307–18
[82] Tang Z, Zhang D, Cui W, Zhang H, Pang W and Duan X 2016 Fabrications, applications and challenges of solid-state nanopores: a mini review *Nanomater. Nanotechnol.* **6** 35
[83] Ha D, Hong J, Shin H and Kim T 2016 Unconventional micro-/nanofabrication technologies for hybrid-scale lab-on-a-chip *Lab Chip* **16** 4296–312
[84] Duan C, Wang W and Xie Q 2013 Review article: fabrication of nanofluidic devices *Biomicrofluidics* **7** 026501
[85] Leulmi Pichot S, Bentouati S, Ahmad S S, Sotiropoulos M, Jena R and Cowburn R 2020 Versatile magnetic microdiscs for the radio enhancement and mechanical disruption of glioblastoma cancer cells *RSC Adv.* **10** 8161–71
[86] Knapitsch A, Auffray E, Barbastathis G, Chevalier C, Hsieh C H and Kim J G *et al* 2016 Large scale production of photonic crystals on scintillators *IEEE Trans. Nucl. Sci.* **63** 639–43
[87] Abbott J, Ye T, Krenek K, Gertner R S, Ban S and Kim Y *et al* 2020 A nanoelectrode array for obtaining intracellular recordings from thousands of connected neurons *Nat. Biomed. Eng.* **4** 232–41

[88] Melissinaki V, Gill A A, Ortega I, Vamvakaki M, Ranella A and Haycock J W *et al* 2011 Direct laser writing of 3D scaffolds for neural tissue engineering applications *Biofabrication* **3** 045005

[89] Prabhakaran M P, Vatankhah E, Kai D and Ramakrishna S 2015 Methods for nano/micropatterning of substrates: toward stem cells differentiation *Int. J. Polym. Mater. Polym. Biomater* **64** 338–53

[90] Kang K, Cho Y and Yu K J 2018 Novel nano-materials and nano-fabrication techniques for flexible electronic systems *Micromachines* **9** 263

[91] Hsieh C Y, Chen J S, Tsou W A, Yeh Y T, Wen K A and Fan L S 2009 A biocompatible and flexible RF CMOS technology and the characterization of the flexible mos transistors under bending stresses *Proc. IEEE Int. Conf. Micro. Electro. Mech. Syst.* 627–9

[92] Weiland J D, Walston S T and Humayun M S 2016 Electrical stimulation of the retina to produce artificial vision *Annu. Rev. Vis. Sci.* **2** 273–94

[93] Röck F, Barsan N and Weimar U 2008 Electronic nose: current status and future trends *Chem. Rev.* **108** 705–25

[94] Sitti M and Wiersma D S 2020 Pros and cons: magnetic versus optical microrobots *Adv. Mater.* **32** 1906766

[95] Barends R, Kelly J, Megrant A, Sank D, Jeffrey E and Chen Y *et al* 2013 Coherent josephson qubit suitable for scalable quantum integrated circuits *Phys. Rev. Lett.* **111** 080502

[96] Hollenberg L C L, Dzurak A S, Wellard C, Hamilton A R, Reilly D J and Milburn G J *et al* 2004 Charge-based quantum computing using single donors in semiconductors *Phys. Rev.* B **69** 2–5

[97] Mourik V, Zuo K, Frolov S M, Plissard S R, Bakkers E P A M and Kouwenhoven L P 2012 Signatures of majorana fermions in hybrid superconductor-semiconductor nanowire devices *Science* **336** 1003–7

[98] Torrejon J, Riou M, Araujo F A, Tsunegi S, Khalsa G and Querlioz D *et al* 2017 Neuromorphic computing with nanoscale spintronic oscillators *Nature* **547** 428–32

[99] Potyrailo R A 2016 Multivariable sensors for ubiquitous monitoring of gases in the era of internet of things and industrial internet *Chem. Rev.* **116** 11877–923

IOP Publishing

Nanofabrication
Nanolithography techniques and their applications
José María De Teresa

Chapter 2

Optical lithography

Francesc Perez-Murano, José Ignacio Martín and José María De Teresa

Optical lithography uses optical radiation to define patterns in a photosensitive material (photoresist). Patterning resolution in optical lithography is mainly limited by light diffraction. Decreasing the wavelength of the optical radiation is the main way to improve resolution. Systems based on extreme ultra-violet light achieve a resolution close to 10 nm. In the most common approach optical lithography is performed using optical masks to locally expose the photoresist, allowing for high throughput processing. Greater flexibility and 3D patterning is achieved by laser and two-photon lithography. A further improvement in the resolution of optical lithography without compromising throughput could be achieved by complementing optical lithography with self-assembly methods.

2.1 Concept of optical lithography

Optical lithography is the most employed lithography method. The principle is described in figure 2.1: optical radiation is used to expose a photosensitive material in selected areas. As a result of the exposure, the photosensitive material experiences a change in its physical and chemical properties, varying its solubility with respect to a specific solvent. After the exposure, the latent pattern in the photosensitive material is revealed in the development process, when a specific chemical agent is used to either eliminate the exposed or the non-exposed areas depending on the characteristics of the resist.

As is the case in all lithography methods, resolution is the main figure of merit to describe the performance. Resolution is defined as the minimum feature that can be printed in the resist. In optical lithography, diffraction phenomena limit the resolution. As diffraction is directly dependent on the wavelength of the optical radiation, the shorter the wavelength, the better will be the resolution. Reducing the wavelength of the optical radiation has been one of the ways to improve the resolution of optical lithography, but at the expenses of increasing the process cost.

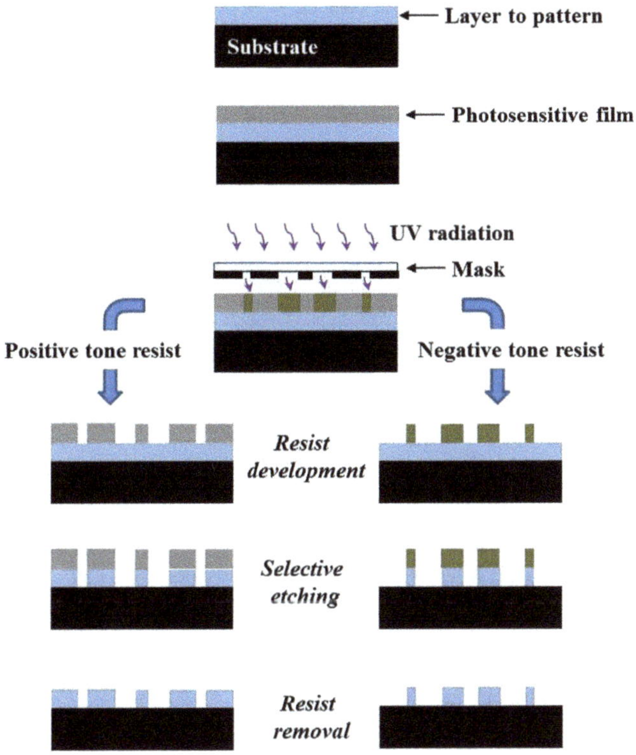

Figure 2.1. Sketch showing the concept of optical lithography.

There are several key features in an optical lithography process, which define the different variations from the global scheme represented in figure 2.1:

- use of an optical mask or direct exposure;
- wavelength of the optical radiation;
- positive or negative tone of the photosensitive material.

The chapter is organized as follows. Section 2.2 is devoted to optical lithography using an optical mask, describing the different exposure modes, the optical radiation used and the types of photosensitive materials. In section 2.3, optical lithography by direct write exposure without a physical mask is reviewed. In section 2.4, the concept of directed self-assembly is introduced.

2.2 Optical lithography using optical masks

2.2.1 Optical lithography exposure modes and their resolution

The most common way to perform an optical lithography process is by means of an optical mask. An optical mask consists of a substrate (glass or quartz, for example) which is transparent for the wavelength of the optical radiation used for the exposure. On this substrate, non-transparent areas are defined, conforming the patterns that will be transferred to the substrate. The mask is located between

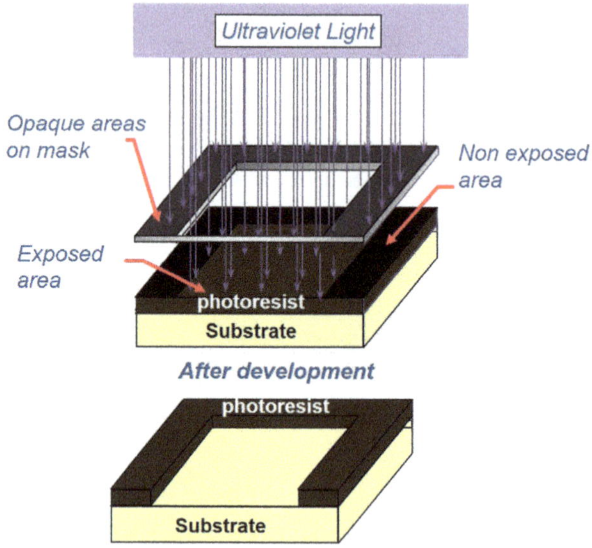

Figure 2.2. General scheme of optical lithography using an optical mask.

the optical radiation source and the sample coated with the photosensitive material. When light crosses the optical mask, the non-transparent areas will block the light and, in consequence, some regions of the photosensitive material will be exposed and some areas not (see figure 2.2).

The scheme depicted in the figure is simple, but the practical implementation is much more complex. In order to achieve micrometric and sub-micrometric resolution the overall exposure process needs to be carefully considered. Specially, an accurate control of the energy intensity delivered to the photosensitive material is crucial to obtain the desired final pattern.

As indicated above, the photosensitive material changes its physical and chemical properties upon exposure. For each photosensitive material, there is an optimal quantity of energy that induces the change of the photosensitive material properties, which in turn depends on the wavelength of the optical radiation and on the thickness of the photosensitive material layer. If the energy delivered is less than optimal, the change in the solubility of the material will not be sufficient, so the contrast between the exposed and unexposed regions will be weak. If the energy delivered is too large, then the photosensitive material may experience degradation or other effects different than the expected change of solubility.

To achieve an accurate control of the energy delivered, the following aspects must be fulfilled:

(1) the wavelength of the optical radiation has to fit with the physical and chemical properties of the photosensitive material;
(2) the intensity of the optical source has to be known and stable;

(3) all the exposed areas have to receive the same amount of energy delivered;
(4) the position of the optical mask has to be fully parallel to the photosensitive material.

Attending to these requirements, figure 2.3(a) presents the three general modes of an optical lithography process: contact, proximity and projection. Figure 2.3(b) shows an image of an extreme ultraviolet (EUV) projection lithography system. The optical projection system is made of multiple mirrors instead of lenses, as this is more appropriate for EUV radiation (see section 2.2.3). Figure 2.3(b) reveals the complexity of these systems, which are only appropriate for high-volume manufacturing activities due to their high costs.

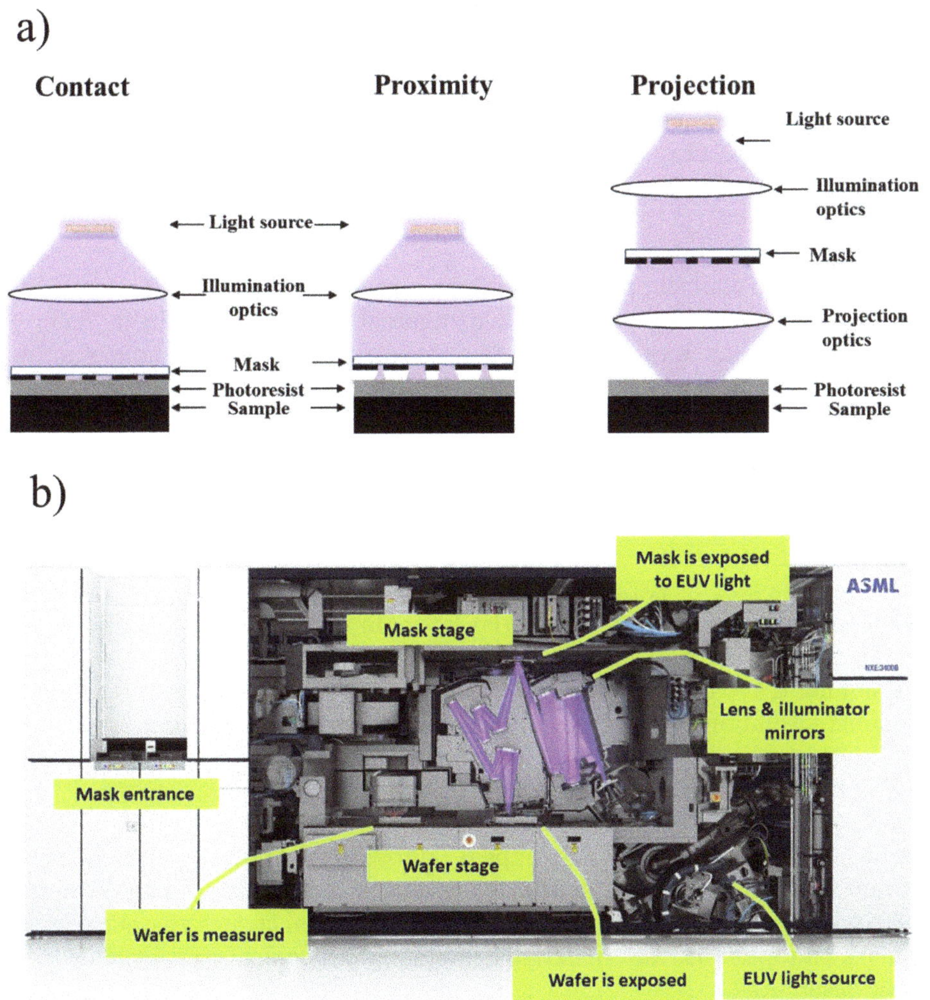

Figure 2.3. (a) Optical lithography modes using an optical mask: contact, proximity and projection. (b) Photograph of an extreme-UV optical lithography system. The main elements are labelled.

2.2.1.1 Contact optical lithography

In contact optical lithography, the mask is placed in mechanical contact with the photosensitive material. In this way, the deterioration in resolution is minimized, as the distance between the apertures in the mask and the exposed volume in the resist is minimized. An approximate calculation of the resolution, R, in this case is the following:

$$R = (d \cdot \lambda)^{1/2} \tag{2.1}$$

where d stand for the thickness of the resist and λ is the wavelength of the optical radiation.

If we consider a common example in a usual microfabrication laboratory, a typical thickness of the photosensitive material is $d = 1$ μm, and the wavelength of the optical radiation in the near UV range, $\lambda = 435$ nm. Then, the minimum feature size that can be printed is $R = 0.7$ μm.

Masked optical lithography in contact presents the drawback of inducing mask damage. The accumulation of defects in the mask means that the mask cannot be used for many exposures, drastically decreasing its lifetime. Consequently, the use of this method should be avoided.

2.2.1.2 Proximity optical lithography

An alternative to contact optical lithography is proximity optical lithography. In this case, the mask and the photosensitive material are separated by a finite gap, preventing any mechanical contact. As a consequence the useful lifetime of the mask is increased, but at the expense of decreasing the resolution. An empirical relation for the resolution in proximity optical lithography is the following:

$$R = k(g \cdot \lambda)^{1/2} \tag{2.2}$$

where g is the distance between mask and the photosensitive material and k is an empirical constant which stands for additional detrimental effects that degrade resolution, like lack of parallelism between the mask and the sample, or the finite thickness of the photosensitive material layer. In consequence, k is always larger than 1.

Assuming a typical situation, $k = 1.1$, $g = 20$ μm and $\lambda = 435$ nm, the best attainable resolution in proximity optical lithography is $R = 3.2$ μm.

2.2.1.3 Projection optical lithography

Projection optical lithography overcomes some of the resolution limitations occurring in contact and proximity optical lithography (such as the damage by mechanical contact and diffraction) by inserting an optical system between the mask and the photosensitive material. The optical system creates an optical image of the mask on top of the resist material, eliminating mechanical contact while keeping a good resolution.

Resolution in projection optical lithography is calculated as in the resolution in any optical system, but including the specificities of the fact that a photosensitive material is exposed. In practice, the resolution is calculated attending to the following semi-empirical relation:

$$R = k_1 \frac{\lambda}{NA} \tag{2.3}$$

where NA is the numerical aperture of the optical system and k_1 is known as the technological constant which stands for all the improvements engineered to enhance the resolution. Considering moderate values of the resolution in optical lithography like $k_1 = 0.66$, $\lambda = 435$ nm, $NA = 0.7$, an important improvement in resolution is obtained: $R = 0.4$ µm.

Projection optical lithography was introduced in the 1970s [1]. Since then, improvements in multiple aspects have allowed for a continuous enhancement of resolution. Resolution in projection optical lithography (i.e., expression (2.3)) cannot be decoupled from the need to keep the exposure in focus. Depth of focus (DOF) stands for the range of vertical distance over which the broadening of the image becomes acceptable. DOF for projection optical lithography can be estimated through the semi-empirical expression:

$$\mathrm{DOF} = k_2 \frac{\lambda}{(NA)^2} \tag{2.4}$$

where k_2 is a constant that, similarly to k_1, depends on technological and process parameters. Comparing expressions (2.3) and (2.4), it is observed that *DOF* decreases when the resolution of the optical system improves. This imposes important requirements for resist thickness and wafer surface planarization.

2.2.1.4 Improvements in projection optical lithography

Projection optical lithography is the method used to obtain high resolution. Figure 2.4 shows the evolution of lithography systems used over the years to improve the resolution.

In the following, the different advances introduced in projection optical lithography will be described:

(a) Reduction optical lithography: step-and-repeat and step-and-scan systems

Introducing a reduction (or demagnification) factor in the optical system presents several advantages: (i) if the reduction factor is n, the features on the optical mask will be n times larger than the exposed features, relaxing the requirements to fabricate the masks. (ii) As the field of exposure decreases, it becomes easier to keep the image in good focus, which is critical when the DOF decreases for high-resolution systems. Masks used in combination with reduction optical lithography are usually referred to as optical reticles (see figure 2.5).

The introduction of an optical reduction factor implies that the whole wafer area will not be exposed in a single illumination shot. In consequence, reduction optical lithography is performed in combination with step-and-repeat and step-and-scan systems in stepper equipment. In the step-and-repeat

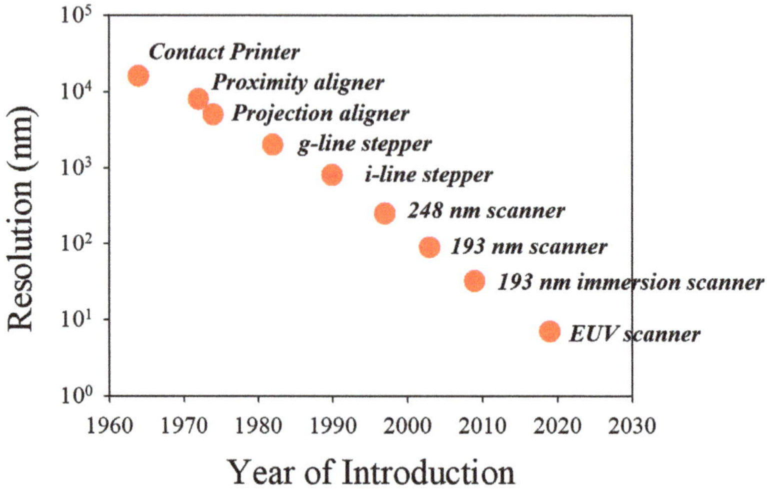

Figure 2.4. Evolution of the optical resolution during the years. Data extracted from [2].

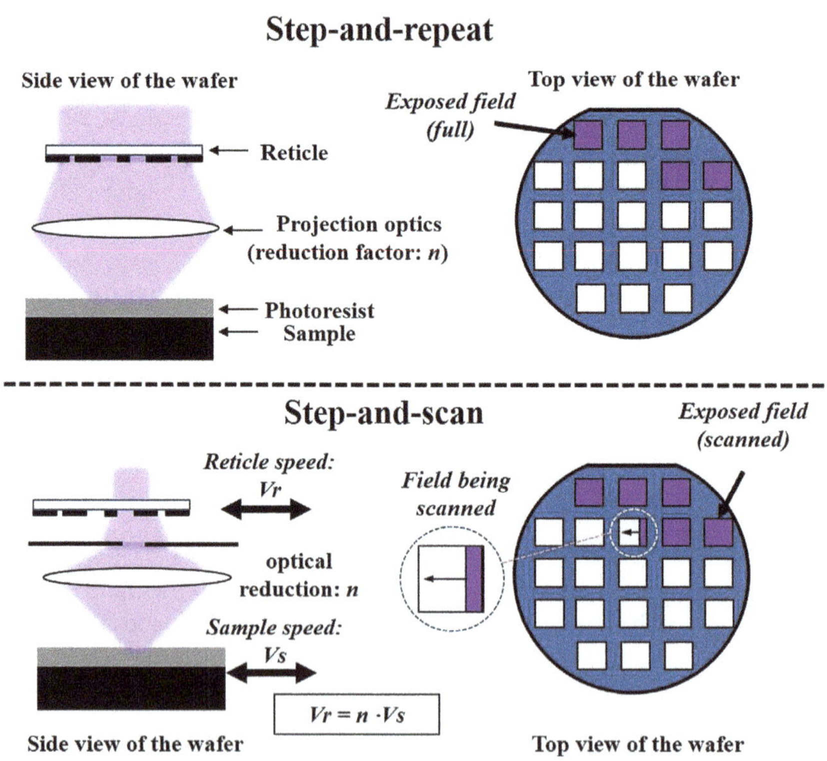

Figure 2.5. Step-and-repeat and step-and-scan systems. In step-and-repeat, the reticle is fully illuminated during the field exposure. In step-and-scan, the reticle is scanned by the laser. In consequence, sample and reticle are moved in a synchronous way, with a scanning speed ratio equal to the reduction factor of the optical system. Adapted from [3].

system, all the desired area of the reticle is exposed in a single shot, then the sample is moved a fixed quantity, and another exposure is performed. Each different exposed area corresponds to a dice or chip.

A further improvement is achieved using step-and-scan systems. While the size of the reticle would be the same for step and repeat and step and scan systems, in the step and scan systems the reticle is only partially illuminated. At the same time, the reticle and sample are being scanned in a synchronous way: for a reduction factor *n*, the scanning speed of the reticle has to be *n* times faster than the scanning speed of the sample. Illuminating a smaller area of the reticle (and, in consequence exposing a smaller area of the wafer) allows for a better focus capability and illumination uniformity.

(b) Decreasing the wavelength of the optical source

A direct implication of expression (2.3) is that resolution improves as the wavelength of the optical radiation decreases. Three different ranges of wavelength are employed: near ultraviolet (NUV), deep ultraviolet (DUV) and previously-introduced EUV. In all cases, light sources of narrow bandwidth are employed: as the optical properties of lenses and the opto-chemical properties of the photosensitive materials depend on the wavelength, it is necessary to use a specific wavelength to achieve an accurate exposure.

- NUV optical lithography

  NUV optical lithography uses optical radiation of 436 nm (g-line), 405 nm (h-line) and 365 nm (i-line). The lines correspond to the optical emission spectra of mercury-arc lamps. The light sources for NUV optical lithography have been obtained from mercury lamps with special optical filters for many years. Presently, UV LED light sources are being employed as they can offer wavelength tunability, optical filters are not required, maintenance is much more simple, offer longer lifetime and are much more eco-friendly. NUV optical lithography is used for optical lithography in the proximity mode and also for step-and-repeat systems.
- DUV optical lithography

  DUV optical lithography systems use optical radiation of 248 and 193 nm, the latter being one of the most used in present systems. KrF and ArF excimer lasers are used as optical light sources for 248 and 193 nm wavelengths, respectively. ArF excimer lasers generate laser light from the stimulated emission of ArF complex created from the ionized gas containing argon and fluorine. Excimer laser operates in a pulsed mode, with typical frequencies of kHz. DUV optical lithography is usually performed in the optical projection mode using step-and-repeat or step-and-scan systems.
- EUV optical lithography

  EUV optical lithography provides the ultimate resolution in mask-based optical lithography using 13.5 nm wavelength. A separated section is dedicated to EUV due to its future importance for the next generation of semiconductor electronics.

(c) Increasing the numerical aperture NA: immersion DUV optical lithography

Increasing the numerical aperture requires increasing the size of lens or reducing the focal distance. Both options present limitations as the need for a minimum distance between the final part of the lens and the resist to account for the sample scanning and also to avoid damaging the lens from the evaporation of material during the exposure.

A solution to it has come from immersion lithography, a key breakthrough for decreasing the attainable resolution of optical lithography, as the presence of a liquid with a refraction index n increases n times the value of NA. In particular, pure water, with $n \approx 1.47$, is suitable for this purpose as it only absorbs ~5% of the incident light. It is usually compatible with the lenses and the photoresist, and it does not contaminate the system.

(d) Decreasing k_1:

As mentioned before, k_1 accounts for adjusting the resolution value from the technological constraints. k_1 has a physical lower limit of 0.25 [4]. Additional difficulties arise when k_1 is decreasing, such as the amplification of errors in masks that cannot be compensated by the demagnification [5] and the need of a very flat topography to compensate for the low depth of focus [6]. Approaching these physical lower limits is achieved by several improvements that cover many aspects of a lithography process:

i. Off-line illumination. Engineering the illumination of the mask improves the resolution because higher NA optical system can be used while maintaining a durable depth of focus. Engineered off-line illumination implies stopping the light from the center along with patterning the optical beam to obtain annular dipole and quadrupole illumination.
ii. Optical proximity correction (OPC). Optical proximity effects are important in dense patterns, and it is caused by the dependence of the actual exposure dose on the geometry of the patterns of the mask. For example, for the same mask illumination, a dense array of exposed lines will receive more dose than a single-exposed line. When features of different geometries and sizes exist, correcting for the proximity effects requires adjusting the geometries, what is known as OPC. For complex structures at the limit of resolution, OPC is complex and requires dedicated modelling.
iii. Phase shifting masks [7]. Diffraction of light through small apertures makes difficult the definition of high-density features. When the optical source is coherent, constructive interference between the light transmitted from two adjacent apertures maximizes the intensity in between the apertures. The underlying concept in phase shifting mask

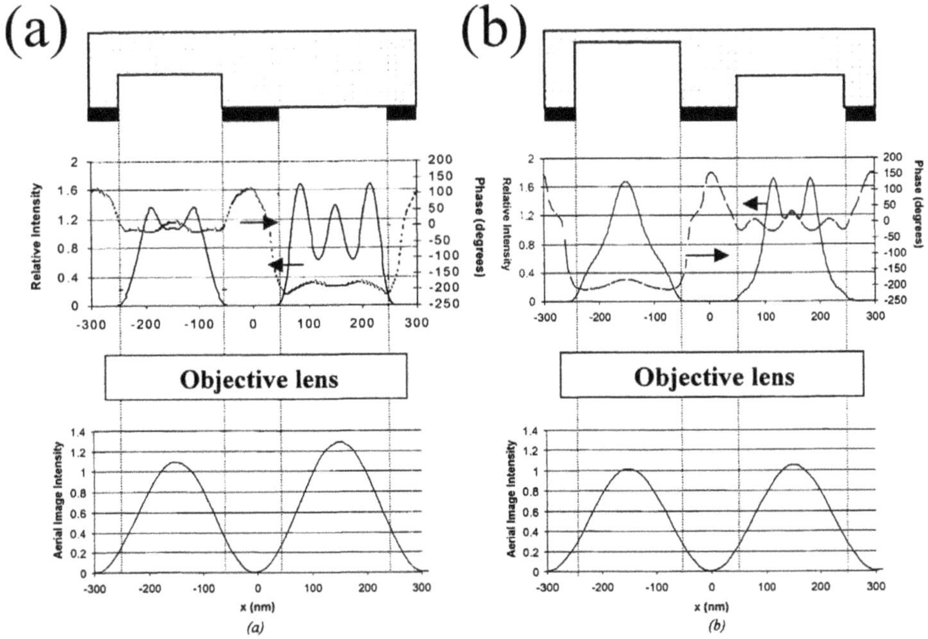

Figure 2.6. Concept of alternating phase shifting mask. Calculation of the in-focus aerial image after passing through a 4X stepper at DUV illumination (248 nm wavelength), and numerical aperture of 0.6 for a simple subtracting alternating phase shifting mask (a) and a dual trench one (b). Figure reproduced from [8], copyright SPIE.

is the variation of the optical path through the mask to add a 180° phase shift to the optical wave, promoting a negative interference in between the apertures. In the case of alternating phase shifting mask, a recession in the apertures allows the control of the phase difference between the optical radiation through the apertures (see figure 2.6).

2.2.2 Photo-resists

Optical lithography is based on the selective exposure of a thin layer of a photosensitive material that is spread over the surface where the patterning is going to be realized. The photosensitive material is called a photoresist, because its main role, after being exposed and revealed, is to protect the surface from the subsequent pattern transfer.

Usually, a photoresist is composed of, at least, the following three components: the resin, a photoactive compound and the solvent. The resin is the polymer that remains on the surface after the exposure and development. The physical–chemical structure of the polymer changes as a consequence of the exposure to optical radiation. The photoactive component controls the photochemical reactions that induce the structural change of the resin. The solvent dissolves the resin in a solution of a certain viscosity. Usually, the photoresist is deposited on the surface by spin coating, forming a thin and very uniform layer of polymer. The viscosity of the solvent along with the spinning speed defines the final thickness of the polymer layer. After deposition, an annealing process is performed to remove the solvent.

The resin and the photoactive compound determine the response of the photoresist to the exposure and development. In the case of a positive tone photoresist, the resin becomes more soluble in the developer, so that the exposure regions are eliminated. For a negative tone photoresist, the effect of the exposure is to decrease the solubility. As a consequence, the exposed areas remain on the surface.

2.2.2.1 Positive resists

For the most used class of positive tone photoresist, the photoactive compound inhibits the dissolution of the resin in the developer before exposure. Upon irradiation, the photoactive compound triggers a chemical reaction that increases the resin solubility [9]. The most used photoresist for NUV optical lithography is based on Novolac resin [10]. Novolac resins are typically produced with formaldehyde, at molar excess of phenol, and an acid catalyst. Typical photo-active compounds belong to the DiazoNaphtoQuinon-sulphonate group (DNQ). Their presence in photoresist reduces the alkaline solubility in one or two orders of magnitude. During exposure, DNQ is converted into a carboxylic acid, causing that the development rate increases by three to four orders of magnitude.

The resists used for NUV optical lithography are not adequate for DUV optical lithography, as they have to face two issues: the lower intensity of the DUV optical radiation and the higher absorbance of NUV resists to DUV radiation. Chemical amplified resists solve the problem of lower optical intensity, increasing the sensitivity of the resist. They are based on the use of photoacid generators (PAG) as photoactive compound. Upon exposure the PAG produces an acid, which catalyzes the breaking of the protecting groups of the resin. As the acid actuates as a catalyzer, the effect on enhancing the solubility of the resin is maintained after the exposure. Concerning the requirement for transparency at DUV wavelength, the resist is no longer based on Novolac but on other polymers like poly (hydroxystyrene).

2.2.2.2 Negative resists

For the most used class of negative tone photoresists, the exposure causes a cross-linking of the polymer molecules, and, consequently, the solubility decreases in the exposed volume. A review about negative resists can be found in [11].

2.2.2.3 Photoresist profile after exposure and development

The final profile of the resist is key to obtain the desired features on the sample after pattern transfer. The resulting profile is a function of the photoresist properties and of the conditions of exposure and development and it is difficult to predict. For the resist properties, it is important to know the thickness, the optical refraction index, the optical absorption and the resist contrast. These properties will define the optimal conditions for exposure and development. Optimal conditions are those allowing a perfectly-vertical profile of the resist. Figure 2.7 schematizes the profiles

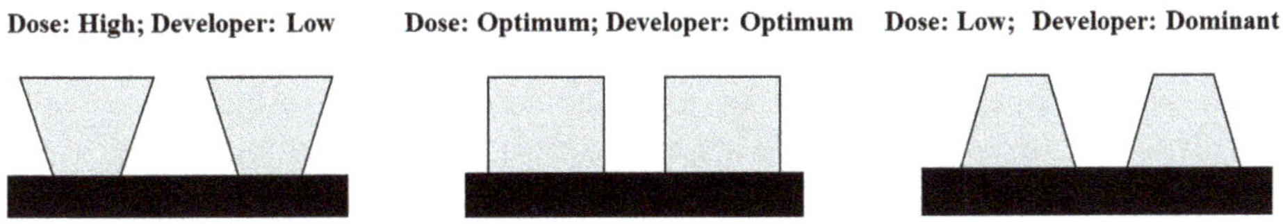

Figure 2.7. Positive photoresist profiles obtained as a function of the exposure (dose of the photoresist exposure) and the development time.

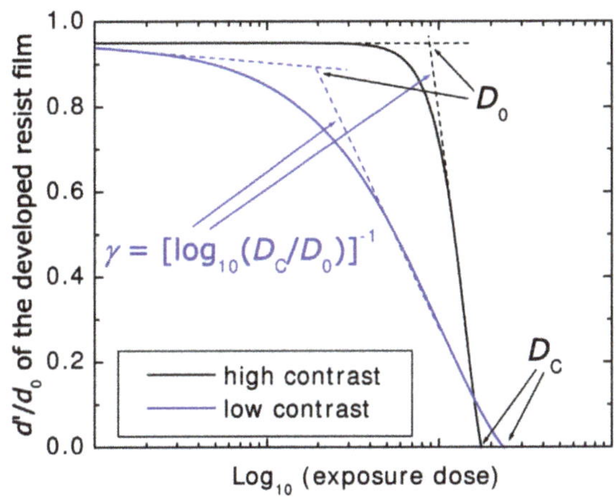

Figure 2.8. Example of the photoresist contrast curve for high contrast and low contrast positive resists. Figure reproduced from [12] with permission of MicroChemicals GmbH.

that can be usually obtained for positive photoresist when the conditions diverge from the optimal ones. Optimal conditions can be obtained from the datasheet of the photoresist.

An additional factor influencing the final profile is the reflection of light at the interface between the resist/sample interface. The additional dose arising from the reflection causes an overexposure and, in consequence, a degradation of the resolution. To avoid this effect, antireflection coatings are usually added to the sample before the resist coating step.

2.2.2.4 Photoresist contrast curve

The contrast curve of a photoresist shows the dependence of the ratio of the remaining photoresist thickness (d'/d_0) as a function of the exposure dose (see figure 2.8).

The photoresist contrast parameter, γ, is defined according to the following relation

$$\frac{d'}{d_o} = \gamma \log_{10}\left[\frac{D_c}{D}\right] \tag{2.5}$$

where D is the dose and D_c is the clearance dose.

2.2.3 Extreme ultraviolet (EUV) optical lithography

As has been properly discussed in previous sections, the resolution of optical lithography is proportional to the wavelength (λ) of the radiation. Thus, as early as in the 1980s decade, it was proposed to use light with $\lambda \sim 10$ nm, that is, more than one order or magnitude smaller than the ultraviolet radiation that has been used in lithography for many decades. Initially, the possible use of this short wavelength light was thought for proximity configuration (see section 2.2.1), in the nominally called soft x-ray proximity lithography. However, it was early noticed [13] that projection could be the appropriate configuration in order to obtain the necessary high-quality lithography results. This technique is called EUV lithography [14, 15] which, finally, after overcoming several challenges associated with the use of small λ radiation, has recently become part of the commercial chip production system.

Those challenges have been related with each part of the lithography process: the light source, the optical system, the mask, and the photoresist. Therefore, deep modifications have been implemented in them with respect to those detailed in previous sections, as will be described in the following.

2.2.3.1 EUV sources

As has been discussed in section 2.2.1, any light source used for optical lithography, and, therefore, EUV radiation, must fulfil several requirements, related to its stability, the homogeneity over the exposed area, and a high-enough intensity level to transform the photosensitive resist.

Thus, in order to produce light with those characteristics and $\lambda \sim 10$ nm, it is necessary to generate a high-energy plasma (typically in the order of $\sim 10^5$ K), as other alternatives like synchrotron radiation are not very feasible for the solid angles needed in wafer production. In any case, the production of a high-energy plasma requires a special preparation. First of all, it must be considered that almost all the matter absorbs the radiation at these wavelengths, so that it is mandatory to produce and propagate the light in a system with the highest possible vacuum. Then, there are two methods to generate this kind of dense plasmas: (i) by a quick pulsed electrical discharge between two metallic electrodes in the presence of a flow of a low pressures gas or a jet of small liquid droplets, which is called discharge produced plasma (DPP); (ii) by focusing a highly energetic laser on the flow of liquid droplets, which is called laser produced plasma (LPP). Nowadays, it is considered that LPP method is more appropriate for chip lithography production lines due to its better power scalability, while lower-cost DPP sources are planned for complementary applications like mask inspection where low-power EUV radiation is used.

In general, to produce the required high-energy plasma, the presence of multi-charged positive ions is necessary. Thus, xenon plasmas with typically Xe^{10+} ions and, moreover, tin plasmas with ions ranging from Sn^{8+} to Sn^{13+} are generated to get an EUV emission that can be very strong in a narrow band centered close to $\lambda = 13.5$ nm. It is important to note that, in addition to the plasma production by LPP method, the interaction of the laser beam with the tin droplets also results in tin vapour and particulates that could quickly contaminate or, even more, coat the collector of the optical system that is close to the location where the plasma is generated. Therefore, it is mandatory to implement a contamination mitigation mechanism within the plasma production device (like the use of hydrogen as a buffer gas in the chamber to efficiently decelerate the fast plasma ions and chemically react with them, or the presence of a magnetic field in order to guide and discharge the ions along the magnetic lines of force) to avoid a too frequent replacement of the optical collector interrupting the operation of the lithography system.

Then, in the LPP source of an EUV lithography system (see a sketch in figure 2.9), a stream of Sn droplets is injected in an ultra-high-vacuum vessel from a heated reservoir with highly-purified tin. The droplets have a diameter size of ~30 μm and fly at a speed larger than 50 m s^{-1}, the flow rate being several tens of thousands of droplets per second. This droplet flow is monitored with high-speed cameras in order to ensure as much as possible the targeting of the droplets with the focused spot of the laser pulses. The droplets are usually irradiated with a high-power CO_2 laser (~50 kW) that works, accordingly with the droplet flow, at several tens of kHz. Actually, in order to increase the efficiency of the LPP process, they are irradiated twice; first, the droplet receives a pre-pulse that modifies its shape from a sphere-like geometry to a flattened disk; subsequently, the main laser pulse hits the deformed droplet. It favours the conversion of the laser power into the plasma generation that produces the EUV light. After it, the EUV radiation is collected by an ellipsoidal mirror, focused in an intermediate focus, and, from there, transported out of the

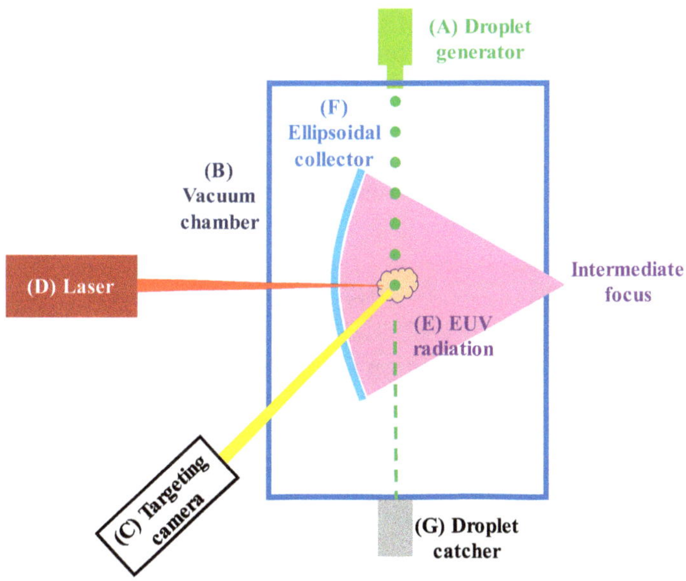

Figure 2.9. Sketch of a LPP EUV-source. A stream of Sn droplets is injected from a generator (A) into the vacuum chamber (B). The droplet position is monitored with cameras (C) to tune the targeting with the laser pulse (D). It produces the plasma with EUV emission (E). With an ellipsoidal collector (F), this EUV radiation is focused to transport it to the exposition area of the lithography system. In front of the droplet generator, a droplet catcher (G) takes the droplet waste.

source chamber by an optical system whose main characteristics are described in the following. Finally, it must be noted that the conversion efficiency of the energy is very low, approximately of about 5%; however, it can be enough to reach an output power level of ~250 W.

2.2.3.2 EUV optical system

From the optical point of view, the fact that the radiation with $\lambda \sim 10$ nm is absorbed for any solid material implies that the refractive optics (based on quartz or, even, CaF_2 lens) used in NUV or DUV stepper lithography systems no longer apply for the short-wavelength EUV light. Instead, a reflective optics architecture must be configured in order to transport the light and illuminate the wafers.

In this sense, although for $\lambda < 30$ nm the reflectance at normal incidence of any natural material is only of ~1% as the refraction index is very close to 1, multilayer structures made of two materials with very different absorption coefficients can result in much larger reflectance values. Thus, mirrors based on multilayer structures are designed for EUV lithography, similarly to the case of x-ray telescopes in astronomy. In particular, Mo/Si multilayers with the appropriate thicknesses and a large number of Mo/Si bilayers (see sketch in figure 2.10) can provide reflectance values up to 70% approximately for EUV radiation with $\lambda = 13.5$ nm, that is, the one produced by the source of the lithography system [16]. Typically, the number (N) of Mo/Si bilayers is of the order of 50. In general, for a constructive interference close to normal incidence between the beams reflected at the different multilayer interface, the thickness (t) of the Mo/Si bilayer must fit $t = \lambda/2$, that is $t \approx 6.8$ nm. To optimize the reflectance, the thickness of the Mo layers is around a little more than half of the thickness of the Si layers. Moreover, it is completely necessary that the deposition of the layers is as perfect as possible in order to avoid losses at the interfaces due to imperfections such as roughness or interdiffusion. Thus, the deposition of additional ultrathin layer barriers of boron carbide at each interface results in a significant reduction of the interdiffusion and compound formation at the Mo/Si interfaces.

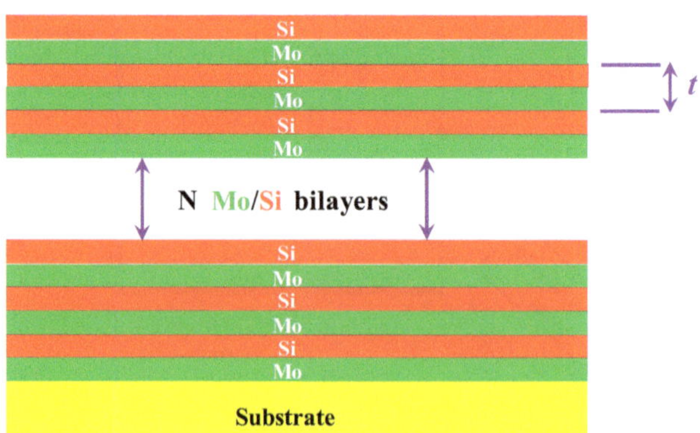

Figure 2.10. Sketch of the Mo/Si multilayer structure of the EUV mirrors. The number of bilayers is, typically, $N \sim 50$. The thickness of each bilayer must fit $t \approx \lambda/2$. To protect the multilayer from oxidation, a thin capping layer of Ru is usually grown on top.

Also, as the dispersion of the light due to roughness (σ) has, approximately, a dependence ~ $(\sigma/\lambda)^2$, considering the short EUV wavelength, the surfaces must present a very high quality smoothness, with a roughness of only 0.1 nm (root mean square). This requirement must be fulfilled along the complete curved surface of all the mirrors of the optical system. This is a huge challenge taking into account that, for example, the ellipsoidal collector of the EUV source has a large diameter of about 600–650 mm to allow the collection of the EUV light emitted by the source in a solid angle of ~5–5.5 steradians.

From the point of view of the architecture of the optical system, after the EUV radiation has been redirected by the ellipsoidal mirror collector to the intermediate focus aperture, the light arrives to the illumination optics part where, after several reflections, it is focused on the EUV mask element and, from there, it is projected to the wafers. In total, the number of multilayer mirrors where the EUV is reflected from the source to the wafer in an EUV lithography system can be 12. In each reflection, the incidence angle with respect to the mirror normal is small, about ~5°–6°. The large number of reflections (n) implies that, although the reflectance in each multilayer mirror is ~70%, the portion of the EUV light that arrives at the wafers from the radiation source is small, $(0.7)^n$ ~ 0.01–0.02, that is, of only a few percent.

On the other hand, considering the resolution of the lithography system (see equation (2.3)), the size of the multilayer mirrors together with their spatial configuration resulted in low values of the numerical aperture (NA $\approx$ 0.25) in the first EUV lithography tools. Increasing the surface size of the mirrors has led to higher values of the NA and, therefore, to an improvement of the lithography resolution. Thus, at the end of the 2010s decade, the first EUV lithography commercial systems were shipped having NA = 0.33, for a 13 nm resolution.

2.2.3.3 EUV lithography masks

Similarly to the other optical elements, due to the short wavelength of the radiation, the mask objects for EUV lithography must work in reflection, so that they present a complex structure that is very different from the masks used in NUV or DUV optical lithography (see a sketch in figure 2.11).

First, as more than 30% of the EUV incident radiation is absorbed in the process, the substrate of the mask must be of a material with a low thermal-expansion coefficient, such as TiO_2-doped SiO_2 amorphous glass, in order to prevent the distortion of the image when the mask is heated by the energy of the absorbed light. On top of the substrate, a Mo/Si multilayer with ~50 bilayers is deposited to act as a reflector for the EUV 13.5 nm light, in the same way as in other lenses of the optical system. This multilayer is capped with a thin layer of a protection material, like Ru, in order to prevent oxidation as well to stop the subsequent etching process that will be done on top to define the desired pattern. For this purpose, a film of an EUV absorber material such as tantalum nitride (TaN) or tantalum boron nitride (TaBN) is grown on the protected multilayer. Also, it is usual to deposit a DUV antireflection coating (ARC), for example a Ta oxide, to improve the inspection contrast. Then, the absorber/ARC is patterned by electron beam lithography and an etching process to define the desired design to be transferred to the wafer. Thus, the

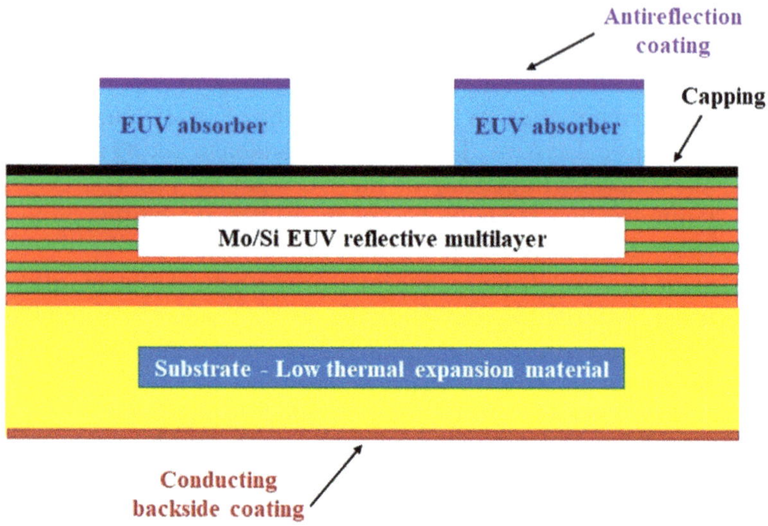

Figure 2.11. Sketch of the cross section of the structure of a EUV lithography photomask.

absorber patterned regions act as a 'black' area for the transmission of the EUV light, while the free Mo/Si multilayered regions work as if they were 'transparent' for the EUV radiation propagation. Finally, in the backside of the substrate, a coating thin layer of a conducting material like CrN is deposited to help the electrostatic fitting to the mask stage.

It is important to note that, for the short wavelength of the EUV light, any tiny defect in height causes a local difference in the reflection and, as a consequence, a phase defect appears in the pattern transferred to the wafer. Therefore, it is mandatory that the substrates are extremely smooth and highly defect-free before the deposition of the Mo/Si reflective multilayer, as later it would not be possible to repair them. Only defects associated with the absorber patterning can be relatively straightforwardly fixed, but it is not the case for the defects related with a height inhomogeneity in the Mo/Si multilayer; in this latter case, it is only possible to detect the defect area to avoid the use of that region for patterning. On the other hand, as any material absorbs EUV radiation, after the fabrication of the mask it cannot easily be protected with a pellicle like in the case for other optical UV lithography masks. Then, without this protection, any dirt or contamination particle that appears is directly in contact with the patterned reflective mask. Although it could be carefully cleaned, it would imply a shortening of the mask lifetime.

Finally, another problem is related with the 3D aspect ratio of the mask; in this sense, it must be considered that, for high performance, the thickness of the absorber layer must be, at least, ~50 nm. Thus, due to the reflective optics configuration, there is a finite angle of incidence of the EUV radiation with respect to the substrate normal; this angle is small, of ~6°, but results in shadowing effects at the edges of the patterned structures that must be taken into account when the mask pattern is defined.

2.2.3.4 Photoresist for EUV lithography

The use of EUV radiation also has implications in the characteristics required for the photoresist [15]. First, after all the losses in the radiation power along the system, the dose at the wafer is typically less than half of the one used in DUV lithography for similar exposition times, requiring a resist with a low clearance dose to change its solubility.

On the other hand, considering the chemical process itself, as discussed in section 2.2.2, in the case of DUV lithography it is necessary to increase the sensitivity of the resist by using PAG photoactive compounds that catalyse the breaking of the groups of the resin, improving its solubility. For EUV radiation, the process is different. Now, the energy of the EUV photons is much higher. As a consequence, they are not absorbed by functional groups but by individual atoms. Thus, the immediate result of the photon absorption is not a molecule in an excited state, but the emission of a highly energetic electron from the individual atom. This emitted electron then interacts with other atoms and functional groups when travelling across the resist, with the possibility of secondary electron generation. Both, primary and secondary electrons, can trigger the PAG catalytic action. Finally, this fact helps the sensitization process of the resist as only a small fraction of the radiation is absorbed directly by the PAG molecules, but they are still activated by the electrons coming from other atoms in the whole resist layer. Conversely, the effect of the secondary electrons can result in a noise contribution to the line-edge roughness. Actually, each EUV absorbed photon changes the solubility of a certain volume of the resist, which is called *blur*. A higher resist resolution implies a smaller blur, but smaller blur indicates lower number of reactions and, therefore, slower photospeed, so that there is a trade-off between line roughness and photospeed.

Besides, in order to avoid the collapse of close patterned structures with high aspect ratios as well as to work with a lower depth of focus, it is convenient that the thickness of the resist layer is not larger than ~20–30 nm. For these values, it is a real challenge to reach the appropriate light absorbance level. Then, it is necessary the use of resists based on metallic nanoparticles or organometallic compounds, given the high absorbance of heavy atoms.

Despite all the difficulties, very high resolution has been reached with EUV lithography and, nowadays, there is commercial equipment shipped by ASML company which, with a numerical aperture of 0.33 in the optical system and a dose of 20 mJ cm^{-2}, is able to produce 125 wafers (300 mm in diameter, with 96 dies) per hour. It has become competitive as it simplifies the multi-patterning immersion process associated with high resolution DUV lithography, and, therefore, reducing the relative cost per patterned pixel. With a further improvement in some of the parameters (an enhancement of the conversion efficiency of the EUV radiation, an increase in the numerical aperture of the optical system, the implementation of some phase-shift strategies in the mask, better tuned resist characteristics, …) it is expected that in the next years the EUV resolution can be further moved to reach the 7 or 5 nm node values.

2.3 Maskless optical lithography

2.3.1 Direct laser lithography

The high resolution and throughput of photomask-based optical lithography come at a great cost due to the complexity and high price of steppers as well as the long fabrication time and cost of masks. In 1989, a new method was proposed to expose a photoresist without the need of optical masks [17]. This method, called direct write lithography (DWL), relies on a laser beam, which can be focused down to 1 μm or less, and an XY-moving stage holding the resist-covered sample, as sketched in figure 2.12. In DWL, the light produced by the laser exposes the resist in a sequential way across the full area of the sample. Thus, in DWL the total exposure time is much longer than in conventional optical lithography using a photomask. The simple scheme shown in figure 2.12 is indeed more complicated in commercial equipment in order to enhance the writing speed. In the first publication on DWL by Rensch *et al* [17], the lens system included a rotating polygon mirror that allowed a fast raster scan of the laser beam along one direction, being in practice a laser scanner. Another approach to produce a fast raster scan of the laser beam along one direction is to use an acousto-optic modulator. More recently, other approaches have been introduced in DWL technology. For example, the use of a spatial light modulator (SLM) placed after the laser allows the dynamic projection of a given pattern on a working area with typical linear length of a few hundreds of microns.

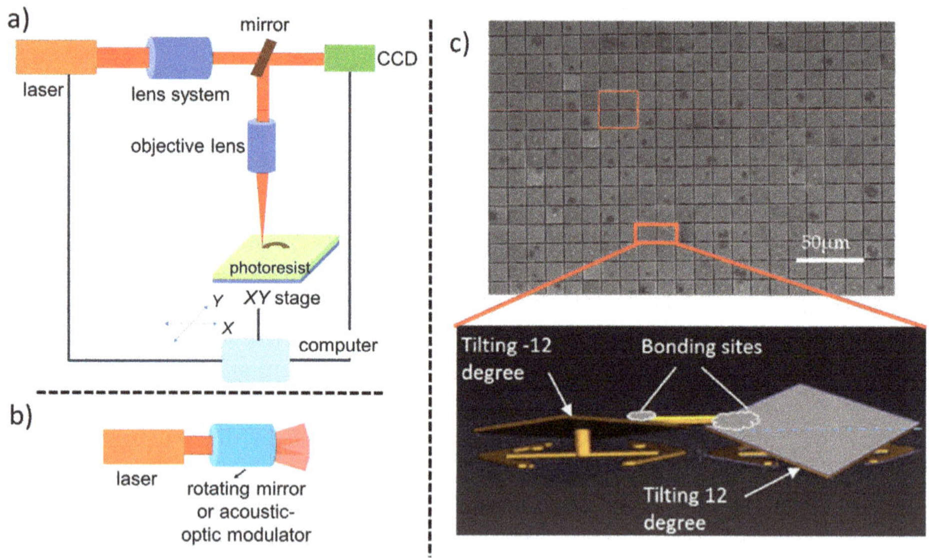

Figure 2.12. (a) Sketch of DWL. Light from a laser beam is projected through the lenses and the mirror reflection onto the photoresist in order to produce its exposure. A computerized XY stage allows scanning the full sample area sequentially. The CCD camera allows imaging the sample. (b) In some DWL systems, the laser is raster scanned along one direction using a rotating mirror or an acousto-optic modulator. (c) In some DWL systems, an array of micro-machined mirrors acts as a dynamic mask, projecting a pattern on a small area (typically <1 mm^2). Adapted and reprinted from [19] MDPI.

Once the targeted pattern has been imprinted in the resist on a small area (<1 mm^2), the stage is moved to the next working area. An SLM can be considered a dynamic mask that can be changed from a certain working area to the next one. SLMs can be based on liquid-crystal technology or on an array of micro-machined mirrors (sometimes known as digital light processor, DLP, or digital mirror device, DMD) [18]. Nowadays, both concepts, that of a well-focused laser beam scanning across the sample and that of an SLM projecting a dynamic mask, can be found in commercial DWL equipment.

The resolution achievable in patterns exposed by DWL is ultimately limited by the light diffraction, thus, of the order of the wavelength of the laser. Typical laser wavelengths used in DWL are around 400 nm. As a consequence, the same or similar photoresists to the ones commonly used in NUV optical lithography are used in DWL. Processes have been developed using either positive resists or negative ones. In figure 2.13, two examples of 2D structures patterned by DWL are shown. In figure 2.13(a), representative SEM micrographs of high aspect ratio (AR, up to 40) microstructures after DWL exposure, post-exposure bake (85 °C) and development are shown. In this example, an epoxy negative-tone resist and a laser wavelength of 413 nm have been used [20]. In figure 2.13(b), SEM and AFM images of a process for the fabrication of x-ray diffraction gratings are shown. The top SEM micrograph

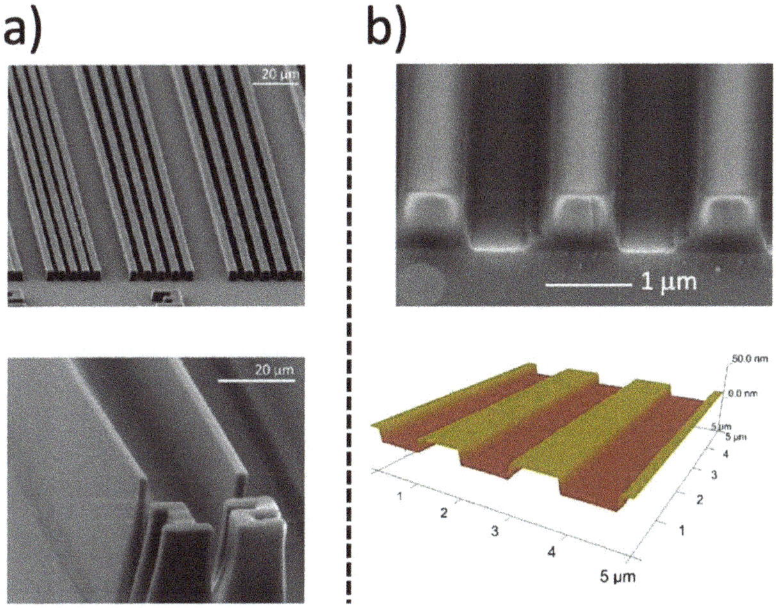

Figure 2.13. (a) SEM micrographs of high aspect ratio microstructures fabricated by DWL. The bottom image contains microstructures with widths of 1 and 2 µm and height of 40 µm. Reproduced from [20]. Copyright IOP Publishing. Reproduced with permission. All rights reserved. (b) Fabrication of x-ray diffraction gratings by DWL with 600 lines/mm. The top image is an SEM micrograph of the resist after exposure and development whereas the bottom image is an AFM measurement of the structure after O_2 plasma etching and resist removal. Reprinted from [21], copyright SPIE.

displays the 550 nm thick photoresist (S1805) on Si after exposure with line periodicity of 600 lines/mm (line width ~0.8 μm) and development [21]. The bottom AFM image has been taken in the same area after oxygen plasma, resist removal and cleaning steps.

In general, the advantages of DWL compared to mask-based optical lithography are the cheaper equipment and the absence of physical masks, making the technique affordable and flexible. On the other hand, the disadvantages are the long resist exposure times and the worse resolution when compared to projection optical lithography described in the previous sections. DWL has been a commonly used technique in last years for the following applications:

- fabrication of optical masks and nanoimprint templates [22];
- fabrication of MEMS [20];
- fabrication of micro-optics components like diffraction gratings, waveguides and microlenses [23];
- fabrication of microfluidic channels [24];
- fabrication of electrical contact pads and micro-contacts [25];
- patterning of magnetic or superconducting materials [26];
- fabrication of sensors [27].

Interesting developments in DWL have brought about the fabrication of non-flat structures, for instance exploiting the gray-scale lithography capabilities of the technique. In gray-scale DWL, each pixel is addressed individually, providing the desired light intensity (typically up to 1000 gray levels are used). As a result, the resist thickness after exposure and development can be tailored from zero up to the maximum value. This is very useful to fabricate certain structures such as micro-lens arrays, MEMS or other mechanical devices containing several heights, blazed diffraction gratings, light diffusers and reflectors, microfluidic channels, etc. Moreover, 3D structures have been achieved by one-photon absorption DWL when working in an ultra-low absorption regime [28]. In this case, at each incidence point the laser is focused at different heights with different light intensity. However, the two-photon absorption technique to be described in the next section is more appropriate for the fabrication of 3D structures as it allows a higher lateral resolution.

DWL commercial equipment can be purchased from companies such as Heidelberg Instruments or Durham Magneto Optics Ltd. A great variety of DWL products are sold, from tabletop equipment for exposure of small substrates with resolution down to 0.5 μm to large equipment that is capable of exposure on a substrate size of 140 cm × 140 cm with writing resolution down to 0.75 μm. Such large equipment is suitable for photomask fabrication to be subsequently applied for the fabrication of large display panels, LEDs or touch screens. Interestingly, in the same equipment, DWL can be combined with a scanning-probe-lithography (SPL) module, providing the large-area lithography capabilities of DWL and the high-resolution capabilities of SPL [29]. Moreover, the mix and match concept can be used to take advantage of the complementarity of DWL with other techniques such as electron beam lithography or nanoimprint lithography [30].

Another strategy derived from DWL is the realization of interference laser lithography, in which the initial laser beam is split into two beams, which are driven to interference on the photoresist at an angle 2Θ. We show two different approaches used to generate interference patterns on the photoresist, as shown in figure 2.14. In the first approach, see figure 2.14(a), the light from a laser beam is split into two laser beams, which are driven to interference on the photoresist. In this configuration, the modulation of the light intensity profile is given by $\lambda/(2*\sin\Theta)$, producing line-shape patterns with the same periodicity. If the sample is rotated 90° and a second exposure is carried out, a chess-like pattern is produced. In the second approach, a double iris at the exit of a laser generates two coherent laser beams, which are driven to interference on the photoresist. In the sketch shown in figure 2.14(b), one can see that besides the angle 2Θ formed by the laser beams, it is possible to modify the α and β angles, which define the sample position with respect to the light incidence angle. The 2D square structure displayed in figure 2.14(c) shows periodicity of 1 μm and has been fabricated with $\Theta = 9.35°$ and $\lambda = 325$ nm with two consecutive exposures with $\alpha = 0°$ and $\alpha = 90°$ whilst keeping $\beta = 0°$. On the other hand, the 3D square structure displayed in figure 2.14(d) shows periodicity of 2 μm and has been fabricated with $\Theta = 7.38°$ and $\lambda = 514$ nm with three consecutive exposures with $(\alpha,\beta) = (60°, 0°), (0°, 30°)$ and

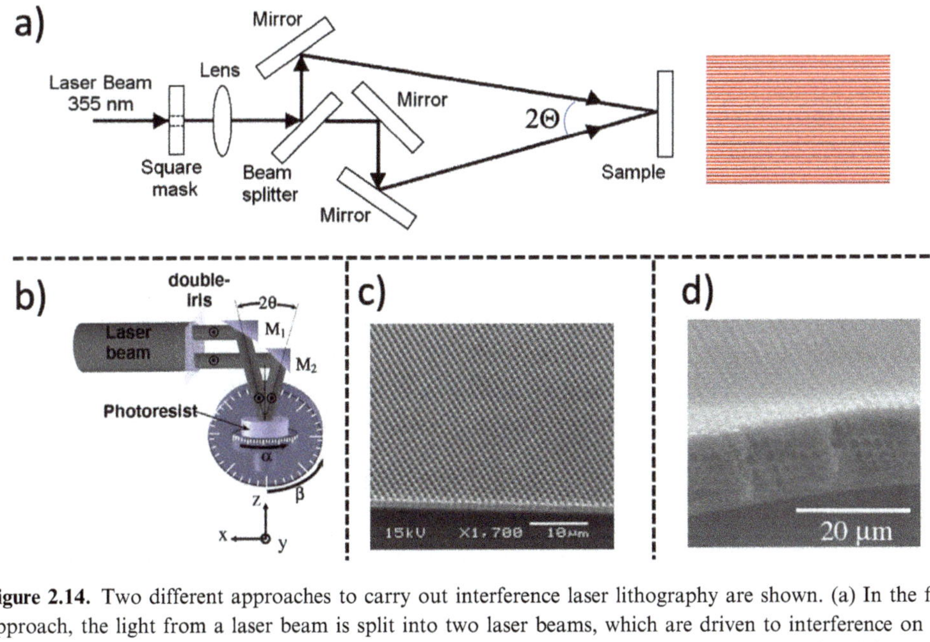

Figure 2.14. Two different approaches to carry out interference laser lithography are shown. (a) In the first approach, the light from a laser beam is split into two laser beams, which are driven to interference on the photoresist. In this configuration, the obtained line periodicity is given by $\lambda/(2*\sin\Theta)$. Reproduced from [33]. Copyright (2007), with permission from Elsevier. (b) In the second approach, a double iris generates two laser beams, which are driven to interference on the photoresist. (c) An example of a 2D structure fabricated following the strategy shown in (b) and using the angle values mentioned in the text. (d) An example of a 3D structure fabricated following the strategy shown in (b) and using the angle values mentioned in the text. Adapted with permission from [34]. Copyright The Optical Society.

(180°, 30°). Other more complex interference laser lithography schemes have been developed [31], which have even achieved sub-100 nm resolution [32].

2.3.2 Two-photon lithography

The process resolution limit imposed by the laser wavelength in DWL (based on the underlying one-photon absorption mechanism) can be overcome by two-photon absorption lithography, as proposed by Wu *et al* [35]. For a given photoresist, the number of photons absorbed per laser pulse and per molecule (n) is given by the following formula:

$$n = \frac{p_0^2 \delta}{\tau f_p^2}\left(\frac{\mathrm{NA}^2}{2\hbar c\lambda}\right)^2 \tag{2.6}$$

where p_0 is the time-averaged laser power, δ is the two-photon absorption (TPA) cross section, τ is the pulse duration, f_p is the pulse repetition rate, NA is the lens numerical aperture, and λ is the laser wavelength [36]. Due to the dependence of n with $(p_0)^2$, the two-photon absorption is non-linear, which can be used to generate significant absorption only in the very small volume of the photoresist where the laser is focused. The TPA phenomenon is sketched in figure 2.15 and was theoretically predicted by Maria Göppert-Mayer in 1931 [37] and experimentally demonstrated with the advent of lasers [38]. At the laser focusing point, the infrared laser light intensity is high enough to generate a substantial TPA, which provides an energy equivalent to that of a UV photon with twice the energy of a single infrared photon. When the TPA takes place, it will trigger the chemical reaction that will change the solubility of the photoresist in the developer. As the laser focus can be displaced in X, Y and Z with respect to the substrate, complex 3D structures can be fabricated with resolution beyond the diffraction limit. As will be shown later, 3D structures with resolution better than 100 nm are doable by TPA lithography.

The important elements in TPA exposure are the following ones:

- The laser: a popular one is a Ti:sapphire laser with $\lambda \sim 800$ nm, typical pulse width of tens to hundreds of femtoseconds and repetition rates ~80 MHz, with average output power around one watt. Nevertheless, continuous-wave (532 nm wavelength) lasers have also been employed for producing 3D structures by TPA lithography [42].
- The resists: it can be a positive or a negative resist, but most of the processes rely on negative resists. The negative resists consist of a pre-polymer resin containing a photo-initiator that can be excited by a TPA (equivalent energy to one UV photon). The excited photo-initiator starts a chemical reaction that leads to the resist photo-polymerization around the exposed regions. Several resists have been employed for TPA lithography [36]. One popular negative resists for TPA lithography is SU8, which can be cast on films up to 500 μm thickness and can yield structures with high aspect ratio [43]. Positive resists are useful in certain applications like in the fabrication of 3D microfluidic channels [44].

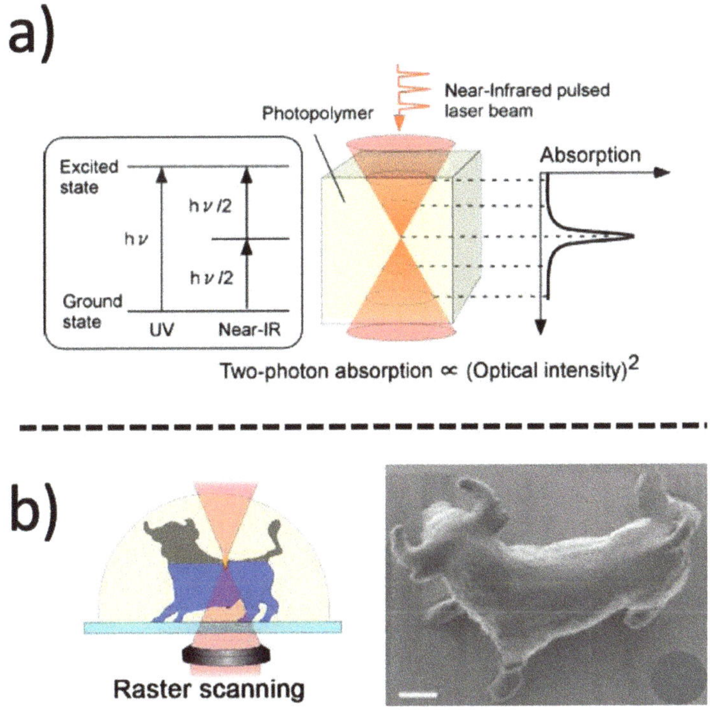

Figure 2.15. (a) Mechanism of resist exposure by two-photon absorption. In a very small area of the resist, the intensity of a well-focused laser can be high enough to produce the simultaneous absorption of two infrared photons, with energy equivalent to that of a single UV photon. This effect brings about the exposure of the UV-sensitive resist in a very small volume. Reprinted from [39]. Copyright John Wiley & Sons. (b) The displacement of the laser focus inside the photoresist along the X, Y, and Z coordinates allows the precise 3D exposure of the resist, producing a 3D structure after resist development. The scale bar is 2 μm. Reprinted by permission from Springer Nature [40]. Copyright (2001) and from [41], Copyright (2010), with permission from Elsevier.

- The post-exposure steps. Each resist needs an appropriate developer, and, in many cases, an additional post-baking step is required.

Once the TPA exposure and development has revealed a 3D polymer structure, two routes can be followed. In the first route, the polymer is functional and can be straightforwardly used for an application. This is the case when the polymer stands for structural, mechanical, magnetic or optical properties. A few examples are shown in figure 2.16. In figures 2.16(a) and (b), the polymer constitutes a 3D scaffold structure around which cell growth can proliferate. In figure 2.16(c), the polymer is structured as a mechanical cantilever resonator. In figure 2.16(d) the hybrid nature of the photoresist (polymer plus magnetic nanoparticles) provides magnetic properties to the micro-swimmers. In figure 2.16(e), the fabricated ring polymer, coupled to a laser fiber, can be used for ultrasound detection.

In the second route, the polymer is not the functional material and another step is required to grow the material of interest either inside or around the 3D-structured

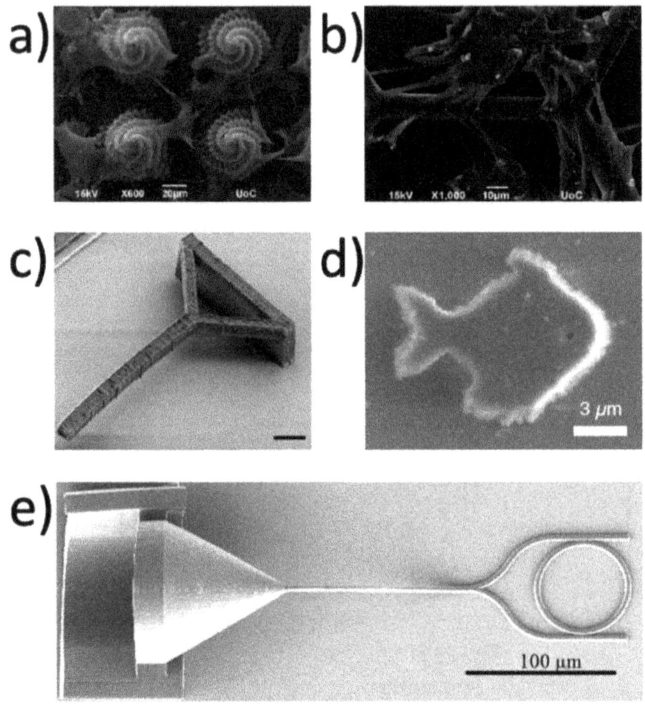

Figure 2.16. Examples of 3D structures imprinted on a photopolymer by TPA lithography: (a) Sea-shell 3D polymer structures patterned by TPA lithography and growth of neural cells after 7 d of culture. Reproduced from [45]. Copyright IOP Publishing. Reproduced with permission. All rights reserved. (b) 3D polymer scaffolds patterned by TPA lithography and growth of bone marrow mesenchymal stem cells after 2 d of culture. Reprinted from [46], Copyright (2017), with permission from Elsevier. (c) High-aspect-ratio cantilever obtained from TPA lithography on an acrylic resin. Reprinted from [36]. (d) Micro-swimmer fabricated by TPA lithography whose composition is a nanocomposite of a polymer and magnetic nanoparticles so that it feels magnetic forces. Reprinted from [47], copyright John Wiley & Sons. (e) Fabrication by TPA lithography of a 3D polymer micro-ring resonator integrated with a fiber laser. Reprinted with permission from [48]. Copyright The Optical Society.

polymer. In figure 2.17, two examples of this are shown in which positive resists are used. In both cases, after resist coating, exposure and development, holes with a 3D structure are created in the resist. An electrodeposition or electroplating process is capable of filling in these holes with the metal of interest. For the success of this step, the substrate material should be conductive and the transparent indium tin oxide (ITO) material is an ideal choice. In the last step, the resist is plasma or wet etched. On the other hand, if negative resists are used, the TAP exposure produces a polymeric 3D scaffold structure that can be coated with a functional material by means of a conformal coating technique such as atomic layer deposition, sputtering, chemical vapour deposition, etc. For example, strong, lightweight and recoverable 3D ceramic nanolattices have been grown following this approach [49].

The resolution of TPA lithography is essentially governed by the volume of the region exposed with high intensity by the laser beam (voxel), which will change its

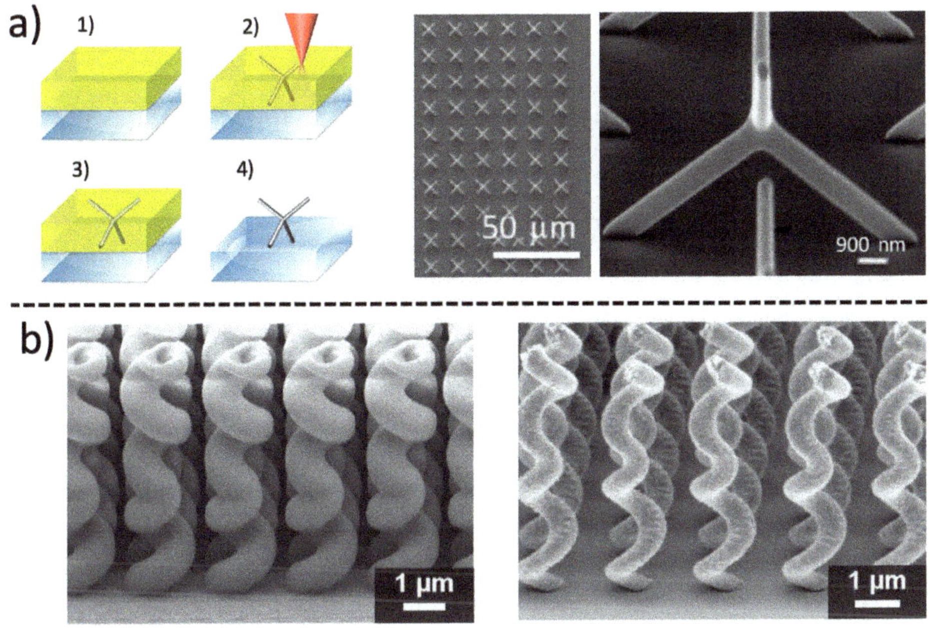

Figure 2.17. Two examples of TPA lithography with subsequent growth of the functional material. (a) On the left side, the four steps to create four connected magnetic nanowires: positive photoresist coating (1), photoresist exposure and development (2), electrodeposition of the magnetic material (3), resist etching and cleaning (4). On the right side, the result of the process is shown. Reprinted from [50], copyright (2018) with permission from Springer. (b) On the left side, polymer templates fabricated by TPA lithography are shown. On the right side, the result of the process after gold electrodeposition and polymer etching is shown. Reprinted from [51], copyright John Wiley & Sons.

solubility in the developer. Important parameters determining this volume will be the spatial distribution of the laser intensity in the photoresist (determined by the quality of the laser and the optical lenses used to focus the laser beam on a small point), the two-photon absorption cross section of the resist and the dimensions and spatial distributions of the photoinitiators, pre-polymer and polymer molecules. Sub-100 nm resolution is possible using photoresists such as for example SCR500, SU8 and ORMOCER [52]. By means of optimization processes, certain structures such as nanorods have been obtained with 30 nm resolution in SU8 [53] and 15 nm in SCR500 [52]. In the latter case, see figure 2.18(a), such ultrahigh resolution has been achieved by means of a repolymerization process in which a nanorod receives directly a low light dose (not enough to be later revealed in the development process) but indirectly (via close exposure areas) receives an additional dose so that the structure is revealed in the development process. In addition, post-processing steps such as plasma etching and pyrolysis can further reduce the structure dimensions, as shown in figure 2.18(b).

In general, the advantages of TPA lithography with respect to related techniques is the high resolution, the capability for 3D patterning and the absence of photomasks. On the other hand, the main disadvantage of the technique is the long exposure times due to its sequential nature.

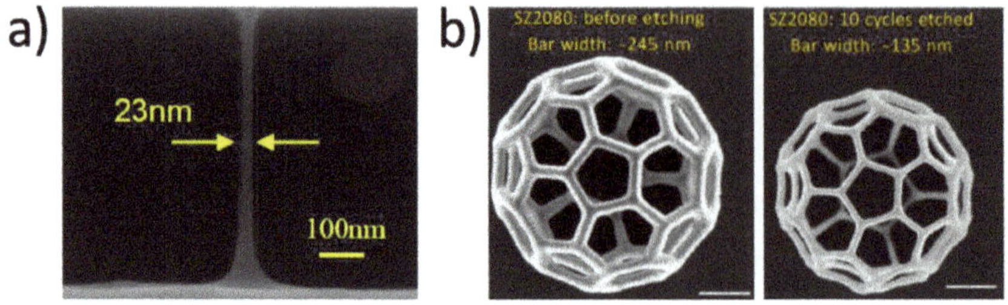

Figure 2.18. (a) Sub-25 nm resolution obtained by TPA lithography on suspended fibers of SCR500 resist by repolymerization. Reprinted from [52] with the permission of AIP Publishing. (b) Reduction of the bar width in a buckyball via a two-step process (the scale bar is 1 μm): (1) TAP exposure of SZ2080 resist and development and (2) ten cycles of an oxygen plasma etching process. Reprinted from [54]. Copyright (2018), with permission from Elsevier.

The main applications of TPA lithography nowadays are in the following domains:

- Micro- and nano-optics related with photonic crystals: micro-lenses, Fresnel lenses, non-planar or suspended components of waveguide structures, linear couplers, Y splitters, Mach-Zender interferometers, diffraction gratings, metamaterials, etc [36, 55, 56].
- Micro-mechanics for the fabrication of strong ultralight structures [49], cantilevers, nanotweezers, microrotors, MEMS, etc [39].
- Micro-fluidics for the fabrication of channels and integration in sensors [56].
- Biology for the fabrication of microneedles and 3D scaffolds for studies on tissue growth [57].
- Electronics for the fabrication of conductive materials, microinductors or microcoils [36, 39].
- Magnetic materials with 3D spin textures [50].

A few companies commercialize TPA lithography equipment, such as Nanoscribe and Multiphoton Optics. The existing equipment can address 3D exposure patterns with either large or small features, from the millimeter range to the sub-200 nm range.

2.4 Directed self-assembly (DSA)

Optical lithography presents the limitation of the maximum resolution achievable, if throughput and reasonable cost of ownership is maintained at levels compatible with the progress of the semiconductor industry. Present lithography-based alternatives to optical lithography do not guarantee their incorporation into the high volume manufacturing chain because of their performance in terms of low throughput or high defectivity.

A change of paradigm is to introduce self-assembly and bottom-up concepts: the use of small building blocks to realize patterns (see chapter 1). In this case, the resolution is dictated by the size of the building blocks. Combination of self-assembly with lithography to position and align the building blocks at the desired location is known as directed self-assembly (DSA).

DSA can be implemented in multiple ways, involving different kinds of nanometer scale objects as building blocks: carbon nanotubes, silicon nanowires, single molecules, DNA origami, among others.

One of the DSA methods that have caught more interest by the industry has been the DSA of block co-polymers, for several reasons such as: its good compatibility with the materials and methods of the semiconductor industry, its simplicity in terms of processing, its inherent high throughput and the demonstration of sub-5 nm resolution. At the time of writing this chapter, DSA is considered as a future complementary method to EUV optical lithography, aiming to improve line edge roughness and decrease process complexity for the most advanced manufacturing nodes in integrated circuit manufacturing.

In this section, the principles of DSA of block copolymers are presented, along with recent examples of strategies for alignment of the block copolymers by combination with optical lithography. One of the principal problems to date is the insufficiently low defect density for the integration of block copolymer lithography in high volume manufacturing processes [58].

2.4.1 Principles of DSA of block copolymers

Block copolymers are macromolecules constituted by two or more chemically different polymer chains covalently bonded together to one single macromolecule [59]. The chemically different blocks can either be merged in a linear or in a branched manner. In that sense, a great variety of different block copolymer structures may be thought of.

Driven by the repulsive force between chemically different molecules, the chains self-assemble in periodic structures in the size range between few nanometers and few tens of nanometers predominantly consisting of alternating phases of only one component A or B (figure 2.19). A quantification of the repulsive interaction between two blocks is given by the Flory–Huggins interaction parameter χ, based on the works of Flory and Huggins [60, 61]:

$$\chi_{AB} = [\varepsilon_{AB} - 1/2(\varepsilon_{AA} + \varepsilon_{BB})]^{*} Z / k_B T \tag{2.7}$$

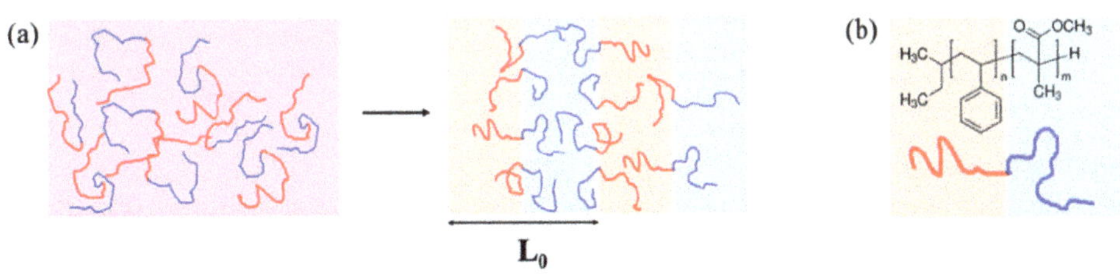

Figure 2.19. Microphase separation in block copolymers: (a) difference between molecular order in a diblock copolymer in disordered state (left) and the microphase separated state (right), (b) molecular architecture of a PS-b-PMMA (polystyrene-block-poly(methyl methacrylate)) block co-polymer, with PS on the left side (red colour) and PMMA on the right side (blue colour).

where Z is the number of nearest neighbors of a unit and ε_{ij} is the interaction energy between two monomers i and j.

In general, the microphase separation in block copolymers is achieved in two different ways [62]. On the one hand, thermal annealing can be used to heat the block copolymers above their glass transition temperature, enhancing their chain mobility in this way [63] and inducing microphase separation. Solvent annealing, on the other hand, is the second strategy used to induce the self-assembly of block copolymers [64]. The selectivity of the solvent plays a role in the orientation of the block copolymer domains with respect to the substrate [65–67].

The most studied block copolymer architecture is A-b-B block copolymers, which show a well-studied phase behavior, as shown in figure 2.20. In the phase diagram, χN, where N is $N_A + N_B$, with N_A and N_B being, respectively, the number of monomers of polymer A and polymer B, and χ is the Flory–Huggins interaction parameter, is represented as a function of f, which is the ratio N_A/N [68][69]. The theoretical phase diagram of A-b-B diblock copolymers has been determined by means of Monte Carlo simulations in [70] (see figure 2.20).

A symmetric block copolymer self-assembles in lamellar structures (denoted L), whereas upon increasing the fraction of one polymer, f_a, the material will first pass to a gyroid morphology and then to cylinders in a matrix of material B, before it will finally self-assemble in spheres in a matrix of material B (see figure 2.20(b)). Microphase separation occurs for all materials that are above the order–disorder transition line in the phase diagram. The curve indicating the order–disorder transition in the phase diagram is parabolic with a global minimum at $\chi N = 10.4$ for perfectly symmetric block copolymers. Based on that, microphase separation for PS-b-PMMA block copolymers at 160 °C is *a priori* precluded for materials consisting of less than 251 monomers. The smallest reported full pitch (total length

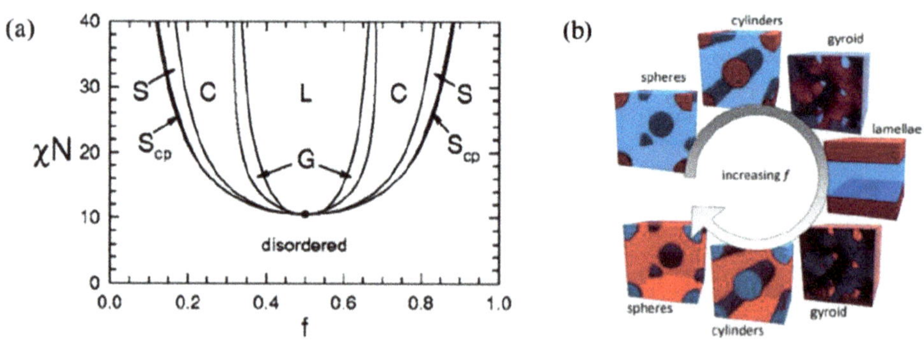

Figure 2.20. Phase behavior of diblock copolymers. (a) Phase diagram of diblock copolymers according to Monte Carlo simulations. Reprinted from [70], with permission from AIP publishing. The parameter f (x-axis) indicates the weight fraction of the chain of one of the blocks, for example PS. The term χN (y-axis) represents the product of the Flory–Huggins interaction parameter χ and the chain polymerization N. Lamellar self-assembly is expected in the areas denoted 'L' and the cylindrical morphology in the areas denoted 'C'. Less frequently used morphologies are the spherical and the gyroid phase, denoted 'S' and 'G', respectively. (b) Sketches of the block copolymer self-assembly morphologies upon changing the relative fraction, f. Reprinted from [71], Copyright (2007) with permission from Elsevier.

of one repetition unit) obtained for PS-b-PMMA is 17.5 nm [72] or 19 nm [73], although the utility of sub-22 nm pitch PS-b-PMMA materials for pattern transfer processes is questioned by some authors [74].

For the energy minimization process in block copolymer thin films, in general four energy components are taken into account [50] (see figure 2.21(a)), namely (1) the energy stored in the chain stretching/compression other than the radius of gyration, (2) the A–B interface curvature, which contributes with a negligibly low amount of free energy, (3) the interface free energy between the blocks A and B and (4) the surface energy at the block copolymer–air and the block copolymer–substrate interface. The self-assembly morphology is always such that the sum of the four free energy components is minimal. For the self-assembly of thin films in unconfined states (e.g., on energetically homogenous, non-patterned substrates) the chain will stretch such that pattern pitch minimizes the chain extension and the total surface energy between the blocks A and B.

In the case of lamellar block copolymers, the decision about whether the material self-assembles in vertical or horizontal lamellae (see figure 2.21(b)) depends on the interface energy between the substrate, the material properties of the two blocks and their interaction energy with the surface. Mansky *et al* have found that the energy at the block copolymer-substrate interface of PS-b-PMMA can be tuned by using PS-r-PMMA random copolymer brushes [75]. Random copolymer brushes whose monomer ratio is similar to that of the block copolymer in question can be used as energy neutralization layers and promote upright standing features. If one of the blocks, however, wets the random copolymer brush layer in an energetically clearly favorable manner, the block copolymer is assembled in horizontal features.

2.4.2 Strategies for alignment of block copolymers

The absence of long-range order and placement accuracy precludes, however, any lithography application of the material in the semiconductor industry in its as-spin-coated and annealed state. It is therefore crucial to induce long-range order in a controlled manner. Guiding patterns that are usually defined by top-down

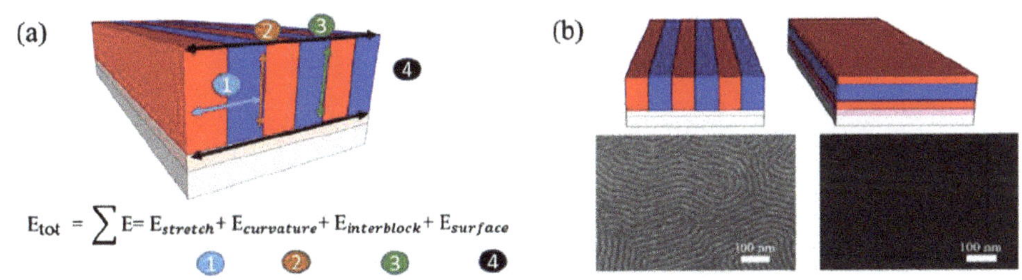

Figure 2.21. Free energy in block copolymer thin films. (a) Sketch providing an overview of free energy contributions to the total free energy in a block copolymer thin film, (b) vertically (left) and horizontally (right) aligned block copolymer thin films. When the interaction energy of the surface is similar for the two blocks, vertically aligned lamellae are obtained, which appear as fingerprint patterns in SEM images. If one of the blocks wets preferentially the surface, then the lamellae self-assemble horizontally, without generation of any surface pattern.

lithography are required to induce long-range order in the material. In the following we will briefly refer to the most commonly used techniques to direct the self-assembly of block copolymers: graphoepitaxy and chemoepitaxy, as it is sketched in figure 2.22.

2.4.2.1 Graphoepitaxy

Directing the self-assembly of block copolymers with topographical patterns is referred to as graphoepitaxy. The technique aims at fabricating topographical structures in such a way that they energetically favour the block copolymer to self-assemble in the desired structures, as depicted in figure 2.22(a). The energetic advantage of ordered (e.g. directed) self-assembly in the created trenches with respect to a disordered self-assembly is in the preferential wetting of guiding pattern walls by one of the blocks. Furthermore, the creation of defects in block copolymers assembled in commensurate trenches imposes a free energy penalty, which converts the self-assembly in lamellae parallel to the guiding pattern walls into the minimum-free-energy state. The probability to form defects increases with the distance between the guiding pattern features because of the finite correlation length of the block copolymers far away from the order-inducing surface.

This principle has been successfully employed by a number of authors, as for example reported in references [76–78]. The best results are obtained when the width of the confining trench is exactly or close to an integer multiple of the natural pitch of the block copolymer [79]. A suitable guiding pattern design permits, for example, to direct the self-assembly of block copolymers in device-oriented features [80] and induce long-range order in sub-10 nm half-pitch block copolymers [81] and sub-5 nm half-pitch liquid crystals [82]. Another widely-used approach to make use of cylindrical block copolymers is contact hole shrink applications [83].

2.4.2.2 Chemoepitaxy

The chemoepitaxy approach relies on the introduction of chemical patterns into an underlying substrate (see figure 2.22(b)). A lithographic process is applied to locally

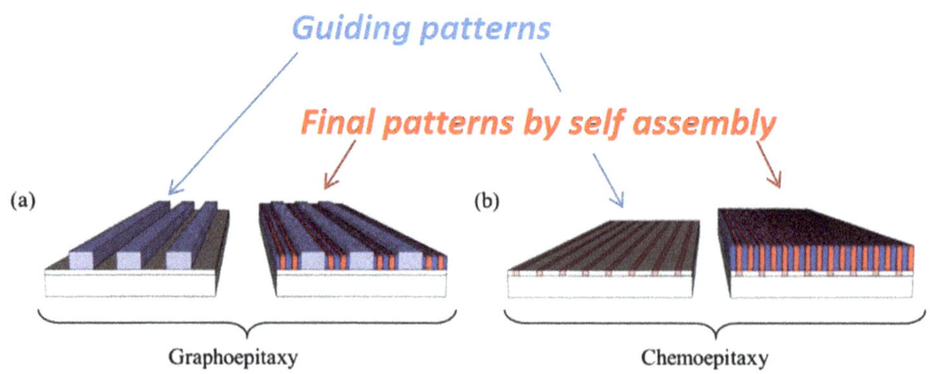

Figure 2.22. Scheme of most common techniques to direct the self-assembly of block copolymers: (a) graphoepitaxy; (b) chemoepitaxy.

modify the interface free energy between the block copolymer and the underlayer in such a way that certain regions are more attractive to one of the blocks than to the other. A characteristic feature of the chemoepitaxy approach is that the guiding pattern must have a pitch close to η-times larger than the natural pitch of the block copolymer. The factor η is an integer referred to as density multiplication factor of the guiding pattern. The critical dimension of the guiding pattern features has to correspond to the half-pitch of the block copolymer [84] or 1.5 times the block copolymer pitch [85].

Several materials can be used as background layers to create the guiding patterns: self-assembled monolayers [86], HSQ resist [87] and end-grafted homopolymers or random copolymer brushes [84]. The most widely used technique consists of using end-grafted homopolymers or random copolymer brushes (see figure 2.23).

Different approaches are used to create chemical guiding patterns. Ruiz *et al* reported a viable technique based on the use of EBL and oxygen plasma functionalization for cylindrical block copolymers with pattern density multiplication of $2L_0$ [89]. Later on, Cheng *et al* developed a technique to increase the density multiplication factor of a lamellar PS-*b*-PMMA BCP, based on 193 nm immersion lithography [87]. In order to achieve the defectivity level required by the industry, Liu and Nealey in 2011 developed the so-called LiNe process. In this process, two materials with different chemical affinities are employed to create the chemical guiding pre-patterns.

As shown in the scheme from figure 2.23, the process flow starts by spin-coating a cross-linkable polystyrene (X-PS) on a substrate and after a lithography step, the material is etched. Then, a random hydroxyl terminated poly(styrene-b-methyl methacrylate) (PS-r-PMMA-OH) brush is spin-coated and grafted to the substrate between the X-PS guiding stripes, creating thus a chemical pattern. The X-PS guiding stripes are more affine to PS domains while the PS-r-PMMA background stripes have equal affinity for both domains [90].

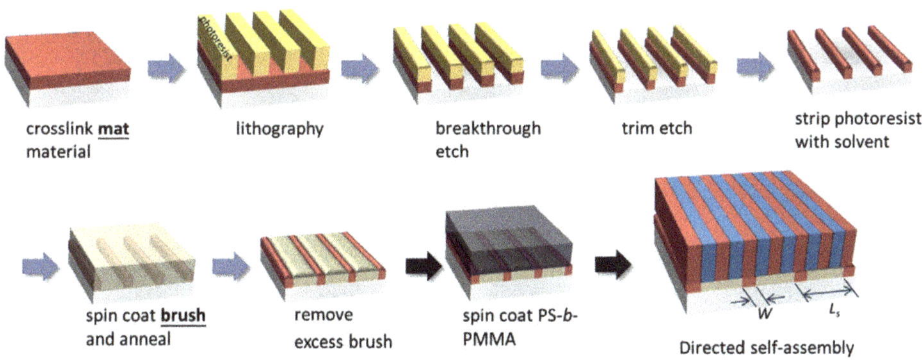

Figure 2.23. Sketch of DSA based on chemical epitaxy using an alternate pattern of two polymers. Reprinted with permission from [88]. Copyright (2011) American Chemical Society.

Another common path to manipulate the surface free energy of a PS-r-PMMA random copolymer brush layer is the combination of a lithography step with its exposure to UV light or an oxygen plasma [89].

2.4.3 Applications

The periodic structures defined in that way can be used as bottom-up templates for nanofabrication processes that require very high resolution, such as bit patterned media for hard disk drives [91–93], integrated circuit finFETs [80], contact holes [83] and patterning of 2D materials [94].

2.5 End-of-chapter summary

Optical lithography is the most employed lithography method. Optical radiation is used to expose a photosensitive material (photoresist) in selected areas. Because of the exposure, the photosensitive materials experience a change in their physical and chemical properties, varying their solubility with respect to a specific solvent. If the solubility of the photoresist increases upon exposure, the resist is named positive resist. Conversely, when the exposed photoresist areas remain after development, the photoresist is called negative.

In optical lithography, diffraction phenomena limit the resolution. As diffraction is directly dependent on the wavelength of the optical radiation, the shorter the wavelength, the better will be the resolution.

The most common way to perform an optical lithography process is by means of an optical mask. An optical mask consists of a transparent substrate, on which opaque areas define the patterns that will be transferred to the substrate. Depending on the position of the mask with respect to the sample, three modes of optical lithography exist: contact, proximity and projection.

Projection optical lithography is the way to obtain the better resolution. Over the years, multiple tricks have been implemented to improve the resolution: decreasing the wavelength, increasing the numerical aperture, engineering the illumination, optical proximity correction, phase shifting masks, among others.

Extreme ultraviolet optical lithography is a particular case of optical lithography in which the wavelength of the optical radiation is 13.5 nm. Major challenges have been solved to make it practical: the type of sources, the optical lithography systems and the materials for the masks and the photoresists.

An alternative approach to optical lithography consists in exposing the photoresist without using masks (mask-less optical lithography). The two main methods to perform mask-less optical lithography are direct write laser and two-photon lithography.

Direct write lithography relies on a laser beam, which can be focused down to 1 μm or less, and an XY-moving stage holding the resist-coated sample. The resolution achievable in patterns exposed by this technique is ultimately limited by the light diffraction, thus, of the order of the wavelength of the laser. Typical laser wavelengths used are around 400 nm. As a consequence, the same or similar photoresists to the ones commonly used in NUV optical lithography are used.

Another derived strategy is the realization of interference laser lithography, in which the initial laser beam is split into two beams, which are driven to interference on the photoresist at an angle.

The process resolution limit imposed by the laser wavelength can be overcome by two-photon-absorption lithography. The two-photon absorption is a non-linear process, which can be used to generate significant absorption only in the very small volume of the photoresist where the laser is focused. As the laser focus can be displaced in X, Y and Z with respect to the substrate, complex 3D structures can be fabricated with resolution beyond the diffraction limit.

Pushing the limits of optical lithography towards sub-10 nm resolution without decreasing throughput will require the combination with self-assembling methods based on small building blocks. Combination of self-assembly with lithography to position and align the building blocks at the desired location is known as directed self-assembly. One of the directed self-assembly methods that have caught more interest by the industry has been the directed self-assembly of block co-polymers. Driven by the repulsive force between chemically different molecular chains which are covalently bonded, thin films of block copolymer molecules self-assemble in periodic structures in the size range between a few nanometers and a few tens of nanometers. The self-assembled film consists of alternating phases of the chains that form the block co-polymer. For a di-block copolymer, large area lamellar or cylindrical patterns are formed just by spin coating the solution and a short thermal annealing. Long-range order in a controlled manner of these patterns can be achieved by pre-defining chemical (chemoepitaxy) or topographical (graphoepitaxy) patterns on the surface. Applications of these technologies have been demonstrated in hard disk drives, integrated circuits and 2D materials.

2.6 Student's corner

- In a good approximation, we can assume that the value of the technological constant, k_1, has been decreasing linearly over the years, with a constant decrease rate of 0.017/year. If the value of k_1 was 0.8 in 1980, indicate which information can be obtained about the rest of the parameters that define the resolution in projection optical for the six cases indicated in figure 2.4.
- The parameter γ determines the contrast of a photoresist after exposure and development, and, in consequence, it is a parameter that influences the resolution. Evaluate the interplay between photoresist contrast and optical exposure. For this, indicate the extension of the dimensions of a pattern for three photoresists with $\gamma = 1$, 5 and 10, and assuming that the profile of the optical exposure at the edges of the exposed area follows a Gaussian like distribution with $\sigma = 1$ µm (case 1) and $\sigma = 3$ µm (case 2). Make the analysis for a positive photoresist.
- In an EUV lithography system the light is reflected in 12 multilayered mirrors from the intermediate focus of the EUV source to the exposed wafer. The reflectance of each mirror is 0.68, the exposure dose of the used photoresist is 40 mJ cm^{-2}, each die is 50 mm × 30 mm in size, and the exposition time is 1.5 s.

If the energy conversion efficiency at the light source is 4%, estimate the power consumption at the EUV source.

- In the case that you had to fabricate a prototype of a planar circuit including superconducting and/or magnetic materials, with patterning resolution of 0.8 μm, would you choose NUV optical lithography, EUV optical lithography, laser direct write lithography or two-photon-absorption lithography? Why?
- If you were working in the MEMS domain and needing to fabricate cantilevers with aspect ratio (height/width) of 20 and widths down to 2 μm resolution, propose a lithography fabrication process based on optical lithography, including the used technique, the photoresist tone (positive or negative) and thickness, the post-exposure steps, etc.
- The following images correspond to SEM images of self-assembled patterns of PS-b-PMMA block copolymer, where the darker areas correspond to the PMMA phase and brighter areas to the PS phase. The block copolymer film has been deposited on a silicon oxide surface and the thickness is around 40 nm. In the three cases, a neutralization surface layer has been formed on the surface before the block copolymer deposition to create the optimal neutralization. Indicate:

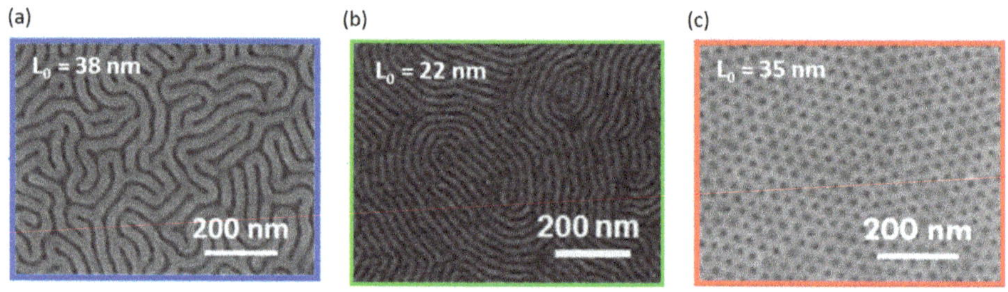

i. Which images correspond to lamella forming block copolymers and which images correspond to cylindrical forming block copolymer?
ii. Are the block copolymer molecules oriented parallel or perpendicular with respect to the surface plane?
iii. What can be said about the presence of a neutralization layer on the surface of the substrate?
iv. Which is the position of each block copolymer in the phase diagram of block copolymers (figure 2.20(a))?

References

[1] Bruning J H 2007 Optical lithography: 40 years and holding *Opt. Microlithogr. XX* **6520** 652004

[2] Hutcheson G D 2018 Moore's law, lithography, and how optics drive the semiconductor industry *Proc. SPIE* **1058303** 501

[3] Levenson H J 2004 *Principles of Lithography* second edition (Bellingham, VA: SPIE)

[4] Lin B J 1990 Methods to print optical images at low-k 1 factors *Opt. Microlithogr. III* **1264** 2–13

[5] Wong A K, Ferguson R A and Mansfield S M 2000 The mask error factor in optical lithography *IEEE Trans. Semicond. Manuf.* **13** 235–42

[6] Kang H 2009 Novel assist feature design to improve depth of focus in low k1 EUV lithography *Proc. SPIE* **7520** 752037

[7] Levenson M D, Viswanathan N S and Simpson R A 1982 Improving resolution in photolithography with a phase-shifting mask *IEEE Trans. Electron Devices* **29** 1828–36

[8] Gordon R L, Mack C A and Petersen J S 1998 Design and analysis of manufacturable alternating phase-shifting masks *18th Annu. BACUS Symp. Photomask Technol. Manag.* **3546** 606

[9] Lin Q 2007 Properties of photoresist polymers *Physical Properties of Polymers Handbook* ed J E Mark (New York: Springer) pp 941–55

[10] Fink J K 2018 Phenol/formaldehyde resins *React. Polym. Fundam. Appl.* 225–53

[11] Shaw J M, Gelorme J D, LaBianca N C, Conley W E and Holmes S J 1997 Negative photoresists for optical lithography *IBM J. Res. Dev.* **41** 81–94

[12] Ed. MicroChemicals GmbH 2000 MicroChemicals Application Notes 2000: Exposure—Basics of Microstructuring (Ed. MicroChemicals GmbH)

[13] Kinoshita H 1989 Soft x-ray reduction lithography using multilayer mirrors *J. Vac. Sci. Technol.* B **7** 1648

[14] Bakshi V 2018 *EUV Lithography* (Bellingham, VA: SPIE Press)

[15] Wurm S, Kaiser W, Dinger U, Müllender B, La Fontaine B, WoodII O R and Neisser M 2020 EUV lithography *Microlithography, Science Technology* ed B W Smith and K Suzuki (Boca Raton, FL: CRC Press)

[16] Bajt S, Alameda J B, Barbee T W Jr, Clift W M, Folta J A, Kaufmann B B and Spiller E A 2001 Improved reflectance and stability of Mo/Si multilayers *Proc. SPIE* **4506** 65–75

[17] Rensch C, Hell S, Schickfus M V and Hunklinger S 1989 Laser scanner for direct writing lithography *Appl. Opt.* **28** 3754

[18] Van Kessel P F, Hornbeck L J, Meier R E and Douglass M R 1998 A MEMS-based projection display *Proc. IEEE* **86** 1687–704

[19] Jiang C, Hu D and Lu Y 2016 Digital micromirror device (DMD)-based high-cycle torsional fatigue testing micromachine for 1D nanomaterials *Micromachines* **7** 49

[20] Cadarso V J, Pfeiffer K, Ostrzinski U, Bureau J B, Racine G A, Voigt A, Gruetzner G and Brugger J 2011 Direct writing laser of high aspect ratio epoxy microstructures *J. Micromech. Microeng.* **21** 017003

[21] Voronov D L, Diez S, Lum P, Hidalgo S A, Warwick T, Artemiev N and Padmore H A 2013 Fabrication of x-ray gratings by direct write maskless lithography *Adv. X-Ray/EUV Opt. Components VIII* **8848** 88480Q

[22] Ivan M G, Vaney J-B, Verhaart D and Meinders E 2009 Direct laser write (DLW) as a versatile tool in manufacturing templates for imprint lithography on flexible substrates *Altern. Lithogr. Technol.* **7271** 72711s

[23] Korolkov V P, Nasyrov R K and Shimansky R V 2006 Zone-boundary optimization for direct laser writing of continuous-relief diffractive optical elements *Appl. Opt.* **45** 53–62

[24] Singh A, Scotti G, Sikanen T, Jokinen V and Franssila S 2014 Laser direct writing of thick hybrid polymers for microfluidic chips *Micromachines* **5** 472–85

[25] Zhou S, Wang R, Han J, Wang D, Li H, Gan L and Zhai T 2019 Ultrathin non-van der Waals magnetic rhombohedral Cr_2S_3: space-confined chemical vapor deposition synthesis and Raman scattering investigation *Adv. Funct. Mater.* **29** 1805880
[26] Vallès F, Palau A, Rouco V, Mundet B, Obradors X and Puig T 2018 Angular flux creep contributions in $YBa_2Cu_3O_{7-\delta}$ nanocomposites from electrical transport measurements *Sci. Rep.* **8** 5924
[27] Torres Alonso E *et al* 2018 Graphene electronic fibres with touch-sensing and light-emitting functionalities for smart textiles *npj Flex. Electron.* **2** 25
[28] Do M T, Nguyen T T N, Li Q, Benisty H, Ledoux-Rak I and Lai N D 2013 Submicrometer 3D structures fabrication enabled by one-photon absorption direct laser writing *Opt. Express* **21** 20964
[29] Rawlings C *et al* 2018 Fast turnaround fabrication of silicon point-contact quantum-dot transistors using combined thermal scanning probe lithography and laser writing *Nanotechnology* **29** 505302
[30] Steen S *et al* 2006 Hybrid lithography: the marriage between optical and e-beam lithography. A method to study process integration and device performance for advanced device nodes *Microelectron. Eng.* **83** 754–61
[31] Xia D, Ku Z, Lee S C and Brueck S R J 2011 Nanostructures and functional materials fabricated by interferometric lithography *Adv. Mater.* **23** 147–79
[32] Smith H I 2001 Low cost nanolithography with nanoaccuracy *Phys. E Low-Dimens. Syst. Nanostruct.* **11** 104–9
[33] Lasagni A, Holzapfel C, Weirich T and Mücklich F 2007 Laser interference metallurgy: a new method for periodic surface microstructure design on multilayered metallic thin films *Appl. Surf. Sci.* **253** 8070–4
[34] Lai N D, Liang W P, Lin J H, Hsu C C and Lin C H 2005 Fabrication of two- and three-dimensional periodic structures by multi-exposure of two-beam interference technique *Opt. Express* **13** 9605
[35] Wu E S, Strickler J H, Harrell W R and Webb W W 1992 Two-photon lithography for microelectronic application *Proc. SPIE* **1674** 776–82
[36] LaFratta C N, Fourkas J T, Baldacchini T and Farrer R A 2007 Multiphoton fabrication *Angew Chemie Int. Ed.* **46** 6238–58
[37] Göppert-Mayer M 2009 Elementary processes with two quantum transitions *Ann. Phys.* **18** 466–79
[38] Kaiser W and Garrett C G B 1961 Two-photon excitation in CaF_2: Eu^{2+} *Phys. Rev. Lett.* **7** 229–31
[39] Maruo S and Fourkas J T 2008 Recent progress in multiphoton microfabrication *Laser Photonics Rev.* **2** 100–11
[40] Satoshi K, Hong-Bo S, Tomokazu T and Kenji T 2001 Finer features for functional microdevices *Nature* **412** 697–8
[41] Zhang Y L, Chen Q D, Xia H and Sun H B 2010 Designable 3D nanofabrication by femtosecond laser direct writing *Nano Today* **5** 435–48
[42] Thiel M, Fischer J, Von Freymann G and Wegener M 2010 Direct laser writing of three-dimensional submicron structures using a continuous-wave laser at 532 nm *Appl. Phys. Lett.* **97** 2010–2

[43] Teh W H, Dürig U, Drechsler U, Smith C G and Güntherodt H J 2005 Effect of low numerical-aperture femtosecond two-photon absorption on (SU-8) resist for ultrahigh-aspect-ratio microstereolithography *J. Appl. Phys.* **97** 054907

[44] Yu T, Ober C K, Kuebler S M, Zhou W, Marder S R and Perry J W 2003 Chemically amplified positive resists for two-photon three-dimensional microfabrication *Adv. Mater.* **15** 517–21

[45] Melissinaki V, Gill A A, Ortega I, Vamvakaki M, Ranella A, Haycock J W, Fotakis C, Farsari M and Claeyssens F 2011 Direct laser writing of 3D scaffolds for neural tissue engineering applications *Biofabrication* **3** 045005

[46] Chatzinikolaidou M, Pontikoglou C, Terzaki K, Kaliva M, Kalyva A, Papadaki E, Vamvakaki M and Farsari M 2017 Recombinant human bone morphogenetic protein 2 (rhBMP-2) immobilized on laser-fabricated 3D scaffolds enhance osteogenesis *Colloids Surf. B Biointerfaces* **149** 233–42

[47] Nguyen D T T, Del Guercio O, Au T H, Trinh D T, Mai N P T and Lai N D 2018 Optical lithography of three-dimensional magnetophotonic microdevices *Opt. Eng.* **57** 041406

[48] Wei H and Krishnaswamy S 2017 Polymer micro-ring resonator integrated with a fiber ring laser for ultrasound detection *Opt. Lett.* **42** 2655

[49] Meza L R, Das S and Greer J R 2014 Strong, lightweight, and recoverable three-dimensional ceramic nanolattices *Science* **345** 1322–6

[50] Williams G *et al* 2018 Two-photon lithography for 3D magnetic nanostructure fabrication *Nano Res.* **11** 845–54

[51] Kaschke J, Blume L, Wu L, Thiel M, Bade K, Yang Z and Wegener M 2015 A helical metamaterial for broadband circular polarization conversion *Adv. Opt. Mater.* **3** 1411–7

[52] Tan D, Li Y, Qi F, Yang H, Gong Q, Dong X and Duan X 2007 Reduction in feature size of two-photon polymerization using SCR500 *Appl. Phys. Lett.* **90** 071106

[53] Juodkazis S, Mizeikis V, Seet K K, Miwa M and Misawa H 2005 Two-photon lithography of nanorods in SU-8 photoresist *Nanotechnology* **16** 846–9

[54] Seniutinas G, Weber A, Padeste C, Sakellari I, Farsari M and David C 2018 Beyond 100 nm resolution in 3D laser lithography—post processing solutions *Microelectron. Eng.* **191** 25–31

[55] Malinauskas M, Farsari M, Piskarskas A and Juodkazis S 2013 Ultrafast laser nano-structuring of photopolymers: a decade of advances *Phys. Rep.* **533** 1–31

[56] Wu Z L *et al* 2019 Polymer-based device fabrication and applications using direct laser writing technology *Polymers (Basel)* **11** 553

[57] Selimis A, Mironov V and Farsari M 2015 Direct laser writing: Principles and materials for scaffold 3D printing *Microelectron. Eng.* **132** 83–9

[58] Neisser M and Wurm S 2015 ITRS lithography roadmap: 2015 challenges *Adv. Opt. Technol.* **4** 235–40

[59] Bates F S and Fredrickson G H 1999 Block copolymers-designer soft materials *Phys. Today* **52** 32–8

[60] Huggins M L 1941 Solutions of long chain compounds *J. Chem. Phys.* **9** 440

[61] Flory P J 1942 Thermodynamics of high polymer solutions *J. Chem. Phys.* **10** 51–61

[62] Ji S, Wan L, Liu C C and Nealey P F 2016 Directed self-assembly of block copolymers on chemical patterns: A platform for nanofabrication *Prog. Polym. Sci.* **54–55** 76–127

[63] Liu G, Delcambre S P, Stuen K O, Craig G S W, de Pablo J J, Nealey P F, Nygård K, Satapathy D K, Bunk O and Solak H H 2010 Mechanism and dynamics of block copolymer directed assembly with density multiplication on chemically patterned surfaces *J. Vac. Sci. Technol.* B **28** C6B13–9

[64] Jung Y S and Ross C A 2009 Solvent-vapor-induced tunability of self-assembled block copolymer patterns *Adv. Mater.* **21** 2540–5
[65] Bosworth J K, Paik M Y, Ruiz R, Schwartz E L, Huang J Q, Ko A W, Smilgies D M, Black C T and Ober C K 2008 Control of self-assembly of lithographically patternable block copolymer films *ACS Nano* **2** 1396–402
[66] Bang J, Kim B J, Stein G E, Russell T P, Li X, Wang J, Kramer E J and Hawker C J 2007 Effect of humidity on the ordering of PEO-based copolymer thin films *Macromolecules* **40** 7019–25
[67] Guo R, Huang H, Chen Y, Gong Y, Du B and He T 2008 Effect of the nature of annealing solvent on the morphology of diblock copolymer blend thin films *Macromolecules* **41** 890–900
[68] Bates F 1990 Block copolymer thermodynamics: theory and experiment *Annu. Rev. Phys. Chem.* **41** 525–57
[69] Leibler L 1980 Theory of microphase separation in block copolymers *Macromolecules* **13** 1602–17
[70] Matsen M W, Griffiths G H, Wickham R A and Vassiliev O N 2006 Monte Carlo phase diagram for diblock copolymer melts *J. Chem. Phys.* **124** 024904
[71] Darling S B 2007 Directing the self-assembly of block copolymers *Prog. Polym. Sci.* **32** 1152–204
[72] Anastasiadis S H, Russell T P, Satija S K and Majkrzak C F 1989 Neutron reflectivity studies of the surface-induced ordering of diblock copolymer films *Phys. Rev. Lett.* **62** 1852–5
[73] Chevalier X *et al* 2013 Scaling-down lithographic dimensions with block-copolymer materials: 10-nm-sized features with poly(styrene)-block-poly(methylmethacrylate) *J. Micro/Nanolithogr. MEMS MOEMS* **12** 031102
[74] Wan L, Ruiz R, Gao H, Patel K C, Albrecht T R, Yin J, Kim J, Cao Y and Lin G 2015 The limits of lamellae-forming PS-b-PMMA block copolymers for lithography *ACS Nano* **9** 7506–14
[75] Mansky P, Liu Y, Huang E, Russell T P and Hawker C 1997 Controlling polymer-surface interactions with random copolymer brushes *Science* **275** 1458–60
[76] Yoshida A, Yoshimoto K and Ohshima M 2016 Effect of wall potential on morphology of symmetric diblock copolymers in nanotrench *Jpn. J. Appl. Phys.* **55** 06GE01
[77] Han E, Kang H, Liu C C, Nealey P F and Gopalan P 2010 Graphoepitaxial assembly of symmetric block copolymers on weakly preferential substrates *Adv. Mater.* **22** 4325–9
[78] Ilievski F and Ross C A 2010 Graphoepitaxy of block copolymers using selectively removable templates *J. Vac. Sci. Technol.* B **28** 42–4
[79] Walton D G, Kellogg G J, Mayes A M, Lambooy P and Russell T P 1994 A free energy model for confined diblock copolymers *Macromolecules* **27** 6225–8
[80] Tsai H *et al* 2014 Two-dimensional pattern formation using graphoepitaxy of PS-b-PMMA block copolymers for advanced FinFET device and circuit fabrication *ACS Nano* **8** 5227–32
[81] Borah D, Shaw M T, Holmes J D and Morris M A 2013 Sub-10 nm feature size PS-b-PDMS block copolymer structures fabricated by a microwave-assisted solvothermal process *ACS Appl. Mater. Interfaces* **5** 2004–12
[82] Nickmans K, Murphy J N, de Waal B, Leclère P, Doise J, Gronheid R, Broer D J and Schenning A P H J 2016 Sub-5 nm patterning by directed self-assembly of oligo(dimethylsiloxane) liquid crystal thin films *Adv. Mater.* **28** 10068–72

[83] Yi H, Bao X Y, Tiberio R and Wong H S P 2015 A general design strategy for block copolymer directed self-assembly patterning of integrated circuits contact holes using an alphabet approach *Nano Lett.* **15** 805–12
[84] Fernández-Regúlez M, Evangelio L, Lorenzoni M, Fraxedas J and Pérez-Murano F 2014 Sub-10 nm resistless nanolithography for directed self-Assembly of block copolymers *ACS Appl. Mater. Interfaces* **6** 21596–602
[85] Evangelio L, Fernández-Regúlez M, Borrisé X, Lorenzoni M, Fraxedas J and Pérez-Murano F 2015 Creation of guiding patterns for directed self-assembly of block copolymers by resistless direct e-beam exposure *Proc. SPIE* **9423** 942326
[86] Kim S O, Solak H H, Stoykovich M P, Ferrier N J, de Pablo J J and Nealey P F 2003 Epitaxial self-assembly of block copolymers on lithographically defined nanopatterned substrates *Nature* **424** 411–4
[87] Cheng J K, Rettner C T, Sanders D P, Kim H C and Hinsberg W D 2008 Dense self-assembly on sparse chemical patterns: Rectifying and multiplying lithographic patterns using block copolymers *Adv. Mater.* **20** 3155–8
[88] Liu C C *et al* 2013 Chemical patterns for directed self-assembly of lamellae-forming block copolymers with density multiplication of features *Macromolecules* 1415–24
[89] Ruiz R, Kang H, Detcheverry F A, Dobisz E, Kercher D S, Albrecht T R, De Pablo J J and Nealey P F 2008 Density multiplication and improved lithography by directed block copolymer assembly *Science* **321** 936–9
[90] Liu C C, Han E, Onses M S, Thode C J, Ji S, Gopalan P and Nealey P F 2011 Fabrication of lithographically defined chemically patterned polymer brushes and mats *Macromolecules* **44** 1876–85
[91] Hellwig O, Bosworth J K, Dobisz E, Kercher D, Hauet T, Zeltzer G, Risner-Jamtgaard J D, Yaney D and Ruiz R 2010 Bit patterned media based on block copolymer directed assembly with narrow magnetic switching field distribution *Appl. Phys. Lett.* **96** 7–9
[92] Ruiz R, Dobisz E and Albrecht T R 2011 Rectangular patterns using block copolymer directed assembly for high bit aspect ratio patterned media *ACS Nano* **5** 79–84
[93] Griffiths R A, Williams A, Oakland C, Roberts J, Vijayaraghavan A and Thomson T 2013 Directed self-assembly of block copolymers for use in bit patterned media fabrication *J. Phys. D Appl. Phys.* **46** 503001
[94] Bai J, Zhong X, Jiang S, Huang Y and Duan X 2010 Graphene nanomesh *Nat. Nanotechnol.* **5** 190–4

Chapter 3

Electron beam lithography and its use on 2D materials

Vito Clericò, Mario Amado and Enrique Diez

Electron beam lithography (EBL) is a fabrication technique based on a scanning electron microscope (SEM) integrated with a lithographic system for the control of the electron beam and/or the SEM stage. In this chapter, starting from the description of a conventional field emission scanning electron microscope (FE-SEM), we introduce the basic concepts of EBL, its advantages and main limitations. In the second part of the chapter, we report some possible applications in which the EBL technique has a very important role, focused on the fabrication of nanodevices based on 2D materials.

3.1 Introduction

Electron beam lithography (EBL) is a nano-fabrication technique based on the controlled irradiation of a sample covered with a sensitive resist via a focused beam of electrons. EBL is typically performed by a scanning electron microscope (SEM) integrated with a lithographic system for the control of the beam and/or the SEM stage. This technique appeared in the late 1960s arising from the necessity to fabricate submicrometric devices, unreachable to other widespread methods such as optical (ultra-violet) lithography, x-ray lithography or imprinting. A standard EBL process consists essentially of three steps: exposure, development and pattern transfer. The successful outcome of the final process relies on the careful control of these singular, yet cumulative, actions shown in figure 3.1.

The emitter, also known as gun, consists of a thermionic filament or a field emission (FE) source based on a single crystal tip, both of them typically manufactured with tungsten. An FE gun produces a far more coherent and brighter beam compared to the one obtained with the thermionic filament. Throughout this chapter, our focus will be given to systems described as field emission scanning electron microscopes (FE-SEM), which are nothing more than an SEM equipped with an FE gun.

doi:10.1088/978-0-7503-2608-7ch3

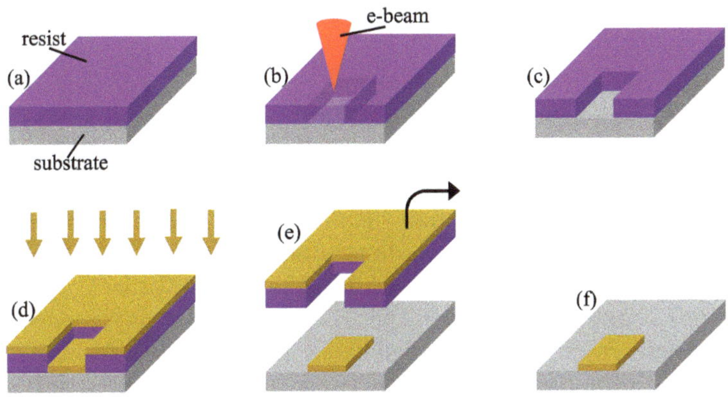

Figure 3.1. Sketch of a standard EBL process to evaporate a thin film only in the desired region of a substrate with a positive resist.

Once the electron-beam (e-beam) is produced, it must be focused to be used as a 'nano-pencil' to write on a thin layer of material sensitive to it (the so-called resist or e-beam resist) that was previously deposited onto the surface on which the final pattern is expected to be transferred to. The resist film can be of positive or negative tone, behaving similarly to how slides and negative films performed in analog photography.

The pattern drawn in the resist can be used either to tailor the material with an etching process or to define electrical contacts by metal evaporation (figure 3.1). The electron beam follows typically a Gaussian-like distribution and the higher the e-beam energy the lower is the wavelength and consequently a minor diameter of the beam, i.e. the size of your 'pencil-tip', is potentially reachable.

The SEM, considered as a microscope, has been primarily built for imaging purposes where the size of the sample area visible when taking the image will depend on the selected magnification. If one wants to inspect nanometric structures, the maximum available magnification will be required, and, therefore, the image will cover a reduced area of the sample. If you keep the magnification at its maximum value and the e-beam of the SEM is used to *write* a pattern, the size of the region where you can *write* (known as writing field, WF) will coincide with the area of the image. Of course, one can decrease the magnification and enlarge consequently the size of the *writeable* area, but it comes at a cost. Generally, a higher area is accompanied by an increase of the e-beam diameter, which reduces extraordinarily the accuracy of the pattern. This is one of the great disadvantages of EBL compared to other lithographic methods: writing accurate small nano-patterns is only possible in a micrometric scale. When we aim repeating a sub-micrometric pattern in a large area, as the electronic industry does in the fabrication of nanoelectronic devices in large wafers, the solution is at hand and well-known from the first textile manufacturers: stitching. In our case, stitching means keeping the WF at its minimum size hence keeping your e-beam size at its minimum too. After drawing the desired pattern in the WF, the physical motion of the stage holding the sample is performed, repeating the process in an adjacent unexposed WF of the same area. Stitching requires an extraordinary accuracy in terms of alignment and positioning (sub-micrometric accuracy indeed) that can be achieved either using interferometric (automatic) stages or by manual alignment using again the SEM and reference marks.

While truly EBL systems are designed to work only for lithographic purposes (integrating very-high-energy electron beams and interferometric stages for high-accuracy stitching), a good imaging SEM can be used for EBL and be good enough for a correct lithographic process. A typical example of the first species are the EBL commercial set-ups sold by companies such as Crestec, Vistec or Raith, well-suited for large-wafer technology or mass scale production, but with the major drawback on their cost, which can rise well over 1.5 M€. Henceforth we will devote the majority of this chapter to EBL using SEMs plus a patterning processor for EBL, rather modest setups compared to dedicated EBL instruments, but nevertheless more than enough for fundamental basic research such as preparing prototypical state-of-the-art quantum devices. There are numerous commercial systems to be purchased almost *off the shelf*, one of the most well-known is the *Elphy+* from Raith patterning system, which can be combined with an SEM Sigma from Zeiss, as we will thoroughly present in this chapter.

As we have already mentioned, in order to produce a final working device we will typically require one or several steps of lithography combined with the auxiliary evaporation of metals or dielectrics as well as any kind of etching (dry or wet) process to define the final geometry of the device. These processes are intimately related; as an example, the selection of a particular resist will be affected by the nature of the material to be processed, the size of the pattern and the required accuracy, also having an effect on the procedure for the etching and/or evaporation.

This chapter will be divided as follows: we will start overviewing a conventional FE-SEM set-up and introducing the basic (common) concepts in EBL as well as its main advantages and limitations. We will show the steps for a successful EBL process and we will briefly introduce electron beam deposition of metals and dielectrics as well as inductively coupled reactive ion etching as two of the processes that might be needed for the final completion of the desired device. We will finally focus on some of most widespread device geometries when working with two-dimensional (2D) materials, as a first approach for the reader to the application of EBL to 2D materials. In the last part we report general applications of EBL as well as new developments of this technique.

3.2 Fundamentals of EBL

3.2.1 Chamber and electron column

The column of the FE-SEM defines the region where electrons are generated, accelerated, focused and deflected. It possesses unique integrated beam booster technology to keep the beam at a high potential in the full length of the column, independent from the landing energy [1]. A schematic view of a typical FE-SEM with Gemini column from Zeiss and its main components is shown in figure 3.2(a).

Electron microscopes use bundled high-energy electrons as the basis of their operation that are generated in the source. These electrons are emitted in a filament presented in position (1) in figure 3.2(a), controlled by the suppressor electrode

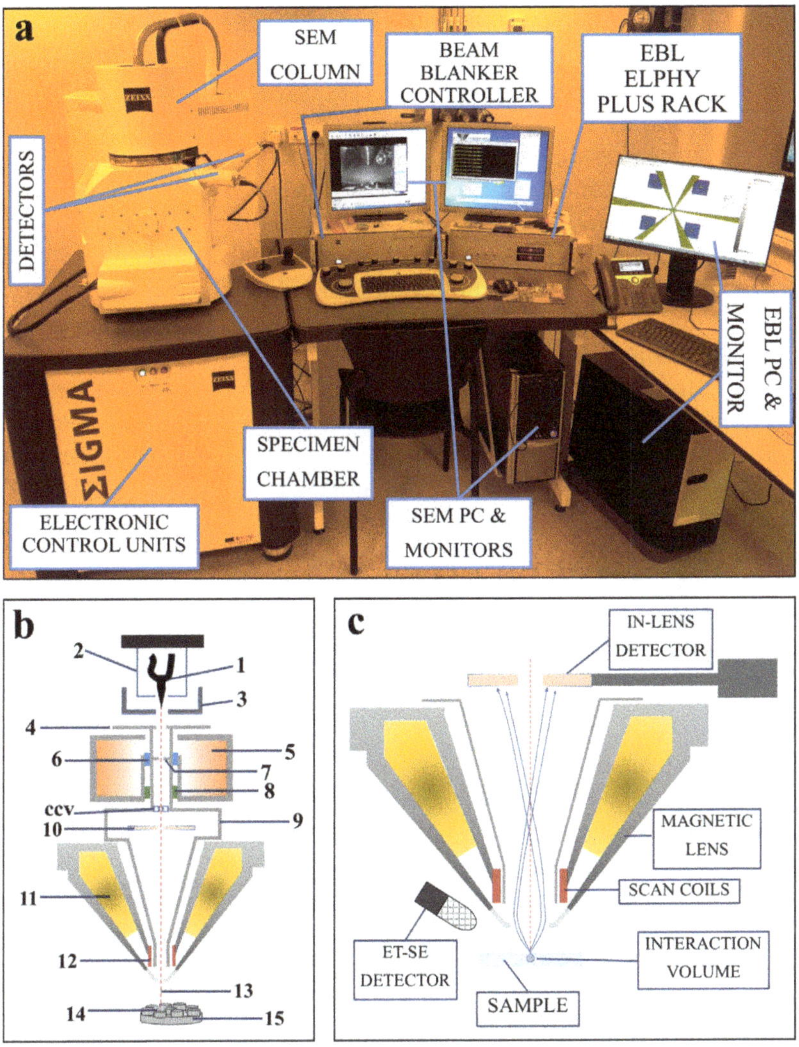

Figure 3.2. (a) Photo of the FE-SEM and the Raith nanolithography system with their main components. (b) Schematic views of FE-SEM: Filament (1), suppressor electrode (2), extractor electrode (3), anode (4), condenser lens (5), gun/aperture alignment system (6), multi-hole aperture (7), stigmator (8), liner tube (9), In-lens detector (10), electromagnetic lens (11), scan coils (12), primary electron beam (13), specimen (14), holder (15), column chamber valve (ccv). (c) Sketch of the positions for the In-Lens and ET-SE detector (2) with respect to the specimen and column.

(2) and collected by the extractor (3) and finally accelerated to the anode (4). The electron beam passes through the borehole of the anode and the condenser (7) and leaves the source before being collimated by the aperture (5).

In EBL, filaments can be composed of different materials, which is tightly related to the diverse working principles and expected properties. Thermionic filaments (W and LaB_6) were commonly used in the pioneer systems, where a moderate vacuum (10^{-6}–10^{-7} mbar) was sufficient. These systems liberate electrons from the well-known thermal electrode emission, that is, by increasing the temperature of the filament until the energy of the carriers overcomes the work function of the material [2].

Schottky sources, also known as thermal field emission sources, represent the evolution of the thermionic ones and are based on tungsten filaments covered by zirconium oxide. They are present in a vast number of the last generations of EBL systems. Their operational temperatures are lowered and thus electron tunneling due to electric-field-induced emission prevails. The combination of sharp tips where the electric field at their apex is locally enhanced (which drastically lowers down the energy barrier) and the ZrO coating that reduces the work function to be overcome increases dramatically the emission current.

Once the electron beam has been produced in the source, it must be focused and guided by the optics present in the column, where electrostatic or magnetic lenses bundle and deflect the e-beam (see position 11 in figure 3.2(b)). Electromagnetic fields will guide the beam of electrons, analogously to geometric classical optics. Similar to optical microscopes where the resolution is limited by the diffraction within the different lenses and depends on the wavelength of the light itself, in an electron microscope the access to high resolution beam spots is allowed when the electron wavelength is shortened.

The wavelength λ is related to the momentum according to the formula:

$$\lambda = \frac{h}{p} \tag{3.1}$$

where h is the Planck constant and p is the momentum of the electrons. Because the momentum p is determined by the accelerating voltage V where

$$eV = \frac{p^2}{2m}; p = \sqrt{2meV}$$

with e being the electron charge and m the electron mass. Therefore, the wavelength of propagating electrons at a given accelerating voltage V can be determined by

$$\lambda = \frac{h}{p} = \frac{h}{\sqrt{2meV}}$$

Since the mass of an electron is $m = 9.1 \times 10^{-31}$ kg, the charge $e = 1.6 \times 10^{-19}$ C and the Planck constant $h = 6.626 \times 10^{-34}$ J s we obtain:

$$\lambda = \frac{1.228}{\sqrt{V}} 10^{-9} m = \sqrt{\frac{1.5}{V}} \text{nm} \tag{3.2}$$

Resolution in microscopy is limited by diffraction effects to about $\lambda/2$. Therefore formula (3.2) allows us to obtain the electron's wavelength (and therefore the nominal resolution of our electron microscope) for different voltages used in the SEM. For a typical value of the accelerated voltage (5 kV) wavelength is 0.017 nm without considering relativistic effects. We can include relativistic effects straightforwardly including the Lorentz factor [3] but the correction for voltages below

300 kV is small so usually to obtain the nominal resolution we have restricted ourselves to formula (3.2). However, in practice, other effects such as the aberration of the lenses (spherical and chromatic), the spread of the primary electron beam inside the resist and the generation of secondary electrons around the incidence point are limiting factors compared to diffraction effects, as will be later discussed.

Multiple apertures are mounted in the vacuum tube. They can well serve both for interrupting the beam or for limiting it (reducing the amount of beam hitting the specimen). For imaging processes, the standard aperture used in the majority of the SEMs is 30 microns, but especially for EBL different apertures (typically 7.5, 10, 20, 30, 60 and 120 microns) are accessible in any commercial system permitting one to achieve a good compromise between resolution and exposure time.

SEM is a scanning microscope in which the scanning of the sample surface takes place thanks to the deflection of the electron beam during their crossing through the scan coils (or deflector plates) (see position 12 in figure 3.2(b)). Electron beam is deflected in X or Y direction by the magnetic field (or electric field in the case of deflector plate) that is created in the scan coils, so the electron beam starts to scan in a raster way over a rectangular area of the sample surface with very high precision. The size of the scan field on the sample is chosen according to the magnification. The magnification is given by the ratio between the dimensions of the field on the SEM display and the real size of the field on the sample.

Since the SEM screen has a fixed size, higher magnifications are obtained reducing the size of the field on the sample.

With respect to a conventional SEM, an FE-SEM has a field emission gun that allows the use of highly focused and low-energy electron beams. This feature entails an improved spatial resolution and the possibility to work at very low potentials (0.01–5 kV). The use of a low potential minimizes the charging effect and reduces the possibility to damage sensitive samples such as 2D materials.

The interaction between the beam and the specimen will give us back crucial information about the sample (position, topography, size and shape, etc). When the incident electron beam interacts with a specimen, it will partially transfer its energy to electrons in the vicinity of the sample's surface, which eventually can escape from the specimen. They are coined secondary electrons (SEs) if they have low energy (less than 50 eV) and backscattered electrons (BSEs) when they are more energetic (up to the energy of the primary beam). SEs and BSEs can be used for imaging purposes and for the exposure of the resist in the EBL process.

The electrons leaving the sample can be collected using different detectors in the SEM. In-Lens detectors and Everhart–Thornley detectors (figure 3.2(c)) are the most typical ones in commercial systems. They work with secondary electrons which can be divided into two different kinds:

- SE1: They are directly generated when the sample is hit by the beam and leave the surface in the vicinity (up to a few nm) of the impact point of the incident beam.
- SE2: They are generated after multiple scattering inside the specimen and leave the sample at a certain distance (up to a few μm) from the impact point of the incident beam.

In-Lens detectors (figure 3.2(b)) are commonly placed inside the electron column of the microscope and collect SEs and BSEs that are emitted from the surface of the sample, providing high-quality topography and compositional images of the surface with a rather low applied energy ($\leqslant$ 5 kV), which is useful for non-conductive samples.

Everhart–Thornley (ET) detectors (collecting SEs) are located in a lateral position (ET-SE detector is shown on the bottom left of figure 3.2(c)) and are designed to operate at working distances between 5–12 mm. ET detectors are mainly used for topography imaging and are more sensitive when used to scan tilted samples.

A column chamber valve (ccv in figure 3.2(b)) separates the gun from the chamber where the specimen is placed. With the use of an ion-getter pump (IGP), the typical pressure is in the range of 10^{-10} mbar where the gun is located. The vacuum in the main chamber for the sample is obtained thanks to turbo pumps with a typical pressure of 10^{-6} mbar. There are several options to place and position the sample in the different available holders. They can be fixed either by pressure clamps, directly screwed or glued using special double-sided conductive tape. Sample holders typically host the different standards that are required for the calibration of the current, focus and astigmatism.

A controlled three-dimensional movement is essential in a modern EBL system. Access to numerous samples by accurate calibrated electromechanical displacements and the possibility to tilt the sample holder *in situ* are beneficial both for lithographic and imaging purposes. In the system presented in this chapter, the instrument includes three-axis (X, Y, Z) fully-motorized rotary stage with tilted option to rotate along the Z-axis from 0° to 90°. Samples with diameters up to 4 inches are allowed in the system discussed in this chapter.

3.2.2 Basic and lift-off EBL processes

A schematic view of the typical process for fabrication, in which EBL has a fundamental role, is shown in figure 3.3(a). Firstly, the spin-coating of the resist on the sample takes place (see section 3.3.1 for details). The used resists are sensitive to the electron beam, so during the EBL process part of the exposed resist becomes sensitized (step 1). In this way, after the developing process (step 2), some part of the resist is removed, and the uncovered sample can be removed by etching process (dry or wet etching), meanwhile the covered sample is protected by the resist, that acts as mask (step 3). Finally, the residual resist is eliminated in the last step 4, in which an immersion of the sample in resist removal (typically acetone) takes place.

In figure 3.3(b) another common fabrication process is shown. In this case the sample is not previously present on the substrate, but it is grown after the EBL process. After exposure (step 1) and developing process (step 2), part of the sample is grown directly on the substrate or on the resist mask (step 3). Finally, the sample is defined through the so-called lift-off process, where an immersion in resist removal (typically acetone) removes the resist mask and the material on it (step 4), defining the rest of the sample on the substrate.

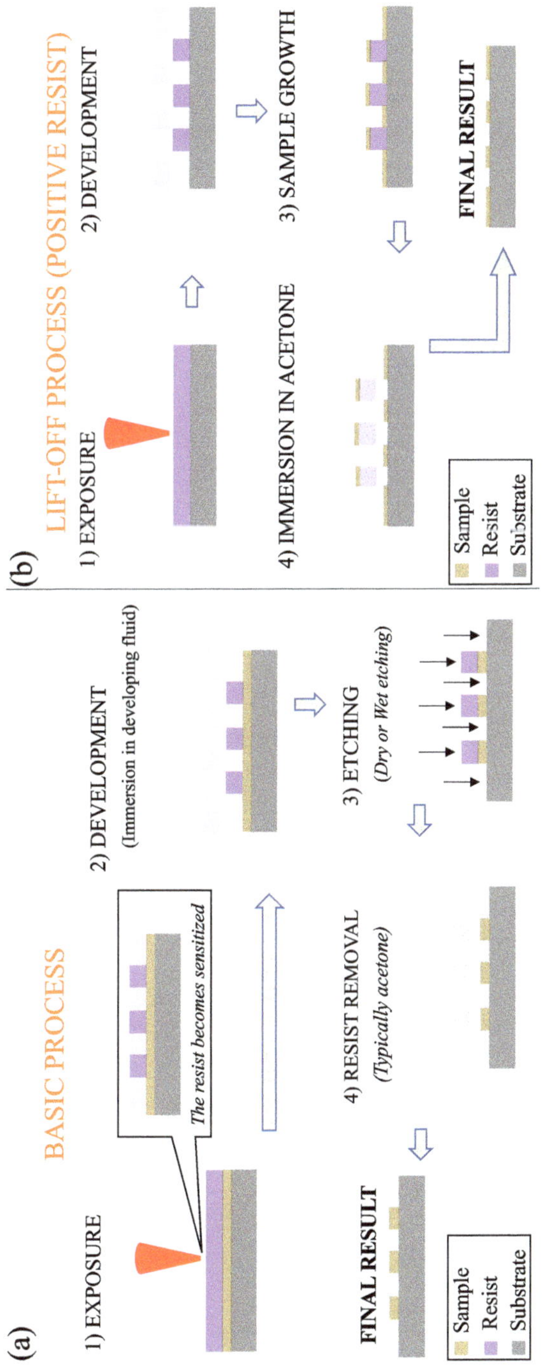

Figure 3.3. (a) Basic nanofabrication process by steps. (b) Lift-off procedure.

Some common examples of these processes will be discussed later in detail, in particular the dry-etching process with ICP-RIE and evaporation with e-beam will be described in section 3.4.

3.2.3 Resolution

Real resolution in an EBL process depends on different parameters/effects. Here we report the main important contributions:

- **The Beam size (d)**

  SEM resolution strongly depends on the SEM beam diameter. Its size, apart from the diffraction limit, is determined by the electromagnetic lenses used to focus the electron beam on the sample. The theoretical effective beam size (d) is given by the formula [4]:

$$d = (d_g^2 + d_s^2 + d_c^2 + d_d^2)^{\frac{1}{2}}$$

  In which $d_g = d_f/M$ is the virtual source contribution (with d_f the source size and M the magnification). $d_S = 0.5C_S\alpha^3$ (with C_S the spherical aberration term and α the beam convergence angle) is the distortion term related to the spherical aberrations. $d_c = C_c\alpha\Delta E/V$ is the chromatic aberration contribution (with C_c chromatic aberration term and α the beam convergence angle, ΔE the energy spread of electron beam and V the acceleration voltage). $d_d = 0.6\lambda/\alpha$ is the diffraction limit contribution. Schematic representation of spherical and chromatic aberrations is shown in figure 3.4(a) in analogy with the optical glass lenses.

- **Beam-resist interaction and main limitations; proximity effect**

  When an electron beam collides with the resist, the interaction strongly depends on the acceleration voltage as schematically shown in figure 3.4(b). This interaction is complex and influenced by many factors, but in general, resolution is typically worse at low voltage because the aberrations of the microscope are greater. Working at high voltage is normally preferred to achieve higher resolution, but there is also another important effect that has

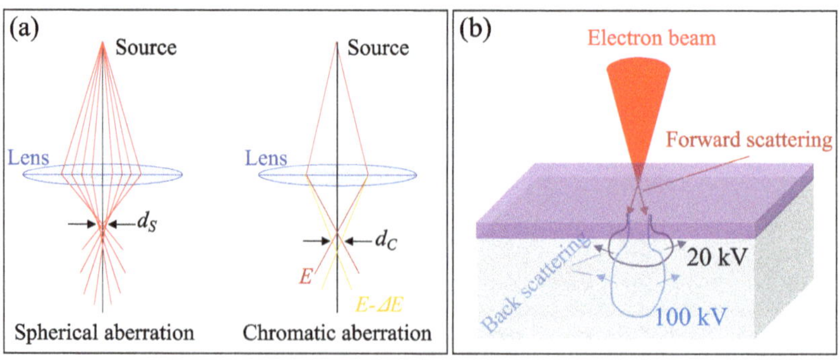

Figure 3.4. (a) Analogy with glass lenses for spherical and chromatic aberration. (b) Schematic representation of electron beam collisions in an EBL process.

to be taken into consideration for the interaction between beam and resist: the proximity effect (PE).

PE relates to the resist outside the targeted area of exposure that unavoidably receives a non-zero electron dose. In fact, when the electron beam collides with the resist or the substrate, the generation of forward (with small angles) and backscattering electrons (with large angles), as shown in figure 3.4(b), causes the exposure of resist far away from the original incident point. In particular, backscattered electrons, with high energy, high scattering angle and large travel lengths generate secondary electrons along their paths, which holds major responsibility for exposure far from the incident point. PE affects more specifically submicron structures, causing a nonuniform exposition distribution between pattern elements or inside a pattern element [5]. The nonuniformity in the exposure between pattern elements might arise from a poor selection of the pattern geometry in the CAD file or from the reduced (limiting) space and size of the adjacent features to be exposed. In fact, smaller elements receive less real dose than the larger ones, and also isolated elements receive less dose than packed elements. The nonuniformity in the exposure inside a single element is instead related to the difference in the exposure between the center of the feature with respect to the edges and the selection of the writing pattern within the single element. Scanning (thus writing) along parallel lines until the completion of the polygon is the most common pattern but diagonal or radial scans can also be performed when necessary.

In the schematic representation of the PE in figure 3.4(b), it is clear that the use of a higher beam voltage implies drastic changes in the PE effect owing to the variations in the interaction volume of the electron beam in the sample and in the forward scattering angle of the primary beam [6]. Also, the nature of the substrate itself has a lot of influence on the interaction volume, thus, on the backscattering process.

There are different ways to reduce PE, such as an appropriate choice of the resist. In fact, the use of a thinner resist layer often reduces the forward scattering. In addition, the use of multilayer resists, in which the upper layer is used for patterning and the lower layer is used to reduce the backscattering processes can be advantageous [7]. Moreover, a high-contrast resist (i.e. HSQ) is typically less sensitive to PE.

Another possibility for the complete or partial correction of the PE error consists in introducing element fragmentation with different doses in the CAD file features, choosing the best scan direction [8] and changing shape/dimension of the element. It can be done manually by the user (but it requires a high level of expertise and it is very time consuming) or the PE correction can be computationally calculated by the EBL software, basically simulating the effect due to backscattering electrons, whose contribution is higher at high voltage.

To reduce the backscattering contribution, it is also possible to perform a background-dose PE correction (known as GHOST) [9, 10]. It consists in writing, through the use of defocused beam, the inverse of the intended image (after the primary exposure), so that a constant background level of dose in all the regions is obtained. Apart from the longer exposure time due to the use of two lower dose scans, the technique is characterized by a low contrast in the feature definition.

3.2.4 Scan modes, writing field, DAC, stitching, beam stability

In scanning beam systems there are two main writing strategies for lithographic purposes and they strongly depend on the beam. Standard Gaussian (round) beams address the desired pattern and their features pixel by pixel and it can be done either by a raster scan or a vector scan. The entire field is scanned with the raster scan method and the beam is blanked where exposure is not required. Instead, a vector scan system scans only over the required positions and the beam is blanked as it moves between different scanning places (figure 3.5(a)) [4].

Both modes have their advantages and drawbacks; as an example, a raster scan would be slower than a vector one but the possibility of the scan to exceed the limits of the pattern would be higher in a vector scan.

As mentioned in section 3.1, another important concept in the EBL process is the WF, which is the largest area exposed without the stage movement and it depends on the SEM magnification. Each WF is composed by a finite number of steps in the exposure (exposure elements (EXELs)), determined by the deflection of the beam according to a certain value of an applied potential. This potential is controlled by a PC through a card for digital to analogic conversion (DAC), connected with the deflection system of the SEM (figure 3.5(b)). For real systems, for a DAC of 16 bits for each axis for example, the WF is subdivided in a writing grid of $2^{16} \times 2^{16} = 65\,536 \times 65\,536$ EXELs.

Thus, WFs with lower size permit performing the EBL process with higher resolution. For example, the nominal resolution with a WF of 100 μm is 2 nm[1]. WFs are chosen by the user with the aim to achieve the required resolution minimizing the stage movements. When the pattern size (or Working Area (WA)) needs the use of

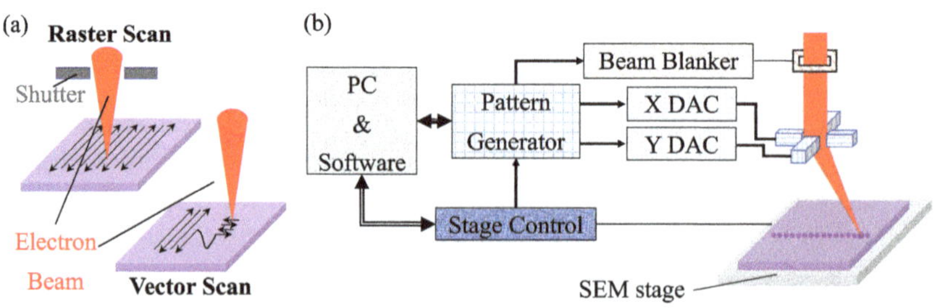

Figure 3.5. (a) Raster scan and vector scan modes. (b) A schematic EBL set-up.

[1] The nominal resolution of an EBL process depends also on the selected SEM aperture (beam current). Using a WF of 100 μm, a resolution of 2 nm can be achieved when the aperture is set to 7.5 μm.

more WFs, the fields are stitched through the stage movement. The accuracy of this stitching depends on the precision of the stage movement. Typically, interferometric stage guarantees nanometric stage movement, meanwhile mechanical stages are characterized by hundreds of nanometers error during the stitching process.

During the exposition, acceleration voltage and beam current are fixed. The real dose is measured in terms of current per unit area and it is determined, as well as the writing speed, by the time in which the beam remains in each EXEL. This time is called area dwell time.

Hereafter, the used nanocontroller is the Elphy-Plus, a system from *Raith* company which makes it possible to produce micro and nano structures by means of a Gaussian electron beam vectorial writing in connection with the FE-SEM. The hardware assembly is built into a 19' rack consisting of a digital control unit and a separate thermo controlled analog electronics box. It has X-Y beam shift inputs and a video out to a monitor (BAS out). The electronics units are completely remote controlled by a PC data system via an integrated small computer system interface (SCSI). Apart of the ELPHY Plus rack, a nanolithography system should present a beam blanker unit that will operate to avoid any exposure when the beam is in transit from one pattern element to another (figure 3.5(b)).

The beam blanker hardware is installed in the SEM column. It consists of sets of parallel plates where one is grounded and the other is connected to the high voltage end of the system. When the voltage is zero, the beam of electrons is allowed to pass through a spray aperture and so they can reach the sample. Conversely when a non-zero voltage is applied, electrons are deflected, and the aperture does not permit the electrons to reach the sample. The beam blanker control unit is a high voltage power supply, accepting a TTL voltage from the ELPHY Plus rack.

The ELPHY Plus software includes all the features needed to expose micro and nano structures (WFs, doses, etc) and it is integrated with a GDSII based editor compatible with ASCII, DXF, CIF and CAD formats. The generation and edition of patterns is graphic and based on the superposition of different layers where the different elements are drawn. In these layers, diverse elements will be designed for multilevel exposure where beam intensity, dose, exposure time, focus, writing field can be modified and unique to each layer.

3.2.5 Resists

The resist has a central role in every lithographic process. It is typically used as sacrificial layer and deposited on the main working material through a spin-coating process (see section 3.3.1) [11]. Typically, the thickness of an EBL resist can measure between 20 nm up to a micrometer. Resists are either organic or inorganic polymeric materials dissolved in solvents and they can be classified in two different classes according their tone: positive or negative tone. After the exposition on the desired areas of the resist and the developing process, the exposed resist remains on (negative tone) or leaves (positive tone) the surface of the substrate ([figure 3.6(a)). The exposition to the electron beam affects the molecular composition of the resist, in particular chain scission in positive tone

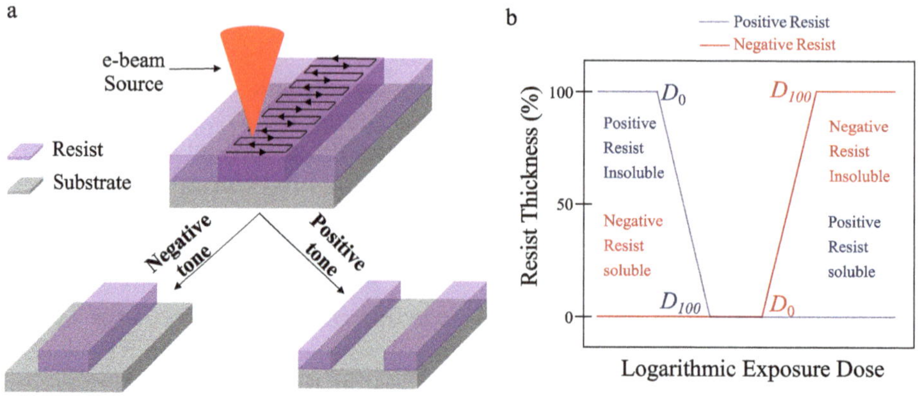

Figure 3.6. (a) Negative and positive tone resists after electron beam exposition and developing process. (b) Remaining resist thickness in positive (dark blue) and negative resist (red) [13].

resists and cross-linking in negative tone resist. The choice of a resist is typically taken according to three principal characteristics: resolution, contrast and sensitivity. Resolution is defined by the minimum feature size (or minimum feature distance) that can be resolved. The contrast is the capability in differencing between exposed and unexposed areas. The sensitivity defines the alteration of the resist with respect to the electron beam doses [12]. Contrast and sensitivity are strictly correlated and expressed according to the equation [13]:

$$\gamma = \frac{1}{\log \frac{D_{100}}{D_0}}$$

where γ is the contrast value, D_0 is highest dose exposition that does not affect the resist, D_{100} is the minimum dose for a complete exposition of the resist. Even if it is desirable to have high contrast and high sensitivity resist, an increase of contrast results in the worsening of the sensitivity and vice versa. A typical plot of the remaining resist thickness versus the exposure dose is shown in figure 3.6(b) for positive and negative resists. Apart from the electron energy, the type and concentration of the developer solvent, the developing time and temperature affect the contrast and sensitivity.

In table 3.1, we report some common resists used for an EBL process and their developers:

In section 3.3.4 we will compare two different EBL processes, using positive (PMMA) and high negative (HSQ) resists, highlighting the main limitations and problems related to a typical EBL process.

3.2.6 Cross-linking problem

Cross-linking might appear when the energy of the electron beam is much higher than the resist ionization energy. Through this process, chemical reactions occur in the resist, bringing about unwanted molecular bonds. The crosslinked resist cannot be easily removed from the sample surface. Although a crosslinked resist can be used

Table 3.1. Comparison among the most frequently used e-beam resists [12, 14].

Risist	Tone	Resolution(nm)	Sensitivity(μC cm^{-2})	Developer
PMMA	Positive	10	100.0	MIBK:IPA
EBR-9	Positive	200	10.0	MIBK:IPA
PBS	Positive	250	1.0	MIAK:2-pentanone 3:1
ZEP	Positive	10	30.0	Xylene: p-dioxane
AZ5206	Positive	250	6.0	KLK PPD 401
SML	Positive	15	60	7:3 IPA:DI Water
SU-8	Negative	25	5.0	EC solvent
COP	Negative	1000	0.3	MEK: ethanol 7:3
SAL-606	Negative	100	8.4	MF312:DI water
HSQ	Negative	<5 nm	500.0	TMAH:DI-water
NEB	Negative	<10 nm	12.0	CD26
Calix[n]arene	Negative	15	115.0	MIBK

for some purposes, as an alternative dielectric film for example [15], it is commonly an undesired effect. Actually, cross-linking is frequently not an effect that appears during the EBL exposure step[2], but it is rather a 'post-exposure' problem related to a second fabrication step, typically the dry-etching process or during metal evaporation. In fact, as we will explain in the next sections, during an ICP-RIE or e-beam-evaporation process, some electrons are generated and if they reach the resist mask, they can crosslink it. The final result is a permanent resist mask after a dry etching process, or in case of evaporation problems during the lift-off process.

3.3 Steps for a successful EBL process

In this section the most common steps in an EBL process will be presented. After introducing basic concepts about spin coating, we will present a process where the WA is larger than the size of the writing field (WA>WF). This is typical when working with a wafer without alignment marks. Afterwards, we will consider pre-structured samples (with pre-deposited alignment markers) in which we face the inverse scenario, WA<WF. Both processes are carried out with the use of a standard SEM stage, in which the movement on the sample takes place with a typical stage error movement of 1 μm, far from the desired accuracy in a nanolithography process. We will therefore present a common self-alignment strategy that will serve to arrive at a final writing accuracy of some tens of nanometers. Two different kinds of resists will be used, both leading to two unwanted, though common, effects when performing EBL: the crosslinking of the resist and the PE.

[2] Indeed, crosslinking the resist requires a rather high dose. For example, for PMMA resist the dose necessary for the crosslinking is roughly ten times the standard one for a correct lithography, hence, crosslinking will mainly happen because of indirect effects.

During the EBL processes we used different WFs. All the WFs must be calibrated beforehand, so a correspondence between the applied voltage and the beam deflection in the (X,Y) plane is found. For the WF alignment, we used a sample (Chessy sample) with calibrated distances. During the alignment procedure, different marks are overlaid on the Chessy image. We move each mark to the correct position and, at the end of this procedure, the WF becomes corrected and the calibration values are stored in the pattern generator hardware.

Different SEM magnifications will imply a change in the maximum area covered by the WF. All the processes described in this section were carried out with FE-SEMs where a fixed beam accelerating voltage of 15 kV was set.

3.3.1 Spin coating

An EBL process should start with a thorough cleaning of the substrate that will depend on the nature of the sample. As a standard procedure, substrates undergo a two-stage cleaning. They are firstly rinsed in acetone in order to remove organic impurities followed by ethanol or isopropanol washing, which removes contaminated acetone before it can form streaks on the surface. The cleaning process might be carried out in a sonic bath cleaner that will help to remove precipitates on the surface. It is nevertheless worth mentioning that such a cleaning processes will strongly vary from sample to sample.

A clean uniform substrate is ready for the very first step of every lithography process, i.e. the spin coating. This process consists in covering the surface of the sample with some drops of the chosen resist, via standard pipetting technique, that will eventually turn into a uniform film using the spin coater. The final thickness of the film depends on the chemical composition of the resist, revolutions per minute and total time of the spin coating process as well as the nature of the surface and the size of the sample. Before the exposition, the resist must be baked by placing the samples onto hot plates with a given (desired) temperature. The definition and dose of the whole EBL process will strongly depend on the temperature and time of this baking process.

In this section we focus the attention on one of the most common resists used in an EBL process: the polymethylmethacrylate (PMMA). PMMA represents also the best candidate for EBL on 2D materials, especially when graphene is involved [16], mainly due to its easy treatment and to the chemical components of the resist itself and its developer, a mixing of methylisobutylketone (MIBK) and isopropanol. In fact, they are less aggressive than other resists and respective developer; moreover, residual PMMA can be removed with an annealing process at relative low temperatures after the fabrication process. PMMA is typically dissolved in anisole or chlorobenzene and its concentration in the solvent will affect the final thickness of the film [17].

The quality and reproducibility of an EBL process strongly depends on the condition of the resist coating, in particular its homogeneity. For this reason, a preliminary check with a filtered optical microscope is convenient. Figures 3.7(a)–(c) display optical images of 1 × 1 cm wafers consisting of 300 nm thick SiO_2 on Si that

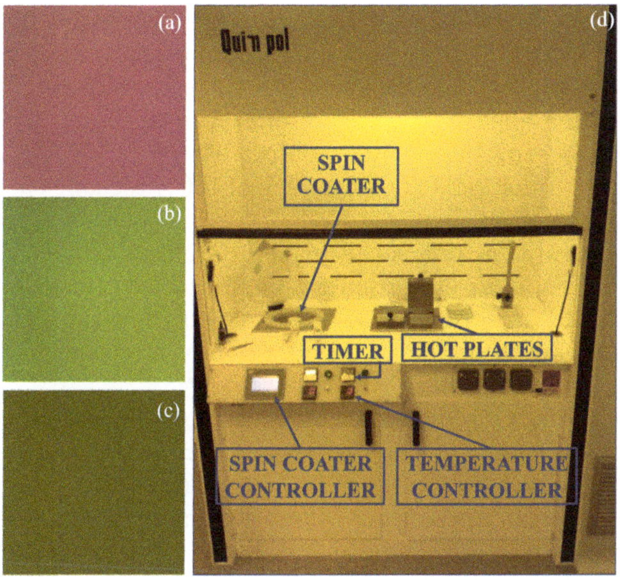

Figure 3.7. Optical images of PMMA coatings with different concentrations on sacrificial SiO_2/Si wafers: (a) PMMA2%, (b) PMMA4% and (c) PMMA8%. (d) A typical fume hood for EBL process. Including a LabSpin spin coater, two hot plates, deionized water supplier and sink.

has been coated with the same homemade PMMA[3] (molecular weight 996 K) but with different concentrations in chlorobenzene: (a) PMMA 2% spun at 4000 rpm for 60 s, (b) PMMA 4% spun at 4000 rpm for 60 s, (c) PMMA 8% spun at 3000 rpm for 60 s. The optical contrast (after baking each of them at 160 degrees for 10 min) is different, corresponding to the thickness (a) 70 nm, (b) 220 nm, (c) 800 nm.

Spin coaters are usually installed in a fume hood with an independent exhaust system that guarantees the extraction of toxic gases. We have used a manual spin coater LabSpin 6 from SÜSS MicroTec (maximum speed of 8000 rpm) and two hotplates (figure 3.7(d)). In a spin-coating process, the resist is deposited by using the spinner on a substrate with a uniform thickness of hundreds of nanometers. Finally, the resist is baked on the hot plate. We also have an ultrasonic cleaner and a magnetic hotplate, which can we employed for cleaning, lift-off and wet etching purposes.

3.3.2 EBL on a bare wafer (WA>WF)

In this section we describe an EBL process on a bare sample in which a precise alignment on a particular position of its surface is not essential. This is a good option when it is necessary to define arrays of structures with nanometric sizes. As an example, we describe the EBL process to create an array (1 mm^2) of thousands of nanoresonators with lateral sizes of 100×200 nm and a separation of 1 μm centered

[3] For the homemade PMMA, we will always refer to PMMA 996 K from now onwards unless stated otherwise.

on a GaAs substrate of 1 × 1 cm^2. In this case an aperture of 7.5 μm and a WF of 100 μm were used, resulting in a nominal resolution of 2 nm. Actually, the resist selected for this process, a commercial PMMA2% resist (Allresist© ARP679.02) has far worse resolution than the nominal one. The standard CAD file for the lithography process shows rectangular shapes of 100 × 200 nm with a selected dose of 230 μC cm^{-2}, repeated for a whole WF of 100 × 100 μm. Therefore, this EBL process is performed in the limit WA>WF.

The process starts with three small manual scratches or indentations on three different borders or corners of the sample similar to the shown in figure 3.8(a). A scratch helps for the correction of SEM parameters (focus, aperture and astigmatism), since the dielectric nature of the resist on a bare sample will bother their adjustment.

Firstly, the stage must be adjusted through the global coordinates. In this way, we can unambiguously define the exact correspondence between the physical coordinates (X,Y) of the stage (SEM stage coordinates) and the (U,V) coordinates of the sample. The global origin can be taken (and it is commonly taken) in the left-bottom corner of the sample and it is essential to compensate the difference between the (X,Y) and (U,V) origins. A second global point is taken after a controlled movement on the SEM X-axis (for example 1 cm) so that the angle between the two systems can be adjusted as well[4]. From this moment it is possible to move the SEM stage using the (U,V) coordinates because the internal electronics in the system will use a roto-translation matrix to this end.

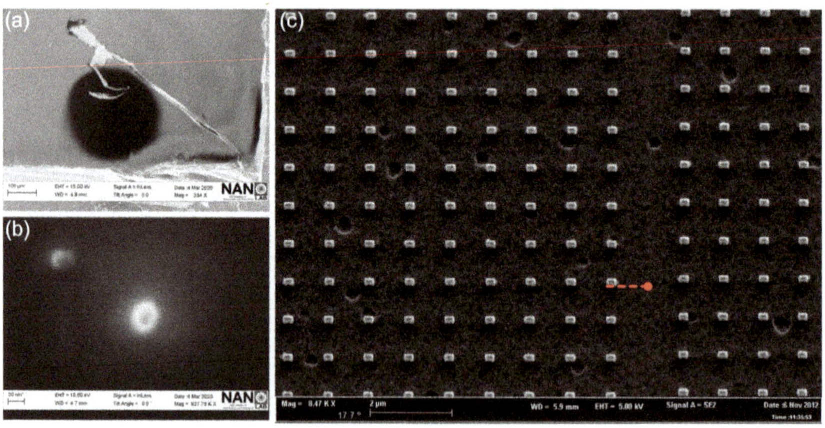

Figure 3.8. Several micrographs of the WA>WF process showing a scratch made on the bottom-right corner of a coated sample (a). Spot grown on PMMA, used for optimization of the exposition parameters (b). Final array of nanoresonators obtained through EBL without markers (c).

[4] Here the axes of the sample are parallel to the axes of the stage, because we do not care about the orientation of the array with respect to the sample orientation. In section 3.3.3, we will show a case in which we need to carry out an angle correction for the compensation of the difference.

The next step consists of defining an array (for example, of 1 mm^2), ideally in the center of the sample and in which positioning scratches are ideally separated 5–10 mm[5]. A further correlation between the coordinates in the sample (U,V) and coordinates in the CAD file (u,v) must be found. Firstly, the software needs to define the 'three points alignment' adjustment of the bare sample. We can take the bottom-left corner scratch as first (U,V) point (0,0) mm, moving physically with the SEM stage along the borders of the sample (so avoiding unwanted exposition of the sample). Then, we choose the bottom-right corner scratch as second (U,V) point, it is ideally positioned at (10, 0) mm. The third (U,V) point at (0, 10) mm is ideally chosen at the top-left corner scratch of the sample. In each point we correct the focus. In the case of PMMA resist, a useful trick can be used to fix correctly the focus (the scratch can have a different WD with respect to the resist coating) and check the quality of the SEM parameters. In fact, it is possible to create a spot (burned area) of very small size on PMMA resist. This process takes place only when the focus is approximately the correct one, so we improve the focus repeating the process in another position (figure 3.8(b)). Its shape gives us information on the astigmatism parameter, since a circular spot means a good correction of astigmatism, whereas an elongated shape is a signature of an uncorrected astigmatism. A good defined circular spot represents also a good candidate for the aperture adjustment through the wobble adjustment, in which the electron beam is displaced back and forth across the focus and it is corrected when the image of the spot 'blinks' in the same position. Once a good spot is obtained, all the parameters are definitively corrected. For a very precise EBL, in particular for the focus correction[6], the 'spot trick' can be applied for all the three-point alignment. Once the alignment is completed, it is necessary to move the stage on the Faraday Cup and measure the beam current that will serve us to calculate the real doses during the exposition. Therefore, after the selection of the WF of 100 μm, we drag and drop the CAD file in the so-called *positioning list*, we select the layer, step size, central dose, WA with the same dimension and position of the WF, (U,V) pattern position and we calculate the dwell time for the chosen step size[7]. For the first WA, the center of the patterning is fixed at (U,V) coordinates close to the center of the sample. We repeat the import of the CAD file in the positioning list as a matrix (10 rows × 10 columns)[8] and, for each one, the patterning center is shifted 100 μm in U and/or V, in order to obtain a total matrix of 1 mm × 1 mm. The time for each patterning exposure (fixing the resolution

[5] We have supposed an ideal distance between adjacent scratches of 10 mm for simplicity (and only in one axis-direction). This will vary between different processes and substrates.

[6] We typically need to correct the focus at different points of the sample. Through Elphy Plus software it is also possible to read in the focus value together with the coordinates for all three alignment points. So, we move the stage to all the three points and re-adjust the focus on each point before reading in the coordinates. In this way the focus will be changed for each digitally addressed (U,V) location.

[7] The step size is the distance between two adjacent exposed points, the dwell time is the time for the exposition at each point.

[8] In this way, the WA will be larger than the selected WF at the end of the EBL process. It is also known as multiple patterning process and it is largely used for dose array matrix (in which doses can be changed along U or V directions).

at 2 nm and the correspondent dwell time) is estimated around 15 min. So, the whole EBL process (without considering the loss of time due to the movement of stage) requires 25 h of exposure. Such a long exposure time is a major limitation is most conventional systems. Therefore, when and where possible, the combination of EBL with another faster lithographic technique like optical lithography [18] would be preferred.

Once the exposure is finished, the last step for the EBL process is to develop the film. Here, we will perform it under cold environment (10°C) for 90 s and rinsing the sample in commercial developer AR60050 from Allresist, with a subsequent stop for 30 s in room-temperature isopropanol.

The electronic micrograph shown in figure 3.8(c) shows the final array of nanoresonators (after evaporation and ICP-RIE processes to be later described) [10]. One of the main limitations of EBL, stitching, can be noticed. Such misalignment/error arises between two adjacent WAs due to the stage movement. In figure 3.8(c) the position of the array on the right is shifted both in (X,Y) with respect to the correct one (red circle indicates the correct separation between the two matrices of array). In this case the shift is lower than 1 μm, but it depends on the SEM stage precision (it could be also some micrometers). Stitching can be completely overcome with the use of an interferometric stage, in which the movement of the stage has precision of a few nm.

3.3.3 EBL in a wafer with alignment markers (WA<WF)

Here we report the process for an e-beam lithography with the use of pre-patterned alignment markers. It is predominant when working with (sub)micrometric-sized 2D-materials such as exfoliated flakes [20, 21], when patterning source-drain contacts in nanowires [22] or carbon nanotubes [23] nanotubes or when several self-aligned steps of EBL are required.

Here we describe one example in which a flake of graphene is first contacted by EBL and through a second EBL process we define the shape of the flake into a Hall bar or a constriction. This is a typical example in which the use of EBL represents the best available option provided that the features in the CAD files are not so small that a very high EBL resolution is required. In fact, it is well known that the fabrication process greatly affects the quality of graphene and PMMA (and its developer) represents a very good choice in terms of resist because different techniques can be easily employed to remove eventual contaminants [24–26].

First of all, it is necessary to design the final area to be exposed in the CAD file and for this reason it is necessary to take accurate distances and orientations of the graphene flake to the closest markers. This measurement can be very accurate through SEM imaging (and preparing the CAD file directly on SEM image) or with a rather lower resolution by using an optical microscope with sufficiently high zoom precision (precision lower than 500 nm in the best available microscopes). The use of optical microscopes is widespread for two main reasons. First, they avoid damaging the sample or causing its propertie to deteriorate because of the unprotected interaction with the electron beam. Secondly, because typical flakes are larger than 5 μm width/length and the optical resolution is sufficient. During the preparation of the CAD file, in this case we need to define the features with different layers for each exposure (contact, mask and pads) with respect to the position of the markers, in which their position is defined in the centers of a square drawn with another layer. In this process the square side is 20 μm[9].

Once the distances are taken and the CAD file is ready, the exposure process can take place. Three different apertures and three different WFs were chosen in this case. We used the aperture of 60 μm, the WF of 1000 μm (nominal resolution 200 nm) and a dose of 350 μC cm^{-2} to define the large metallic pads in this first EBL step. For the exposure of the contacts from the pads to the sample, the following parameters were used: aperture of 10 μm, WF of 500 μm (nominal resolution 7.8 nm) and a dose of 300 μC cm^{-2}. An aperture of 7.5 μm and WF of 100 μm (nominal resolution 2 nm) were used for the exposure of the mask (and a dose of 270 μC cm^{-2}) and for the bar or constriction (dose of 180 μC cm^{-2}) that will define the final shape of the flake after the subsequent reactive ion etching step. We have used home-made resist PMMA4% in chlorobenzene for the first and second steps and homemade resist PMMA2% for the final one. Both resists were baked at 160°C for 10 min.

As the different EBL processes are quite similar, we will describe the one for the exposure of the contacts. In this case, the knowledge of the orientation of the (U,V) coordinates of the sample is essential. Therefore, it will be necessary to align the system with the available markers (figure 3.9(a)). We choose two markers at the border of the sample (far enough to the flake to avoid undesired exposure) and rotate the SEM stage until a perfect alignment is achieved between the two markers (figure 3.9(b)). Similar to the EBL without markers, the global parameters (origin and angle) are

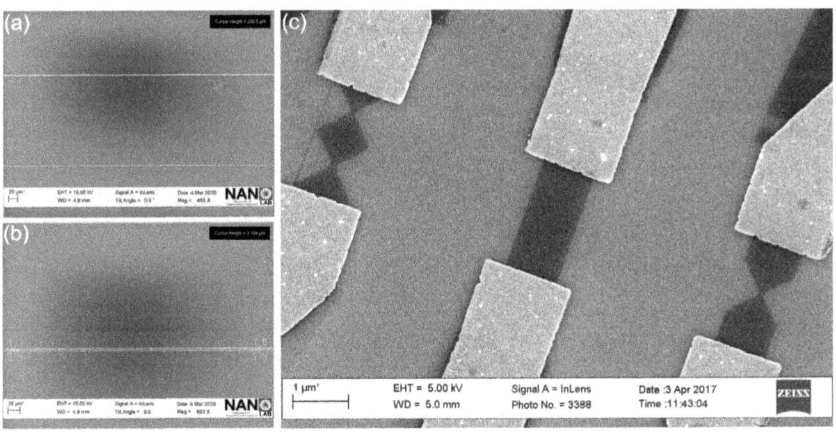

Figure 3.9. SEM micrographs displaying the orientation of the markers before (a) and after (b) the rotation of the SEM stage for alignment. Final graphene device fabricated with two EBL processes through the use of markers (c).

[9] Fours markers are typically used for the alignment. All the squares centered in the markers have to be contained in the chosen WF. Therefore, the distance between them must be lower than the WF size.

fixed using two different markers in this case, as well the SEM parameters and the three-points alignment for the exact correspondence between (X,Y) coordinates of the SEM stage, the (U,V) coordinates of sample and the (u,v) coordinates of the CAD file. Instead of points or scratches we will now use three of the makers for the local alignment. In fact, the distance between markers is well known as well as their (U,V) local coordinates with respect to the position of the graphene flake. After measuring the beam current in the Faraday Cup, we open a position and drag and drop the CAD file, this time selecting the layer for exposure (i.e. layer for contacts) as well as the layer of the four markers around the flake of graphene. Before exposing the sample, the program opens several windows with the chosen dimensions (in our case 20 × 20 μm, as previously stated) where the markers must be found to inspect and correct *in situ* the misalignment. By centering each marker in the center of its window we correct the error intrinsic to the displacement of the stage[10] and by repeating this procedure several times before the real exposure, a nanometric alignment can be achieved.

Once the different lithographic steps and subsequent evaporations, and etching processes are completed we obtain geometries shown in figure 3.9(c). Nanometric alignment was achieved between contacts and the graphene bar/constrictions [27]. The total exposure time was roughly 1 h, quite reasonable for a standard EBL process and suited for its implementation when working with graphene and other 2D materials (see section 3.5).

3.3.4 Two different resists: polymethylmethacrylate (PMMA) versus higher resolution hydrogen silsesquioxane (HSQ)

Typically, in a standard low-energy EBL set-up, and for the vast majority of resists, it is difficult to define shapes with a precision below 25 nm. PMMA resist, in spite of its versatility, suffers from this limitation. As described in section 3.2.5, there are few resists capable of resolution below 25 nm. One of them is hydrogen silsesquioxane (HSQ) resist. HSQ permits the definition of few-nm structures, even at low-energy EBL [28], as well as lines and space with a resolution better than 10 nm [29].

In this section, we compare two different EBL processes with two different resists. A commercial positive low-resolution PMMA (ARP679.02 from Allresist©) resist and a commercial negative high-resolution HSQ resist (XR-1541-004 Dow Corning) are compared. For both cases, we want to define a 'bone' structure with features lower than 100 nm using the same beam voltage of 15 kV. For the first process, with ARP679.02 resist, the same recipe described in section 3.3.2 is used. For the HSQ process, we have developed a higher contrast recipe. In detail, after a cleaning step in hot acetone and isopropanol and a pre-bake at 150°C for 5 min, we use a plasma cleaner process for 3 min before the coating. For spin coating we use 4000 rpm for 1 min with a baking at 80°C for 4 min. In the exposure, the dose for HSQ is 2000 μC cm^{-2}. The developing process is carried out rinsing tetramethylammonium hydroxide

[10] Before exposure, the SEM stage moves (with a micrometric error in the movement) to the chosen (U,V) coordinates for the patterning.

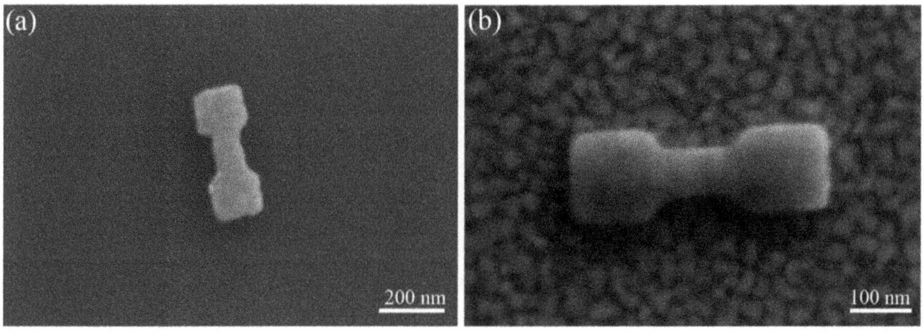

Figure 3.10. SEM images of EBL processes to obtain a bone structure using either PMMA (AR679.02) resist (a) or HSQ resist (b).

at 25% in DI-water (TMAH25%) at 80°C and stopped in flowing DI-water for 1 min. The final thickness of HSQ with this treatment is 50 nm.

In figures 3.10(a) and (b) the final result of these two processes are displayed. The very high contrast and very high resolution of the HSQ structure in comparison with the PMMA one is very clear from the two SEM images[11]. In spite of its very high resolution, HSQ has some limitations and problem in comparison with PMMA resist. In fact, it needs a previous treatment of the substrate before coating, it is very sensitive to the time between coating and exposure and it needs very high doses during the electron exposure (an order of magnitude larger compared to PMMA). Another limitation is related with its developer (for high contrast recipe). In fact, the high concentration of TMAH and the temperature during development are too aggressive for many types of devices (i.e. 2D materials devices).

3.4 ICP-RIE and E-beam evaporation

As we mentioned in section 3.1, in order to produce a final working device we typically require several EBL steps combined with (Joule, sputtering or e-beam) metal deposition and wet or dry-etching to define the device geometry.

3.4.1 ICP-RIE

In this section we will restrict ourselves to describing one of the most common dry-etching techniques for 2D-based systems: the ICP-RIE (inductively coupled plasma-reactive ion etching). There are different dry-etching techniques of this kind: RIE, ICP-etch, planar etch and deep reactive ion etch. Even though the above techniques need a plasma source, the main difference amongst them lies on the configuration of the electrodes for the plasma generation.

A plasma is defined as a medium of positive and negative particles. It is formed when a gas becomes ionized due to radio frequency (RF) energy, microwave

[11] The two SEM image are taken at different magnification, nevertheless the imperfection of the SEM image with lower magnification corresponds to the PMMA process (figure 3.11(a)) are evident in comparison with the higher magnification SEM image corresponds to the HSQ process (figure 3.11(b)).

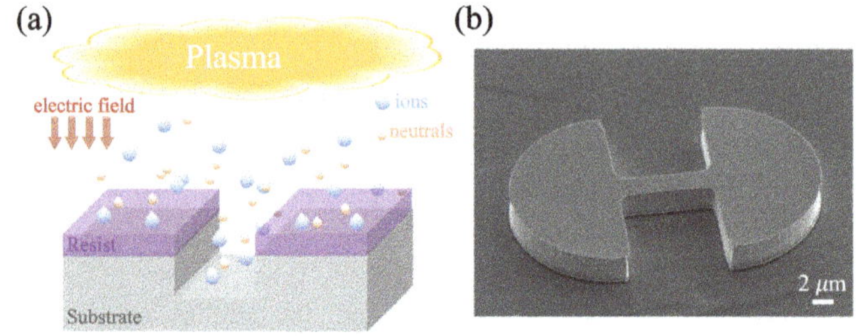

Figure 3.11. (a) Typical plasma etching process. (b) SEM image of GaAs structure etched through an ICP-RIE process with chlorine gas.

radiation or electromagnetic (EM) fields with an energy that overcomes the binding one for the electrons present in the atoms or molecules in the gas. This ionization energy varies from one gas to another. As an example, in a commercial system PlasmaPro 100 Cobra from Oxford Instruments, the energy ranges from 11.6 eV for Cl_2 to 15.7 eV for Ar. The use of the plasma permits performing high-temperature chemistry at low temperatures where slight variations in the different parameters (gas mixture, temperature, pressure or plasma power) will imply anisotropic or isotropic etched profile. Optical or EBL resist as well as previously deposited metal laid on the surface of the sample will play the role of the mask. They must be previously defined by any kind of lithographic process, where the uncovered areas of the sample will be effectively etched while the remaining ones, protected by the mask, will be left untouched. Remnant mask will be typically removed after the etching process (see figure 3.11(a) for a sketch of the whole process).

In an RIE process, a plasma is created in the system at low pressure by applying a strong RF field. Once the gas has begun to ionize, the free electrons strike neutral gas molecules generating more electrons and ions. The interaction of these electrons with the gas molecules could be multiple, such as atom ionization and dissociation of gas molecules leading to formation of radicals and electron impact excitation. The ions created in this process are extracted from the plasma by applying an electric field and impact on the sample, removing (etching) material from the sample of interest.

Plasma can interact with the surface of the sample during the etching in different ways, producing ion bombardment (sputtering), radical chemical reaction and ion assisted chemical reaction. The etching processes can be isotropic or anisotropic. The processes with a pure chemical etching (similar to the wet etching) do not need ion bombardment for reaction and in this case there is no directionality on the etching. For anisotropic etching (normally the desired one), the process needs ion bombardment, so the reaction can take place and the obtained etching has a vertical profile. In an RIE, as well as an ICP-RIE process, chemical and physical reactions take place.

Compared to RIE, in an ICP-RIE there is a physical separation between the plasma generation (created by means of inductive coils) and the acceleration of the resulting ions. This separation permits ICP-RIE a higher control during the overall etching process with respect to the standard RIE, including a higher confinement

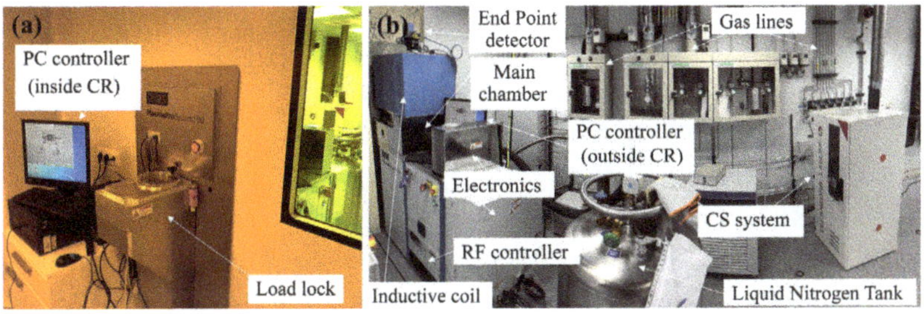

Figure 3.12. (a) View of the loadlock and ICP-RIE PC controller situated in a clean room environment. (b) Main body of the ICP-RIE placed in an adjacent technical room.

and higher ion flux uniformity with a more intense plasma and thus a deeper etching. A typical vertical etching on a GaAs heterostructure through ICP-RIE is shown in figure 3.11(b).

The Oxford Instruments PlasmaPro 100 Cobra system (see figure 3.12) is able to perform high-quality etching through a plethora of different materials (dielectric, metals, semiconductors, organics, etc) in wafers up to 100 mm (4') diameter. A typical set of gases for the ICP-RIE processes consists of O_2, Ar, H_2, CH_4, SF_6, C_4F_8, CHF_3, Cl_2, BCl_3 together with N_2 and He for venting purposes. This system is equipped with a cryogenic module allowing to work down to −120°C to perform cryo-etching. This option plays a crucial role in the fabrication of graphene nanoconstrictions and other nanodevices where low edge roughness is crucial. Therefore, we will devote a whole sub-section in this chapter to this process applied to 2D material heterostructures. As shown in figure 3.12(a), only the load-lock (for mounting/unmounting the samples) and one of the two controller PCs are within the premises of the clean room. The main body of the equipment (plasma chamber, gas lines, pumps, nitrogen tank, clean solution scrubber system for the disposal of the toxic waste) are placed in an adjacent technical room (figure 3.12(b)).

3.4.2 E-beam metal evaporation

A standard e-beam evaporator for the deposition of metals (Au, Cr, Ti, Al, Ni, Co, Pt) and dielectrics (SiO_2, Al_2O_3, HfO_2) is shown in figure 3.13(a). It is a prototype designed with the help of Tecnovac (Spain) equipped with a load-lock vacuum chamber isolated from the main chamber with a manual guillotine valve. The samples are loaded/unloaded using a mechanical valve. In the main chamber, a vacuum in the low 10^{-9} mbar can be achieved while in the load-lock it is possible to achieve 10^{-5} mbar in approximately 40 min.

In an e-beam evaporation process, a high current passing through a tungsten filament (see figure 3.13(b)) permits obtaining an electron beam which is accelerated in a gun[12] and focused onto the evaporation material using either permanent magnets

[12] The one presented in this chapter is equipped with a commercial Telemark 7 keV e-beam gun.

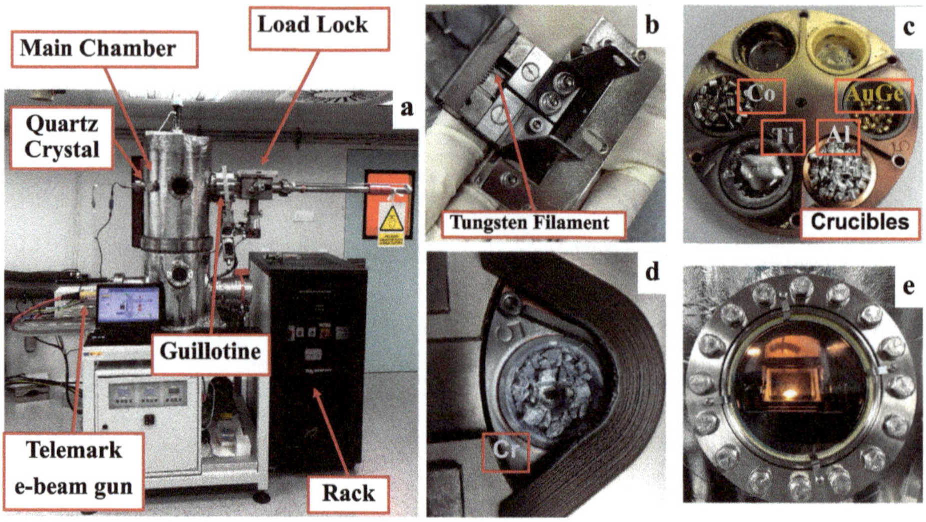

Figure 3.13. (a) E-beam evaporator designed by Tecnovac, with Telemark e-beam gun. (b) E-beam source model 244. (c) 6-pocket crucible carousel. (d) Detailed reinforced graphite crucible containing Cr. (e) View of a e-beam evaporation from an external window thanks to an internal periscope.

or electromagnets. These systems have typically a carrousel with different pockets where the crucibles contain the materials initially shaped in pellets, ingots or rods (figures 3.13(c) and (d)). By a controlled 'XY Sweeping', the e-beam circulates along the whole surface of the materials which are homogeneously evaporated when brought to their melting or sublimation points.

The combination of long distance between the material and the sample (over 1 m) and the operation at very low pressures ensures a high control on the evaporation rate (less than 1 $Å\ s^{-1}$) which is monitored by a quartz crystal controlled by a TTL controller and creating high homogeneity thin films on the samples but with the major drawback of a high consumption of materials. A typical evaporation process, in which the metal is heated up as the emission current is increased until its conversion to a gaseous state, is shown in figure 3.13(e).

3.5 EBL applied to 2D materials

2D materials [30–32], a family that encompasses an enormous selection of natural and artificial compounds with almost all the elements of the periodic table represented, are single-layer-thick systems arranged in a crystalline manner. They were widely believed to be thermodynamically unstable and, therefore, unsuitable to exist naturally in the 2D form but just embedded in the bulk in layered materials. The isolation of a single layer of graphene back in 2004, obtained by mechanical cleavage using the scotch tape method [33], represented the source of a sudden cascade of discoveries that rapidly enlarged the 2D-family. Nitrides, oxides, chalcogenides, halides, carbides (among others) have been exfoliated directly from layered 3D materials or artificially produced in the lab since then. Most of these materials are binary compounds but single layered (such as graphene or phosphorene) or ternary ($CuSbS_2$) compounds. The wide variety of materials and compounds has implied an enormous assortment of electronic properties present that includes semimetals, metals, semiconductor and insulators with enormous potentiality to be used in nanodevices and optoelectronic devices.

The majority of the devices described in this section have in common the use of exfoliated graphene materials from the bulk by means of mechanical cleavage (for example graphene, h-BN, MoS_2 or $MoSe_2$). Exfoliation of a bulk material aiming to obtain a single layer of them represents the first step for the bottom-up approach in the generation of 2D-material-based devices (while the growth of artificial 2D-layers such as hexagonal boron nitride would be the top-down approach). The mechanical exfoliation [33, 34] consists in a peeling force applied to a piece of graphite (or any other layered bulk material) by an adhesive tape that has been placed on its surface. By exerting such (normal) force it is possible to overcome the Van der Waals attraction dividing drastically the pristine bulk material into thinner aggregates. Subsequent repetitions of such mechanical exfoliation would eventually lead to the reduction of the material into single-layer flakes. Once obtained, the peeled one-layer-thick flakes are ready to be transferred onto (typically) sacrificial substrates. SiO_2 on Si has been widely used to this end where thickness of the insulating barrier (usually 90 or 290 nm) is crucial for the identification of the flakes via optical contrast [35, 36] or other non-invasive techniques such as micro-Raman analysis [37–39], ellipsometry [40], etc. Once the flakes are positioned and identified on the desired substrate they can be processed further.

As graphene and the majority of the 2D materials are typically of a semi-metallic or semiconducting behavior, they have been extensively processed following the standard roadmap for III-V semiconductor compounds. We will show a thorough and detailed study of some of the most common nanofabrication processes involving 2D materials in which one or several steps of EBL must be performed. We will concentrate our efforts in the most typical structures and geometries found in these low-dimensional systems that might serve the reader as the first steps into the generation of eventually more complicated devices.

3.5.1 Hall bars in graphene heterostructures

In this section we will show how to fabricate a Hall bar device in a hBN/graphene/hBN/graphite heterostructure [41]. The encapsulation of graphene in hBN guarantees a complete isolation of the graphene flake from the external environment [42], so the fabrication of high-quality graphene devices can be achieved. Furthermore, the group of Klaus Ensslin [43] reported seminal results showing that the use of a graphite backgate can improve the performances of encapsulated graphene devices.

Before a standard exfoliation of the single flakes on SiO_2, the substrate was cleaned with acetone (ultrasonic bath for 3 min) then isopropanol (or ethanol) for 30 s and finally annealed in our rapid thermal process (RTP) furnace (AS-One 100-Annealsys) for 15 min at 400°C in vacuum. For exfoliation of graphene and graphite, we previously put a thick natural graphite piece on a Scotch Magic tape where it was folded onto itself. We transferred the thinner graphite pieces onto another adhesive tape with lower force adhesion, Ultron tape (1009R-60). In the last step we transferred different graphene and graphite layers on the Si/SiO_2 substrate through the peeling process. The use of Ultron tape (instead of the direct use of Magic tape) guarantees a minor contamination of the graphene flake and of the substrate. Typically, single layer (SL) or bilayer (BL) graphene flakes are chosen. A graphite flake with a thickness between 10 and 20 nm is also selected.

On the other hand, the hBN flakes have to be subsequently obtained for the fabrication of graphene-hBN based heterostructures. Particular attention in the choice of the bottom hBN flake is requested [44] because the thickness should be around 35–40 nm and well measured to avoid a problem during the etching process (avoiding short-circuited graphite with ohmic contacts). Ultra-pure bulk crystals of hBN used for exfoliation were bought from HQ Graphene. The exfoliation process was similar to graphene, but for hBN we use only Ultron tape (1009R-60) and the peeling occurs on a hot plate at 100°C after heating the substrate for 1 min. In this way, relatively big flakes with a minimum lateral size of 30 μm and a thickness between 10 and 40 nm were obtained with less tape contaminant. The residual contaminant of the tape was definitively removed after a standard cleaning process (acetone and isopropanol) and an annealing at 350°C with Ar atmosphere in the RTP furnace.

After exfoliation, a preliminary characterization was obtained by optical contrast of the graphene, graphite and hBN flakes. The number of layers for graphene were confirmed by MicroRaman measurements [45, 46], and profilometer measurements gave us the accurate information regarding the thickness and quality (roughness) of hBN and graphite flakes. After choosing the flakes, the fabrication of the graphene/hBN heterostructures takes places. Two different dry transfer techniques based on polydimethylsiloxane (PDMS) stamps and polycarbonate (PC) film [47] or polypropylene carbonate [PPC] [48] are used. In the following we describe the PPC technique in detail [41].

Fabrication steps for an hBN/graphene/hBN/graphite heterostructure

1. **Preparation of the PDMS stamp.** A mixture 1:10 of the bicomponent SYLGARD® 184 was placed (after removal of air bubbles with various steps of vacuum-vent in a bell jar) in a Petri dish and baked at 75°C for 4–5 h.
2. **Preparation of the PC film on microscope glass slide. Preparation of the PDMS stamp with the PPC film.** A square piece of 0.8 cm^2 of PDMS is cut and placed in a glass slide. The surface of the PDMS stamp is functionalized in an atmosphere of O_2 through ICP-RIE (see section 3.4.1). A solution of polypropylene previously dissolved in anisole (concentration of 9%) is spun on the PDMS stamp and baked for 15 min at 100°C in vacuum. After its preparation, PPC film on PDMS can be also used as stamp for longer time (typically 1 week).
3. **Placing the PPC film and PDMS stamp on a glass slide.** A small piece (9 mm^2) of PDMS and PPC stamp is cut and stuck with transparent glue on a glass slide. Everything is ready for the transfer process.

4. **Pickup of the hBN (bottom flake).** The stamp is placed over the Si/SiO_2 substrate with the bottom hBN and by using the micrometer stages and the optical microscope, the top hBN flake (figure 3.14(a) 1A) is located and positioned under the stamp. Then, the SiO_2 substrate approaches the PDMS stamp through the z-movement stage until a contact area appears (figure 3.14(a) step 2A). If the contact area is too far from the flake, the tilt stage can be used to increase it. The temperature of the substrate is increased up to 60°C and the PPC will extend and cover the whole flake. Turning off the heater the SiO_2 substrate is removed with the z-movement allowing to pick up the bottom hBN flake by the stamp. In this step the flake has been transferred on the PPC film(figure 3.14(a) 3A).
5. **hBN/graphite.** Then the graphite flake (figure 3.14(b)) on Si/SiO_2 is aligned with (X,Y) stages and approaches the hBN bottom flake until a contact area (between SiO_2 substrate and stamp) appears. Tilt in case the bubble is too far from the flake (4A, 5A in figure 3.14(a)). The bottom hBN flake is deposited on the graphite flake. The temperature of the substrate with the

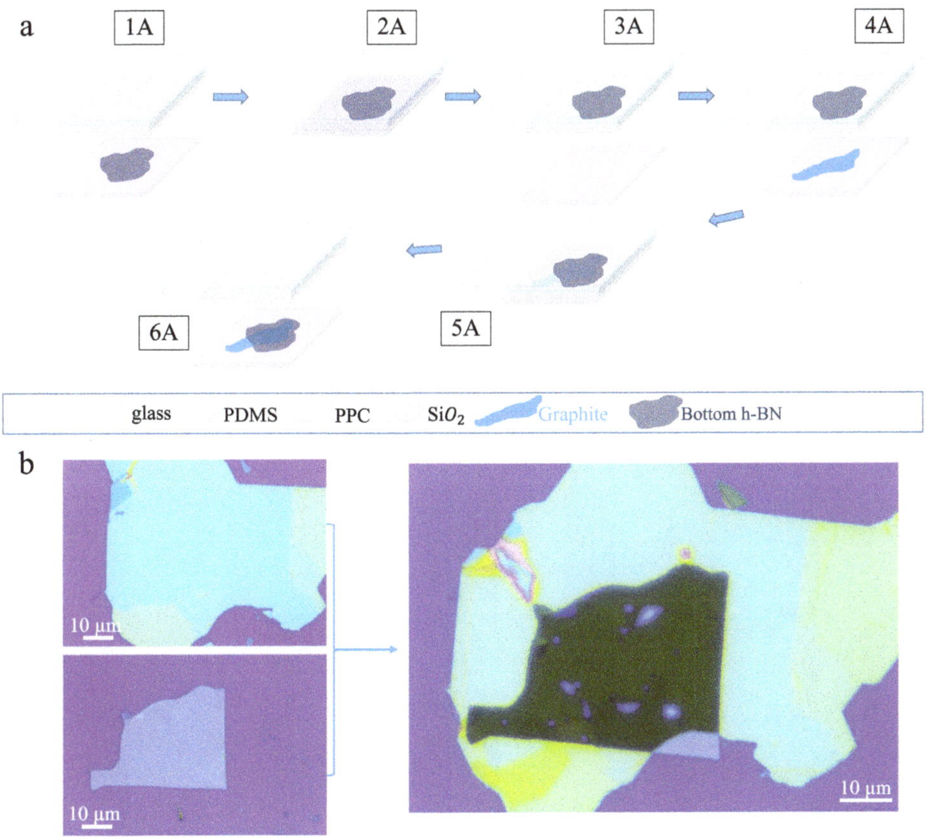

Figure 3.14. (a) Transfer steps process for hBN/Graphite heterostructure. (b) Optical images of bottom hBN (top left), graphite (bottom left) and their correspondent heterostructure.

graphite is increased up to 150°C which permits the PPC film to melt and enhancement of the adhesion between graphite and hBN. Therefore, the hBN bottom flake is deposited on the graphite flake reaching the above temperature (figure 3.14(a) 6A).

6. The result is a heterostructure of graphite/hBN on a Si/SiO_2 substrate (see figure 3.14(b)). The heterostructure was cleaned in anisole, acetone and isopropanol and annealed at 350°C in Ar in our RPT for 15 min to remove the contamination before the next steps of the transfer process.
7. **hBN/graphene**. Using a new stamp, we pick up the top hBN flake from the SiO_2 substrate approaching the PDMS stamp, extending the contact area with the temperature (until 60°C) and removing the substrate from the stamp through z-movement (figure 3.15(a) 1B, 2B). Once the top hBN flake stays on the PPC film (3B in figure 3.15(a)), we align and approach with (x,y,z) stage the graphene flake on SiO_2 with the top hBN flake (figure 3.15(a) 4B). After the appearance of the contact area (5A in figure 3.15(a)), it is increased with the temperature at 150°C, then through the melting of the PPC film and removing the substrate with z-movement, the top hBN flake is released on

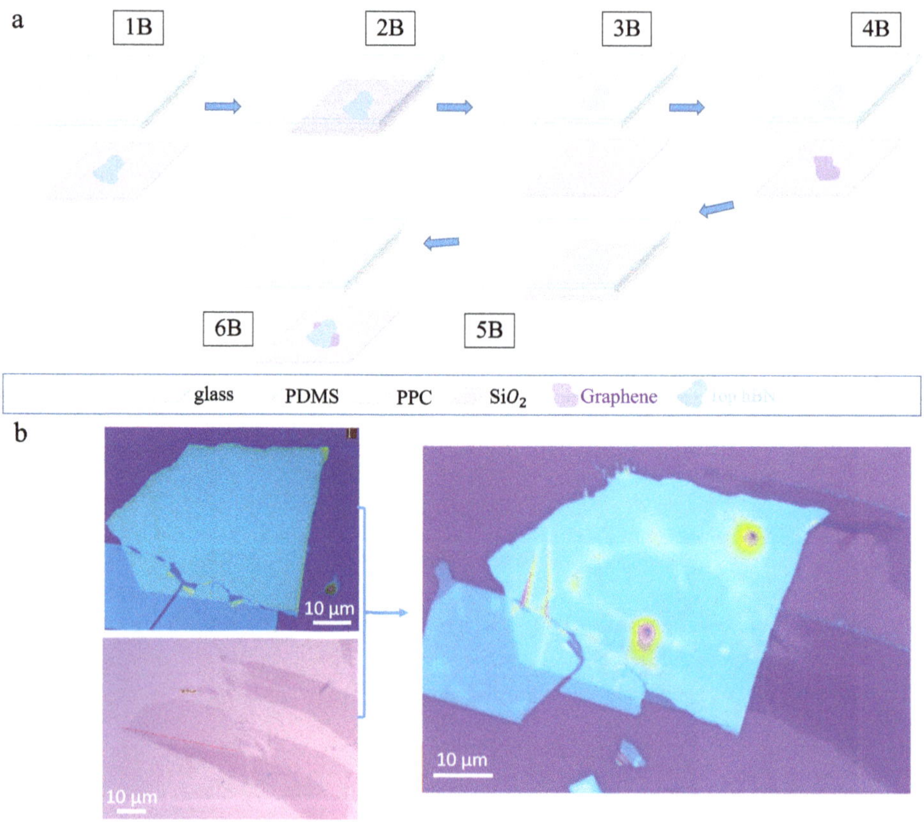

Figure 3.15. (a) Transfer steps process for hBN/graphene heterostructure. (b) Optical images of top hBN (top left) graphene (bottom left) and their correspondent heterostructure.

the graphene flake (6B in figure 3.15(a)). Now the heterostructure of graphene/hBN (top) was obtained and cleaned in anisole, acetone and isopropanol (figure 3.15(b)). So the heterostructure of graphene/hBN (top) is ready to be picked up and placed on the previously prepared heterostructure of graphite/hBN (bottom).

8. **Graphite/hBN/graphene/hBN.** First the pickup of the heterostructure graphene/hBN(top) takes place with a new stamp. Introducing the SiO_2 substrate to the PDMS stamp, extending the contact area with the temperature (until 60°C) and removing the substrate from the stamp through z-movement. After this step the heterostructure of graphene/hBN(top) is transferred on the PPC film (steps 1C, 2C, 3C in figure 3.16(a)). Then the alignment and approach of the heterostructure of graphite/hBN (bottom) on SiO_2 substrate with the heterostructure of graphene/hBN takes place until we obtain a contact area (4C, 5C in figure 3.16(a)). Increasing the temperature (150°C), the heterostructure of graphene/hBN (top) is transferred on the heterostructure of graphite/hBN(bottom) (6C in figure 3.16(a)).
9. **Cleavage and cleaning of the substrate**. This is the best moment for the cleavage of the substrate because the area with the heterostructure is easily recognizable and the heterostructure is still protected by PPC film. The whole heterostructure is so cleaned in anisole, acetone and isopropanol and annealed in our RTP furnace at 350°C in Ar atmosphere removes contamination and PPC residual from the heterostructure and from the substrate.

Once the hBN/graphene/hBN/graphite heterostructure is obtained, the fabrication by EBL of the final device is ready to be tackled. During this process a lot of blisters or air bubbles were trapped within the heterostructure. It is important to choose areas free of bubbles since it would reduce the quality of the encapsulated graphene. The best area was chosen through optical microscope and Raman spectroscopy (figure 3.16).

EBL Fabrication steps of hBN/graphene/hBN/graphite device in Hall bar geometry.

1. **First EBL for the definition of the markers.** For high-quality devices, it is better to use wafers without pre-patterned markers to avoid contamination due to a preliminary process (in particular evaporation). So, firstly the definition of marker arrays is necessary. For this aim, we prepare a CAD file in which there are markers with a separation of 80 μm. The dose of each marker is 270 $\mu C\ cm^{-2}$. After the coating and baking of the homemade PMMA4% resist, the first EBL process for the markers definition takes place. In this case we used a WF of 1000 μm and the aperture of 30 μm with a nominal resolution of 6 nm and the carried process is very similar to the process described in section 3.3.2. This time the WA is the whole sample, so typically 4 × 4 mm, but the size of the sample depends on the dimensions of the sample holder. After exposure, a shorter developing time (30 s) in 1:3 MIBK: IPA is chosen to avoid damage of the resist.
2. **Measuring distances and preparation of the CAD file.** Once the markers are defined on the PMMA, we need to measure accurately the distances between

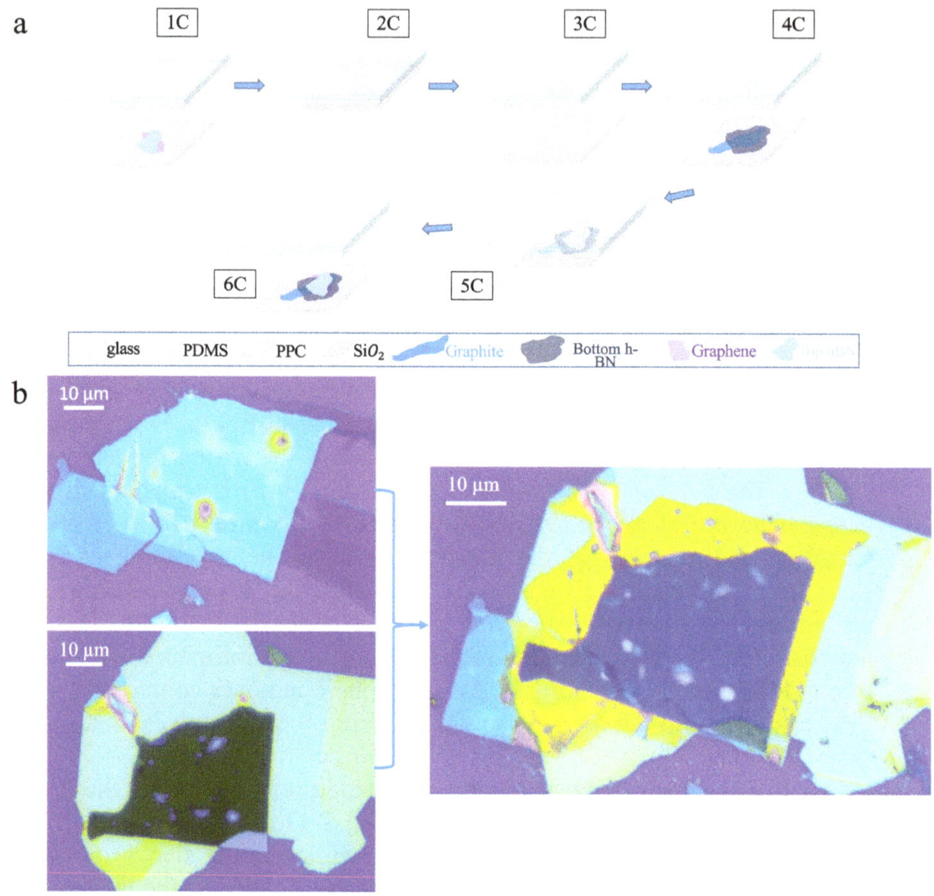

Figure 3.16. (a) Final steps of the dry transfer method for a h-BN/graphene/hBN/graphite heterostructure. (b) Optical images of graphene/top hBN structure (top left) and the graphite/bottom hBN (bottom left) one and the combination of both resulting in the whole heterostructure (right).

the flake and the closer markers, obtained by optical microscopy and taking into account the lower optical contrast of graphene flake due to the presence of the hBN underneath it. In the CAD file, we define the mask Hall Bar (first EBL process) and then add layers for the metallic pads and graphite contacts (second EBL process). The preparation of the CAD file in heterostructures with graphite bottom gate is particularly complex, since it is very important to avoid eventual cracks in the heterostructure.

3. **Second EBL process for definition of an etching mask.**

   The second EBL process is very similar to the section 3.3.3, the main difference in this case is the lower dose for the Hall bar mask (250 μC cm^{-2}) due to the presence of hBN flakes. The presence of the dielectric nature of the hBN could provoke charging effects and for this reason a smaller dose has to be used. After a developing time of 2 min, the Hall bar mask is defined (first step in figure 3.17(a)).

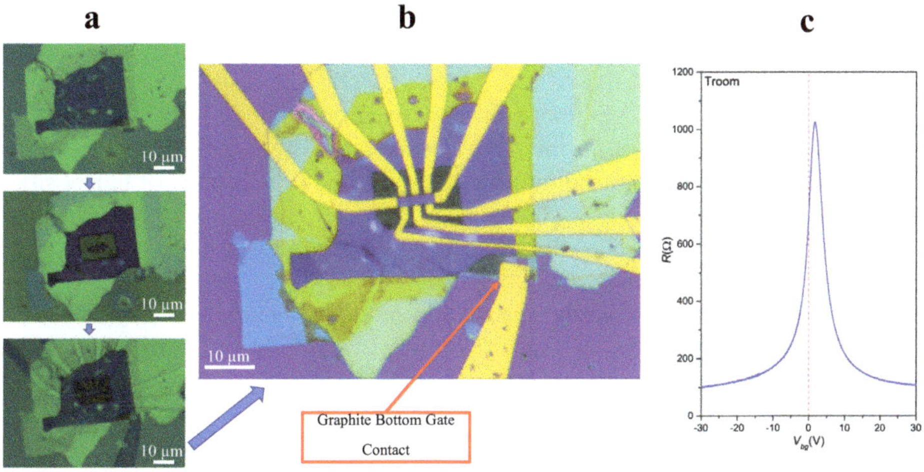

Figure 3.17. (a) From the top to the bottom: definition of the Hall Bar through EBL, etching of the Hall Bar through ICP-RIE process, definition of the contacts (and contact for the graphite bottom gate) with a second EBL. (b) Final devices after evaporation of Cr/Au and lift-off process. (c) Four probe resistance in function of the back-gate voltage at room temperature for an encapsulated graphene device.

4. **Shaping the heterostructure into a Hall bar and definition of the pyramidal profiles in encapsulated graphene.**
   Through ICP-RIE we etched the encapsulated graphene in the Hall bar geometry and created the appropriate etching profile for side-contacts in SF_6 atmosphere with an etching rate of 3 nm s^{-1} on hBN. In our ICP-RIE recipe, we used 40 sccm of SF_6, a power in RIE mode of 75 W at 6 mbar of pressure and 10 °C of temperature for typically 12 s. A controlled etching rate gives us the possibility to etch only the top hBN flake avoiding the full etching of the bottom hBN (see second step in figure 3.17(a)). In this case, the etching does not have a vertical profile, as typically desired in a standard etching process, but it is pyramidal [42]. To avoid possible oxidation of the atomic exposed (to the external environment) graphene layer, it is important to minimize the time between the etching and the following nanofabrication steps. It is important to note that, after this process, the markers pattern is transferred to the substrate. In fact, the silicon oxide is also partially etched (etching rate ~1.5 nm s^{-1}) with this recipe.
5. **Annealing process for residual resist and bubbles removal.**
   After the etching process, the sample is cleaned in acetone and isopropanol and annealed through RTP in Ar atmosphere at 380°C. In this way, contamination of the resist and residual bubbles from the heterostructure are removed. In fact, it's well known that an annealing (below 500°C) reduces the number of bubbles encapsulated in the graphene heterostructure [49].
6. **Third EBL process for definition of a new array of markers.**
   The pattern of markers transferred on the substrate in the first lithography are not properly visible by SEM through a new resist coating, but their

presence is still well visible by optical contrast (also through a new layer of coating). For this reason, we repeat the markers array with the same EBL process described in the point 1.

7. **Adaptation of the CAD file to the new markers.**
   Without the use of an interferometric stage, the new markers are slightly shifted with respect to the old markers (and the graphene Hall bar position). The measurement of this shift can be properly measured by optical microscope and the CAD file (in particular contact and pads) is so adapted with respect to the new markers' position.
8. **Fourth EBL process for the fabrication of edge contacts.**
   Then, after a new deposition of the PMMA resist, with a new nanolithography process, we define the ohmic contacts and the backgate contact (see third step in figure 3.17(a)). This process is again very similar to the process in section 3.3.3 apart from the reduction of the dose at 270 $\mu C\ cm^{-2}$ for the ohmic contacts.
9. **Evaporation of Cr/Au and lift-off process.**
   Finally, after an evaporation of Cr (10 nm) and Au (50 nm) and the subsequent lift-off process in acetone, the final sample is ready to be characterized (figure 3.17(b)). In this case, a controlled and low evaporation rate is strictly necessary. In fact, only with low evaporation rate and pyramidal profile previously obtained, is it possible to achieve a good contact between the atomic exposed layer of graphene and the evaporated metals, in particular the adhesion layer metal (Cr) [50].

Using this fabrication method, it is possible to obtain edge contacts with a contact resistance lower than 1 $k\Omega\ \mu m^{-1}$. A typical characterization of an encapsulated device at room temperature is shown in figure 3.17(c), where we report a four-probe resistance in function of the back-gate voltage. Encapsulated graphene devices have very sharp and narrow charge neutrality point (CNP) in comparison with standard graphene devices on SiO_2 substrate and the 'Dirac peak' is typically very close to zero (see red dashed line in figure 3.17(c)). Typical mobilities[13] in encapsulated graphene are higher than 100 000 $cm^2\ (Vs)^{-1}$ at room temperature [41]. For comparison, in graphene on SiO_2 substrate, the mobility is always lower than 20 000 $cm^2\ (Vs)^{-1}$ also at low temperatures and with chemical treatments to reduce the influence of the substrate [27].

3.5.2 Quantum point contacts defined by cryo-etching

It is important to highlight the cryo-etching process due to its novelty applied to graphene [51]. Cryo-etching has been widely used in the fabrication of quasi-2D

[13] The carrier mobilities are obtained through the fit of the resistance in function of the carrier density with the use of the field effect formula, at zero magnetic field. The carrier density is directly calculated by the gate voltage [41].

structures achieving an unprecedented control of the edge-definition. It was successfully implemented in silicon-based devices, where it was possible to control and define low-roughness sidewalls, observe the absence of sidewall scalloping [52], and generate samples with a high aspect ratio and cleaning [53] in nanometric-size structures [54]. The extension of the cryo-etching technique to graphene [51] allowed us to build graphene nanoconstrictions or quantum point contacts (QPCs) with very low sidewall roughness.

EBL Fabrication of a QPC in an h-BN/graphene/hBN heterostructure

1. **First EBL: bar definition and definition of pyramidal profile.** A bar (instead of a Hall Bar) was defined on the graphene heterostructure by EBL, using PMMA as a mask. The etching for the edge contacts is obtained by a dry etching processing with the same recipe of the previous section. A standard cleaning process and a rapid thermal annealing was performed at 350 °C in Ar atmosphere for 15 min to remove eventual small non-visible bubbles and the PMMA remnants.
2. **Second EBL: Edge contacts.** A second EBL is performed for the definition of the edge contacts after the deposition of Cr/Au through e-beam evaporator and lift off process.
3. **Third EBL: Cryo-etching.** We used PMMA but with lower concentration (2%) in chlorobenzene (resist thickness 70 nm), a smaller aperture (7.5 μm with a dose of 185 μC cm^2) and a WF of 100 μm (using closer markers with a separation of 70 μm previously defined in the second EBL). In this way a nominal resolution of 2 nm can be obtained. From a technological point of view, the choice of a constriction with V-shape, in contrast to an 'abrupt' constriction or a ribbon, as well as the use of a thinner resist, permit the reduction of the proximity effect, as explained in section 3.2.3. The cryo-etching process was performed at −110°C in a controlled Ar/SF_6 (10/40 sccm) atmosphere at $P = 6$ mbar for 18 s, $P = 75$ W. The use of cryogenic temperatures and Ar allows us to obtain QPCs with very low roughness. Ar gas, instead of the more common O_2 gas for a cryo-etching recipe, reduces the possibility of passivation to a minimum. The etching rate of the cryo-etching recipe is 2 nm s^{-1}.

Figure 3.18 shows a few SEM images of a typical encapsulated graphene QPC etched by means of the cryo-etching technique (magnification of 450k) (a) and by standard room temperature etching process (magnification of 300k) (b). The upper panels display the original SEM micrographs while the lower part colored SEM images help to visualize the borders of the constrictions. The higher roughness of the second image is evident at the edges, in comparison with the very well-defined cryo-etched structure. Two devices are shown in the SEM image of figure 3.18(c), where two different nanoconstrictions are defined. The upper one has a width of ~200 nm, whereas the lower one has a width of ~100 nm. The high quality of these QPCs allowed almost ideal ballistic transport. Therefore, the cryo-etching technique paves the way for the fabrication of more complex graphene-based nanostructured prototypical devices with truly ballistic transport of high-mobility carriers [41].

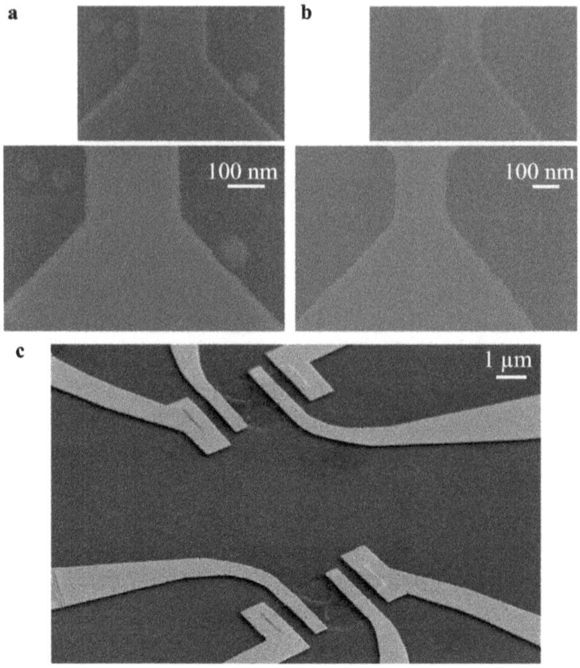

Figure 3.18. (a) SEM image of a nanocontact etched by cryo-etching ICP-RIE. (b) Micrograph of a similar constriction etched using standard (no cryo) ICP-RIE. A noticeable reduction of the roughness along the edges can be found when using low-temperature RIE. (c) Tilted SEM image of typical devices with nanocontacts in encapsulated graphene. Reproduced with permission of Springer.

3.5.3 Dual-gated graphene FETs

Here we will describe the fabrication process of terahertz detectors based on dual-gated graphene field-effect transistors (FETs), using encapsulated graphene [55, 56]. Thanks to its high carrier mobility, frequency-independent absorption, gapless or controllable gap according to the number of layers (monolayer or bilayer) graphene is a very interesting material for the development of detectors working in the terahertz region of the electromagnetic spectrum, still severely lacking in terms of solid-state devices.

Fabrication of a dual-gated FET in a h-BN/graphene/hBN heterostructure

1. **Fabrication of the heterostructure.** By using a similar procedure to that described in section 3.5.1, we obtain a hBN-graphene-hBN heterostructure.
2. **First EBL: bar definition.** For the bar definition, a nanolithography and etching process similar to the Hall bar definition described in section 3.5.1 was carried out. After the deposition of the resist, a bar is defined by means of the lithography process (step 2 in figure 3.19) and etched with ICP-RIE in SF6 atmosphere (step 3 in figure 3.19).
3. **Second EBL: edge contacts.** Edge contacts were fabricated after annealing in Ar atmosphere at 350°C, through a second nanolithography process (step 4

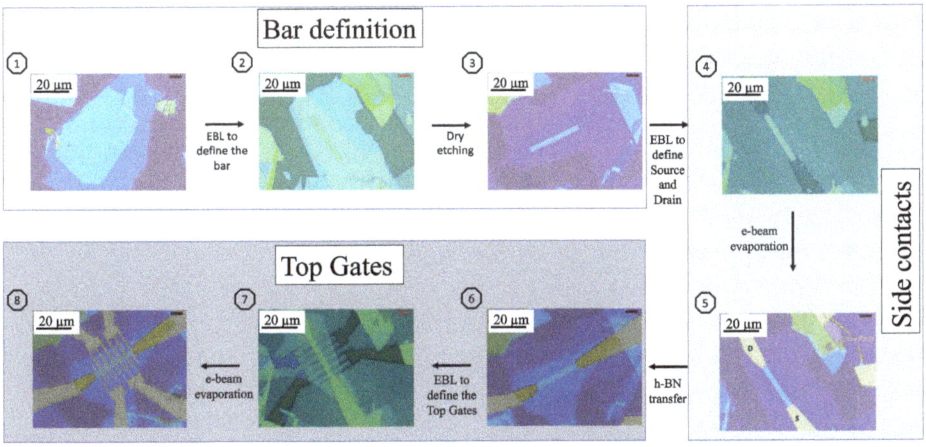

Figure 3.19. Fabrication process of FETs in encapsulated graphene can be divided into three different steps: bar definition, fabrication of the side contacts and top gates.

in figure 3.19) and subsequent evaporation of Cr/Au (5–40 nm) and lift off process (step 5 in figure 3.19).

4. **Transfer of a new hBN flake.** Before the definition of the top gates it is necessary to cover the device with a dielectric film to avoid preparing new edge contacts instead of gates! In our clean room we do not have equipment (i.e. atomic layer deposition, sputtering, etc) for dielectric deposition, which would be the first option. On the other hand, a few-layer hBN flake could be a very good choice to obtain the higher performances of the device. For this reason, we transferred a new flake of hBN on the device (step 6 in figure 3.19).
5. **Third EBL: Top gates.** The top gates were defined by another EBL process (step 7 in figure 3.18) and evaporation and lift-off of Cr/Au (5/40 nm) (step 8 in figure 3.18). The final device, an asymmetrical dual grating gate FET based on graphene (ADGG-GFET), is shown in the final (eighth) step of figure 3.19.

A multi-finger dual-gate topology improves considerably the performance of THz detectors based on plasma waves, as demonstrated theoretically and experimentally. The use of asymmetric fingers FETs based on graphene instead of a large antenna tuned for THz frequency has a technological advantage with respect to a conventional antenna in terms of space (see [55] and references therein for more details).

3.5.4 Hall bars in other 2D materials

In 2005, Novoselov *et al* [57] demonstrated that, beyond graphene, it is possible to obtain other 2D materials by mechanical exfoliation. They are named Van der Waals or 2D materials (hBN, MoS_2, $MoSe_2$, $MoTe_2$, $NbSe_2$, $Bi_2Sr_2CaCu_2O$) [58], highlighting their very small thickness in comparison with the other two dimensions. Starting from a bulk 3D material, it is possible to obtain a 2D material through mechanical exfoliation. This is possible because layered materials show strong in-plane bonds, but weak van-der-Waals coupling between layers. Transition metal dichalcogenide (TMD) materials attracted particular interest among the 2D materials [59]. They are thin semiconductors composed by a transition metal atom (M) and a chalcogen atom (X). In this section we describe the fabrication by EBL of hall bars in $MoSe_2$ and $MoTe_2$, as well as their preliminary characterization. They have a direct band gap between 1 and 2 eV and represent excellent candidates for application in electronics and optics.

$MoSe_2$ devices

We started with the mechanical exfoliation of the material from the bulk material (grown by HQ Graphene). The bulk material was firstly peeled with Magic tape and later transferred in the blue tape, where we peeled 5–6 times before placing the blue tape on a SiO_2 substrate (it was previously cleaned in acetone and isopropanol and treated with plasma cleaner for 5 min).

The flake is obtained and identified by optical microscope (inset in figure 3.20(a)) and a first characterization was made by profilometer (figure 3.20(a)), before the Raman measurements (figure 3.20(b)). For example, for the flake of the inset of figure 3.20(a), the measured thickness was 5 nm and considering the thickness of a monolayer flake (9.5 Å), we can estimate a five-layer $MoSe_2$ flake. This was confirmed by Raman spectroscopy, comparing another 5 nm flake in literature [60] with which we found similarities. There is a prominent A1G peak on the material also in bulk as well as other peaks (E_{1G}, $E^1{}_{2G}$, $B^1{}_{2G}$). We tried to transfer with the PPC method (see section 3.5.1 for details) this flake on pre-patterned contacts, but the flake was damaged, as presented in figure 3.20(c).

The device worked but we noticed a large electrical instability. For this reason, we performed a different fabrication method on another 4L-$MoSe_2$ flake. In this case we used EBL on ARP67904 (coating at 4000 rpm, bake 160°C for 10 min) and evaporation of Ti and Au (5 and 35 nm). The final device is shown in figure 3.20(d).

We measured the drain current as a function of the gate voltage from −30 to 40 V at fixed drain voltage (V_d = 2 V) (figure 3.20(e)) and the behavior of the current is typical of an n-type semiconductor, as expected [61]. We fabricated also a 1L-$MoSe_2$ device (shown in the inset of figure 3.20(f)) but it did not work probably due oxidation. We estimated the gap with photoluminescence (PL) measurement through Raman, as shown in figure 3.20(f). The measured gap was 1.57 eV [62].

$MoTe_2$ devices

The mechanical exfoliation of $MoTe_2$ is similar to that in the $MoSe_2$ material. The only difference consists in peeling 3–4 times with the Magic tape and placing a stripe of the blue tape and peeling only once with the blue tape. In fact, this material was much harder to be exfoliated than other 2D materials. We also tried to transfer flakes with thicknesses of approximately 20 nm (optical image in the inset of figure 3.21(a)) on pre-patterned contacts. The transfer process was successful as shown measurements were electrically unstable.

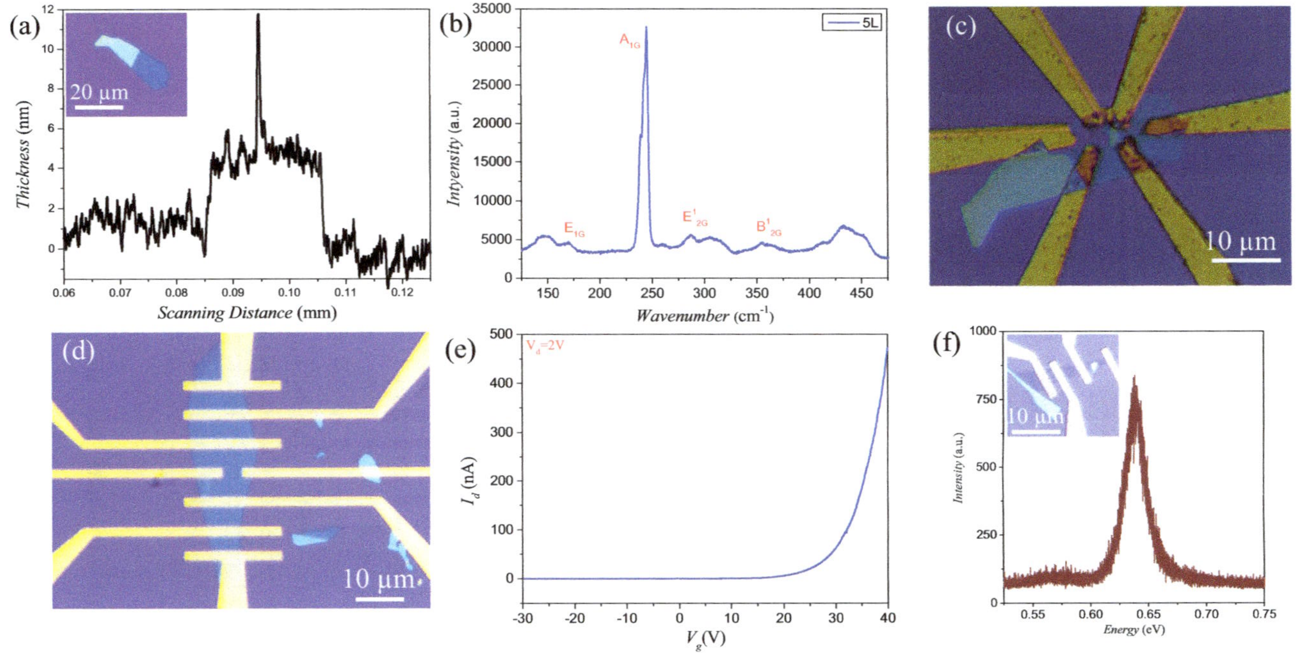

Figure 3.20. (a) Profilometer scan of the exfoliated $MoSe_2$ with a thickness of 5 nm, inset. The inset shows its optical image. (b) Raman spectrum of the $MoSe_2$ flake that was transferred onto pre-patterned Cr/Au contacts (c). Optical image of a Hall-bar-like device fabricated on a 4 nm thick $MoSe_2$ (d). (e) Drain current versus gate voltage sweep with a fixed drain voltage (2 V). (f) Photoluminescence spectrum of a monolayer $MoSe_2$ flake with its optical image in the inset.

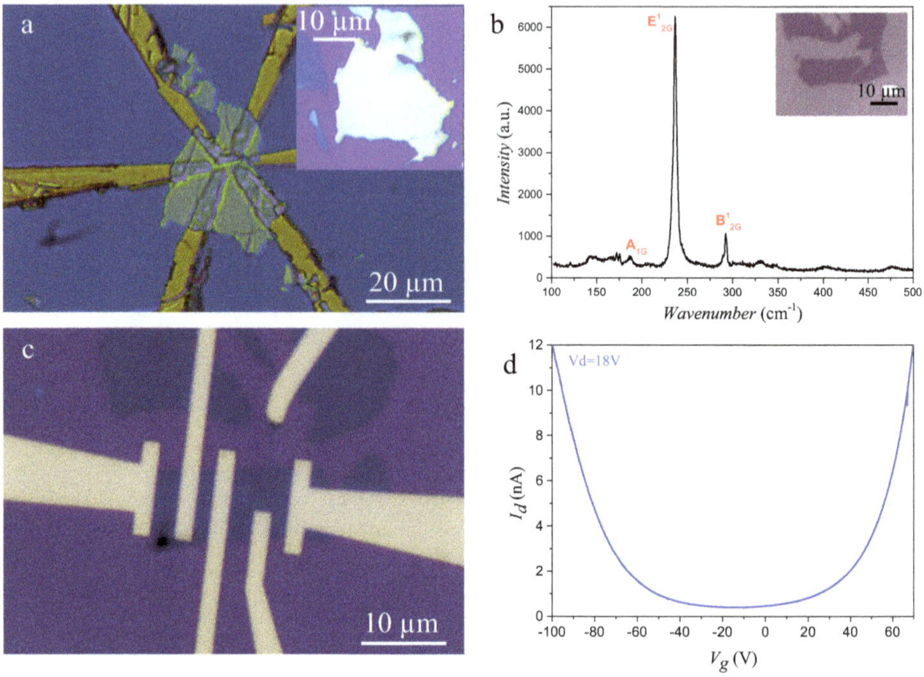

Figure 3.21. (a) Optical image a flake of MoTe2 (inset) transferred on pre-patterned contacts. (b) Raman spectra of a 5L-MoTe2. (c) Hall-bar-like device on the 5L-MoTe2 flake. (d) Drain current versus gate voltage sweep with a fixed drain voltage (18 V).

For this reason, we fabricated another device with the same process described in the previous subsections. The final device was designed on a quintuple layer flake (optical image in the inset of figure 3.21(b)) whose thickness was characterized by Raman spectroscopy (as shown in figure 3.21(b)) where a 0.17 ratio between the characteristic peaks $B_1{}^{2G}$ and $E_1{}^{2G}$ confirmed the number of layers in the sample [63]. After the EBL process, evaporation of Ti and Au (5 and 35 nm) and lift-off, the final device is shown in figure 3.21(c) (one contact was broken during the lift-off process). A preliminary measurement of the drain current (I_d) as a function of the gate voltage (V_g) at fixed drain voltage (V_d = 18 V) (figure 3.21(d)) shows the characteristic ambipolar behavior of this material [64].

We also tried to transfer flakes with thicknesses of approximately 20 nm (optical image in the inset of figure 3.21(a)) on pre-patterned contacts. The transfer process was successful as shown measurements were electrically unstable.

3.6 General applications and new developments in EBL technique

EBL plays a central role for the top-down nanofabrication approach. Its range of applications in different fields is really wide, as shown in figure 3.22.

EBL technique is largely used for the contact definition of (sub)micrometic-sized structures (nanowires, nanotubes, etc). A typical example is shown in figure 3.23(a), where metal contacts are defined on cobalt nanowires. The EBL process is carried

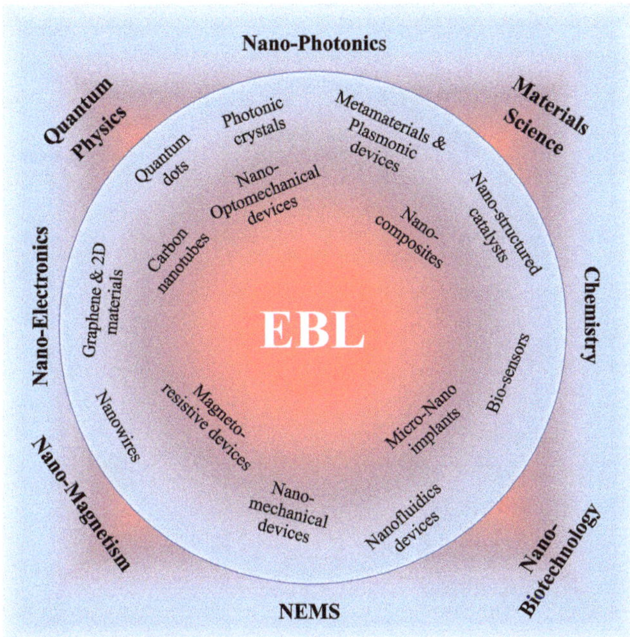

Figure 3.22. Range of applications of EBL for different fields.

out after the identification of the cobalt nanowires position randomly deposited on the substrate. In this case the position of the single nanowire can be taken through SEM and the design for the CAD directly prepared on the SEM image, achieving in this way an alignment of the contacts with nanometric precision.

EBL is also used in the mask fabrication for optical and x-ray lithography [65]. The fabrication of photomasks is similar to a wafer fabrication. For optical lithography mask a quartz/glass substrate is covered with a chrome film and electron beam resist. The resist is subsequently exposed through an EBL process and used as mask for the etching of the chrome film in the targeted areas. Features of the order of ~1 μm and line control of ~50 nm is required for an optical lithography mask. The process for x-ray lithography masks is very similar. The main differences are the minimum feature size (~100 nm) and line control (~20 nm) required and the materials used for the mask (Si, SiN, SiC, diamond with Au or W as metallic mask).

In the previous sections we introduced some of the limitations of the EBL technique. New developments of EBL aim to overcome the problems related to the technique itself. One of the major limitations is the difficulty in exposing a large area due to the limited scan range of the electron beam (the WF) in a reasonable exposition time. For example, the problems related to defective stitching between different WFs was already addressed in section 3.2.4. This problem occurs even when one wants to define features (for example a line) bigger than the WF size, and it can be partially solved with the use of an interferometric stage with high resolution, but a residual stitching error (~10 nm) is always present. The use of

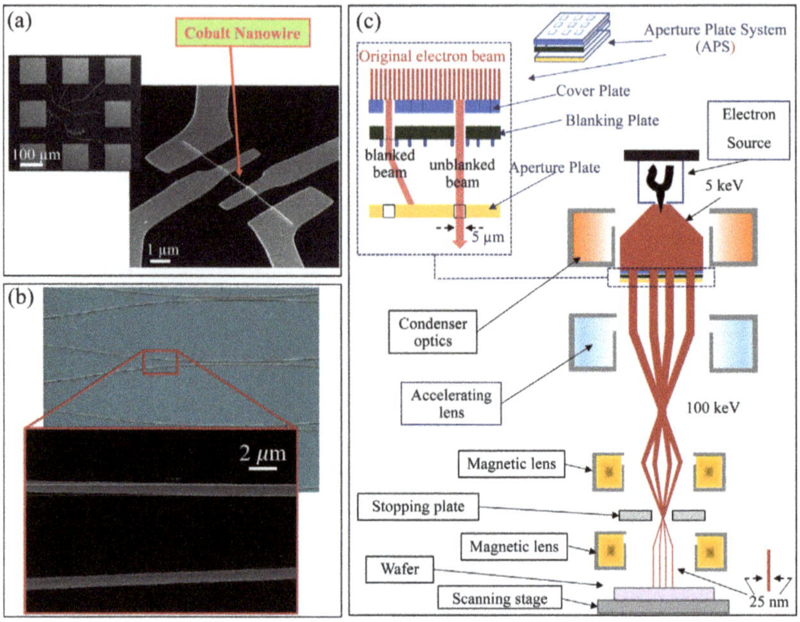

Figure 3.23. (a) Electrical contacts on cobalt nanowires fabricated through the EBL process. Permission by J M De Teresa (unpublished). (b) Long optical waveguides exposed with the fixed-beam-moving-stage (FBMS) approach [67]. (c) Schematic representation of a multibeam system with the aperture plate system [68]. Reprinted with permission from *Acta Physica Polonica* A.

the fixed-beam-moving-stage (FBMS) strategy addresses this problem [66]. In figure 3.23(b) optical waveguides are defined with this technique [67]. The waveguides (length of 2.5 mm and width of 700 nm) are exposed continuously, maintaining electron beam and dose fixed during the stage movement, thus overcoming any stitching error.

But probably, the main important limitation, especially for a use of the EBL technique in the industry, is related with its low throughput due to the long exposition time to cover large areas. Multi-beam systems for parallel writing are much faster than standard EBL system and potentially competitive with the optical lithography. The basic idea is schematically shown in figure 3.23(c). A low-energy electron beam (5 keV) is generated by a single electron source. Various electrons beams are obtained and selected by a programmable aperture plate system (APS), basically composed of a cover plate (for the division of the original beam), a blanking plate (for the selection of the beams) and an aperture plate. Different beams with a typical size of 5 μm are so accelerated up to 100 keV and focused through an electrostatic and magnetic lens on the wafer with a final beam size of 25 nm. The scanning of the beams on the wafer takes place through the movement scanning stage. This technique was developed for the semiconductor industry, and some companies (Mapper Lithography in Delft, IMS Nanofabrication GmbH in Wien) base their business on this new technology.

3.7 End-of-chapter summary

In this chapter, we have provided the reader with the basic knowledge and understanding of the indispensable processes in and around EBL, with a final scope of their application for the generation of fully working 2D-material-based devices. After reviewing the core concepts of an FE-SEM nanolithography set-up (section 3.2) we have introduced the fundamental (common) concepts for a typical EBL in the following section. Section 3.3 has been devoted to introducing the steps for an EBL process as well as the auxiliary, though necessary, processes essential to obtaining low-dimensional devices, i.e. plasma etching and e-beam metal evaporation (section 3.4).

In section 3.5 we have combined all the knowledge previously acquired and given a thorough and complete step-by-step study in the generation of some selected devices based on graphene and related 2D materials: Hall bars, quantum point contacts and FETs. We want to stress once more that the different recipes will definitely vary between facilities, set-ups and devices to be fabricated but we do believe that this chapter will be a good starting point for neophytes in the technique. We encourage Master and PhD students that are taking their first steps in EBL lithography to adapt and optimize these recipes at their convenience.

In section 3.6 we introduce general applications of EBL and new developments for the EBL technology, thought to overcome its own limitations. In the Student's corner we introduce briefly some other applications of EBL in 2D-materials and provide the reader with some introductory and seminal papers to continue their study further. We have included a few questions for a student self-assessment where they will challenge their acquired knowledge. The authors would like to thank the reader for any feedback that could be remitted to improve this text.

3.8 Student's corner

So far, we've dedicated this chapter to electron beam lithography using a specific set-up but trying to introduce general concepts of this powerful technique. We have thoroughly explained how to obtain some devices in nanoelectronics such as Hall bars and FETs while the generation of quantum point contacts with smooth walls has been introduced because of the unusual combination of EBL and cryo-etching.

Nevertheless, the student can go deeper in the application of EBL to 2D materials. For that, here below we mention other devices and geometries relevant in nano-electronics. Therefore, we do encourage the reader to take some time to go through the proposed examples and literature.

- Nano-etched suspended graphene can be obtained by a combination of scanning electron microscope (in this case present in a focused ion beam system) with a selective gas injection needle that tailors pre-deposited chemical vapour deposition (CVD) graphene into line-cuts with a resolution down to ~15 nm, whilst the edge roughness stays below 5 nm [69].

- Graphene nanoelectrode arrays can be obtained solely by the use of an EBL set-up combined with a standard voltage ramp apparatus. After the spin coating of MicroChem 950 PMMA,A5 with a speed of 4000 rpm for 45 s, a CVD graphene sheet was patterned as follows: a 5 nm width dash line between each pair of electrodes was designed on top of the graphene in order to obtain 50 μm length windows creating a pattern of numerous nanocontacts that are etched by standard oxygen RIE. Finally, electroburning technique was carried out by applying a slow voltage ramp to the device until the graphene links burn [70]
- Generation of unusual systems such as edge-free Corbino graphene geometry where concentric circular electrodes with no physical edge connecting the inner and outer electrodes is defined by a series of self-aligned EBL in combination with the van-der-Waals dry transfer technique [71]. Graphene nanoribbons (GNRs) represented the first benchmark attempts to study real ballistic transport and the effect of the edges in low-dimensional systems. GNRs have been largely generated by EBL typically using negative tone resists such as HSQ [72] or MIBK-diluted HSQ [73] in order to attain samples with extreme length-to-width ratios, much larger than present in QPCs.
- Josephson junctions represent a novel technological paradigm with various applications in nanoelectronics. They consist of a normal (metallic or semiconducting) region known as weak link coupled to superconducting leads harnessing dissipationless electrical current. Graphene-based Josephson junctions have been possible due to the generation of EBL-designed (sub)micrometric devices. Such junctions could be as simple as two bare superconducting electrodes in direct contact to the graphene sheet via edge contacts [74] or to more complicated fabrication process involving a first EBL step on a single PMMA layer plus the deposition of the superconducting leads followed by a second (more complex) lithography with a PMMA/HSQ mask combined with an RIE process for the definition of the final geometry of the system [75].

Concept checking questions

Choose the correct answer among the different options (more than one correct answer is possible):

- An FE-SEM allows working at low potentials which is crucial when dealing with…
  - A. Metallic films;
  - B. Semiconducting films;
  - C. Dielectric films.
- If you need to take an SEM image for topographic information at a rather high tilting angle (>45°), what would you choose?
  - A. It is not possible to take tilted SEM images as the SEM configuration is fixed (beam, detectors, etc).
  - B. In-Lens detector is the best choice.
  - C. ET detector is the best choice.

- For an EBL process, which of the following statements are true?
  - A. After the exposure and development processes, the positive tone resist is removed in the exposed areas.
  - B. After the exposure and development processes, the negative tone resist is removed in the exposed areas.
  - C. Development determines the tone of the resist.
- Choose the WF with the nominal higher resolution.
  - A. 100 μm
  - B. 500 μm
  - C. 1 mm
- What does the concentration of PMMA in solvent affect?
  - A. The homogeneity of the coating.
  - B. The thickness of the resist.
  - C. The optical contrast of the resist.
- If you need to expose an array (larger than 1 mm) of the same nanostructured pattern, which of the following approaches are the most suited for a correct lithography?
  - A. Expose the whole WA in a unique (U,V) center.
  - B. A smaller area (according to the used WF) CAD file is prepared and the whole WA is exposed thanks to the creation of a matrix of CAD elements with shifted (U,V) centers.
  - C. The whole CAD file is prepared, but we translate the WA exposure at different (U, V) centers.
- How can we overcome the stitching problem?
  - A. Through the use of an interferometric position stage.
  - B. Through the use of a micrometric position stage.
  - C. Through the use of markers.
- Suppose that during an alignment process with markers, their centers are not visible in the windows that are opened by the EBL program. How would you proceed?
  - A. I would repeat the 3-point alignment and probably I would need to repeat also the global origin and angle correction.
  - B. I would change the dimensions of the markers square in the CAD file (I would make them bigger so that I can see their center in the next four markers alignment).
  - C. I would repeat the alignment process on the four markers as many times as necessary. In the first alignment, I would center each window in the visible part of the marker so that in the subsequent alignments I will finally identify the center of the markers.
- What are the advantages of the HSQ resist with respect to the PMMA?
  - A. Versatility.
  - B. Resolution.
  - C. Lower exposition time due to the lower used doses.

- How would you reduce the PE problem through fragmentation?
  A. In an isolated feature by reducing the dose at the center and increasing the dose at the edges.
  B. In an isolated feature by increasing the dose at the center and decreasing the dose at the edges.
  C. In packed structures by increasing the dose between neighbouring features.
- How would you solve a lift-off problem due to the crosslinking of the resist?
  A. Using thinner resists.
  B. Using multilayer resists.
  C. Understanding if crosslinking is related first to certain positions of the position of the sample in the evaporator.

Acknowledgements

VC, MA, and ED from University of Salamanca acknowledge financial support from the *Agencia Estatal de Investigación* of Spain (Grants MAT2016-75955, and PID2019-106820RB-C22) and the *Junta de Castilla y León* (Grant SA256P18), including funding by ERDF/FEDER. We are also thankful to Juan Antonio Delgado, Yahya M Meziani, Enrique Velázquez, Mercedes Velázquez, Jorge Quereda, and Adrián Martín Ramos for their help and valuable comments on this manuscript.

References

[1] Zhou W, Apkarian R, Wang Z L and Joy D 2006 *Fundamentals of Scanning Electron Microscopy (SEM)* (New York: Springer) pp 1–40

[2] Abbas A 2013 Nanofabrication using electron beam lithography: novel resist and applications *PhD Thesis* University of Waterloo

[3] Alymani A and Lemine O M 2012 *FE-SEM Characterization of Some Nanomaterial* (London: InTech) pp 463–72

[4] Rius Suñé G 2008 Electron beam lithography for nanofabrication *PhD Thesis* Universitat Autònoma de Barcelona

[5] Chang T H P 1975 Proximity effect in electron-beam lithography *J. Vac. Sci. Technol.* **12** 1271–5

[6] Pala N and Karabiyik M 2016 Electron beam lithography (EBL) *Encyclopedia of Nanotechnology* ed B Bhushan 1st ed. (Berlin: Springer)

[7] Owen G 1990 Methods for proximity effect correction in electron lithography *J. Vac. Sci. Technol.* B **8** 1889

[8] Demers H, Poirier-Demers N, Phillips M R, de Jonge N and Drouin D 2012 Three-dimensional electron energy deposition modeling of cathodoluminescence emission near threading dislocations in GaN and electron-beam lithography exposure parameters for a PMMA resist *Microsc. Microanal* **18** 1220–8

[9] Kostelak R L, Kung E H, Thomson M G R and Vaidya S 1988 Evaluation of the proximity effect and GHOST correction technique for submicron electron beam lithography at 50 and 20 kV *J. Vac. Sci. Technol.* B **6** 2042–7

[10] Watson P G, Berger S D, Liddle J A and Waskiewicz W 1995 A background dose proximity effect correction technique for scattering with angular limitation projection electron lithography implemented in hardware *J. Vac. Sci. Technol.* B **13** 2504
[11] Mc Cord M and Rooks M 1997 Electron beam lithography *Handbook of Microlithography, Micromachining, and Microfabrication* vol 4 (San Francisco, CA: SPIE) ch 2
[12] Gangnaik A S, Georgiev Y M and Holmes J D 2017 New generation electron beam resists: a review *Chem. Mater.* **5** 1898
[13] Zheng Z 2008 Nanofabrication by charged beams *Nanofabrication Principles, Capabilities and Limits* (New York: Springer) ch 3
[14] Chen Y 2015 Nanofabrication by electron beam lithography and its applications: a review *Microelectron. Eng.* **135** 57
[15] Maex K 2003 Low dielectric constant materials for microelectronics *J. Appl. Phys.* **93** 8793
[16] Ferrari A C *et al* 2015 Science and technology roadmap for graphene, related two-dimensional crystals, and hybrid systems *Nanoscale* **7** 4598–810
[17] Zhou W, Wang W and Lin W 2007 E-beam nanolithography integrated with scanning electron microscope *Scanning Microscopy for Nanotechnology: Techniques and Applications* (New York: Springer) ch 6
[18] Carbaugh D J, Pandya S G, Wright J T, Kaya S and Rahman F 2017 Combination photo and electron beam lithography with polymethyl methacrylate (PMMA) resist *Nanotechnology* **28** 455301
[19] Clericò V, Masini L, Boni A, Meucci S, Cecchini M, Recchia F A, Tredicucci A and Bifone A 2014 Water-dispersible three-dimensional LC-nanoresonators *PLoS One* **9** e105474
[20] Özyilmaz B, Jarillo-Herrero P, Efetov D and Kim P P 2007 Electronic transport in locally gated graphene nanoconstrictions *Appl. Phys. Lett.* **91** 192107
[21] Winter A *et al* 2018 Lateral heterostructures of two-dimensional materials by electron-beam induced stitching *Carbon* **128** 106–16
[22] Paajaste J, Amado M, Roddaro S, Bergeret F S, Ercolani D, Sorba L and Giazotto F 2015 Pb/InAs nanowire Josephson junction with high critical current and magnetic flux focusing *Nano Lett.* **15** 1803–8
[23] Pitner G, Hills G, Llinas J P, Persson K-M, Park R, Bokor J, Mitra S and Wong H-S P 2019 Low-temperature side contact to carbon nanotube transistors: resistance distributions down to 10 nm contact length *Nano Lett.* **19** 1083–9
[24] Dan Y, Lu Y, Kybert N J, Luo Z and Johnson A T C 2009 Intrinsic response of graphene vapor sensors *Nano Lett.* **9** 1472–5
[25] Cheng Z, Zhou Q, Wang C, Li Q, Wang C and Fang Y 2011 Toward intrinsic graphene surfaces: a systematic study on thermal annealing and wet-chemical treatment of SiO_2-supported graphene devices *Nano Lett.* **11** 767–71
[26] Fan J, Michalik J M, Casado L, Roddaro S, Ibarra M R and De Teresa J M 2011 Investigation of the influence on graphene by using electron-beam and photo-litography *Solid State Commun.* **151** 1574
[27] Clericò V, Delgado-Notario J A, Saiz-Bretín M, Hernández Fuentevilla C, Malyshev A V, Lejarreta J D, Diez E and Domìnguez-Adame F 2018 Quantized electron transport through graphene nanoconstrictions *Phys. Status Solidi a* **215** 1701065
[28] Guerfi Y, Carcenac F and Larrieu G 2013 High resolution HSQ nanopillar arrays with low energy electron beam lithography *Microelectron. Eng.* **110** 173–6

[29] Grigorescu A E, van der Krogt M C, Hagen C W and Kruit P 2007 10 nm lines and spaces written in HSQ, using electron beam lithography *Microelectron. Eng.* **84** 822–4

[30] Zeng M, Xiao Y, Liu J, Yang K and Fu L 2018 Exploring two-dimensional materials toward the next-generation circuits: from monomer design to assembly control *Chem. Rev.* **118** 6236

[31] Das S, Robinson J A, Dubey M, Terrones H and Terrones M 2015 Beyond graphene: progress in novel two-dimensional materials and van der Waals solids *Annu. Rev. Mater. Res.* **45** 1

[32] Novoselov K S, Andreeva D V, Ren W and Shan G 2019 Graphene and other two-dimensional materials *Front. Phys.* **14** 13301

[33] Novoselov K S, Geim A K, Morozov S V, Jiang D, Zhang Y, Dubonos S V, Grigorieva I V and Firsov A A 2004 Electric field effect in atomically thin carbon films *Science* **306** 666

[34] Yin M and Shen Z 2015 A review on mechanical exfoliation for the scalable production of graphene *J. Mater. Chem.* A **3** 11700

[35] Blake P and Hill E W 2007 Making graphene visible *Appl. Phys. Lett.* **91** 063124

[36] Roddardo S, Oingue P, Piazza V, Pellegrini V and Beltram F 2007 The optical visibility of graphene: interference colors of ultrathin graphite on SiO_2 *Nano Lett.* **9** 2707

[37] Ferrari A C *et al* 2006 Raman spectrum of graphene and graphene layers *Phys. Rev. Lett.* **97** 187401

[38] Malard L M, Pimenta M A, Dresselhaus G and Dresselhaus M S 2009 Raman spectroscopy in graphene *Phys. Rep.* **473** 51

[39] Berkdemir A *et al* 2013 Identification of individual and few layers of WS_2 using Raman Spectroscopy *Sci. Rep.* **3** 1755

[40] Weber J W, Calado V E and van den Sanden M C M 2010 Optical constants of graphene measured by spectroscopic ellipsometry *Appl. Phys. Lett.* **97** 091904

[41] Clericò V 2020 Fabrication and characterization of quantum materials: graphene heterostructures and topological insulators *PhD Thesis* University of Salamanca http://nanotech.usal.es/publications/

[42] Wang L *et al* 2013 One-dimensional electrical contact to a two-dimensional material *Science* **342** 614

[43] Overweg H *et al* 2018 Electrostatically induced quantum point contacts in bilayer graphene *Nano Lett.* **18** 553

[44] Britnell L *et al* 2012 Electron tunneling through ultrathin boron nitride crystalline barriers *Nano Lett.* **3** 1707

[45] Ferrari A C 2007 Raman spectroscopy of graphene and graphite: Disorder, electron–phonon coupling, doping and nonadiabatic effects *Solid State Commun.* **143** 47

[46] Neumann C *et al* 2015 Raman spectroscopy as probe of nanometre-scale strain variations in graphene *Nat. Commun.* **6** 8429

[47] Zomer P J, Guimarães M H D, Brant J C, Tombros N and van Wees B J 2014 Fast pick up technique for high quality heterostructures of bilayer graphene and hexagonal boron nitride *Appl. Phys. Lett.* **105** 013101

[48] Pizzocchero F, Gammelgaard L, Jessen B S, Caridad J M, Wang L, Hone J, Bøggild P and Booth T J 2016 The hot pick-up technique for batch assembly of van der Waals heterostructures *Nat. Commun.* **7** 11894

[49] Frisenda R, Navarro-Moratalla E, Gant P, Pérez De Lara D, Jarillo-Herrero P, Gorbachev R V and Castellanos-Gomez A 2018 Recent progress in the assembly of nanodevices and

van der Waals heterostructures by deterministic placement of 2D materials *Chem. Soc. Rev.* **47** 53

[50] Jain A, Szabó Á, Parzefall M, Bonvin E, Taniguchi T, Watanabe K, Bharadwaj P, Luisier M and Novotny L 2019 One-dimensional edge contacts to a monolayer semiconductor *Nano Lett.* **19** 6914

[51] Clericò V, Delgado-Notario J A, Saiz-Bretín M, Malyshev A V, Meziani Y M, Hidalgo P, Méndez B, Amado M, Domínguez-Adame F and Diez E 2019 Quantum nanoconstrictions fabricated by cryo-etching in encapsulated graphene *Sci. Rep.* **9** 13572

[52] Defforgea T, Song X, Gautier G, Tillocher T, Dussart R, Kouassi S and Tran-Van F 2011 Scalloping removal on DRIE via using low concentrated alkaline solutions at low temperature *Sens. Actuators* A **170** 114

[53] Cadarso V J, Chidambaram N, Jacot-Descombes L and Schift H 2017 High-aspect-ratio nanoimprint process chains *Microsyst. Nanoeng.* **3** 17017

[54] Liu Z, Wu Y, Harteneck B and Olynick D 2013 Super-selective cryogenic etching for sub-10 nm features *Nanotechnology* **24** 015305

[55] Delgado-Notario J A 2019 Silicon- and graphene-based FETs for THz technology *PhD Thesis* University of Salamanca http://nanotech.usal.es/publications/

[56] Delgado-Notario J A, Clericò V, Diez E, Velázquez-Pérez J E, Taniguchi T, Watanabe K, Otsuji T and Meziani Y M 2020 Asymmetric dual-grating gate graphene FET for detection of terahertz radiations *APL Photonics* **5** 066102

[57] Novoselov K S, Jiang D, Schedin F, Booth T J, Khotkevich V V, Morozov S V and Geim A K 2005 Two-dimensional atomic crystals *PNAS* **30** 102

[58] Geim A K and Grigorieva I V 2013 Van der Waals heterostructures *Nature* **499** 419

[59] Manzeli S, Ovchinnikov D, Pasquier D, Yazyev O V and Kis A 2017 2D transition metal dichalcogenides *Nat. Rev. Mater.* **2** 17033

[60] Roy A, Movva H C P, Satpati B, Kim K, Dey R, Rai A, Pramanik T, Guchhait S, Tutuc E and Banerjee S K 2016 Structural and electrical properties of $MoTe_2$ and $MoSe_2$ grown by molecular beam epitaxy *ACS Appl. Mater. Inter.* **2** 7396

[61] Mouafo L D N, Godel F, Froehlicher G, Berciaud S, Doudin B, Kamalakar M V and Dayen J-D 2017 Tuning contact transport mechanisms in bilayer $MoSe_2$ transistors up to Fowler–Nordheim regime *2D Mater.* **4** 015037

[62] Tonndorf P *et al* 2013 Photoluminescence emission and Raman response of monolayer MoS_2, $MoSe_2$, and WSe_2 *Opt. Express* **21** 4908

[63] Huang H *et al* 2016 Highly sensitive visible to infrared $MoTe_2$ photodetectors enhanced by the photogating effect *Nanotechnology* **27** 445201

[64] Lin Y-F *et al* 2014 Ambipolar $MoTe_2$ transistors and their applications in logic circuits *Adv. Mater.* **26** 3263

[65] Yoshioka N 2005 Optical masks: an overview *Handbook of Photomask Manufacturing Technology* 1st edn (Boca Raton, FL: Taylor & Francis Group) ch 6

[66] Khodadad I 2017 Electron beam lithography using fixed beam moving stage *J. Vac. Sci. Technol.* B **35** 051601

[67] Kirchner M and Kahl M 2009 Proc. of the III National Conference on Nanotechnology NANO (Warsaw, 2009) *Acta Phys. Pol.* A 116 S-198

[68] Chalom D, Klikovits J, Geist D, Hudek P, Eder-Kapl S, Daneshpanah M, Laske F, Eyring S and Roeth K-D 2015 *Proc. SPIE* **9658** 965805

[69] Sommer B, Sonntag J, Ganczarczyk A, Braam D, Prinz G, Lorke A and Geller M 2015 Electron-beam induced nano-etching of suspended graphene *Sci. Rep.* **5** 7781

[70] Gu G, Su D, Jia C, Ren S and Guo X 2018 Building nanogapped graphene electrode arrays by electroburning *RSC Adv.* **8** 6814

[71] Zeng Y, Li J Y A, Dietrich S A, Ghosh O M, Watanabe K, Taniguchi T, Hone J and Dean C R 2019 High-quality magnetotransport in graphene using the edge-free corbino geometry *Phys. Rev. Lett.* **122** 137701

[72] Han M Y, Özyilmaz B, Zhang Y and Kim P 2007 Energy band-gap engineering of graphene nanoribbons *Phys. Rev. Lett.* **98** 206805

[73] Hwang W S, Tahyl K, Li X, Xing H, Seabaugh A C, Sung C Y and Jena D 2012 Transport properties of graphene nanoribbon transistors on chemical-vapor-deposition grown wafer-scale graphene *Appl. Phys. Lett.* **100** 203107

[74] Borzenets I V *et al* 2016 Ballistic graphene Josephson junctions from the short to the long junction regimes *Phys. Rev. Lett.* **117** 237002

[75] Calado V, Goswami S, Nanda G, Diez M, Akhmerov A R, Watanabe K, Taniguchi T, Klapwijk T M and Vandersypen L M K 2015 Ballistic Josephson junctions in edge-contacted graphene *Nat. Nanotechnol.* **10** 761

IOP Publishing

Nanofabrication
Nanolithography techniques and their applications
José María De Teresa

Chapter 4

Focused electron beam induced deposition

Javier Pablo-Navarro, Soraya Sangiao, César Magén and José María De Teresa

In this chapter, the physical principles, applications and perspectives of focused electron beam induced deposition (FEBID) are overviewed. FEBID is a single-step direct-write nanolithography technique relying on a nanometre-size electron beam to decompose the molecules of a precursor gas delivered onto a substrate. The process is carried out inside a scanning electron microscope with a gas injector system. The shape and particular dimensions of the resulting nanodeposit can be tailored from the micron scale down to the nanometre scale. The absence of resists makes FEBID extremely adaptable to virtually any kind of substrate material or topography, including cantilevers and tips. The availability of precursors enables the growth of a wide range of functional materials, whose physical properties (conducting, insulating, mechanical, superconducting, magnetic, optical) can be tailored to a great extent, with relevant applications in nanoelectronics, nanophotonics, nanosensing and nanomechanics. Being an additive and site-selective technique, FEBID offers unique possibilities as a 3D nano-printer and prototyping of complex nanoarchitectures.

4.1 Introduction

The possibility of fabricating micro- and nanostructures paves the way for systems at the nanometre scale with novel capabilities and better performance. In this regard, micro- and nanolithography techniques are common methods used to produce structures by transferring a specific pattern to a sample. These nanoscale objects are fabricated following two different approaches: the bottom-up and the top-down strategies (see chapter 1 for more details). The bottom-up approach refers to architectures based on self-assembly or self-organization processes, such as the electrochemical deposition and the block co-polymers technology with promising applications and high tunability in terms of the size, shape and periodicity of arrays of objects [1]. The top-down methods are basically rooted in the deposition of thin films onto which the motifs are patterned by lithography techniques [2, 3]. Several

doi:10.1088/978-0-7503-2608-7ch4

types of top-down nanometre scale patterning lithography methods have been developed: electron beam lithography (EBL) [4, 5]; focused ion beam lithography [6]; near-field scanning optical lithography [7], laser lithography [8], holographic or interferometric lithography [9, 10], extreme ultraviolet lithography [11], x-ray lithography [12], zone plate array lithography [13]; stamping techniques such as micro-contact printing [14, 15], nano-imprint [16] and step and flash imprint lithography [17]; and others such as nanografting [18] or dip-pen lithography [19].

However, although these nanofabrication techniques have a great potential for a wide range of applications, they also present some limitations such as the need of an additional layer prior to the fabrication, the use of masks and lift-off processes, material implantation, sputtering and deposition processes, the effect of elevated temperatures for the fabrication, the inability to grow deposits with complex materials and shapes in 3D or the use of several steps for the growth of the final nanostructure, among others. In this context, focused electron beam induced deposition (FEBID) emerges as a direct-write fabrication approach for nanosynthesis of functional materials. The method is similar to chemical vapor deposition, but, rather than on thermal dissociation, it is based on the electron-impact decomposition of gas molecules that are continuously refreshed at the area of growth. In FEBID, the deposition is confined to the area where the primary electron beam impinges and to the surrounding area where secondary electrons are generated. This technique has already demonstrated to serve as a robust nanofabrication method for electronic [20], magnetic [21], photonic [22], and mechanical [23] devices.

4.2 Principles of FEBID

FEBID is a single-step nanolithography technique based on the decomposition by a finely-focused electron beam of the molecules of a precursor gas delivered and adsorbed onto the surface of the substrate, which gives rise to the deposition of a solid material [24–27]. Before the consolidation of this technique, the potential of electron beam induced processing for patterning was already hinted in the 1970s by the local deposition of contaminants while performing scanning electron microscopy (SEM) [28]. When the electron beam scans a region of the surface of the substrate, a few-nanometre layer of material starts to cover the scanned area. This phenomenon is produced by the decomposition of hydrocarbons present in the vacuum chamber as residual gases, and adsorbed on the surface.

FEBID was formally introduced in 1984 by Matsui and Mori [29, 30], who injected different precursor gas molecules inside the vacuum chamber on purpose, producing micro- and nanostructures with different functionalities. Once the precursor gas molecules are delivered locally near the surface of the substrate, some of them are dissociated by the electron beam. The non-volatile part of the molecule is deposited, whereas the volatile fragments are pumped out of the working chamber. Nowadays, this method is capable of nanometre-scale resolution for the growth of 2D [31] and 3D [32] architectures. The physical and chemical properties of the deposit will depend on its obtained geometry and composition, which will be determined by the complex interactions between the electron beam, precursor gas molecules, substrate and the nascent structure [33, 34]. For this reason, understanding the influence of the growth parameters on these interactions is a key point to develop proper approaches for the fabrication of nanostructures with the optimum targeted properties (figure 4.1).

The list of most relevant growth parameters in FEBID includes:

- the energy of the primary electrons, which is determined by the beam acceleration voltage;
- the electron beam current, defined as the electron flow reaching the sample surface;
- the dwell time, which identifies the lapse of time for which the electron beam is held still on a particular position or pixel;
- the pitch, which is the distance between two neighbouring pixels;
- the scan direction or patterning strategy, associated with the trajectory of the beam within the pattern [34, 35];
- the number of loops, which is the number of single complete scans of the established pattern;
- the refresh time, which is the period of time between two consecutive loops, allowing the replenishment of precursor gas molecules in the deposition area.

In addition, the pattern dimensions and geometry, the chamber base pressure, the precursor gas type, its temperature and flux, the position of the gas injector system (GIS) nozzle, the substrate material and its temperature are also decisive variables for the fabrication of nano-objects by FEBID.

However, the simultaneous control of all the parameters playing a role in the growth process of nanostructures is a daunting task [24], not only for the high number of variables, but also because some of them depend on the others. With the goal of learning how these parameters affect the growth operation and handle them, the dissociation process caused by the electron beam deserves special attention. Among the many distinct electron–molecule interaction processes [36], the elastic scattering, vibrational and electronic excitation, dissociative electron attachment,

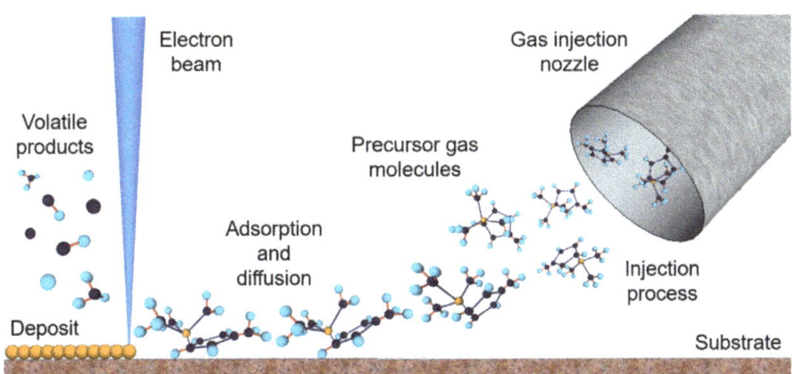

Figure 4.1. Sketch of the FEBID process. Example with the $(CH_3)_3Pt(CpCH_3)$ molecule.

neutral and bipolar dissociation, and dissociative ionization can be mentioned. Given this situation, it may be thought that the beam–molecule interaction is too complex to have a thorough comprehension of the mechanism at molecular scale. In fact, few experimental [37] and theoretical studies [38] in this regard can be found in literature. However, simulations of the fabrication process by continuum models and approaches of growth geometries by Monte Carlo method have already been performed to clarify the intricacies of the FEBID growth process [39].

4.2.1 Electron–matter interaction

Since the basic principles of FEBID are closely related to SEM, whose basics are explained in section 3.2.1, the physical events emerging from the electron–matter interaction will be first introduced and correlated with SEM imaging. First, the elastic collisions give rise to a broadening of the primary beam which degrades the lateral resolution. In addition, inelastic collisions lead to other physical events such as the emission of x-rays and Auger electrons, phonon scattering, or cathodoluminescence (visible light). Also, the electron–sample interaction generates secondary electrons (SE) and backscattered electrons (BSE), both being used for imaging purposes. The SE are generated by the inelastic scattering of the primary beam with the valence electrons in the outer shells of the sample atoms. They are emitted with typical energies lower than ~50 eV, having mean free paths of few nanometres. Since only SE emitted near the surface can escape and be detected, they are mainly used to obtain images of topographic contrast [40, 41]. On the other hand, the BSE are generated by the quasi-elastic scattering of the primary beam with the atoms. Thus, the energy lost by these electrons is very small, but the momentum transfer is large, allowing the electrons to be scattered at high angles. As BSE are high-energy electrons, they can reach the surface from deep regions of the sample, not giving accurate topographic information. However, they provide images with compositional contrast given that the scattering cross section of the BSE increases with the atomic number, Z. The maximum resolution of an SEM image is mainly determined by the primary electron beam voltage and current, but the interaction phenomena of the electron beam with the sample is also relevant.

Regarding the physical phenomena involved in the growth of structures by FEBID, Fowlkes *et al* noted that the most significant contribution to the dissociation of the precursor gas molecules arises from the primary electrons (PE) and their generated secondary electrons (SE_I), with lower relevance from BSE and their generated secondary electrons (SE_{II}) [42]. Later, Smith *et al* stressed the important role played by forward-scattered electrons (FSE) in the growth of 3D nanostructures as well as the dependence of their effect on the primary beam energy and the resulting growth regime [43]. Moreover, Arnold *et al* evidenced that SE_{II} are responsible for a halo of deposited material around the scanned area in 2D structures, affecting the final deposition resolution [44]. A similar study by Schmied *et al* showed various competing effects in the case of 3D deposits, with significant broadening of the deposit upon decreasing primary electron beam energies [45]. The FEBID process is thus complex and, in general, the precursor

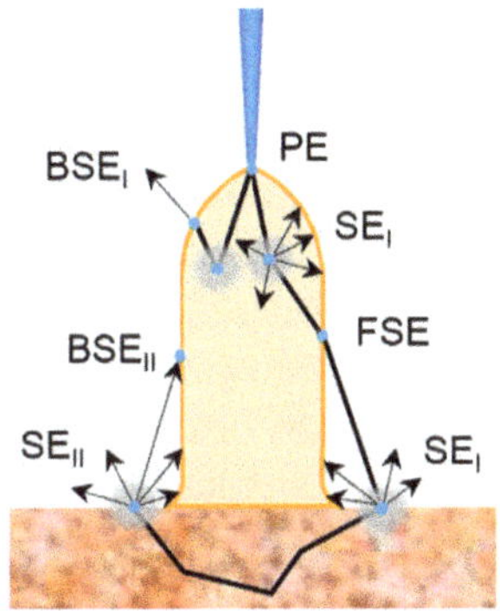

Figure 4.2. Types of electron emission contributing to the FEBID process. Adapted from Harald Plank, Graz University of Technology, Austria, 2020.

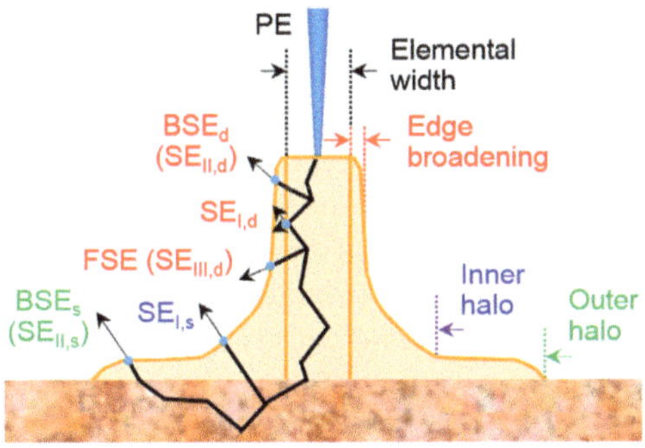

Figure 4.3. Broadening effects induced by the electron–matter interactions. The *d* and *s* subscripts refer to electrons emerging from the deposit and substrate, respectively. Adapted from [45] with permission of Hindawi.

dissociation is determined by high-energy electrons (mainly arising from PE and FSE) as well as by low-energy electrons (mainly SE_I and SE_{II}) (figure 4.2).

In brief, due to the complex electron–matter interaction, different broadening effects with respect to the primary beam width can be identified in FEBID, as represented in figure 4.3 [44]:

- The edge broadening, in the range of a few nanometres, which is mainly caused by the SE generated in the growing structure by the PE, the BSE and the FSE.
- The inner halo broadening, in the range of a few tens of nanometres, which is mainly caused by the SE_I generated by the PE in the substrate and reaching the substrate surface.
- The outer halo broadening, up to a few hundreds of nanometres, which is mainly caused by the SE_{II}, generated by the electrons backscattered by the substrate, that reach the substrate surface.

Regarding the inner and outer halo broadening, the substrate atomic number, *Z*, and the electron beam voltage, *V*, have a decisive influence:

- A higher *Z* value leads to a smaller interaction volume of the electron beam in the substrate, thus reducing the outer halo.
- A higher *V* produces two effects. First, the primary beam is narrower due the lower aberrations of the electron column lenses, thus diminishing the primary irradiation spot and the inner halo produced by the SE_I. Secondly, a larger interaction volume of the electron beam with the substrate gives rise to SE_{II} emitted at further distances, thus enlarging the outer halo.

4.2.2 Monte Carlo method towards FEBID simulations

The Monte Carlo method can be used to comprehend the FEBID growth by simulating the electrons trajectories and precursor gas dynamics and dissociation [32, 43]. The interaction between the electron beam and the atoms of the substrate and of the growing structures can be modelled via a strong interaction mechanism leading to a multiple electron scattering process. The Monte Carlo method typically implies two assumptions. First, the electron trajectory only depends on elastic events, modelled with a relativistic-screened Rutherford scattering cross section. Secondly, the electron loses energy continuously as a function of the path length in the solid [25]. A large number of Monte Carlo scattered electron trajectories and SE generation are simulated. After parametrizing the electron trajectories and calculating the scattering events, the motion and energy distribution of the electrons through a material can be obtained as a function of the primary beam conditions and substrate material [36], as illustrated in figure 4.4. SE generated close to the surface

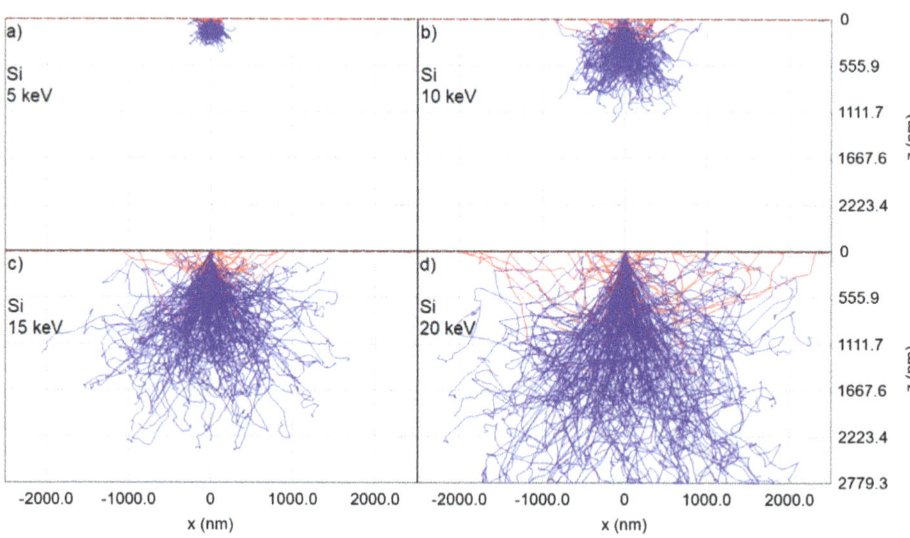

Figure 4.4. Monte Carlo simulations using CASINO software [46] of electron trajectories in Si for a PE beam with incident energy in the range from 5 to 20 keV in normal incidence, with SE and BSE in blue and red, respectively. The number of simulated trajectories is 1000 at each energy.

with sufficient energy to escape are candidates to produce precursor dissociation, thus, deposit growth.

4.2.3 Single precursor species model

In order to shed light on the physical and chemical phenomena behind the deposit growth, the single precursor species continuum model of FEBID is enlightening. Considering the precursor adsorbate density, n, and the radial coordinate with respect to the electron beam position, r, the radially symmetric equation for the adsorption rate is written as follows [47]:

$$\begin{aligned}\frac{\partial n(r,\,t)}{\partial t} &= sJ\left[1 - \frac{n(r,\,t)}{n_{ML}}\right] - \frac{n(r,\,t)}{\tau} \\ &+ D\left[\frac{\partial^2 n(r,\,t)}{\partial r^2} + \frac{1}{r}\frac{\partial n(r,\,t)}{\partial r}\right] - \sigma_d f(r) n(r,\,t)\end{aligned} \tag{4.1}$$

where the adsorption, desorption, diffusion and dissociation terms can be sequentially identified in the right-hand side of the equation. Firstly, in the adsorption term, s is the sticking coefficient, J the precursor flux, n_{ML} the complete area density which corresponds to that of a monolayer, and $1 - n/n_{ML}$ the fraction of surface sites available for adsorption. This model assumes a weak precursor–substrate interaction and neglects the interaction amongst the adsorbed precursor gas molecules. Secondly, the desorption term is defined as the adsorbate density divided by the average residence time of the precursor gas molecules on the substrate surface, τ. Thirdly, the diffusion term denotes the spreading of the precursor gas molecules on the surface and is proportional to the diffusion constant, D. Finally, the dissociation term models the depletion of n induced by the electron-beam-induced decomposition of the molecules, which is proportional to the energy-integrated dissociation cross section, σ_d, and the electron flux distribution per unit area, $f(r)$, with the Gaussian shape:

$$f(r) = \frac{I/e}{2\pi a^2}\exp\left(-\frac{r^2}{2a^2}\right) \tag{4.2}$$

where I is the electron beam current, e the electron charge and a the standard deviation of a Gaussian distribution.

Additionally, the local deposition growth rate $\mathcal{R}(r)$ can be defined as

$$\mathcal{R}(r) = \frac{V\sigma_d f(r)}{t_D}\int_0^{t_D} n(r,\,t)dt \tag{4.3}$$

where V is the volume of the non-volatile deposit resulting from the decomposition of one precursor gas molecule and t_D is the dwell time of the electron beam at each point. Neglecting the diffusion term—which is a good approximation for a wide range of D and t_D values [24]—, the analytical solution of equation (4.3) is reduced to equation (4.4) for a pulsed irradiation (*on* and *off* irradiation at each point):

$$\mathcal{R}(r) = V\sigma_d f(r)\left[sJ\left(\frac{1}{k_r} - \frac{1}{k_d}\right)\frac{1 - \exp(-k_d t_D)}{k_d t_D} + \frac{sJ}{k_d}\right] \tag{4.4}$$

where k_d and k_r are, respectively, constants related to the available precursor at a given point when the electron beam is *on* (depletion period) and when the electron beam is *off* (replenishment period):

$$k_d = \frac{sJ}{n_{ML}} + \frac{1}{\tau} + \sigma_d f \tag{4.5}$$

$$k_r = \frac{sJ}{n_{ML}} + \frac{1}{\tau} \tag{4.6}$$

A rough interpretation of the growth rate can be obtained as a function of t_D. On the one hand, for relatively small t_D, typically on the order of nanoseconds, $\mathcal{R}$ increases as a function of f. This growth regime is the *electron-limited regime* because the growth is restricted by the inability of the electron beam to decompose all the available precursor molecules on the surface (figure 4.5). On the other hand, in the case of large t_D, the dissociation rate exceeds the gas precursor replenishment rate, giving rise to the *precursor-limited regime*.

Thus, the balance between the availability of precursor molecules on the growth area and the electron beam current will determine the regime in which the growth occurs, affecting not only the growth rate but also the composition of the nanostructures [48, 49] and the sample morphology [34]. This balance can be also modified through the variation of the precursor flux reaching the growth area [48], which can be tuned by several means such as varying the GIS temperature, changing the GIS valve opening, modifying the distance between the GIS and the substrate, etc.

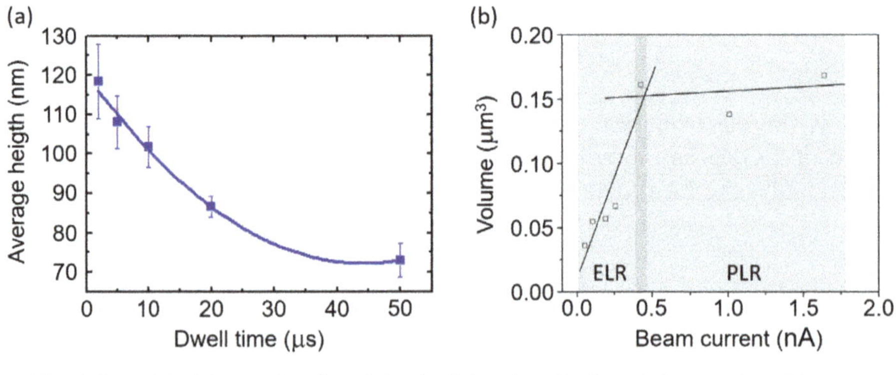

Figure 4.5. (a) Deposit height as a function of the dwell time for a fixed total electron dose; (b) representation of the deposit volume as a function of the electron beam current, indicating the electron-limited regime (ELR) and the precursor-limited regime (PLR). (a) Adapted with permission from [48]. Copyright 2011 American Chemical Society. (b) Adapted from [49] with permission of de Gruyter.

It is also worth mentioning that the precursor molecule sticking coefficient (or equivalently, the precursor molecule residence time) is temperature dependent. Thus, an increase in the growth temperature entails a lower residence time of the precursor molecules, giving rise to lower growth rates [50, 51]. This is particularly relevant in the growth of 3D structures, where heat dissipation is less efficient [52]. On the other hand, if significant heating of the growing deposit occurs, the decomposition of the precursor gas molecules is more efficient when the reached temperatures are close to that of the thermal decomposition of the precursor [53–55]. However, if the deposit temperature exceeds the decomposition temperature of the precursor, a spontaneous and uncontrolled thermal decomposition of the molecules takes place.

4.2.4 General features of FEBID precursors, growth parameters and applications

The specific characteristics of the precursor are of utmost importance concerning the physical and chemical properties exhibited by the FEBID deposit and its subsequent applications [56]. The precursor material needs to be suitably chosen following various recommendations, as discussed in reference [56].

The chemical stability is one of the first requirements to fulfil. The precursor material is stored inside a GIS container and delivered to the working chamber through a narrow nozzle that is placed in proximity to the substrate growth area. Although the precursor reaches the substrate surface in the gas phase, it can be stored in solid, liquid or gas phase. The transfer from the GIS container to the working chamber should be free from undesired chemical reactions or premature decomposition. Sometimes, the lifetime of the precursor is too short or polymerization reactions occur inside the container itself.

- The gas flux needs to be precisely controlled and requires precursor materials with appropriate vapour pressure values. Rough approximations suggest that this variable should range between 0.01 and 50 mbar. The two main knobs to control the gas flux are the automatic or manual valve, which allows the releasing of the gas into the chamber, and the GIS temperature, which can be tuned in commercial equipment up to a few tens of degrees.
- In order to obtain high-purity deposits mainly containing the element of interest, small precursor molecules are preferred, given that long molecules are more prone to produce non-volatile residues inside the deposit. Along this line of thought, precursor molecules based on carbonyl groups, phosphine groups and halogen groups are normally used.
- A large sticking coefficient of the precursor molecule, closely linked to its residence time, favours a high growth rate. If the sticking coefficient is very low, the technique becomes too slow. On the other hand, a large sticking coefficient implies that the precursor molecules remain for a certain period of time (a few minutes typically) adsorbed on the substrate surface, which can hinder further sample processing.

- The precursor molecule should not be hazardous (toxic, flammable, corrosive or carcinogenic) to minimize health risks during precursor handling (normally performed under glove-box environment) and to avoid chamber and vacuum-system contamination.

In table 4.1, the most important FEBID process parameters are listed. Finding an optimum value for each parameter will ensure an efficient process, but, at the same time, it can be a Herculean task. The general notions presented in this chapter can be applied for the fabrication of any type of structure, offering a starting working point. However, an optimization or refinement process should be performed for each particular deposit shape and precursor material.

Due to the physical principles of FEBID described in this section, this technique presents unique advantages with respect to other nanolithography techniques:

- FEBID is a direct-write nanolithography technique which implies that only one lithography step is required, in sharp contrast to other competing techniques such as EBL. Thus, FEBID can be useful in device prototyping as a 2D (or 3D) nano-printer.
- FEBID does not require the use of resists. This implies that FEBID is competitive when the use of resists is cumbersome, such as in the case of non-flat substrates or substrates whose physical/chemical properties are sensitive to resist residues.
- FEBID can achieve extremely high lateral resolution. Using an SEM, 3 nm dots have been grown. More generally, sub-100 nm resolution is routinely achieved, in various cases reaching 10 nm resolution.

Table 4.1. Main FEBID tunable variables, and their impact on the growth parameters and the deposit properties.

Tunable variable	*Growth parameter*	*Deposit property*
E-beam acceleration voltage	Interaction volume	Resolution, growth rate
E-beam current	Beam diameter	Resolution, composition
Dwell time	Deposition time	Growth rate, composition
Refresh time	Gas replenishment	Growth rate, composition
Pitch	Pattern geometry	Shape, growth rate
Scan direction	Pattern geometry	Shape
Number of loops	Deposition time	Thickness
Chamber base pressure	Residual gases	Composition
Precursor gas type	Deposition elements	Composition, growth rate
Precursor gas flux	Molecules injected	Growth rate, composition
Gas injection system position	Gas direction and gas flux	Growth rate, shape
Substrate material	Interaction volume	Growth rate, resolution, charging effects
Substrate temperature	Sticking coefficient	Growth rate, composition

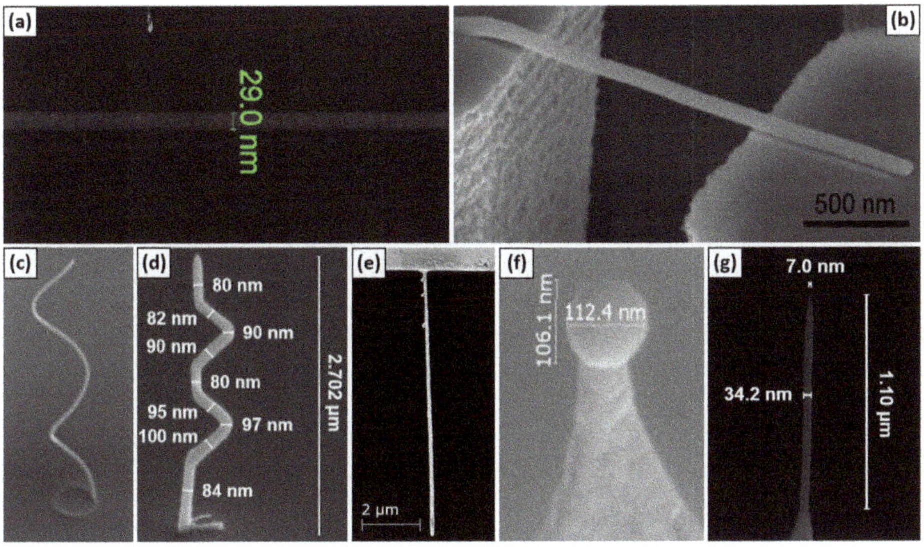

Figure 4.6. Examples of FEBID deposits that exploit the advantages of the technique: (a) high-resolution Co nanowire grown on a Si_3N_4 membrane, (b) suspended Co nanowire grown between two pads [57], (c) two-loop Co nanowire, (d) Co@Pt core–shell nanowire built with various bends, (e) Co nanowire with an aspect ratio around 100 grown on a Si substrate, (f) Co nanosphere grown on a cantilever and (g) Fe nanowire grown on the tip of an atomic force microscope cantilever [58]. (a) Reprinted with permission from [48]. Copyright 2011 American Chemical Society. (c) Reprinted with permission from [59]. Copyright 2013 Springer Nature. (d) Reprinted with permission from [60], copyright IOP Publishing. (e) Reprinted with permission from [61]. Copyright 2020 American Physical Society. (f) Reprinted from [62] with permission of Hindawi.

- FEBID can fabricate 3D objects, which are hard to obtain with competing nanolithography techniques. Recently, specific software has been developed to apply this 3D growth capability to create complex 3D architectures. Please, refer to section 1.4.5 for further details.
- Site selectivity and tunability in terms of material type and composition by modifying the growth parameters are two key advantages of FEBID with respect to other nanofabrication techniques.

In figure 4.6, particular FEBID examples exploiting these advantages are collected.

4.3 FEBID precursors and deposited materials

In this section, we will first review the most frequent precursors used in FEBID and the properties of the resulting as-grown deposits. Secondly, various strategies to tune the properties of the FEBID deposits will be revised. Such strategies, applied during or after the FEBID growth, are mainly aimed at enhancing the purity of the deposits.

4.3.1 Precursors and properties of as-grown materials

The growth of FEBID nanostructures with either metallic, magnetic, photonic, superconducting or structural properties has been explored using a wide range of precursors [63]. One of the potential issues of this synthesis technique is the existence of chemical impurities in the deposit due to incomplete dissociation of precursor molecules. In addition, the structures also contain different levels of contaminants originating from the residual gases existing in the working chamber [64]. Nevertheless, purity can be normally improved by a precise control of the large number of parameters involved in the growth [65], which in some cases enables the fabrication of deposits with a high content of the desired element(s) [48, 66–69]. In the following, the **main precursors** used in FEBID, their as-grown composition and other physical properties are listed.

- *Platinum.* The Pt deposits are among the most common structures grown by FEBID. Typically, these objects are presented in the form of conducting Pt–C nanocomposites where the metallic particles are embedded in a carbonaceous matrix, using $(CH_3)_3Pt(CpCH_3)$ precursor gas [70] and, to a lesser extent, $(CH_3)_3PtC_5H_5$ [71]. Integrated generally by 80 at. % of C and 20 at. % of Pt, and with a resistivity value around 10^4 μΩ cm, the significantly higher growth rate of this material in comparison with the rest of FEBID materials makes it very useful for the standard lamellae preparation for transmission electron microscopy.
- *Tungsten.* W deposits grown with $W(CO)_6$ precursor yield to maximum metallic contents of 50 at. %, showing low electrical resistivity values around 270 μΩ cm [72]. The deposits are composed by an insulating matrix where a disordered array of crystalline metallic nanoparticles (1–5 nm) are embedded. One of the most interesting properties of these W-based granular deposits relies on the capability to cover the full line between insulating to practically metallic behavior as a function of the growth conditions [73]. WF_6 molecule, one of the first precursor gases which gave rise to pure metallic structures under ultra-high vacuum conditions [42, 74], is also currently used.
- *Carbon.* The organometallic nature of a great number of precursors induces the deposition of different levels of C material as a side effect. To purposely grow C deposits, $C_{10}H_8$ molecule can be used as standard precursor gas for patterning transfer into the substrates with high lateral resolution [75].
- *Cobalt.* Metallic Co structures with polycrystalline microstructure, purity higher than 90 at. % and resistivities around 40 μΩ cm were reported in the last decade using $Co_2(CO)_8$ [48, 76]. These structures in the form of nano-squares revealed a very well-defined dependence of the composition with the precursor gas flux. At low precursor gas flux, i.e., in the precursor-limited regime, C and O coming from the molecule and the residual species in the chamber are deposited. On the other hand, at high precursor gas flux, i.e., in the electron-limited regime, the molecules are refreshed too quickly to allow a proper decomposition by the electron beam, thus incorporating non-metallic

material. Finally, there is an intermediate precursor gas flux range where an optimized balance between the precursor flux and the electron beam current is reached, where the decomposition takes place under the most favourable and advantageous conditions. Other molecules such as $Co(CO)_3NO$ have also been used for the fabrication of deposits with 50 at. % of Co, showing non-magnetic behaviour and higher resistivity values, around 6.3×10^3 μΩ cm at room temperature [77].

- *Iron.* Fe deposits have been grown with metallic compositions above 75 at. % with homonuclear $Fe(CO)_5$ [78] and $Fe_2(CO)_9$ [79] precursors. In these amorphous Fe:O:C nanostructures, resistivity values below 100 μΩ cm were found at room temperature.
- *Gold.* The maximum Au purity limit of 30 at. % cannot be overcome for the as-deposited materials based on $C_7H_{10}AuF_3O_2$ [80, 81]. Nano-objects in the shape of nanowires present very high resistivity values of around 5×10^6 μΩ cm at room temperature. Also, other Au precursors such as PF_3AuCl [82] or Au(CO)Cl [83], and even precursor molecules such as $Au(CH_3)_2(C_2ClF_3O)$ have been tested to improve the metallic content in combination with new approaches, as explained in the next section.
- *Silver.* Planar Ag deposits with purity around 75 at. % have been fabricated utilizing the precursor gas $AgO_2(CH_3)_2C_4H_{10}$ [84]. The results indicate that the use of low doses (30 nC μm^{-2}) gives rise to the formation of Ag crystals, whereas high doses (10^4 nC μm^{-2}), which produce larger C atomic percentages, result in Ag nanoparticles embedded into a carbonaceous matrix.
- *Silicon* and *silicon oxide.* Si deposits based on Si_5H_{12} precursor gas [85], or amorphous SiO_2 deposits with additional amounts of C and O contaminants have been obtained from the $Si(OC_2H_5)_4$ precursor [86, 87], the most commonly used gas for SiO_2 deposition. In addition, the shapes and architectures arising from this material also demonstrate the geometrical versatility by the growth of 2D and 3D nanostructures, e.g., in the form of a hand-fan architecture [88].
- *Lead.* Granular Pb deposits, with a 46 at. % Pb composition, based on the dissociation of $(CH_3CH_2)_4Pb$, also exhibit metallic behavior [89]. These deposits present a resistivity value of 190 μΩ cm at room temperature.
- Other less frequent precursors: *ruthenium* $C_7H_9RuC_7H_9$ precursor has been used to fabricate deposits where the as-deposited Ru composition is nearly 10 at. %, the rest being C [90]. *Nickel* precursors such as $Ni(C_5H_4CH_3)_2$ or $Ni(PF_3)_4$ [91] and *copper* precursors such as $CuC_{16}O_6H_{26}$ [92], with resistivities around 10^6 μΩ cm, have also been used to grow structures by FEBID.

Bimetallic precursors. Using heteronuclear precursor molecules that contain two metallic elements, it is possible to grow deposits of alloyed materials. A few examples are collected here below:

- *Cobalt-iron.* Co-Fe bimetallic compounds have been explored through the use of the heteronuclear precursor $HCo_3Fe(CO)_{12}$ [93–95]. Co_3Fe alloy deposits with 80 at. % metallic content were obtained, with a resistivity value of 43 μΩ cm at room temperature, comparable to those obtained in Co-FEBID [76] and Fe-FEBID [78] nanostructures.

- *Ruthenium-iron*. The carbonyl-based complex $H_2FeRu_3(CO)_{13}$ [96] has led to deposits with maximum metallic contents of around 26 at. %, displaying a lowest resistivity of 2500 μΩ cm, close to the metallic conduction regime. The different behavior between $HCo_3Fe(CO)_{12}$ and $H_2FeRu_3(CO)_{13}$ precursors points to the relative thermal stability of the decarbonylated intermediates with the CO ligands. In the first one, the CO groups are thermally unstable in the intermediate compound at room temperature, whereas in the second one the CO ligands are stable and can be integrated into the structure [96].
- *Silicon-manganese*. Si-Mn alloys can be fabricated by using the $SiH_3Mn(CO)_5$ precursor [97]. In this case, amorphous Mn_2Si grains are embedded in a carbonaceous matrix, with relative compositions of approximately 35 at. % of Mn and 18 at. % of Si, the rest being C and O. As has been observed in other FEBID materials, whereas the composition has been found to be independent of the electron beam voltage, the total metallic composition was improved from ~40 at. % up to ~53 at. % by increasing the electron beam current one order of magnitude, as well as the Mn/Si ratio, indicating a loss of Si in the fabrication process. These deposits present a quasi-metallic temperature-dependent behavior, with a resistivity value of 1.5×10^4 μΩ cm at room temperature.

4.3.2 Improving properties

The low purity and the poor crystallinity are key limiting factors for capitalizing the full potential of FEBID. In most cases, the purity of the deposits is not enough to ensure a metallic behavior or the good performance of the magnetic, optical or superconducting properties. To face this limitation, diverse purification protocols have been developed to increase the purity of FEBID deposits, the general strategy being a multi-step synthesis approach [64]. In the following, details are provided on the purification methods and their degree of success when applied to various materials.

Purification methods. Various successful methods have been employed in order to increase the purity of the element of interest, in general a metallic one, and to decrease the unwanted elements, either originating from the precursor molecules or from the chamber contamination. The most successful ones are as follows:

- *In situ* [98, 99] and *ex situ* [100–102] post-annealing treatments at high vacuum and under controlled reactive gas atmospheres [103]. This procedure is performed inside an SEM or TEM equipment using a heating stage and controlling very precisely the temperature, the ramp rate and the annealing time after stabilization.
- The use of substrates at high temperatures during growth [54, 104]. In addition to the preservation of the deposit geometry, the metallic purity of

the structures when using this method is enhanced, sometimes at the cost of a lower growth rate.

- Post-growth electron beam irradiation [105, 106] and ion beam irradiation [107]. Both strategies point towards a strong improvement of the electrical resistivity of the as-grown deposits.
- Laser-assisted heating during deposition [108, 109]. The support of a synchronized pulsed laser gives rise to laser-assisted electron-beam-induced deposition (LAEBID), which has been demonstrated to produce thermal desorption of the reaction by-products leading to an increase of the metallic content.
- Post-growth Joule heating upon injection of high electric currents [110]. This method has also been employed to promote a different microstructure, e.g., from as-grown nanogranular crystal phases to a polycrystalline one when increasing the current.
- Supersonic jet delivery of precursor gas [111]. The high kinetic energy of the inert carrier gas jet stimulates the desorption of the decomposition residues, inducing an enhancement of the metallic purity. The highly-localized precursor delivery also increases the deposition yield.
- Combinations of all previous methods [112–115]. The exploration of post-growth heating annealing in combination with gas atmospheres and electron beam irradiation has been performed in order to improve the metallic content of the as-deposited structures.

Purified materials.
Some important materials are discussed hereafter:

- *Platinum.* In order to increase the metallic content during growth, existing strategies lie in the use of more than one gas during the deposition. The prime example consists of Pt by using $(CH_3)_3Pt(CpCH_3)$ and O_2 injection simultaneously, where metallic contents near 100 at. % and resistivity down to 60 μΩ cm have been found at room temperature [112]. In fact, it is interesting to highlight that the tuning of the precursor gas temperature leads to different fluxes. In these experiments, the resistivity was reduced from 88 μΩ cm to 60 μΩ cm by decreasing the GIS temperature from 40 °C to 28 °C, obtaining pure structures. Also, an additional knob was used, based on the improvement of the purification of the Pt structures by increasing the O_2/precursor flux ratio (always above 1×10^4) or increasing the beam current for a fixed flux ratio. The PE, SE and BSE are quite likely to ionize O_2, prone to react with C residues into volatile species. Finally, it is also worth mentioning the astounding success achieved in growing virtually-pure Pt deposits by post-growth electron irradiation in O_2 atmosphere [103].
- *Gold.* The growth of Au structures with metallic purity higher than 90 at. % has been performed by a water-assisted process in combination with Au $(CH_3)_2(C_2ClF_3O)$ [116]. The use of water as an oxidative enhancer during the deposition has enabled the formation of highly pure Au nanowires with resistivity of 8.8 μΩ cm and associated current density of 715 kA mm^{-2}.

The atomic Au composition can be adjusted by modifying the water flux. It has been found that, although the total thickness of the deposit drops with increasing water addition, the relative contribution of Au to the structure volume increases. It is suggested that water favors the oxidation of the carbonaceous part of the deposits as well as the residual gas compounds in the chamber, improving the efficiency of the Au nanofabrication process since a higher number of adsorption sites are available for the Au precursor gas molecules. Also, other studies support the possibility of fabricating functional Au plasmonic nanostructures by electron beam irradiation in water vapor atmosphere [117].

- *Other metallic materials*:

  In order to improve the electrical properties, different experiments in deposits based on the $SiH_3Mn(CO)_5$ precursor gas revealed that the low-energy electron irradiation allows the tuning of the conductivity. Specifically, recent experiments show the dependence of the resistivity with the irradiation dose. Deposits undergoing 44 nC μm^{-2} dose show resistivity values around 1.5×10^4 $\mu\Omega$ cm at room temperature, this parameter decreasing with the irradiation dose [97]. Using an electron irradiation of 1000 nC μm^{-2} the resistivity reaches 1.9×10^3 $\mu\Omega$ cm, and further irradiation gives rise to a saturation in the conductivity. However, even though this value is two orders of magnitude lower than the one obtained for FeRu [96] and CoSi [118], it is still almost 50 times larger than the CoFe-FEBID one [93].

- *Magnetic materials*:

  Deposits based on $Co(CO)_3NO$ precursor gas were performed on substrates at low temperatures to understand the decomposition process [119]. Firstly, the NO and part of the CO ligands are decomposed, giving rise to a decarbonylated intermediate. Then, the remaining CO groups can be decomposed by the electron beam, increasing the C content of the structure while Co oxidized atoms turn up. At higher temperatures, the thermal reactions lead to C removal, therefore improving the metallic purity. In subsequent studies, it has been demonstrated that the deposition yield increases with the substrate at 130 °C, nevertheless maintaining their high resistivity of 6.3×10^3 $\mu\Omega$ cm at room temperature with a metal-insulator behaviour. The Co composition is always at 50 at. %, after 400 °C annealing, the resistivity decreases two orders of magnitude down to 63 $\mu\Omega$ cm, showing a clear metallic performance. In addition, some of the initial CoO fcc nanometric grains were transformed into Co hcp grains, with sizes between 10–15 nm, which facilitate the connection between Co grains after the thermal treatment, eventually enhancing the electrical properties [77, 120].

  On the other hand, $Co(CO)_3NO$-based as-grown nanowires with moderate atomic content of Co, ranging 50%–55%, present non-metallic and non-magnetic behavior even at low temperatures [120]. However, after performing an annealing treatment at 400 °C in vacuum conditions, the resistivity of the nanostructures decreases down to 62 $\mu\Omega$ cm, corresponding to a metallic behavior at low temperature, retrieving the ferromagnetic properties.

In the deposits formed from $HCo_3Fe(CO)_{12}$, compositional analysis showed that carbonyl groups, and to a lesser extent the O, are removed from the deposit by *ex situ* high-vacuum post-growth annealing process. The atomic composition increases from Co:Fe:C:O = 41:13:26:20 for the as-deposited nanostructures up to 52:17:22:9 after the thermal treatment. In addition, the amorphous C microstructure also changes into graphite crystals increasing the temperature up to 300 °C, and the semiconducting behavior turns into the metallic one, where the resistivity is reduced by around two orders of magnitude to 65 μΩ cm [95].

The widely-used purification strategies have been generally applied to 2D deposits, but also purification of 3D objects has been attempted in vertical Co and Fe nanowires based on $Co_2(CO)_8$ and $Fe_2(CO)_9$ [121, 122]. The 3D case presents specific problems: for instance, the architecture stability can be compromised due to the high-volume shrinkage occurring in low metallic content deposits, eventually leading to a substantial modification of the object shape or even the collapse of the 3D structures [123] (figure 4.7).

- *Superconducting materials*:

  The fabrication of superconducting structures at the nanoscale has been generally tackled by focused ion beam induced deposition (FIBID) [124], even though the physical damage and ion doping generated by this method would prompt considering FEBID as an optimal approach. However, significantly fewer studies have focused their attention on FEBID superconducting materials because of the lack of reproducibility and difficulties in

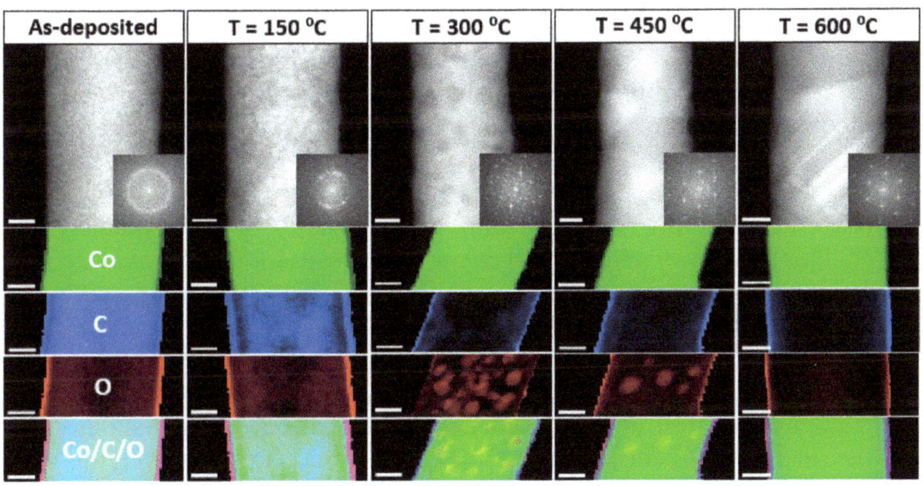

Figure 4.7. Scanning transmission electron microscopy images of as-deposited and annealed 3D Co nanowires under a post-growth purification process (first row of images). In the insets, the corresponding fast-Fourier transform of the images reveals the increase of the crystallinity as a function of the temperature. Chemical maps (following rows) show the spatial distribution of the relative composition of Co, C and O, increasing the metallic content from ~72 at. % for the as-deposited nanowire to ~93 at. % for the 600 °C-annealed one. Scale bars are 20 nm. Adapted with permission from [121]. Copyright 2018 American Chemical Society.

the growth conditions. Nevertheless, Makise *et al* reported for the first time the fabrication of MoC nanowires using $Mo(CO)_6$ in combination with H_2O. In these samples, superconductivity is suggested to be present due to the resistivity decrease below 10 K [125]. Additionally, as previously mentioned, Winhold *et al* showed the growth and characterization of Pb nanostructures using $(CH_3CH_2)_4Pb$ with metallic behavior, with the resistivity dropping to 16 μΩ cm at 9 K, very close to the superconducting transition temperature (7.2 K) [89]. Finally, Sengupta *et al* reported the growth of W nanowires with critical temperature of 2 K [72].

Multi-precursor deposition. The use of more than one precursor gas has allowed the synthesis of binary metallic alloys with outstanding properties. Some examples are considered below:

- $[CoFeSi_2]_n$ multilayers can be prepared by alternating different precursors using a multi-channel GIS [126], where $HCo_3Fe(CO)_{12}$ and $Fe(CO)_5$ are combined simultaneously to obtain deposits with a Co/Fe ratio of 2. Also, by depositing a Si-based dielectric layers with Si_5H_{12} in between the CoFe ones, $CoFeSi_2$ Heusler compounds are grown.
- *Fe–Si* binary compounds have been fabricated by the alternating $Fe(CO)_5$ and Si_5H_{12} as precursor gases. The combination of post-growth annealing and electron-beam irradiation in Fe/Si multilayers has led to FeSi and Fe_3Si metallic binary compounds by the intermixing of Fe and Si atoms [85].
- *Au–Fe* nanoalloys were grown under consecutive deposition of Au and Fe using $C_7H_{10}AuF_3O_2$ and $Fe(CO)_5$ precursor gases, respectively [127].
- *CoSi–C* binary alloys were synthesized using $Co_2(CO)_8$ and Si_5H_{12} precursor gases simultaneously, with the presence of CoSi and $CoSi_2$ nanoclusters embedded in a carbonaceous matrix in the final deposit. The resistivity can be modulated between 7×10^3 μΩ cm and 2.8×10^4 μΩ cm controlling the Si_5H_{12} flux or, in other words, the Si/Co ratio [118].
- *CoPt–C* structures arisen from the combination of $Co_2(CO)_8$ and $(CH_3)_3CH_3C_5H_4Pt$ present a total metallic purity of approximately 29 at. %. This quite low relative metal composition leads to resistivity values around 10^5 μΩ cm. As a result of the growth, the deposits were composed by CoPt nanoparticles embedded in a carbonaceous matrix. Following an electron irradiation approach, the low purity and high resistivity of the nanostructures based on $Co_2(CO)_8$ and $(CH_3)_3CH_3C_5H_4Pt$ can be improved. The CoPt nanoparticles turn into face-centered tetragonal $L1_0$ nanocrystallites, while the resistivity is reduced from 10^5 μΩ cm to 10^3 μΩ cm. Moreover, this specific system, which is superparamagnetic in the as-grown state, becomes ferromagnetic at room temperature after the irradiation treatment. Along this line, the accurate control of the electron dose permits the adjustment of the electrical and magneto-transport properties [128].
- *FePt–C* nanorods were prepared using $Fe(CO)_5$ and $(CH_3)_3(C_5H_5)Pt$ precursor gases. These FePt–C deposits in the form of nanowires were submitted

to post-growth annealing treatments at 600 °C for 120 min, leading to a crystallization process. Several crystalline FePt alloy nanoparticles with face-centered tetragonal microstructure and surrounded by a C shell constitute the composition of the nanostructures [129].

- *PtSi–C* alloys with $CH_3CpPt(CH_3)_3$ and $Si(SiH_3)_4$ precursors have been grown [130]. The composition can be progressively tuned by varying the precursor flux. Although the highest metallic content was found to be ~36 at. % Si, ~13 at. % Pt, ~47 at. % C and ~4 at. % O, the minimum resistivity was achieved for a sample with metallic content of ~42 at. %, when the Si/Pt ratio was 1.5. However, even at ~20 K this sample did not show a metallic behavior.
- *Co–C* deposits were obtained using $Co_2(CO)_8$ and $C_8H_{17}OH$ precursors. Utke *et al* showed that in multi-precursor FEBID deposition, the compositional variation is driven by the adsorption and supply of the adsorbates, together with the electron beam impact efficiency [68]. Thus, the control of the purity in metallic nanostructures to achieve customized physical properties was shown in this material. By means of a two-adsorbate model, the tuning of deposit composition by pulsed electron beams process was demonstrated.

4.4 Applications

FEBID is becoming a key lithographic technique in nanotechnology and materials science [131, 132]. Its unique capability for the growth of different types of materials has been exploited in a broad range of applications, such as integrated circuit edit and mask repair [133], creation of electrical contacts [134], growth of magnetic nanowires [135], fabrication of plasmonic nanostructures [117], photodetection [136] or gas sensing [137]. The last 35 years have witnessed the increasing interest of the scientific community and industry in the development of advanced nano-objects grown by FEBID for various applications, which can be classified attending to the material's nature. Following this criterion, the nanostructures can be classified in metallic, superconducting, magnetic, photonic and plasmonic, structural, etc.

From all the applications arising from table 4.2, we are going to single out the most relevant ones today:

- *Mask repair*. Highly-challenging critical features are required in the lithography and even very small defects should be minimized during the fabrication. One of the most demanding FEBID-based processes is photolithography mask repair, allowing the removal of surplus mask material and addition of the missing material. Binary, attenuated phase-shift or extreme ultraviolet and nanoimprint lithography masks can be repaired overcoming charging or contamination issues and ensuring cleaning lifetime [133]. For example, protective Ru films have been used for nanoscale repairs by deposition and post-purification process based on electron irradiation in combination with O_2 atmosphere [90]. Pt deposits based on $(CH_3)_3Pt(CpCH_3)$ have been used for lithography mask repair [141, 142].

Table 4.2. FEBID materials and applications examples.

Type of material	*Property of interest*	*Applications*
Metallic	Conductivity	Electrical contact to nano-objects [138]
Superconducting	Null electrical resistance, superconducting vortices	Electrical contacts to nano-objects, fluxtronics [125]
Magnetic	Magnetization, magnetic forces, domain walls	Magnetic sensing, actuation and storage [61]
Photonic and plasmonic	light–matter interaction	Optical antenna, filters [22]
Structural	Mechanical properties	Nanotools, mechanical support, templating, nanoactuators, tips [139]

- *Tips for scanning probe microscopy.* 3D FEBID deposits have been used for general scanning probe microscopy (SPM) operation modes, such as conductive-atomic force microscopy (AFM), electrostatic force microscopy, Kelvin force microscopy or scanning thermal microscopy. High-resolution capabilities were achieved by Pt–C, W, Au-based tips and their purified versions thanks to the narrow diameters and especially due to the small and sharp tip ends (<10 nm) [143–145]. In this line, the growth of magnetic Co and Fe nanowire tips has also shown its potential application in magnetic force microscopy (MFM) [58]. The superior behaviour of FEBID tips with respect to standard tips has recently opened the door to the next generation of commercial MFM tips. These tips work properly in ambient conditions as well as in liquid environment, behaving appropriately in terms of mechanical stability, resolution and sensitivity [145]. Besides, tetrapod structures have proven to be an efficient method for localized thermal sensing, where the fast temperature response enables dynamic sensing [146]. In addition, the fabrication of nano-spheres on top of AFM cantilevers has shown powerful sensitivity in ferromagnetic resonance force microscopy as they minimize the remanent effects when measuring magnetic hysteresis [62]. In the field of nano-optics, scanning near-field optical microscopy (SNOM) [147], tip-enhanced Raman scattering nano-probes [148] and AFM-based infrared spectroscopy [149] have considered the use of FEBID pillars as potential basic elements.
- *Electrical contacts to nano-objects.* Due to the low resistivity values presented by some metallic materials grown by FEBID, their implementation in nanoelectronic devices is a significant step forward. The growth of reliable metallic connections for electrical characterization is a routine method performed by FEBID. Commonly, Au [150], W [151] and Pt [152] contacts have been used, minimizing the damage caused by other techniques such as FIBID. In the case of superconducting materials, FEBID nanostructures were found interesting only in MoC and $Pb_xC_yO_\delta$ materials [89]. In the latter,

since the resistivity value at room temperature is seven times larger than the bulk Pb, this compound is a potential candidate for the fabrication of metallic contact electrodes.

- *Nanophotonic applications.* The FEBID community has focused much attention on optical applications. These can be divided according to the dimensional extension of the nanostructures. In the case of planar or 2D nano-objects, photonic nanostructures have been fabricated, for example by Au in the form of disks, or nanorods to be used as antennas [22]. In this sense, a comparison between FEBID and other lithography techniques such as EBL results in a similar qualitative behavior. W and SiO_x elements can also be used in photonic crystals for infrared filters [153] and plasmonic antennas for SNOM [154]. Even Ag nanostructures have been grown for plasmonic applications in the visible range [84]. On the other hand, 3D nanodeposits with strong plasmonic activity can be fabricated by Pt–C [155] and pure Au, the latter acting as nanoantennas or optical filters [65, 117].
- *Magnetic applications (beyond tips).* Magnetic nanostructured materials attract particularly great interest because of their possibilities to be implemented in future spintronic nanodevices. Specifically, they are potential candidates for fast and low-power electronics based on domain wall conduit for storing and handling information [156–158]. In addition, Co magnetic nano-objects grown by FEBID have also been used for the fabrication of electronic devices, e.g., nano-Hall sensors [20, 48]. The step towards 3D nanostructures has given rise to nanoactuators controlled by the application of external magnetic fields [21], artificial spin ice systems [159], complex architectures based on hook-shaped nanowires aimed for domain wall conduit devices [60] or structures where the control of the magnetic chirality is possible [160]. Other materials such as Mn–Si nanostructures have proven to be appropriate systems to study helimagnetism and skyrmionics [97], opening a new pathway for application development. On the other hand, FEBID can also be used for studies devoted to biomolecule immobilization, such as the case of CoNi nanoparticles attached to the substrate using a 30 keV electron beam [161].
- *Nanotools and other structural applications.* FEBID versatility has allowed the development of strategies aimed at fabricating and customizing nanotweezers, enabling their manipulation by applying a voltage [139], as shown in figure 4.8(e). Also, non-magnetic 3D scaffolds have been grown by FEBID, and subsequently covered by a thin film of evaporated magnetic material to bring functionality [162]. In addition, FEBID Fe particles have been used as catalysts for the site-specific synthesis of C nanotubes, which can give rise to the production of novel nanocircuits or biosensors [163], Pt deposits have been used for the fabrication of DNA tunneling detectors [164], and systems based on Cu have been built and characterized, showing their applicability as mechanical devices [23].

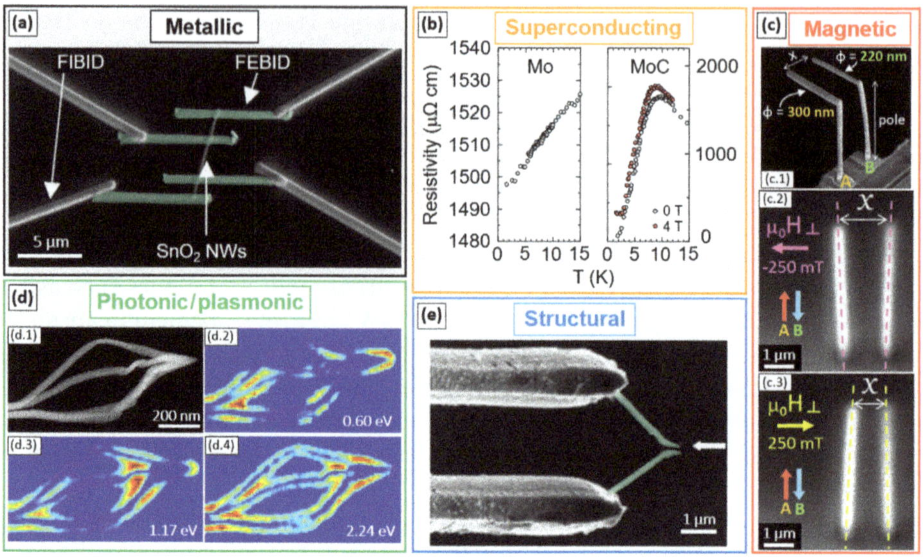

Figure 4.8. Examples of applications based on (a) metallic contacts aiming to protect the SnO_2 nanowire from ion damage, (b) superconducting crystalline Mo and amorphous MoC nanowires, respectively, (c) remote magnetomechanical nanoactuators based on Co nanowires, (d) free-standing Au plasmonic architectures and (e) structural carbonaceous materials used as nanotweezers with a gap of 25 nm. (a) Adapted with permission from [140]. Copyright 2006 Elsevier. (b) Adapted with permission from [125]. Copyright 2015 Springer Nature. (c) Adapted with permission from [21]. Copyright 2016 Wiley-VCH Verlag GmbH & Co. KGaA, Weinheim. (d) Reprinted with permission from [117]. Copyright 2017 American Chemical Society. (e) Adapted with permission from [139], copyright IOP Publishing.

4.5 Additional FEBID strategies and novel developments

Leveraging from the FEBID general framework presented so far, other FEBID-related approaches have been explored. Here, we will focus our attention on six derived strategies: (i) FEBID assisted by an electrically-biased patterned metallic structure, (ii) FEBID under cryogenic conditions, called cryo-FEBID [165], (iii) growth after electron-beam induced surface activation, called EBISA [166], (iv) FEBID in liquid environment [167], (v) combined use of FEBID and atomic layer deposition (ALD) [168], and (vi) the subtractive nanofabrication method known as focused electron beam induced etching (FEBIE) [169].

- *FEBID assisted by an electrically-biased patterned metallic structure.* This approach arises from the difficulties in the use of FEBID on insulating substrates due to detrimental charging effects which ruin the resolution of the beam. To overcome these issues and also to allow for the *in situ* modulation of the geometry of the structures, an electrically-biased metallic structure patterned on the surface of the substrate is used, giving rise to the so-called Architectural Adjustment by Grid Overlay Nanotechnology-Chip (ARAGON-Chip). As shown in figure 4.9, a perforated metallic layer enables the evacuation of the charge produced during the nanostructure growth. On the other hand, the tunable electric fields produce a significant impact on the trajectories of the PE

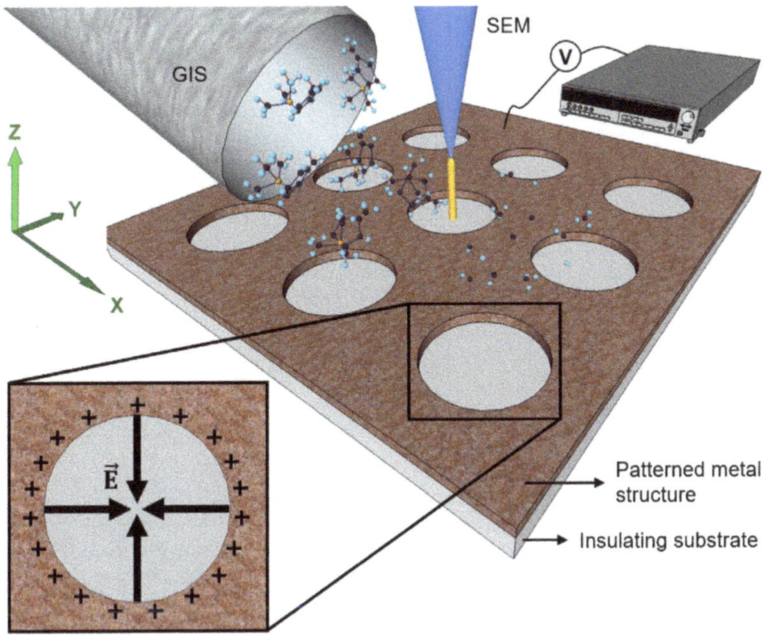

Figure 4.9. FEBID process carried out on an insulating substrate with an electrically-biased patterned metallic structure (ARAGON-Chip). Example with the $(CH_3)_3Pt(CpCH_3)$ molecule. Reprinted with permission from [170], copyright IOP Publishing.

and the generated SE, leading to the dimensional customization of the objects, especially the 3D ones such as vertical nanowires [170].

- *Cryo-FEBID*. This technique relies on the condensation of the precursor material on the surface of the substrate prior to the exposure to the focused electron beam. After exposure, the condensate layer is heated up to a temperature slightly above room temperature (e.g., ~50 °C), to reveal the formation of the deposit in the irradiated areas, whereas the unexposed ones become volatile. The most important advance of cryo-FEBID is the much higher growth rate in comparison with the original FEBID approach. In fact, the growth rate has been found to be up to four orders of magnitude higher compared to room temperature deposition. Owing to the need of a much smaller electron irradiation dose to reveal the deposit, the FEBID process speed is significantly enhanced. The origin of this phenomenon is that in cryo-FEBID the electron irradiation is more likely to break precursor molecules in the thick (a few hundreds of nm) condensed precursor layer than in FEBID, where the precursor molecules are in the gas phase and form an adsorbed monolayer on the substrate (figure 4.10).

Bresin *et al* proposed for the first time the growth of a material by cryo-FEBID using $(CH_3)_3Pt(CpCH_3)$ precursor gas onto a substrate at −155 °C, exposing the condensate and facilitating the desorption of the precursor molecules in non-irradiated regions by heating the substrate to room

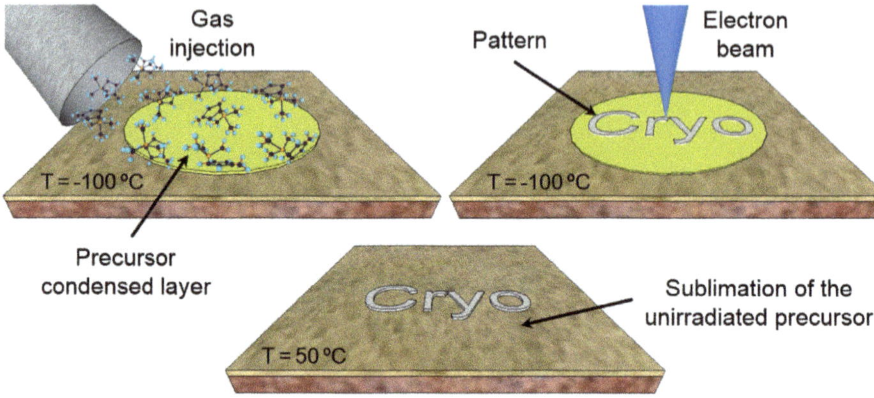

Figure 4.10. Diagram of the cryo-FEBID process. Example with the $(CH_3)_3Pt(CpCH_3)$ molecule.

temperature [171, 172]. In this study, the thickness of the condensed precursor layer was modified between 75 and 300 nm by tuning the GIS temperature, distance to the substrate and valve opening time. The lateral resolution strongly depends on the thickness of the condensed layer. High lateral resolution of 40 nm was reported with a condensed layer of 70 nm in thickness. However, the use of condensed layers around 280 nm worsened the resolution 3.5 times [172]. Moreover, consecutive cryo-FEBID depositions of different layers can be carried out to fabricate 3D structures or to combine different materials and thicknesses. This provides great versatility and opens many possibilities in terms of applications. For instance, mask repair, etching and large-scale patterning processes could benefit from this technique [173], as well as the fabrication of metallic contacts.

- *EBISA.* This technique consists of a previous patterning by focused electron beam over the surface of the substrate at high vacuum conditions with no precursor gases introduced in the chamber. The areas scanned by the electron beam become chemically active for the subsequent decomposition of the precursor material. When the precursor gas is locally delivered onto the substrate surface, autocatalytic growth takes place as the decomposition only occurs at the irradiated areas [174]. Similarly to standard FEBID, the nanostructure size is determined by the electron beam scan as well as the interaction between the beam and the substrate. Different substrates such as SiO_x or TiO_2 are commonly used [175, 176]. Also, additional steps can follow the first deposition by introducing different gases and generating thicker specimens in the EBISA areas (figure 4.11).
- *FEBID in liquid environment.* This strategy consists of the decomposition of solvent molecules by the electron beam, producing solvated electrons and radicals in a domino-like effect [177]. In the standard process, the use of some cells to retain the liquid is the most straightforward way to work under vacuum (figure 4.12).

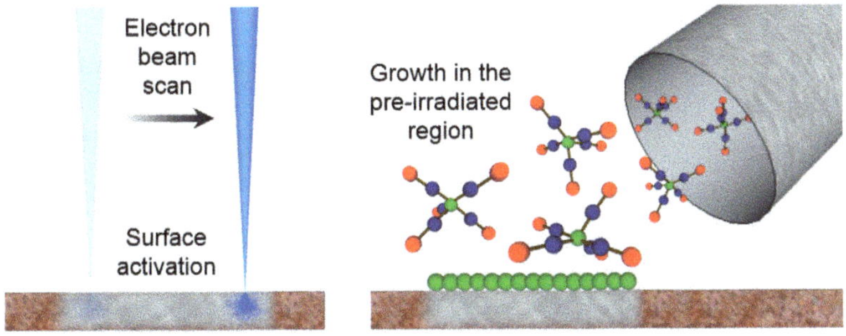

Figure 4.11. Diagram of the EBISA process. Example with the $Fe(CO)_5$ molecule.

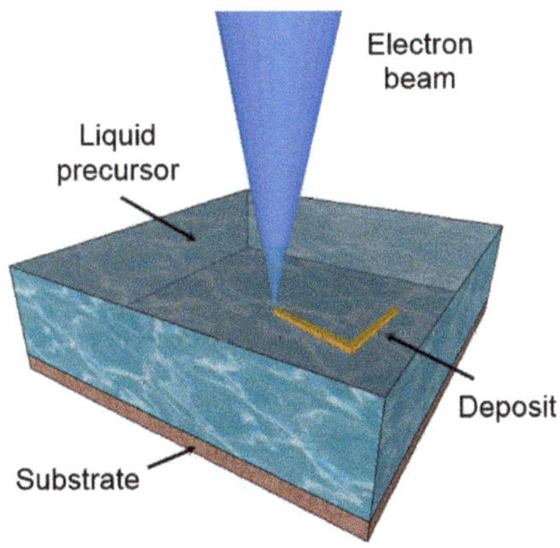

Figure 4.12. Diagram of the FEBID in liquid environment process.

Due to the need of removal of high levels of contaminants in some metallic deposits, liquid-phase processes provide an effective solution to obtain high purity nanostructures. Several materials such as Pt [167, 178], Au [179], Ag [180], Pd [181] or Cu [182] have been used to fabricate nanostructures by this approach. For instance, in the last case, using aqueous solutions of copper sulfate as precursor, with sulfuric acid which induces the elimination of oxygen, deposits of 95 at. % of Cu have been grown [183]. Apart from the high metallic content of the structures, growth rates up to 10^5 times greater than that of standard gas-phase process can be reached.

- *Combination of FEBID and atomic layer deposition.* The combination of FEBID with other lithography or deposition techniques could extend the capabilities of both methods. One of the examples is the conjunction between FEBID and area-selective ALD [168], the so-called direct-write ALD. In several studies, firstly the patterns made of a FEBID seed layer with low Pt contents are deposited, then this metallic part is activated locally for the ALD growth. The method consists in a multi-cycle process involving the alternate surface exposure to two different gas reactants, as shown in figure 4.13 [184]. This cycle approach opens up the possibility to fabricate layers or multilayers

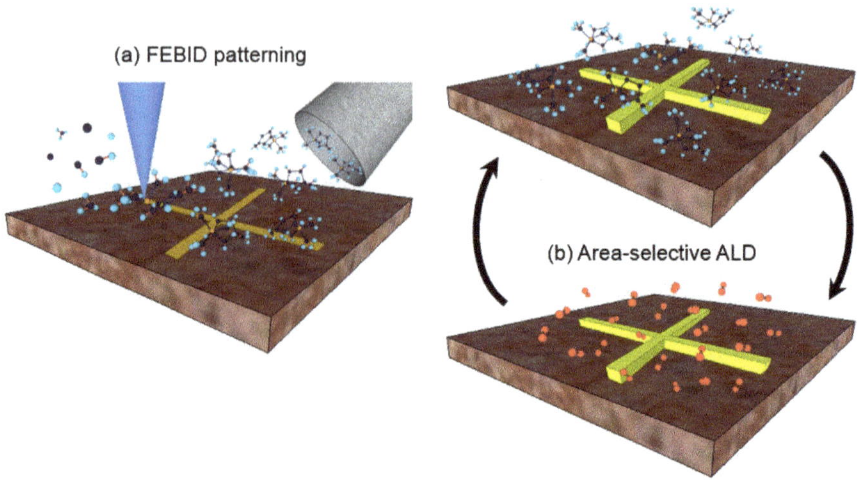

Figure 4.13. Diagram of the direct-write ALD process. Example with $(CH_3)_3Pt(CpCH_3)$ and O_2 molecules.

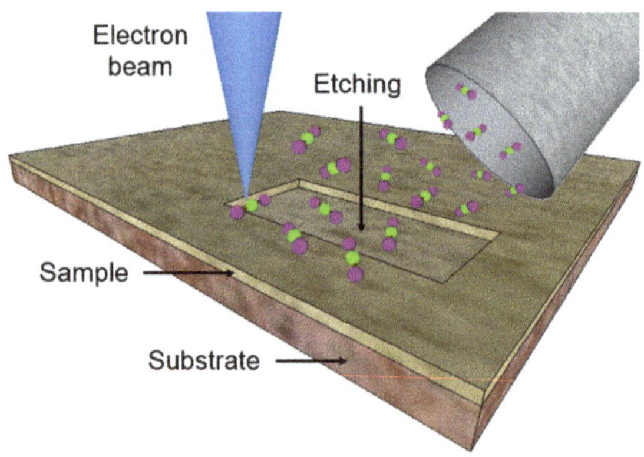

Figure 4.14. Diagram of the FEBIE process. Example with the XeF_2 molecule.

of different materials. For example, selective Pd or Ru deposition on Pt-FEBID seed layer has been demonstrated [185, 186]. Thanks to this innovative strategy, high lateral resolution (down to 10 nm), high chemical purity together with high conductivity, and great thickness control are provided. Moreover, the low electron dose (~10 pC μm^{-2}) and the absence of resists allow patterning on sensitive materials such as graphene or C nanotubes, improving the wetting interaction with respect to other methods [187, 188].

- *FEBIE.* Focused electron beam induced etching consists of a direct-write subtractive nanofabrication technique where the combination of an electron beam and a reactive gas is used to perform chemical dry etching with spatial resolution down to 10 nm [189, 190]. The typical precursor gas molecules used are H_2O, O_2, Cl_2, NH_3, NOCl, SF_6 and XeF_2, among others [169]. In comparison with FIB, the use of an electron beam prevents beam damage and redeposition. This confers the technique a unique behavior for applications such as nanostructures edition [191], photolithographic mask repair [192], upgrading of tips devoted to SPM [193] or fabrication of nanometric pores in membranes [194]. Besides, FEBIE has been widely employed to enhance metallic purity of as-deposited FEBID nanostructures composed of Pt [195], Au [196], Fe_3O_4 [197] or SiO_2 [198] materials (figure 4.14).

4.6 End-of-chapter summary

Over the last decades, FEBID has demonstrated unprecedented capabilities for the growth of customized nanostructured materials. The key virtues of this technique, i.e., a maskless, resist-free direct-writing nanolithography, enable the design of complex motifs on virtually any surface, and an easy adaptation to 3D printing of nanostructures nowadays out of reach for most nanofabrication techniques. This adaptability is sustained on the extensive repertoire of gas precursors available nowadays. Some success cases of FEBID materials are particularly astounding. Platinum precursor is routinely used as structural material, for 3D prototyping and scaffolding as well as for nanocontacting. High purity cobalt has become the standard for ferromagnetic FEBID nanostructures, with proven functionality for scanning probe tip functionalization and sensing. Pure gold can be fabricated by FEBID, with great promise for the design of custom-tailored nanophotonics devices. Finally, further progress on FEBID technology is currently underway. Some exciting developments described in this chapter, including FEBID in extreme environments (cryogenic or liquid media), in combination with other techniques (ALD), reactive etching or autocatalytic FEBID growth, are aimed to boost FEBID performance and throughput, and open new pathways in the competitive market of nanofabrication.

4.7 Student's corner

- A video animation on the FEBID process can be visualized at https://iopscience.iop.org/book/978-0-7503-2608-7.
- Which growth variables have an influence on the lateral resolution of FEBID, and why?
- Which measures would you take to reduce the halo around FEBID deposits?
- Are insulating substrates suitable for FEBID growth?
- Growth rate is one of the key issues of FEBID. For a given electron dose, which growth parameters could you modify to increase the growth rate?

Acknowledgements

The authors acknowledge financial support from the Spanish Ministry of Economy and Competitiveness through the projects MAT2018-102627-T, MAT2017-82970-C2-1-R, MAT2017-82970-C2-2-R, from the Aragon Regional Government (Construyendo Europa desde Aragón) through the project E13_20R, with the European Social Fund funding, from CSIC through the project PIE 202060E187, and from the European's Union Horizon 2020 research and innovation programme under Grant No. 823717-ESTEEM3.

References

[1] Lopes W A and Jaeger H M 2001 Hierarchical self-assembly of metal nanostructures on diblock copolymer scaffolds *Nature* **414** 735–8

[2] Bratton D, Yang D, Dai J and Ober C K 2006 Recent progress in high resolution lithography *Polym. Adv. Technol.* **17** 94–103

[3] Park C, Yoon J and Thomas E L 2003 Enabling nanotechnology with self assembled block copolymer patterns *Polymer* **44** 6725–60

[4] Chou S Y, Wei M S, Krauss P R and Fischer P B 1994 Single-domain magnetic pillar array of 35 nm diameter and 65 Gbits/in^2 density for ultrahigh density quantum magnetic storage *J. Appl. Phys.* **76** 6673–5

[5] Nicolau D V, Taguchi T, Taniguchi H and Yoshikawa S 1999 Negative and positive tone protein patterning on E-beam/deep-UV resists *Langmuir* **15** 3845–51

[6] Morimoto H, Sasaki Y, Saitoh K, Watakabe Y and Kato T 1986 Focused ion beam lithography and its application to submicron devices *Microelectron. Eng.* **4** 163–79

[7] Herndon M K, Collins R T, Hollingsworth R E, Larson P R and Johnson M B 1999 Near-field scanning optical nanolithography using amorphous silicon photoresists *Appl. Phys. Lett.* **74** 141–3

[8] Fischer J and Wegener M 2013 Three-dimensional optical laser lithography beyond the diffraction limit *Laser Photonics Rev.* **7** 22–44

[9] Campbell M, Sharp D N, Harrison M T, Denning R G and Turberfield A J 2000 Fabrication of photonic crystals for the visible spectrum by holographic lithography *Nature* **404** 53–6

[10] Savas T A, Farhoud M, Smith H I, Hwang M and Ross C A 1999 Properties of large-area nanomagnet arrays with 100 nm period made by interferometric lithography *J. Appl. Phys.* **85** 6160–2

[11] Wu B and Kumar A 2007 Extreme ultraviolet lithography: a review *J. Vac. Sci. Technol.* B **25** 1743–61

[12] Miramond C, Fermon C, Rousseaux F, Decanini D and Carcenac F 1997 Permalloy cylindrical submicron size dot arrays *J. Magn. Magn. Mater.* **165** 500–3

[13] Djomehri I J, Savas T A and Smith H I 2015 Zone-plate-array lithography in the deep ultraviolet *J. Vac. Sci. Technol.* B **16** 3426–9

[14] Wilbur J L, Kumar A, Biebuyck H A, Kim E and Whitesides G M 1996 Microcontact printing of self-assembled monolayers: applications in microfabrication *Nanotechnology* **7** 452–7

[15] Cerf A, Thibault C, Geneviève M and Vieu C 2009 Ordered arrays of single DNA molecules by a combination of capillary assembly, molecular combing and soft-lithography *Microelectron. Eng.* **86** 1419–23

[16] Chou S Y, Krauss P R and Renstrom P J 1996 Nanoimprint lithography *J. Vac. Sci. Technol.* B **14** 4129–33

[17] Bailey T, Choi B J, Colburn M, Meissl M, Shaya S, Ekerdt J G, Sreenivasan S V and Willson C G 2000 Step and flash imprint lithography: template surface treatment and defect analysis *J. Vac. Sci. Technol.* B **18** 3572–7

[18] Wadu-Mesthrige K, Xu S, Amro N A and Liu G-Y 1999 Fabrication and imaging of nanometer-sized protein patterns *Langmuir* **15** 8580–3

[19] Piner R D, Zhu J, Xu F, Hong S and Mirkin C A 1999 'Dip-Pen' nanolithography *Science* **283** 661–3

[20] Gabureac M, Bernau L, Utke I and Boero G 2010 Granular Co–C nano-Hall sensors by focused-beam-induced deposition *Nanotechnology* **21** 115503

[21] Vavassori P, Pancaldi M, Perez-Roldan M J, Chuvilin A and Berger A 2016 Remote magnetomechanical nanoactuation *Small* **12** 1013–23

[22] Graells S, Aćimović S, Volpe G and Quidant R 2010 Direct growth of optical antennas using e-beam-induced gold deposition *Plasmonics* **5** 135–9

[23] Friedli V, Utke I, Mølhave K and Michler J 2009 Dose and energy dependence of mechanical properties of focused electron-beam-induced pillar deposits from Cu $(C_5HF_6O_2)_2$ *Nanotechnology* **20** 385304

[24] Huth M, Porrati F, Schwalb C, Winhold M, Sachser R, Dukic M, Adams J and Fantner G 2012 Focused electron beam induced deposition: a perspective *Beilstein J. Nanotechnol.* **3** 597–619

[25] Randolph S J, Fowlkes J D and Rack P D 2006 Focused, nanoscale electron-beam-induced deposition and etching *Crit. Rev. Solid State Mater. Sci.* **31** 55–89

[26] Utke I, Hoffmann P and Melngailis J 2008 Gas-assisted focused electron beam and ion beam processing and fabrication *J. Vac. Sci. Technol.* B **26** 1197–276

[27] van Dorp W F and Hagen C W 2008 A critical literature review of focused electron beam induced deposition *J. Appl. Phys.* **104** 081301

[28] Broers A N, Molzen W W, Cuomo J J and Wittels N D 1976 Electron-beam fabrication of 80-Å metal structures *Appl. Phys. Lett.* **29** 596–8

[29] Matsui S and Mori K 1984 New selective deposition technology by electron beam induced surface reaction *Jpn. J. Appl. Phys.* **23** L706–8

[30] Matsui S and Mori K 1986 New selective deposition technology by electron-beam induced surface reaction *J. Vac. Sci. Technol.* B **4** 299–304

[31] van Kouwen L, Botman A and Hagen C W 2009 Focused electron-beam-induced deposition of 3 nm dots in a scanning electron microscope *Nano Lett.* **9** 5149–2152

[32] Fowlkes J D, Winkler R, Lewis B B, Stanford M G, Plank H and Rack P D 2016 Simulation-guided 3D nanomanufacturing via focused electron beam induced deposition *ACS Nano* **10** 6163–72

[33] Plank H, Smith D A, Haber T, Rack P D and Hofer F 2012 Fundamental proximity effects in focused electron beam induced deposition *ACS Nano* **6** 286–94

[34] Winkler R, Szkudlarek A, Fowlkes J D, Rack P D, Utke I and Plank H 2015 Toward ultraflat surface morphologies during focused electron beam induced nanosynthesis: disruption origins and compensation *ACS Appl. Mater. Interfaces* **7** 3289–97

[35] Winkler R, Fowlkes J, Szkudlarek A, Utke I, Rack P D and Plank H 2014 The nanoscale implications of a molecular gas beam during electron beam induced deposition *ACS Appl. Mater. Interfaces* **6** 2987–95

[36] Utke I, Moshkalev S and Russell P 2012 *Nanofabrication Using Focused Ion and Electron Beams: Principles and Applications* (Oxford: Oxford Univesity Press)

[37] Wnuk J D, Rosenberg S G, Gorham J M, van Dorp W F, Hagen C W and Fairbrother D H Electron beam deposition for nanofabrication: insights from surface science *Surf. Sci.* **605** 257–66

[38] Muthukumar K, Opahle I, Shen J, Jeschke H O and Valentí R 2011 Interaction of $W(CO)_6$ with SiO_2 surfaces: a density functional study *Phys. Rev.* B **84** 205442
[39] Fowlkes J D and Rack P D 2010 Fundamental electron-precursor-solid interactions derived from time-dependent electron-beam-induced deposition simulations and experiments *ACS Nano* **4** 1619–29
[40] Cazaux J 2005 Recent developments and new strategies in scanning electron microscopy *J. Microsc.* **217** 16–35
[41] Zarraoa L, González M U and San Paulo Á 2019 Imaging low-dimensional nanostructures by very low voltage scanning electron microscopy: ultra-shallow topography and depth-tunable material contrast *Sci. Rep.* **9** 16263
[42] Fowlkes J D, Randolph S J and Rack P D 2005 Growth and simulation of high-aspect ratio nanopillars by primary and secondary electron-induced deposition *J. Vac. Sci. Technol.* B **23** 2825–32
[43] Smith D A, Fowlkes J D and Rack P D 2007 A nanoscale three-dimensional Monte Carlo simulation of electron-beam-induced deposition with gas dynamics *Nanotechnology* **18** 265308
[44] Arnold G, Timilsina R, Fowlkes J, Orthacker A, Kothleitner G, Rack P D and Plank H 2014 Fundamental resolution limits during electron-induced direct-write synthesis *ACS Appl. Mater. Interfaces* **6** 7380–7
[45] Schmied R, Fowlkes J D, Winkler R, Rack P D and Plank H 2015 Fundamental edge broadening effects during focused electron beam induced nanosynthesis *Beilstein J. Nanotechnol.* **6** 462–71
[46] Drouin D, Couture A R, Joly D, Tastet X, Aimez V and Gauvin R 2007 CASINO V2.42—a fast and easy-to-use modeling tool for scanning electron microscopy and microanalysis users *Scanning* **29** 92–101
[47] Friedli V and Utke I 2009 Optimized molecule supply from nozzle-based gas injection systems for focused electron- and ion-beam induced deposition and etching: simulation and experiment *J. Phys. D Appl. Phys.* **42** 125305
[48] Serrano-Ramón L, Córdoba R, Rodríguez L A, Magén C, Snoeck E, Gatel C, Serrano I, Ibarra M R and De Teresa J M 2011 Ultrasmall functional ferromagnetic nanostructures grown by focused electron-beam-induced deposition *ACS Nano* **5** 7781–7
[49] Wachter S, Gavagnin M, Wanzenboeck H D, Shawrav M M, Belić D and Bertagnolli E 2014 Nitrogen as a carrier gas for regime control in focused electron beam induced deposition *Nanofabrication* **1** 16–22
[50] van Dorp W F, Hansen T W, Wagner J B and De Hosson J T M 2013 The role of electron-stimulated desorption in focused electron beam induced deposition *Beilstein J. Nanotechnol.* **4** 474–80
[51] Huth M, Porrati F, Gruszka P and Barth S 2020 Temperature-dependent growth characteristics of Nb- and CoFe-based nanostructures by direct-write using focused electron beam-induced deposition *Micromachines* **11** 28
[52] Mutunga E, Winkler R, Sattelkow J, Rack P D, Plank H and Fowlkes J D 2019 Impact of electron-beam heating during 3D nanoprinting *ACS Nano* **13** 5198–213
[53] Hochleitner G, Wanzenboeck H D and Bertagnolli E 2008 Electron beam induced deposition of iron nanostructures *J. Vac. Sci. Technol.* B **26** 939–44

[54] Córdoba R, Sesé J, De Teresa J M and Ibarra M R 2010 High-purity cobalt nanostructures grown by focused-electron-beam-induced deposition at low current *Microelectron. Eng.* **87** 1550–3

[55] Belova L M, Dahlberg E D, Riazanova A, Mulders J J L, Christophersen C and Eckert J 2011 Rapid electron beam assisted patterning of pure cobalt at elevated temperatures via seeded growth *Nanotechnology* **22** 145305

[56] Mulders J J L 2014 Practical precursor aspects for electron beam induced deposition *Nanofabrication* **1** 74–9

[57] Pablo-Navarro J 2020 Development and optimization of 3D advanced functional magnetic nanostructures grown by focused electron beam induced deposition *Dissertation* University of Zaragoza

[58] Jaafar M *et al* 2020 Customized MFM probes based on magnetic nanorods *Nanoscale* **12** 10090–7

[59] Fernández-Pacheco A, Serrano-Ramón L, Michalik J M, Ibarra M R, De Teresa J M, O'Brien L, Petit D, Lee J and Cowburn R P 2013 Three-dimensional magnetic nanowires grown by focused electron-beam induced deposition *Sci. Rep.* **3** 1492

[60] Wartelle A *et al* 2018 Transmission XMCD-PEEM imaging of an engineered vertical FEBID cobalt nanowire with a domain wall *Nanotechnology* **29** 045704

[61] Mattiat H, Rossi N, Gross B, Pablo-Navarro J, Magén C, Badea R, Berezovsky J, De Teresa J M and Poggio M 2020 Nanowire magnetic force sensors fabricated by focused electron beam induced deposition *Phys. Rev. Appl.* **13** 044043

[62] Sangiao S, Magén C, Mofakhami D, de Loubens G and De Teresa J M 2017 Magnetic properties of optimized cobalt nanospheres grown by focused electron beam induced deposition (FEBID) on cantilever tips *Beilstein J. Nanotechnol.* **8** 2106–15

[63] Swiderek P, Marbach H and Hagen C W 2018 Chemistry for electron-induced nanofabrication *Beilstein J. Nanotechnol.* **9** 1317–20

[64] Botman A, Mulders J J L and Hagen C W 2009 Creating pure nanostructures from electron-beam-induced deposition using purification techniques: a technology perspective *Nanotechnology* **20** 372001

[65] Winkler R, Lewis B B, Fowlkes J D, Rack P D and Plank H 2018 High-fidelity 3D-nanoprinting via focused electron beams: growth fundamentals *ACS Appl. Nano Mater.* **1** 1014–27

[66] Utke I, Michler J, Gasser P, Santschi C, Laub D, Cantoni M, Buffat P A, Jiao C and Hoffmann P 2005 Cross section investigations of compositions and sub-structures of tips obtained by focused electron beam induced deposition *Adv. Eng. Mater.* **7** 323–31

[67] Lukasczyk T, Schirmer M, Steinrück H-P and Marbach H 2008 Electron-beam-induced deposition in ultrahigh vacuum: lithographic fabrication of clean iron nanostructures *Small* **4** 841–6

[68] Bernau L, Gabureac M, Erni R and Utke I 2010 Tunable nanosynthesis of composite materials by electron-impact reaction *Angew. Chem. Int. Ed.* **49** 8880–4

[69] Winkler R, Geier B and Plank H 2014 Spatial chemistry evolution during focused electron beam-induced deposition: origins and workarounds *Appl. Phys.* A **117** 1675–88

[70] Botman A, Hesselberth M and Mulders J J L 2008 Improving the conductivity of platinum-containing nano-structures created by electron-beam-induced deposition *Microelectron. Eng.* **85** 1139–42

[71] Tsukatani Y, Yamasaki N, Murakami K, Wakaya F and Takai M 2005 Transport properties of Pt nanowires fabricated by beam-induced deposition *Jpn. J. Appl. Phys.* **44** 5683–6

[72] Sengupta S, Li C, Baumier C, Kasumov A, Guéron S, Bouchiat H and Fortuna F 2015 Superconducting nanowires by electron-beam-induced deposition *Appl. Phys. Lett.* **106** 042601

[73] Huth M, Klingenberger D, Grimm C, Porrati F and Sachser R 2009 Conductance regimes of W-based granular metals prepared by electron beam induced deposition *New J. Phys.* **11** 033032

[74] Klein K L, Randolph S J, Fowlkes J D, Allard L F, Meyer H M III, Simpson M L and Rack P D 2008 Single-crystal nanowires grown via electron-beam-induced deposition *Nanotechnology* **19** 345705

[75] Guan Y, Fowlkes J D, Retterer S T, Simpson M L and Rack P D 2008 Nanoscale lithography via electron beam induced deposition *Nanotechnology* **19** 505302

[76] Fernández-Pacheco A, De Teresa J M, Córdoba R and Ibarra M R 2009 Magnetotransport properties of high-quality cobalt nanowires grown by focused-electron-beam-induced deposition *J. Phys. D Appl. Phys.* **42** 055005

[77] Gazzadi G C, Mulders H, Trompenaars P, Ghirri A, Affronte M, Grillo V and Frabboni S 2011 Focused electron beam deposition of nanowires from cobalt tricarbonyl nitrosyl (Co $(CO)_3NO$) precursor *J. Phys. Chem.* C **115** 19606–11

[78] Lavrijsen R *et al* 2011 Fe:O:C grown by focused-electron-beam-induced deposition: magnetic and electric properties *Nanotechnology* **22** 025302

[79] Córdoba R, Han D-S and Koopmans B 2016 Manipulating the switching in modulated iron nanowires grown by focused electron beam induced deposition *Microelectron. Eng.* **153** 60–5

[80] Belić D, Shawrav M M, Gavagnin M, Stöger-Pollach M, Wanzenboeck H D and Bertagnolli E 2015 Direct-write deposition and focused-electron-beam-induced purification of gold nanostructures *ACS Appl. Mater. Interfaces* **7** 2467–79

[81] Koops H W P, Schössler C, Kaya A and Weber M 1996 Conductive dots, wires, and supertips for field electron emitters produced by electron-beam induced deposition on samples having increased temperature *J. Vac. Sci. Technol.* B **14** 4105–9

[82] Utke I, Hoffmann P, Dwir B, Leifer K, Kapon E and Doppelt P 2000 Focused electron beam induced deposition of gold *J. Vac. Sci. Technol.* B **18** 3168–71

[83] Mulders J J L, Veerhoek J M, Bosch E G T and Trompenaars P H F 2012 Fabrication of pure gold nanostructures by electron beam induced deposition with Au(CO)Cl precursor: Deposition characteristics and primary beam scattering effects *J. Phys. D Appl. Phys.* **45** 475301

[84] Höflich K, Jurczyk J, Zhang Y, Puydinger dos Santos M V, Götz M, Guerra-Nuñez C, Best J P, Kapusta C and Utke I 2017 Direct electron beam writing of silver-based nanostructures *ACS Appl. Mater. Interfaces* **9** 24071–7

[85] Porrati F, Sachser R, Gazzadi G C, Frabboni S and Huth M 2016 Fabrication of FeSi and Fe_3Si compounds by electron beam induced mixing of $[Fe/Si]_2$ and $[Fe_3/Si]_2$ multilayers grown by focused electron beam induced deposition *J. Appl. Phys.* **119** 234306

[86] Frabboni S, Gazzadi G C and Spessot A 2006 Transmission electron microscopy characterization and sculpting of sub-1 nm Si–O–C freestanding nanowires grown by electron beam induced deposition *Appl. Phys. Lett.* **89** 113108

[87] Randolph S J, Fowlkes J D and Rack P D 2005 Effects of heat generation during electron-beam-induced deposition of nanostructures *J. Appl. Phys.* **97** 124312
[88] Gazzadi G C, Frabboni S and Menozzi C 2007 Suspended nanostructures grown by electron beam-induced deposition of Pt and TEOS precursors *Nanotechnology* **18** 445709
[89] Winhold M, Weirich P M, Schwalb C H and Huth M 2014 Superconductivity and metallic behavior in $Pb_xC_yO_\delta$ structures prepared by focused electron beam induced deposition *Appl. Phys. Lett.* **105** 162603
[90] Noh J H, Stanford M G, Lewis B B, Fowlkes J D, Plank H and Rack P D 2014 Nanoscale electron beam-induced deposition and purification of ruthenium for extreme ultraviolet lithography mask repair *Appl. Phys.* A **117** 1705–13
[91] Perentes A, Sinicco G, Boero G, Dwir B and Hoffmann P 2007 Focused electron beam induced deposition of nickel *J. Vac. Sci. Technol.* B **25** 2228
[92] Haverkamp C, Sarau G, Polyakov M N, Utke I, Puydinger dos Santos M V, Christiansen S and Höflich K 2018 A novel copper precursor for electron beam induced deposition *Beilstein J. Nanotechnol.* **9** 1220–7
[93] Porrati F, Pohlit M, Müller J, Barth S, Biegger F, Gspan C, Plank H and Huth M 2015 Direct writing of CoFe alloy nanostructures by focused electron beam induced deposition from a heteronuclear precursor *Nanotechnology* **26** 475701
[94] Kumar T P R, Unlu I, Barth S, Ingólfsson O and Fairbrother D H 2018 Electron induced surface reactions of $HFeCo_3(CO)_{12}$, a bimetallic precursor for focused electron beam induced deposition (FEBID) *J. Phys. Chem.* C **122** 2648–60
[95] Puydinger dos Santos M V, Barth S, Béron F, Pirota K R, Pinto A L, Sinnecker J P, Moshkalev S, Diniz J A and Utke I 2018 Magnetoelectrical transport improvements of postgrowth annealed iron–cobalt nanocomposites: a possible route for future room-temperature spintronics *ACS Appl. Nano Mater.* **1** 3364–74
[96] Kumar T P R, Weirich P, Hrachowina L, Hanefeld M, Bjornsson R, Hrodmarsson H R, Barth S, Fairbrother D H, Huth M and Ingólfsson O 2018 Electron interactions with the heteronuclear carbonyl precursor $H_2FeRu_3(CO)_{13}$ and comparison with $HFeCo_3(CO)_{12}$: from fundamental gas phase and surface science studies to focused electron beam induced deposition *Beilstein J. Nanotechnol.* **9** 555–79
[97] Porrati F, Barth S, Sachser R, Jungwirth F, Eltsov M, Frangakis A S and Huth M 2018 Binary Mn-Si nanostructures prepared by focused electron beam induced deposition from the precursor $SiH_3Mn(CO)_5$ *J. Phys. D Appl. Phys.* **51** 455301
[98] Puydinger Dos Santos M V *et al* 2016 Annealing-based electrical tuning of cobalt-carbon deposits grown by focused-electron-beam-induced deposition *ACS Appl. Mater. Interfaces* **8** 32496–503
[99] Tanaka M, Shimojo M, Takeguchi M, Mitsuishi K and Furuya K 2005 Formation of iron nano-dot arrays by electron beam-induced deposition using an ultrahigh vacuum transmission electron microscope *J. Cryst. Growth* **275** e2361–6
[100] Mitsuishi K, Shimojo M, Tanaka M, Takeguchi M and Furuya K 2005 Resolution in new nanofabrication technique combining electron-beam-induced deposition and low-energy ion milling *Jpn. J. Appl. Phys.* **44** 5627–30
[101] Shimojo M, Takeguchi M, Tanaka M, Mitsuishi K and Furuya K 2004 Electron beam-induced deposition using iron carbonyl and the effects of heat treatment on nanostructure *Appl. Phys.* A **79** 1869–72

[102] Takeguchi M, Shimojo M and Furuya K 2007 Nanostructure fabrication by electron-beam-induced deposition with metal carbonyl precursor and water vapor *Jpn. J. Appl. Phys.* **46** 6183–6
[103] Plank H, Noh J H, Fowlkes J D, Lester K, Lewis B B and Rack P D 2014 Electron-beam-assisted oxygen purification at low temperatures for electron-beam-induced Pt deposits: towards pure and high-fidelity nanostructures *ACS Appl. Mater. Interfaces* **6** 1018–24
[104] Mulders J J L, Belova L M and Riazanova A 2011 Electron beam induced deposition at elevated temperatures: compositional changes and purity improvement *Nanotechnology* **22** 055302
[105] Frabboni S, Gazzadi G C, Felisari L and Spessot A 2006 Fabrication by electron beam induced deposition and transmission electron microscopic characterization of sub-10-nm freestanding Pt nanowires *Appl. Phys. Lett.* **88** 213116
[106] Plank H, Kothleitner G, Hofer F, Michelitsch S G, Gspan C, Hohenau A and Krenn J 2011 Optimization of postgrowth electron-beam curing for focused electron-beam-induced Pt deposits *J. Vac. Sci. Technol.* B **29** 051801
[107] Porrati F, Keller L, Gspan C, Plank H and Huth M 2017 Electrical transport properties of Ga irradiated W-based granular nanostructures *J. Phys. D Appl. Phys.* **50** 215301
[108] Roberts N A, Fowlkes J D, Magel G A and Rack P D 2013 Enhanced material purity and resolution via synchronized laser assisted electron beam induced deposition of platinum *Nanoscale* **5** 408–15
[109] Roberts N A, Gonzalez C M, Fowlkes J D and Rack P D 2013 Enhanced by-product desorption via laser assisted electron beam induced deposition of $W(CO)_6$ with improved conductivity and resolution *Nanotechnology* **24** 415301
[110] Gazzadi G C and Frabboni S 2015 Structural transitions in electron beam deposited Co–carbonyl suspended nanowires at high electrical current densities *Beilstein J. Nanotechnol.* **6** 1298–305
[111] Henry M R, Kim S and Fedorov A G 2016 High purity tungsten nanostructures via focused electron beam induced deposition with carrier gas assisted supersonic jet delivery of organometallic precursors *J. Phys. Chem.* C **120** 10584–90
[112] Villamor E, Casanova F, Trompenaars P H F and Mulders J J L 2015 Embedded purification for electron beam induced Pt deposition using $MeCpPtMe_3$ *Nanotechnology* **26** 095303
[113] Mehendale S, Mulders J J L and Trompenaars P H F 2013 A new sequential EBID process for the creation of pure Pt structures from $MeCpPtMe_3$ *Nanotechnology* **24** 145303
[114] Dobrovolskiy O V, Kompaniiets M, Sachser R, Porrati F, Gspan C, Plank H and Huth M 2015 Tunable magnetism on the lateral mesoscale by post-processing of Co/Pt heterostructures *Beilstein J. Nanotechnol.* **6** 1082–90
[115] Höflich K, Bin Yang R, Berger A, Leuchs G and Christiansen S 2011 The direct writing of plasmonic gold nanostructures by electron-beam-induced deposition *Adv. Mater.* **23** 2657–61
[116] Shawrav M M, Taus P, Wanzenboeck H D, Schinnerl M, Stöger-Pollach M, Schwarz S, Steiger-Thirsfeld A and Bertagnolli E 2016 Highly conductive and pure gold nanostructures grown by electron beam induced deposition *Sci. Rep.* **6** 34003
[117] Winkler R, Schmidt F-P, Haselmann U, Fowlkes J D, Lewis B B, Kothleitner G, Rack P D and Plank H 2017 Direct-write 3D nanoprinting of plasmonic structures *ACS Appl. Mater. Interfaces* **9** 8233–40

[118] Porrati F, Kämpken B, Terfort A and Huth M 2013 Fabrication and electrical transport properties of binary Co-Si nanostructures prepared by focused electron beam-induced deposition *J. Appl. Phys.* **113** 053707
[119] Rosenberg S G, Barclay M and Fairbrother D H 2013 Electron beam induced reactions of adsorbed cobalt tricarbonyl nitrosyl ($Co(CO)_3NO$) molecules *J. Phys. Chem.* C **117** 16053–64
[120] Gazzadi G C, Mulders J J L, Trompenaars P, Ghirri A, Rota A, Affronte M and Frabboni S 2011 Characterization of a new cobalt precursor for focused beam deposition of magnetic nanostructures *Microelectron. Eng.* **88** 1955–8
[121] Pablo-Navarro J, Magén C and De Teresa J M 2018 Purified and crystalline three-dimensional electron-beam-induced deposits: the successful case of cobalt for high-performance magnetic nanowires *ACS Appl. Nano Mater.* **1** 38–46
[122] Pablo-Navarro J, Winkler R, Haberfehlner G, Magén C, Plank H and De Teresa J M 2019 *In situ* real-time annealing of ultrathin vertical Fe nanowires grown by focused electron beam induced deposition *Acta Mater.* **174** 379–86
[123] Puydinger dos Santos M V, Szkudlarek A, Rydosz A, Guerra-Nuñez C, Béron F, Pirota K R, Moshkalev S, Diniz J A and Utke I 2018 Comparative study of post-growth annealing of $Cu(hfac)_2$, $Co_2(CO)_8$ and $Me_2Au(acac)$ metal precursors deposited by FEBID *Beilstein J. Nanotechnol.* **9** 91–101
[124] Sadki E S, Ooi S and Hirata K 2004 Focused-ion-beam-induced deposition of superconducting nanowires *Appl. Phys. Lett.* **85** 6206
[125] Makise K, Mitsuishi K, Shimojo M and Shinozaki B 2015 Microstructural analysis and transport properties of MoO and MoC nanostructures prepared by focused electron beam-induced deposition *Sci. Rep.* **4** 5740
[126] Porrati F, Sachser R, Gazzadi G C, Frabboni S, Terfort A and Huth M 2017 Alloy multilayers and ternary nanostructures by direct-write approach *Nanotechnology* **28** 415302
[127] Shawrav M M, Belić D, Gavagnin M, Wachter S, Schinnerl M, Wanzenboeck H D and Bertagnolli E 2014 Electron beam-induced CVD of nanoalloys for nanoelectronics *Chem. Vap. Deposition* **20** 251–7
[128] Porrati F, Begun E, Winhold M, Schwalb C H, Sachser R, Frangakis A S and Huth M 2012 Room temperature $L1_0$ phase transformation in binary CoPt nanostructures prepared by focused-electron-beam-induced deposition *Nanotechnology* **23** 185702
[129] Che R C, Takeguchi M, Shimojo M, Zhang W and Furuya K 2005 Fabrication and electron holography characterization of FePt alloy nanorods *Appl. Phys. Lett.* **87** 223109
[130] Winhold M, Schwalb C H, Porrati F, Sachser R, Frangakis A S, Kämpken B, Terfort A, Auner N and Huth M 2011 Binary Pt-Si nanostructures prepared by focused electron-beam-induced deposition *ACS Nano* **5** 9675–81
[131] Hirt L, Reiser A, Spolenak R and Zambelli T 2017 Additive manufacturing of metal structures at the micrometer scale *Adv. Mater.* **29** 1604211
[132] Huth M, Porrati F and Dobrovolskiy O V 2018 Focused electron beam induced deposition meets materials science *Microelectron. Eng.* **185-186** 9–28
[133] Bret T, Hofmann T and Edinger K 2014 Industrial perspective on focused electron beam-induced processes *Appl. Phys.* A **117** 1607–14
[134] Sangiao S, Martín S, González-Orive A, Magén C, Low P J, De Teresa J M and Cea P 2017 All-carbon electrode molecular electronic devices based on Langmuir–Blodgett monolayers *Small* **13** 1603207

[135] Utke I, Hoffmann P, Berger R and Scandella L 2002 High-resolution magnetic Co supertips grown by a focused electron beam *Appl. Phys. Lett.* **80** 4792–4
[136] Makise K, Mitsuishi K, Shimojo M and Furuya K 2009 A nanosized photodetector fabricated by electron-beam-induced deposition *Nanotechnology* **20** 425305
[137] Kolb F, Schmoltner K, Huth M, Hohenau A, Krenn J, Klug A, List E J W and Plank H 2013 Variable tunneling barriers in FEBID based PtC metal-matrix nanocomposites as a transducing element for humidity sensing *Nanotechnology* **24** 305501
[138] Dias R J, O'Regan C, Thrompenaars P, Romano-Rodriguez A, Holmes J D, Mulder J J L and Petkov N 2012 Low resistivity Pt interconnects developed by electron beam assisted deposition using novel gas injector system *J. Phys. Conf. Ser.* **371** 012038
[139] Bøggild P, Hansen T M, Tanasa C and Grey F 2001 Fabrication and actuation of customized nanotweezers with a 25 nm gap *Nanotechnology* **12** 331–5
[140] Hernández-Ramírez F, Rodríguez J, Casals O, Russinyol E, Vilà A, Romano-Rodríguez A, Morante J R and Abid M 2006 Characterization of metal-oxide nanosensors fabricated with focused ion beam (FIB) *Sens. Actuators* **B 118** 198–203
[141] Liang T, Frendberg E, Lieberman B and Stivers A 2005 Advanced photolithographic mask repair using electron beams *J. Vac. Sci. Technol.* B **23** 3101–5
[142] Edinger K *et al* 2004 Electron-beam-based photomask repair *J. Vac. Sci. Technol.* B **22** 2902–6
[143] Plank H, Winkler R, Schwalb C H, Hütner J, Fowlkes J D, Rack P D, Utke I and Huth M 2020 focused electron beam-based 3D nanoprinting for scanning probe microscopy: a review *Micromachines* **11** 48
[144] Brown J, Kocher P, Ramanujan C S, Sharp D N, Torimitsu K and Ryan J F 2013 Electrically conducting, ultra-sharp, high aspect-ratio probes for AFM fabricated by electron-beam-induced deposition of platinum *Ultramicroscopy* **133** 62–6
[145] Jaafar M, De Teresa J M, Asenjo A, Pablo-Navarro J, Ares P, Magén C and Gómez-Herrero J 2019 System for an atomic force microscopy *Spain* 2711860 PCT; WO2019/086745
[146] Sattelkow J, Fröch J E, Winkler R, Hummel S, Schwalb C and Plank H 2019 Three-dimensional nanothermistors for thermal probing *ACS Appl. Mater. Interfaces* **11** 22655–67
[147] Sqalli O, Utke I, Hoffmann P and Marquis-Weible F 2002 Gold elliptical nanoantennas as probes for near field optical microscopy *J. Appl. Phys.* **92** 1078–83
[148] De Angelis F *et al* 2010 Nanoscale chemical mapping using three-dimensional adiabatic compression of surface plasmon polaritons *Nat. Nanotechnol.* **5** 67–72
[149] Qian W, Sun S, Song J, Nguyen C, Ducharme S and Turner J A 2018 Focused electron-beam-induced deposition for fabrication of highly durable and sensitive metallic AFM-IR probes *Nanotechnology* **29** 335702
[150] Brintlinger T, Fuhrer M S, Melngailis J, Utke I, Bret T, Perentes A, Hoffmann P, Abourida M and Doppelt P 2005 Electrodes for carbon nanotube devices by focused electron beam induced deposition of gold *J. Vac. Sci. Technol.* B **23** 3174–7
[151] Wilhite P, Uh H S, Kanzaki N, Wang P, Vyas A, Maeda S, Yamada T and Yang C Y 2014 Electron-beam and ion-beam-induced deposited tungsten contacts for carbon nanofiber interconnects *Nanotechnology* **25** 375702
[152] Langford R M, Wang T-X, Thornton M, Heidelberg A, Sheridan J G, Blau W and Leahy R 2006 Comparison of different methods to contact to nanowires *J. Vac. Sci. Technol.* B **24** 2306–11

[153] Koops H W P, Hoinkis O E, Honsberg M E W, Schmidt R, Blum R, Böttger G, Kuligk A, Liguda C and Eich M 2001 Two-dimensional photonic crystals produced by additive nanolithography with electron beam-induced deposition act as filters in the infrared *Microelectron. Eng.* **57-58** 995–1001

[154] Weber-Bargioni A, Schwartzberg A, Schmidt M, Harteneck B, Ogletree D F, Schuck P J and Cabrini S 2010 Functional plasmonic antenna scanning probes fabricated by induced-deposition mask lithography *Nanotechnology* **21** 065306

[155] Esposito M, Tasco V, Cuscunà M, Todisco F, Benedetti A, Tarantini I, De Giorgi M, Sanvitto D and Passaseo A 2015 Nanoscale 3D chiral plasmonic helices with circular dichroism at visible frequencies *ACS Photonics* **2** 105–14

[156] De Teresa J M and Fernández-Pacheco A 2014 Present and future applications of magnetic nanostructures grown by FEBID *Appl. Phys.* A **117** 1645–58

[157] Serrano-Ramón L, Fernández-Pacheco A, Córdoba R, Magén C, Rodríguez L A, Petit D, Cowburn R P, Ibarra M R and De Teresa J M 2013 Improvement of domain wall conduit properties in cobalt nanowires by global gallium irradiation *Nanotechnology* **24** 345703

[158] Fernández-Pacheco A, De Teresa J M, Córdoba R, Ibarra M R, Petit D, Read D E, O'Brien L, Lewis E R, Zeng H T and Cowburn R P 2009 Domain wall conduit behavior in cobalt nanowires grown by focused electron beam induced deposition *Appl. Phys. Lett.* **94** 192509

[159] Keller L *et al* 2018 Direct-write of free-form building blocks for artificial magnetic 3D lattices *Sci. Rep.* **8** 6160

[160] Sanz-Hernández D *et al* 2020 Artificial double-helix for geometrical control of magnetic chirality *ACS Nano* **14** 8084–92

[161] Burbridge D J, Crampin S, Viau G and Gordeev S N 2008 Strategies for the immobilization of nanoparticles using electron beam induced deposition *Nanotechnology* **19** 445302

[162] Sanz-Hernández D *et al* 2018 Fabrication of scaffold-based 3D magnetic nanowires for domain wall applications *Nanomaterials* **8** 483

[163] Sharma R, Moore E, Rez P and Treacy M M J 2009 Site-specific fabrication of Fe particles for carbon nanotube growth *Nano Lett.* **9** 689–94

[164] Ivanov A P, Instuli E, McGilvery C M, Baldwin G, McComb D W, Albrecht T and Edel J B 2011 DNA tunneling detector embedded in a nanopore *Nano Lett.* **11** 279–85

[165] De Teresa J M, Orús P, Córdoba R and Philipp P 2019 Comparison between focused electron/ion beam-induced deposition at room temperature and under cryogenic conditions *Micromachines* **10** 799

[166] Vollnhals F, Drost M, Tu F, Carrasco E, Späth A, Fink R H, Steinrück H-P and Marbach H 2014 Electron-beam induced deposition and autocatalytic decomposition of $Co(CO)_3NO$ *Beilstein J. Nanotechnol.* **5** 1175–85

[167] Donev E U and Hastings J T 2009 Electron-beam-induced deposition of platinum from a liquid precursor *Nano Lett.* **9** 2715–8

[168] Mackus A J M, Mulders J J L, van de Sanden M C M and Kessels W M M 2010 Local deposition of high-purity Pt nanostructures by combining electron beam induced deposition and atomic layer deposition *J. Appl. Phys.* **107** 116102

[169] Schoenaker F J, Córdoba R, Fernández-Pacheco R, Magén C, Stéphan O, Zuriaga-Monroy C, Ibarra M R and De Teresa J M 2011 Focused electron beam induced etching of titanium with XeF_2 *Nanotechnology* **22** 265304

[170] Pablo-Navarro J, Sangiao S, Magén C and De Teresa J M 2019 Diameter modulation of 3D nanostructures in focused electron beam induced deposition using local electric fields and beam defocus *Nanotechnology* **30** 505302
[171] Bresin M, Thiel B L, Toth M and Dunn K A 2011 Focused electron beam-induced deposition at cryogenic temperatures *J. Mater. Res.* **26** 357–64
[172] Bresin M, Toth M and Dunn K A 2013 Direct-write 3D nanolithography at cryogenic temperatures *Nanotechnology* **24** 035301
[173] Martin A A and Toth M 2014 Cryogenic electron beam induced chemical etching *ACS Appl. Mater. Interfaces* **6** 18457–60
[174] Walz M-M, Schirmer M, Vollnhals F, Lukasczyk T, Steinrück H-P and Marbach H 2010 Electrons as 'Invisible Ink': fabrication of nanostructures by local electron beam induced activation of SiO_x *Angew. Chem. Int. Ed.* **49** 4669–73
[175] Walz M-M, Vollnhals F, Rietzler F, Schirmer M, Kunzmann A, Steinrück H-P and Marbach H 2012 Thin membranes versus bulk substrates: investigation of proximity effects in focused electron beam-induced processing *J. Phys. D Appl. Phys.* **45** 225306
[176] Vollnhals F, Woolcot T, Walz M-M, Seiler S, Steinrück H-P, Thornton G and Marbach H 2013 Electron beam-induced writing of nanoscale iron wires on a functional metal oxide *J. Phys. Chem.* C **117** 17674–9
[177] Grogan J M, Schneider N M, Ross F M and Bau H H 2014 Bubble and pattern formation in liquid induced by an electron beam *Nano Lett.* **14** 359–64
[178] Donev E U, Schardein G, Wright J C and Hastings J T 2011 Substrate effects on the electron-beam-induced deposition of platinum from a liquid precursor *Nanoscale.* **3** 2709–17
[179] Schardein G, Donev E U and Hastings J T 2011 Electron-beam-induced deposition of gold from aqueous solutions *Nanotechnology* **22** 015301
[180] Ocola L E, Joshi-Imre A, Kessel C, Chen B, Park J, Gosztola D and Divan R 2012 Growth characterization of electron-beam-induced silver deposition from liquid precursor *J. Vac. Sci. Technol.* B **30** 06FF08
[181] Jesse S, Borisevich A Y, Fowlkes J D, Lupini A R, Rack P D, Unocic R R, Sumpter B G, Kalinin S V, Belianinov A and Ovchinnikova O S 2016 Directing matter: toward atomic-scale 3D Nanofabrication *ACS Nano* **10** 5600–18
[182] den Heijer M, Shao I, Radisic A, Reuter M C and Ross F M 2014 Patterned electrochemical deposition of copper using an electron beam *APL Mater.* **2** 022101
[183] Esfandiarpour S, Boehme L and Hastings J T 2017 Focused electron beam induced deposition of copper with high resolution and purity from aqueous solutions *Nanotechnology* **28** 125301
[184] Mackus A J M, Thissen N F W, Mulders J J L, Trompenaars P H F, Verheijen M A, Bol A A and Kessels W M M 2013 Direct-write atomic layer deposition of high-quality Pt nanostructures: selective growth conditions and seed layer requirements *J. Phys. Chem.* C **117** 10788–98
[185] Mackus A J M, Dielissen S A F, Mulders J J L and Kessels W M M 2012 Nanopatterning by direct-write atomic layer deposition *Nanoscale* **4** 4477–80
[186] Junige M, Löffler M, Geidel M, Albert M, Bartha J W, Zschech E, Rellinghaus B and van Dorp W F 2017 Area-selective atomic layer deposition of Ru on electron-beam-written Pt (C) patterns versus SiO_2 substratum *Nanotechnology* **28** 395301

[187] Mackus A J M, Thissen N F W, Mulders J J L, Trompenaars P H F, Chen Z, Kessels W M M and Bol A A 2017 Resist-free fabricated carbon nanotube field-effect transistors with high-quality atomic-layer-deposited platinum contacts *Appl. Phys. Lett.* **110** 013101

[188] Di Prima G, Sachser R, Gruszka P, Hanefeld M, Halbritter T, Heckel A and Huth M 2017 *In situ* conductance monitoring of Pt thin film growth by area-selective atomic layer deposition *Nano Futures* **1** 025005

[189] Toth M 2014 Advances in gas-mediated electron beam-induced etching and related material processing techniques *Appl. Phys.* A **117** 1623–9

[190] Utke I and Gölzhäuser A 2010 Small, Minimally invasive, direct: electrons induce local reactions of adsorbed functional molecules on the nanoscale *Angew. Chem. Int. Ed* **49** 9328–30

[191] Martin A A, Toth M and Aharonovich I 2015 Subtractive 3D printing of optically active diamond structures *Sci. Rep.* **4** 5022

[192] Lassiter M G, Liang T and Rack P D 2008 Inhibiting spontaneous etching of nanoscale electron beam induced etching features: solutions for nanoscale repair of extreme ultraviolet lithography masks *J. Vac. Sci. Technol.* B **26** 963–7

[193] Roberts N A, Noh J H, Lassiter M G, Guo S, Kalinin S V and Rack P D 2012 Synthesis and electroplating of high resolution insulated carbon nanotube scanning probes for imaging in liquid solutions *Nanotechnology* **23** 145301

[194] Liebes Y, Hadad B and Ashkenasy N 2011 Effects of electrons on the shape of nanopores prepared by focused electron beam induced etching *Nanotechnology* **22** 285303

[195] Wang S, Sun Y-M, Wang Q and White J M 2004 Electron-beam induced initial growth of platinum films using $Pt(PF_3)_4$ *J. Vac. Sci. Technol.* B **22** 1803–6

[196] Mølhave K, Madsen D N, Dohn S and Bøggild P 2004 Constructing, connecting and soldering nanostructures by environmental electron beam deposition *Nanotechnology* **15** 1047–53

[197] Shimojo M, Takeguchi M and Furuya K 2006 Formation of crystalline iron oxide nanostructures by electron beam-induced deposition at room temperature *Nanotechnology* **17** 3637–40

[198] Bishop J, Toth M, Phillips M and Lobo C 2012 Effects of oxygen on electron beam induced deposition of SiO_2 using physisorbed and chemisorbed tetraethoxysilane *Appl. Phys. Lett.* **101** 211605

IOP Publishing

Nanofabrication
Nanolithography techniques and their applications
José María De Teresa

Chapter 5

Focused ion beam induced processing

Pablo Orús, Rosa Córdoba and José María De Teresa

Focused ion beam (FIB) technology makes use of a highly-focused beam of charged ions to perform precise nanopatterning tasks by steering it over a sample, tracing patterns defined by the user. The physical interactions between the ions and the atoms of the sample are exploited for nanostructuring purposes such as material removal or modification, and material deposition (when used in combination with a precursor). Besides, FIBs are also used for imaging and compositional and structural analyses. In this chapter, the instrumentation and technique fundamentals are reviewed, and a selection of application examples is presented.

5.1 Introduction

Foreseen in 1959 by Richard Feynman on *There's Plenty of Room at the Bottom* [1], the continuous need for device miniaturization present in both the industrial and research fields has led to the development of a wide spectrum of techniques in nanofabrication. Since its inception in the 1970s [2–5], the technology of FIBs, which represents the capability to obtain, steer, and focalize an energetic beam of charged ions for different purposes, has established itself as a solid cornerstone within the nanofabrication framework.

Focused ion beam induced processing (FIBIP, although it is common to refer to both the beam and the technique as simply 'FIB') is an umbrella term that encompasses several nanopatterning techniques that share the usage of an FIB as a primary drive for the nanolithography process. These methods exploit the interactions that take place between a focused beam of charged ions and the material it impinges upon. By itself, an FIB can be used to selectively remove material from a surface, and to induce crystallographic and compositional changes on it. In combination with a gaseous precursor capable of adsorbing to the surface of the sample, the FIB can be used to induce a decomposition reaction, with subsequent material deposition. This technique, known as focused ion beam induced deposition (FIBID, IBID), finds application in optical lithography mask repair, circuit editing, and growth of functional materials [6].

doi:10.1088/978-0-7503-2608-7ch5

Depending on the processed material, the ionic species, and the beam configuration, FIBIP provides a working resolution range from up to hundreds of μm down to a single nm. It offers great adjustability and ease of use in terms of patterning of complex geometries at specific locations in a sample.

The object of this chapter is to provide the reader with a basic overview of FIBIP techniques and their applicability. First, the basics of the FIB instruments will be reviewed. After that, the theoretical foundations and experimental details of the technique will be presented. Lastly, several examples of usage of different FIBIP techniques, including milling, modification, and material deposition, will be reviewed.

5.2 The instrument

As the architecture and design of FIB equipment play a pivotal role in FIBIP, the instrument will be described in detail hereafter. FIBIP is performed in either dedicated equipment simply referred to as FIB instruments, or in hybrid microscopes that combine an FIB with a scanning electron microscope (SEM). FIB/SEM microscopes are extensively known by commercial names such as *Dual Beam* (Thermo Fisher Scientific), *Crossbeam* (Zeiss), and *MultiBeam* (Raith). The commercial name *Helium Ion Microscope* or *HIM* (Zeiss), is also commonly used to refer to FIB instruments that make use of He^+ and Ne^+ ions in their columns.

Both FIB and FIB/SEM instruments share two main components: an ion column and a process chamber. The first houses on its upper part an ion source from where the ions are extracted, while the lower part contains a series of elements used to accelerate them, and focus and steer the beam. The chamber is where the processing is performed, and is equipped with a *stage* holding the sample, and several detectors and processing modules, including a gas delivery system used to inject the precursor material into the chamber to perform FIBID, if applicable. Since the interactions of beam ions with contamination and/or air particles are to be avoided, a pumping system is used to maintain vacuum conditions during operation. Typically, at least two levels of vacuum are maintained in the equipment: the operation chamber vacuum (10^{-6} mbar, raised to 10^{-5} mbar during gas delivery), and the ion source vacuum (10^{-8} mbar).

5.2.1 Historical development

In the field of experimental sciences, the achievement of most scientific breakthroughs is intrinsically bound to the development of the specific instrumentation that is required to perform the required experiments. As shall be discussed in the following section, the beating heart of any FIB is the *ion source*—the constituent used to extract the ions from a source material. With ample generosity, the development of ion sources for FIBs can be traced back to the 17th century, when English scientist William Gilbert reported that when an electrically charged fragment of amber was brought close to a water droplet, it would pull some towards it, constituting a conical shape [7]. One of the most significant achievements in the development of FIB instrumentation was the invention of sources based on the ionization of a material in

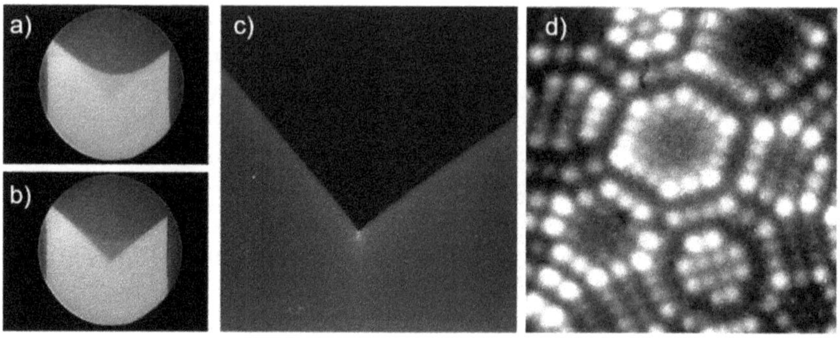

Figure 5.1. (a) and (b) TEM images showing the formation of the Taylor cone at a 20 μm wide Ga-wetted tungsten tip acting as a LMIS. (c) High magnification of the former LMIS, with the emission area visible at the apex of the tip. Reprinted from [9]. Copyright 2004, with permission from Elsevier. (d) Field-ion microscopy image of the tungsten tip utilized by Müller. Each bright spot corresponds to one individual atom in the tip, this image being one of the first ever depicting atomic resolutions. Reprinted from [10] John Wiley & Sons. Copyright Wiley Periodicals, Inc.

liquid form, following the same physical mechanism that made Gilbert's amber piece pull water. These *liquid metal ion sources* (LMISs), initially intended to be used as ion thrusters in space applications [8], would eventually become a main cornerstone in FIB technology (figure 5.1a–c).

The development of actual, usable FIBs based on LMISs would find its beginnings in the 1970s, with Seliger and Orloff being two of the most prominent challengers. Seliger *et al* reported in 1974 the usage of a 60 keV He^+ beam for material modification [2]. The beam had a diameter of 3.5 μm, which would be considered enormous by today's standards. Complementarily, Orloff *et al* reported the successful ionization of gaseous H_2 and Ar to a much more focused beam, with a probe size of 0.1 μm [3, 4]. The first reported usage of FIBIP dates back to 1979, with Seliger *et al* using an improved 57 kV Ga^+ source to mill trenches on an Au film [5].

In parallel to LIMSs, strong effort was directed to the development of ion sources based on the ionization of a material *in gaseous form* [10, 11]. In 1955, Müller *et al* [12] reported the utilization of the so-called *field ion microscope*, based on the ionization of gaseous He. Roughly, the instrument consisted in a glass-blown recipient on which a fine tungsten tip, a fluorescent screen, and gaseous He were placed. By biasing the tip and the screen contrary to each other, field-ionization of the He atoms was achieved at the former, resulting in the He^+ ions being accelerated towards the screen, where the effective emission pattern of the tip would be recorded (figure 5.1d). This experiment marks the very first time in the history of humankind when atomic resolution was achieved—individual tungsten atoms from the tip could be clearly discerned in the fluorescent imprint of the emission pattern.

Field-ionization of a gas thus proved quite a promising approach as a basis for ion sources in FIBs, setting the grounds for what would later be known as the *gas field ion source* (GFIS). However, the rise and consolidation of LMISs somewhat slowed down the research process, since this established approach was also considered, for most applications, more practical and doable at the time. After a series of efforts developing a suitable and repeatable procedure for tip fabrication in the 1990s [13, 14], the atomic trimer-based approach that would definitely stay as the core of GFISs was presented in 2006 by Ward *et al* [15], with a probe size of 0.5 nm and imaging resolution of 1 nm [16]. This technique proved to be commercially viable and is currently present in modern HIMs.

Thanks to all these developments, both LIMS and GFIS-based FIB instruments have become commercially available as of the present day. They can be found at specialized facilities of both commercial companies and research centers, where they can be used either as a supplementary tool (very commonly, for transmission electron microscopy sample preparation), or as the primary means of achieving a patterning goal. FIB equipment manufacturers include the companies Thermo Fisher Scientific, Zeiss, Raith, Orsay Physics, Hitachi, and Tescan.

5.2.2 Ion sources

The *ion source* is the component of an FIB instrument where a source material is ionized to extract the ions that will posteriorly be accelerated in the column, constituting the beam. Depending on the ionization mechanism, several types of ion sources exist. As anticipated, the most commonly used are LMISs [6] and GFISs [17]. While the instruments are typically equipped with a single source capable of providing a certain ionic species, customizable setups from several source materials do exist, both in commercial and home-made forms.

Liquid metal ion source

An LMIS is based on the ionization of a source metallic material in liquid state. It is essentially constituted by a metal reservoir connected to a tungsten needle with a tip radius between 1 and 5 μm, and an extractor electrode (figure 5.2a). The source material, stored in the reservoir, is heated up to its melting point and wets the capillary needle, remaining in place as a consequence of the surface tension of the liquid.

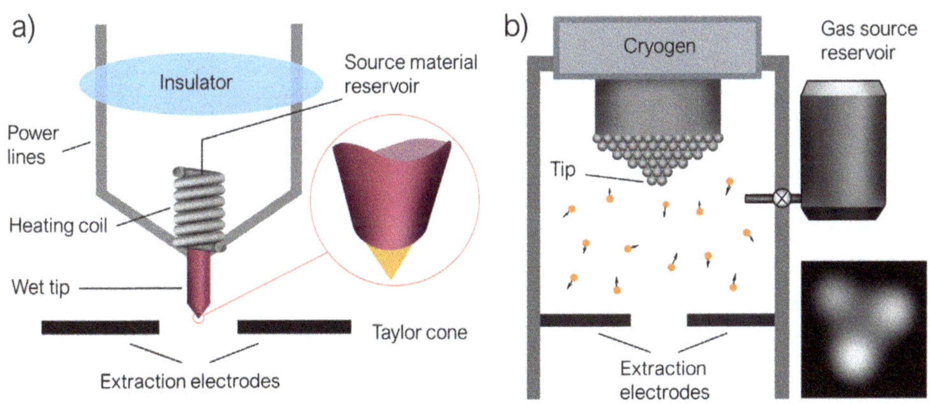

Figure 5.2. (a) LMIS. (b) GFIS. Inset shows a scanning ion microscopy image of the tip edge of a GFIS. Each bright dot corresponds to one of the atoms in the trimer. Reprinted from [18], with permission from Elsevier.

The extractor electrode supplies a high electric field (given by an *extraction voltage*, with values commonly around 10 kV [19]) that induces an electrostatic force on the liquid. As experimentally observed by Gilbert in 1600, the balance between the action of the surface tension of the liquid source material and the electrostatic forces results in the deformation of the liquid drop into a cone shape with a very sharp tip (2–5 nm), referred to as *Taylor cone*. Atoms are then extracted at the point-like tip of the cone by field evaporation through the electric field provided by the extractor voltage, and then accelerated down the FIB column.

Several metals, such as Al, In, Au, or Bi can be used as source materials in LMISs [20], although the most widespread material used in LMISs is undoubtedly gallium, due to its outstanding properties for this purpose, such as low melting point (29.8 °C), low vapor pressure, and long lifetime, among others [6]. Ga^+ LIMSs are commonplace in commercial FIB/SEM microscopes.

The flexibility of FIB technology makes the potential usage of ionic species other than Ga^+ highly appealing due to the technological and scientific interest they may present. However, some pure species present an elevated melting point, making the engineering of the source challenging. This obstacle can be mitigated by using as a source in the LMIS a binary or ternary alloy of the wanted material [21, 22].

The working principle of a metal-alloy based LMIS (sometimes termed LMAIS) is the same as a pure metal LMIS, with a significant difference: metal-alloy LMISs do not emit a single ionic species as LMISs do, and as a consequence require a filtering of the beam before it reaches the substrate. This is achieved by using a *mass separator* (also known as *Wein filter* and *ExB filter*): placed in the column after the source, a series of magnets electromagnetically separate the different beam constituents according to their atomic mass. It is worth noting that the mass separator enables extracting different ionic species from the same alloy.

A myriad of alloys can be employed in metal-alloy LMISs, containing elements that cover about half of the periodic table [22]. Their operating temperatures fall in the 100 °C–1000 °C range. Specific examples include Co–Nd [23] and Au–Si [24], among many others.

Gas field ion source

A GFIS is based on the ionization of a source material in gas form. The main constituent of a GFIS is a 'supertip'—a nanofabricated tungsten crystal whose size is sequentially diminished in a series of terraces down to a single group of three W atoms, or *trimer*. The tip is positively biased against a counter electrode (around 35 kV) and then the source gas of neutral atoms is made flow in its vicinity [25] (figure 5.2b).

Polarized by the electric field, these atoms are attracted to the tip. In close proximity to the trimer, the high electric field results in the appearance of *ionization disks*, i.e., small regions in which electrons from the gas atoms tunnel into the tip, effectively leaving them positively charged. As a consequence, they are immediately accelerated away from the tip and down the FIB column [15, 25]. The source is typically tilted so one of the three atoms in the trimer is aimed down the column [10].

Conventional gases used in GFISs are helium and neon. It is relatively common for GFIS-based instruments to be equipped with reservoirs for both gases, which can be ionized using the same GFISs. As aforementioned, and due to the significant differences they exhibit with conventional Ga^+ FIB instruments, FIB instruments with He^+ (and Ne^+) GFISs are sometimes referred to as helium ion microscopes (HIMs).

Less frequent architectures are capillary-type and impregnated-type GFISs [26]. These are respectively based on ionization occurring at the end of a small capillary through which the gaseous source material is made flow, and on the ionization occurring at the end of a needle joined to a tube inside which the source material is stored.

Other sources

An alternative approach to obtain an FIB is the usage of a *plasma source* (PFIB). These sources, more typically used to obtain broad beams rather than FIBs, operate by creating a plasma of a noble gas and then extracting the ions through an aperture. One of the most relevant advantages of plasma sources is the enhanced amount of beam current they are able to deliver while keeping a relatively small beam spot size below 25 nm (significantly enhancing the removal speed in FIB milling) at the cost of shorter lifetimes and restrictiveness in the number of feasible source materials [27]. Xe, Ne, O, and Ar are amongst the species that are ionized in plasma sources.

Somewhat more 'exotic' designs include *laser ion sources* [28], from which the ions are obtained by laser-induced ionization of a focused atomic beam also obtained using a laser; *electron-impact sources* [29], on which an electron beam induces the ionization of a gas; and *magneto-optical trap sources* [30], among others.

5.2.3 Ion column

The *ion column* is the constituent of FIB equipment on which the ions coming from the source are accelerated, steered, and focalized to the substrate (figure 5.3). One of the reigning parameters of the ion column is the *acceleration voltage*, i.e. the energy acquired by the ions as they are driven down a potential difference in the column.

Apart from the acceleration voltage electrodes and the ion source itself, the column is essentially equipped with two different elements: *electromagnetic* and *electrostatic lenses*, and *apertures*. An aperture is a physical orifice of a fixed size used to define the size of the beam and moderate the number of ions that go down the column. Different beam currents are selected by choosing an appropriate beam-defining aperture.

Electromagnetic lenses are constituted by arrangements of electromagnets in quadrupole or octupole-like dispositions, in such a way that the joint magnetic field they generate radially increases away from the center of the column. The Lorentz force induced by the magnetic field on the passing charged particles deflects those away from the center back onto it, focusing the beam. Conversely, electrostatic lenses are combinations of electrodes that deflect and focus the charged particles in a similar way, but without resorting to magnets and as such, lacking magnetic hysteresis.

A common and relatively simple column setup is based on two main electromagnetic lenses—the *condenser lens*, located near the source, and the *objective lens*,

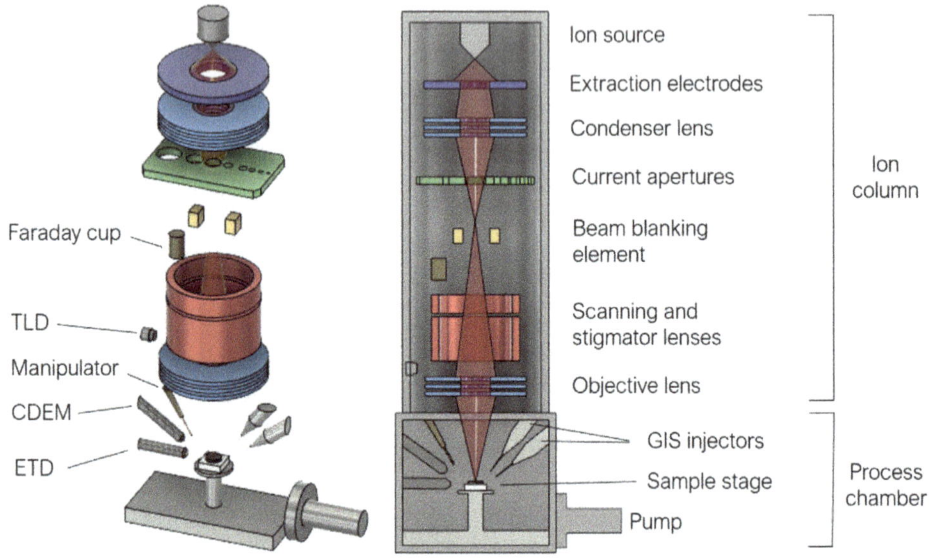

Figure 5.3. Diagram of a generic FIB instrument, depicting the fundamental constituents. The FIB column has been placed in the vertical axis for simplicity, but in dual FIB/SEM instruments it is found angled around 52° with respect to the vertically-placed SEM column. Not to scale.

located near the sample. Depending on the particle optics between the two, four main regimes can be differentiated: semi-convergent and semi-divergent modes, on which the beam respectively converges and diverges towards the objective lens, parallel mode, on which no further beam refocusing is performed before the objective lens, and crossover regime, on which the beam is focused into a point between the two lenses and then diverges into the objective lens.

Other relevant elements of the FIB column are the *beam blanker* element, used to rapidly steer the beam away from the axis when it is required to, and the *Faraday cup*, a conductive cup-shaped device used to precisely measure the ion beam current by quantifying the charge accumulation that occurs on it when the FIB is steered onto it.

5.2.4 Gas injection system

The *gas injection system* (GIS) is the component of FIB instruments used to deliver the precursor materials to the substrate surface during FIBID. The precursor materials are stored in dedicated reservoirs, commonly in solid state, although liquid and gas-state precursors exist. To inject the material into the chamber in gas form, the reservoirs are heated to a certain temperature to sublimate the material, which is then delivered to the chamber through a nozzle placed in close proximity to the sample in terms of lateral distance and height (between 50 and 200 μm in some cases). If there are multiple precursor materials in the system, each may have one individual, dedicated nozzle, or they may all share a common one.

The spectrum of materials that can be grown via FIBID is bound by the availability of a corresponding suitable precursor. Other than obviously containing the element to be deposited, precursor materials must satisfy several requirements to be able to be used in FIBID [31, 32]:

- the vapor pressure of the precursor in the GIS must be high enough to ensure a sufficient flux of precursor is delivered to the chamber;
- precursor molecules must have a sufficient sticking coefficient to the surface to which they adsorb, and a subsequent residence time in that surface large enough to ensure they can be decomposed under irradiation by the beam;
- the precursor material must be non-hazardous to both handling operators and the equipment itself.

Table 5.1 lists some relevant precursor materials.

5.2.5 Process chamber and electron/ion detectors

The process chamber is the main compartment of the instrument. Samples are placed on a moving platform referred to as *sample stage*, which can be maneuvered in all three directions of space, rotated around its axis, and tilted up to a certain angle.

Most systems and modules within FIB instruments, such as the FIB itself, the GIS, or spectroscopy detectors, are calibrated to optimally work at a specific height of the sample stage, called *eucentric height*. For FIB/SEM microscopes, the eucentric height is set at the coincidence point between the two beams and the stage, which are tilted one with respect to the other in an angle typically between 45° and 55° [34]. When tilting the stage while at the eucentric height, the working area remains in focus. Since in most commercial FIB/SEM instruments the SEM column is placed in the vertical direction and the FIB is angled with respect to it, tilting the stage to

Table 5.1. Typical FIBID precursor materials found in literature.

	Precursor material	Short formula
Tungsten	Tungsten hexacarbonyl	$W(CO)_6$
Platinum	Trimethyl(methylcyclopentadienyl) platinum (IV)	$MeCpPtMe_3$
	Trimethyl(cyclopentadienyl) platinum (IV)	$CpPtMe_3$
Gold	Dimethyl gold hexafluoroacetylacetonate	$C_7H_7F_6O_2Au$
Cobalt	Dicobalt octacarbonyl	$Co_2(CO)_8$
Aluminum	Trimethylamine alane	$(CH_3)_3NAlH_3$
Copper	(Hexafluoroacetylacetonate)-copper(I)-(trimethylvinylsilane) [33]	Cu(hfac)TMVS
Silica	Tetraethoxysilane/Tetraethyl orthosilicate/TEOS	$Si(OC_2H_5)_4$
	1,3,5,7-tetramethyl-cyclotetrasiloxane/TMCTS	$C_4H_{16}O_4Si_4$
Carbon	Naphthalene	$C_{10}H_8$
	Phenanthrene	$C_{14}H_{10}$

match the inclination angle of the FIB is mandatory if normal incidence of the FIB is required (as is the case for conventional FIB milling and FIBID).

Most FIBIP procedures, if not all, require (at least to a certain extent) obtaining micro- and nanoscopic images of the sample, either for observation, precise positioning, compositional and structural analysis, etc. These images are retrieved following the conventional principles of scanning electron microscopy, on which, succinctly, pictures are obtained by collecting and analyzing the electrons that the sample emits upon interaction with the beam of charged particles.

The specifics on the origin and nature of the different electrons emitted from the sample will be discussed in a later section. For now, it will suffice to say that mainly two types of emitted electrons are relevant for imaging purposes: *secondary electrons* (SEs), inelastically scattered as a consequence of some loss of energy of the incident particle when it excites an electron in the sample, and *backscattered electrons* (BSEs), exclusive to SEM, which are electrons from the primary beam that are elastically scattered away from the sample after impinging on it. The signal associated with the former mainly provides topographic information, whilst the contrast associated with the latter is mostly related to the atomic number of the elements in the material that is being imaged, thus providing compositional contrast.

FIB/SEM instruments (and some FIB equipment, such as HIMs) are equipped with a series of *electron detectors* that collect these emitted electrons and process their corresponding signal so that an image can be constituted. It is outside the scope of this chapter to go into detail over their working principles and relevance, but a basic overview is provided due to their relevance in most FIB processing techniques.

The most commonly found electron detector is the *Everhart–Thornley* detector (ETD). The ETD is placed inside the chamber, above the sample stage at an angle of about 30°. Its end is covered with a metallic grid named *Faraday cage*, to which a positive bias is applied (variable, up to 250 V). In this way, SEs are attracted towards it, inducing the emission of light by a scintillating material. This signal is further amplified via a photomultiplier and eventually converted into an electric current using a photocathode, from which the image is retrieved. This setup is used to capture mostly SEs, although in SEM, BSEs emerging in the angle at which the ETD is placed can be detected by it as well [35, 36].

The signal retrieved by the ETD also contains second-order SEs or SE-II (i.e., SEs whose emission has not been directly induced by the primary beam, but by backscattered electrons or ions) which worsen the resolution of the image. Some microscopes are equipped with objective lenses whose associated magnetic field is strong enough to reach the sample (provided the working distance is short enough), trapping electrons in an upwards spiral trajectory. Electrons can thus be retrieved *above* the lens using a *through-the-lens* detector (TLD), polarized at high voltage (10 kV). This allows for efficiently separating second-order and backscattered electrons from primary SEs, which comprise the image retrieved in the TLD. The electric signal is also extracted by a scintillator-photomultiplier setup [35, 36].

The ETD and TLD detect BSEs only incidentally, since the energy of emerging BSEs is much higher (in the order of keV) than that of SEs, so BSEs are detected using dedicated detectors. Rather than using a voltage to attract the electrons, these are usually designed as fixed pieces that cover a large area, towards which the BSEs impinge directly. The signal can be obtained either via scintillation, or by inducing electronic excitations in a semiconductor material [35, 36].

The previous detectors are commonplace in SEM, and are typically optimized to detect electrons whose emission has been induced by a focused electron beam (FEB). Although they can be used when the primary beam is an FIB (and indeed are, for instance in HIM imaging [37]), specialized detectors mostly dedicated to FIB imaging are available as well. Widely known as *electron multipliers*, these dynode-based detectors capture the SE signal induced by an FIB and greatly enhance it, allowing for image retrieval. A common example is the *channel detection electron multiplier* detector (CDEM, also standing for *continuous dynode electron multiplier*) [6, 38].

The electric signal coming from the detectors passes through an *image processing board*, which maps the intensity distribution of the electrons coming from the sample and yields a black-and-white image that is visualized on the screen. The contrast can be adjusted by modifying the amplification parameters of the signal, and varies depending on the type of signal that is being processed.

5.2.6 Additional modules

Some FIB and FIB/SEM instruments are equipped with additional modules that are used to either complement the nanofabrication process in some way, or to perform *in situ* additional characterization procedures. These include, but are not limited to, a microprobe station to perform *in situ* electrical measurements on the samples; a nanomanipulator used to handle pieces of extracted substrate material, used in transmission electron microscopy (TEM) sample preparation; spectroscopy modules to determine the composition of the sample via secondary ion mass spectrometry (SIMS) or energy-dispersive x-ray spectroscopy (EDX/EDS) (FIB/SEM systems only, since it needs an electron beam to induce enough counts of the characteristic x-ray emission for each element); a stage temperature controller; and an electron flood gun used to correct charge imbalance.

5.3 Basics of FIB processing

Accelerating the ions down the column provides them with a significant amount of energy that is redistributed as they reach the sample. The energy exchange that occurs at the sample is the driving force for all FIB processing (figure 5.4).

The most common and straightforward application in FIB processing is ion-induced *milling*, through which material is precisely removed from the sample as the FIB scans its surface. Alternatively, ion irradiation may also result in *permanent modification of the material properties* via structural and compositional changes (e.g. amorphization and ion implantation). Materials modification is usually regarded as an unwanted side effect inherent to the nature of the process (referred to as 'ion damage' or similar). However, it can also be exploited for wanted purposes, such as resist exposure for ion beam lithography, in a similar fashion as a focused electron beam selectively sensitizes resist in an electron beam lithography process.

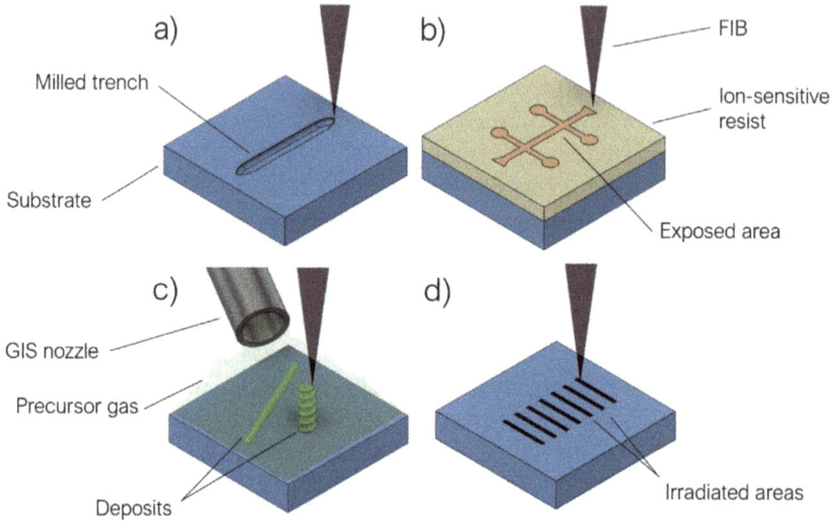

Figure 5.4. Typical applications of FIBIP. (a) Milling, (b) ion beam lithography, (c) FIBID, and (d) material modification by ion irradiation.

When used in combination with a precursor material, ion irradiation results in *material growth*. Deposition occurs as a consequence of several interactions taking place between the ions from the beam with the substrate surface and the precursor molecules.

This section is dedicated to present the relevant interactions taking place during the different processes and the controllable parameters that can be adjusted to optimize it.

5.3.1 Ion–solid interactions

As the beam is scanned over a material, several interactions initiated by the ions from the FIB take place (figure 5.5). They arise as a consequence of the conservative transfer of energy and momentum from the ions to the atoms in the sample.

Depending on the mechanism that governs the energy loss of the impinging ion, the ion–solid interactions can be broadly divided into two categories: *nuclear* and *electronic*. On the former, the interaction is governed by the elastic interaction of the nucleus of the ion with the nuclei of the atoms in the sample, whilst for the second, the interaction takes place between electrons from the ion with those present in the atoms of the material.

The extent up to which these interactions take place is governed by a parameter referred to as *ion range*, which indicates the average distance travelled by an ion until it loses all of its impinging energy. The ion range defines a physical region within the substrate on which the ion irradiation is relevant, called *interaction volume*. This volume is pear-shaped (in most conventional operating conditions) and its bounds are strongly dependent on the ionic species used to perform the irradiation, as well as on the nature of the material that is being irradiated (figure 5.6). Differences in the

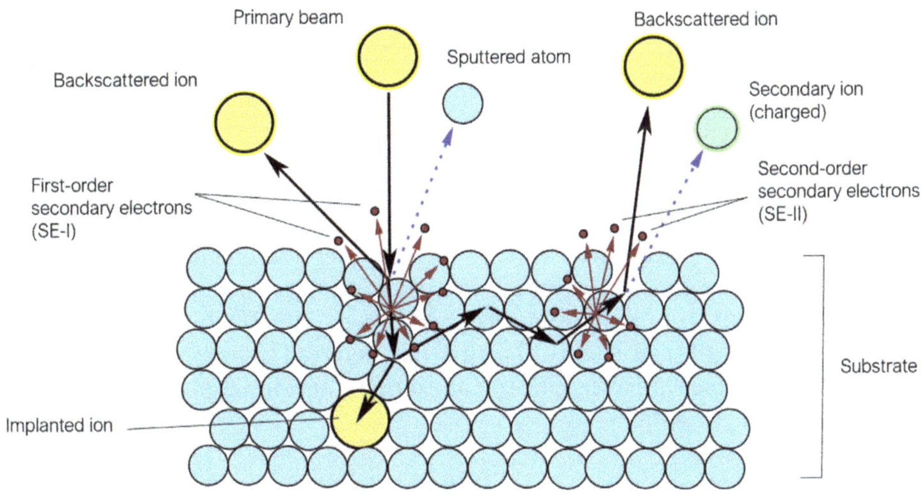

Figure 5.5. Summary of ion–solid interactions in FIB processing.

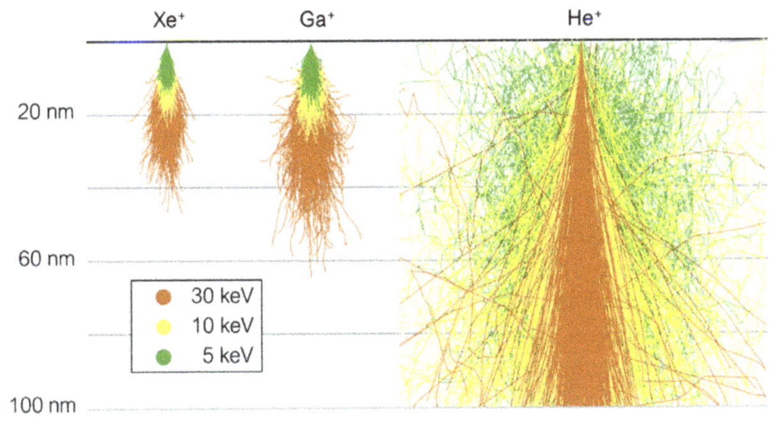

Figure 5.6. SRIM simulations of trajectories in a Si substrate of three ionic species under different acceleration voltages, showing the interaction volume of each.

interaction volume of the ions in the sample lead to substantial changes in the processing and growth of the material [17].

Nuclear energy losses

Ionic nuclear energy losses take place via successions of individual, elastic collisions between the nuclei of the ions with those of the substrate. Incident ions elastically collide with atoms in the substrate, displacing them from their equilibrium position and triggering a *collision cascade*, through which momentum is elastically transferred from one atom in the substrate to a neighboring one. The atomic displacements taking place during the cascade generally result in amorphization of the material.

In a collision cascade, if the momentum transferred to atoms close to the surface of the substrate exceeds their binding energy, they get ejected from the bulk.

This effect is called *sputtering*. Sputtered atoms from the substrate may then redeposit in the vicinity of the point where they were ejected from. Collision cascades can also result in the emission of charged ions from the substrate, referred to as *secondary ions* (SIs), since they are the byproduct of a primary radiation represented by the FIB itself. SIs are used as a compositional fingerprint in SIMS.

Additionally, elastic collisions may result in the ions themselves getting ejected back from the substrate after one or multiple collisions, in a process referred to as *backscattering*. Both sputtering and backscattering are enhanced as the angle between the FIB and the substrate is increased away from normal incidence.

A part of the ions from the primary beam will remain in the bulk of the substrate once they have lost all their energy. This effect is referred to as *ion implantation* and results in the presence of a significant amount of the ionic species on the final composition of the deposit.

Electronic energy losses

Energy from the ions is also transferred to the substrate through electronic interactions. With the energy of the FIB always being set above the ionization potential of the substrate, the energy transfer results in the appearance of excitations in the form of emission of the already mentioned *secondary electrons* as ionization products, emission of electromagnetic radiation, and in the creation of plasmons and phonons.

Analogously to the aforementioned SIs, SEs get their name due to their emission being a consequence of the irradiation with a primary beam, the FIB. As anticipated, depending on where they are extracted from, SEs are termed SE-I, or *first-order*, when they are generated at the local area of irradiation, and SE-II, or *second-order*, when they arise far from it. SE-IIs appear when the impacting ions are scattered away from the irradiation point and induce new excitations in a different region.

SEs are of paramount importance in FIB/SEM microscopy, for the topographic contrast in the images is obtained by processing of their associated signal. In addition, SEs are mainly responsible for the material growth in conventional FIBID.

If the substrate is polycrystalline and the ions happen to impinge in a low-index crystallographic direction, the reduced number of substrate nuclei in these directions results in the ion undergoing much fewer nuclear interactions than it would if it had impinged on a different crystalline direction. This effect, referred to as *crystallographic channeling*, has two main consequences: the appearance of significant contrast in SE images between oriented and non-oriented directions as a consequence of the change in SE yield (oriented directions appear darker); and a reduction on both the milling and material redeposition rates in oriented directions, since the number of nuclear collisions near the surface diminishes as the ions channel (consequently, milling in oriented directions is slower, but sharper) [39].

Numerical models

The computational modelling of ion-substrate interactions, used to gain insight on the expected behavior of a material under irradiation by an FIB, is performed by developing statistical-based approaches named *Monte Carlo simulations*, which have also proven useful for SE emission studies [40, 41]. Their specifics are outside the scope of this chapter, but some of them will be briefly mentioned in this chapter for the sake of completeness.

The program *transport of ions in matter* [42] (TRIM), developed in the 1980s, still remains a relevant and consolidated approach as of the present day. The program defines the trajectories of individual impinging ions by calculating the changes in energy and trajectory that occur at each individual binary collision with the atoms in the substrate, tracing a collision cascade. This approach allows for the extraction of statistically significant conclusions when a large number of paths is simulated. Different conditions can be emulated by modifying the ion's initial energy, or the crystallinity and composition of the substrate, obtaining a reliable approximation of the corresponding interaction volume. TRIM is implemented in an open source software piece named *stopping and range of ions in matter* (SRIM) [43].

More recent and sophisticated approaches to modelling ion behavior inside a material, which are used to account for observed behavior in complex FIBIP scenarios, include the programs *EnvizION* [44–48], which expands TRIM to include the sputtering effects taking place in the surface of the substrate; *IONiSE* [40], which focuses on simulation of SE trajectories; and *TRI3DYN* [49–51], which implements the binary collision approximation for 3D nanostructures.

5.3.2 Operating the beam

Beam scanning

In FIBID and in FIB processing in general, the patterns traced by the beam are constituted by a series of small, identical 'irradiation spots' in close proximity. The FIB is steered upon each point during a fixed amount of time and then moves to the next one (figure 5.7). With some texts referring to each of these points as *pixels*, care must be taken since in general they do not correspond with the digital pixels of a

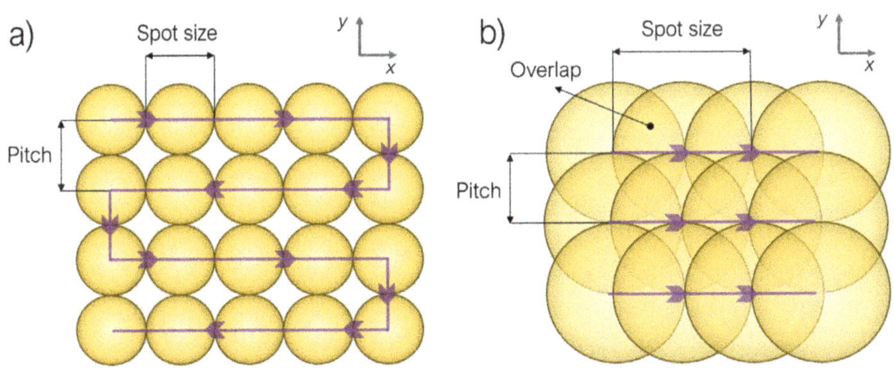

Figure 5.7. Diagram of FIB scanning of a rectangular pattern using different operation parameters. (a) The pattern is traced by successive exposure to the beam of 20 irradiation spots (using a small spot size) with an x-pitch and y-pitch equal to the spot size (equivalent to overlap to 0%), and using a serpentine scan. (b) The design is patterned so as to have 12 irradiation spots and a larger spot size. The x-pitch and y-pitch are set to half the spot size, resulting in a 50% overlap between successive irradiation spots. The scanning is performed in raster mode. Note that if the dwell time is set to the same value in both scanning types, approach (b) will be faster.

bitmap. The scanning of the complete series of points is in most cases repeated multiple times to deliver the required ion dose to the substrate [52].

The shape of each of the irradiation spots is roughly Gaussian [52] with a size ultimately mainly determined by the working current, increasing for higher currents. The scanning is usually set for the individual irradiation spots to overlap with each other, eventually constituting a definite or arbitrary shape fixed by the operator: a line, a curve, a polygon, a digital image in the form of a bitmap, etc.

The shape fidelity and lateral resolution of the technique are given by the dimensions of the interaction volume, which depends on the conditions of each specific FIB system-substrate combined scenario. For milling only, and with flat and conducting substrates, the resolution is mainly limited by the spot size of the beam, associated with the ion beam current: using low ion beam currents, features of around 10 nm can be milled with Ga^+ ions, and down to 0.5 nm with He^+ ions. On the other hand, the resolution will be jeopardized if the substrate contains heavy elements, as they give rise to a strong ion–substrate interaction very close to the substrate surface. For growth purposes, the resolution worsens due to the large lateral range reached by the SEs dissociating the precursor molecules and, in addition, it is eventually bottlenecked by the spot size of the beam. In optimal conditions, the best resolutions of Ga^+ and He^+ FIBID are around 20 and 10 nm, respectively. Other limitations are practical in nature, such as the pattern requiring a reasonable amount of growth time at a specific current.

The most straightforward pattern that can be grown by FIBID is a film-like structure, by repeatedly scanning the FIB over an area while constantly replenishing the precursor. In-plane nanowires and nanowire-like structures are also commonplace when working with FIB. By focusing the beam on a single point under a constant precursor flux, it is possible to grow out-of-plane nanowires, commonly referred to as 'nanopillars'. As will be described in the following sections, FIBID of more complex three-dimensional structures such as nanohelices [53] and nanotubes [54] can also be achieved through specific adjustment of the growth parameters.

Adjustable parameters

Most of the parameters that govern FIBIP can and must be adjusted by the instrument operator to optimize each growth procedure. Setting a growth protocol for a given type of sample, consisting in a systematic tuning of these variables, is commonplace when attempting the growth of a new or challenging geometry.

- The *acceleration voltage* is the potential difference used to accelerate the ions down the ion column. The most commonly used value is 30 kV, although the attainable range extends down to 0.2 kV and up to 60 kV depending on the specific instrument. Generally speaking, higher acceleration voltages yield higher resolution, at the cost of a reduced growth rate.
- The *ion beam current* is a measure of the (ion) charge delivered to the substrate per unit time. It is adjusted by selecting an appropriate aperture and/or modifying the gas pressure at the source if a GFIS is being used. In FIBID, typical ion beam current values usually fall in the pA–nA range. Depending on the equipment, currents as low as 0.1 pA and as high as 2 μA (in Xe PFIB systems [55]) can be delivered.

Table 5.2. Operation parameters in FIBID, with the main features of the processed feature they influence, and their typical values/modes of usage.

Type	Parameter	Definition	Related outputs	Typical range/mode
Beam	Acceleration voltage	Energy acquired by the ions down the column	Interaction volume	30 kV
	Ion beam current	Charge delivered per unit time	Spot size, growth rate	pA to nA
Patterning	Dose/fluence	Received charge per unit area	Growth rate	Variable, measured in charge/area.
	Volume per dose	Achievable growth for a given dose	Growth rate	Variable, measured in volume/charge
	Dwell time	Time spent at each individual pixel	Milling/growth ratio	Nanoseconds, microseconds
	Refresh time	Time spent at each pass OR Wait time between successive passes	Milling/growth ratio	Microseconds
	Pitch/overlap	Physical separation between adjacent pixels	Milling/growth ratio	50% of the spot size
GIS	GIS nozzle positioning	In-plane distance as seen normal to the FIB, and vertical distance to the stage	Amount of available precursor	Hundreds of micrometers
	Precursor temperature	Temperature of the precursor reservoir	Amount of available precursor	Material-dependent
Other	Stage temperature	Temperature of the substrate	Growth mechanism	Variable, most typically room temperature

- The *dose* is a quantification of the charge received by a material per unit area in a given time interval. This word is used somewhat interchangeably with the term *fluence*, the difference between the two being essentially formal in nature (while the dose accounts for the charge actually received by a material, the fluence accounts for the charge passing through a given volume, divided by the cross-sectional area of that volume).
- The *volume per dose* is a parameter that accounts for the volume of material that can be grown/removed per unit charge. It is defined as the ratio between the volume of the processed feature (a hole or a deposit) and the total charge received by the substrate, which can be calculated from the dose.
- The *dwell time* is the time spent by the FIB on each individual irradiation spot during scanning. For milling purposes, high dwell times (i.e. keeping

the beam focused at an individual pixel for an extended period of time) can lead to unwanted redeposition of sputtered material. In FIBID, high dwell times generally produce precursor depletion and, subsequently, milling.

- The *refresh time* is defined as the time required to complete a full scan of the pattern, referred to as a *pass*. It amounts to the dwell time times the number of points in the pattern, plus any extra time added between successive scans. As a note of caution, in some cases the term is used to denote this additional wait time between passes without including the time spent scanning the pattern.
- The *pitch*, also referred to as *step size*, is the physical separation between two adjacent points on a pattern. The pitch is typically set below the spot size of the beam, so the irradiated area overlaps between adjacent spots. A different pitch can be defined in both the x and y directions. The *overlap* is actually a controllable parameter by itself, and in most cases is set to a value of 50% of the spot size.

The chemical composition and morphology of the material can also be tuned via thermally-assisted methods, namely heating the substrate during and after growth (annealing). Performing the deposition above room temperature is found to markedly affect the composition of the material, increasing the metallic content and promoting crystallinity, as observed in FIBID of copper [56, 57] and gold [58, 59].

Annealing is found to have a moderate purifying effect on Pt-FIBID nanowires, slightly increasing the metallic content [60], but with the resistivity of the sample remaining fundamentally unchanged [61]. In addition, rapid annealing treatments are found to have a significant bending effect on freestanding nanostructures [62].

5.4 Milling and materials modification

5.4.1 Milling

Standard milling

The most straightforward application of FIB processing is direct milling. Given the heavy mass of ions compared to electrons, the linear momentum of accelerated ions is high enough to displace and eventually remove atoms from a given material. Under the same acceleration energy, the ratio of the linear momentum of two different charged particles depends on the square root of their mass ratio. In the case of a light ion such as He^+, the ratio of its linear momentum to that of an electron is 85, increasing to 355 for a heavier ion like Ga^+. The impacting ion will produce a collision cascade phenomenon and, as described in section 5.3.1, it will transfer energy and linear momentum to surface atoms, milling the material. The efficiency of this milling process can be quantified through the *sputtering yield* (number of sputtered atoms per number of incident ions) or the *sputter rate* (sputtered volume/ion dose) and depends on various parameters, the main ones being:

- The type of ion: the sputtering yield will increase with the ion mass, given the higher momentum of the impacting ion. For example, the sputtering yield of Si under 20 kV ion irradiation is 0.66 for Ne^+ ions and 2.9 for Xe^+ ions [63].
- The ion energy: in first approximation, the sputtering yield will increase with the acceleration energy given the higher momentum and energy of the impacting

ion. For example, under Ga^+ FIB normal incidence, the sputtering yield of Si at 20 kV is 1.94 [64], increasing to 2.4 at 30 kV [65]. However, at very high ion energies, the sputtering yield decreases again due to deeper ion penetration in the material.

- The angle of the ion with respect to the surface: the ion incidence angle can have a profound influence on the milling rate given that the removal of a surface atom requires a backscattered ion to approach the surface. As an example, in the case of Si, the sputter rate increases from 0.24 $\mu m^3\ nC^{-1}$ for normal incidence (90°) to 1.6 $\mu m^3\ nC^{-1}$ for a 10° incidence angle with respect to the surface, decreasing to zero for parallel incidence (0°).
- The surface binding energy of the atoms in the material to sputter: as a rule of thumb, the lower the binding energy, the higher the sputtering yield.

In table 5.3, the sputter rates of various materials under 30 kV Ga^+ FIB normal incidence are displayed.

The most important aspects to be considered for milling processes are:

- The achievable milling resolution. Given one ion species, a higher ion energy will give rise to higher-resolution milling. The higher resolution is a consequence of a more finely-focused beam occurring at high energy, due to the decrease of the ion column aberrations, and the lower spread of the ion beam in the vicinity of the surface. Comparing the available commercial FIB technologies, Ga^+ FIB columns are widely used and have improved to the point of achieving sub-5 nm resolution [66]. More recently, He^+ FIB and Ne^+ FIB commercial columns can

Table 5.3. Sputter rate for various materials under 30 kV Ga^+ FIB normal incidence.

Material	Sputter rate ($\mu m^3\ nC^{-1}$)
Al_2O_3	0.08
Cr	0.10
Si_3N_4	0.16
C	0.18
Pt	0.23
Si	0.24
SiO_2	0.24
Cu	0.25
Fe	0.29
Al	0.29
Ta	0.32
PMMA	0.4
Ti	0.46
GaAs	0.61
InP	1.2
Au	1.5

provide beam spots of the order of 0.3 and 2 nm respectively, which can be useful for high-resolution milling applications [10].

- The milling throughput. As it increases with the impacting ion mass, Xe^+ FIB columns have been recently commercialized in order to achieve much higher sputter rates than Ga^+ FIB columns [67].
- The damage created in the material around the milling area [68]. Higher ion energies result in damage in larger areas given a deeper penetration in the material. Typical material damage includes amorphization, recrystallization, vacancy creation, alloying, etc. This is why some processes, such as the TEM lamellae preparation, start at high acceleration voltage for obtaining a large milling rate, but they are concluded using a lower acceleration voltage to minimize the damage in the final structure [67].

Generally unwanted side-effects that can occur during FIB milling are:

- Part of the milled material is redeposited in the surrounding areas and can give rise to electrical short-circuit, lack of straight milling profiles, surface roughness, etc. This effect can be sometimes minimized by means of an appropriate patterning strategy such as multi-pass versus single-pass beam scan, left-to-right versus top-down beam scan direction, etc.
- When the milled sample is composed of various materials with different milling rates, FIB milling produces rough milled surfaces, curtain effects, etc. Gas-assisted FIB milling can minimize this drawback (see next section).
- Smoothly milled surfaces can prove challenging to achieve in a single material with various grain orientations due to ion channeling effects along highly-symmetric crystallographic orientations that decrease the milling efficiency. Gas-assisted FIB milling can minimize this drawback (see next section).
- Ion implantation can be detrimental if the physical properties of the milled material are severely affected. The material conductivity, magnetic or optical properties can be dramatically changed.
- The formation of helium bubbles (accumulation of gaseous helium near the surface) when performing He^+ FIB milling on certain materials has been observed.

Despite the abovementioned difficulties, FIB milling is the lithography of choice in several applications requiring material removal given that it is a direct-write technique that does not require the use of resists and is capable of nanometer lateral resolution (figure 5.8). Some relevant applications include:

- Circuit edit in semiconductor industry [69, 70].
- TEM lamellae preparation and cross-sectional FIB-SEM imaging in materials and life sciences [71, 72].
- FIB-SEM 3D tomography [73],
- Small hole drilling in membranes for DNA/protein sequencing and preparation of shadow masks [66, 74, 75].
- Cantilever tip sharpening for scanning probe microscopy [76].
- Atom probe tomography specimen preparation [77].

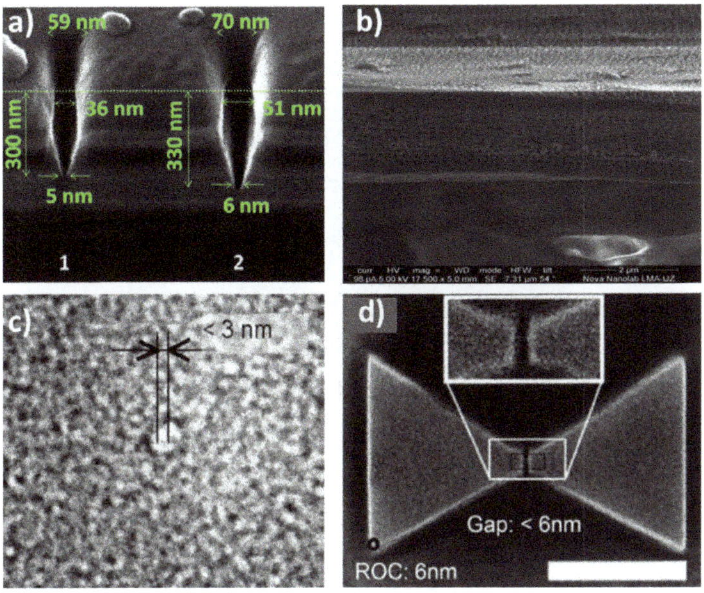

Figure 5.8. (a) 45°-tilted HIM image of the cross-section of a large trench (50 nm × 200 nm) milled by 5 keV Ne^+ FIB through a test sample for circuit edit purposes. Reprinted from [86]. With permission from Elsevier. (b) SEM image after FIB milling of a beetle scale, which allows observing the sub-surface tubular structures producing plasmonic effects. Image credit: Teobaldo E. Torres (LMA), unpublished, collaboration with S Mouchet and O Deparis. Used with permission. For related research, see [87]. (c) TEM image of a nanopore fabricated on a SiC membrane by Ga^+ FIB milling. Reprinted from [88]. Copyright (2009), with permission from Elsevier. (d) 6 nm gap gold nanoantenna fabricated by He^+ FIB milling. Reprinted with permission from [89]. Copyright (2014) American Chemical Society.

- Creation of nanophotonic devices [78].
- Mask repair applications [79, 80], where the technique has to compete with focused electron beam induced etching [81].
- High-resolution milling. Ga^+ FIB milling has been used to produce Fresnel zone plates with resolution down to 15 nm [82, 83], which is of interest to focus x-rays. Remarkably, He^+ FIB milling has been used on gold to fabricate nanoantennas with a resolution of 6 nm [84] and on graphene to create nanoribbons with a resolution of 5 nm [85].

Gas-assisted milling

- In order to enhance the FIB milling rate, the insertion of certain etching gases (Cl_2, XeF_2, I_2, …) onto the area of interest is frequently used [90]. This approach is found in applications requiring high throughput, such as circuit edit. The combination of an etching gas with an accelerated ion beam produces a combined physicochemical etching process that can multiply the milling rate of a factor up to 100, as recently shown [91], although milling rate enhancements up to a factor of 10 are usually reported [65, 90, 92–97].

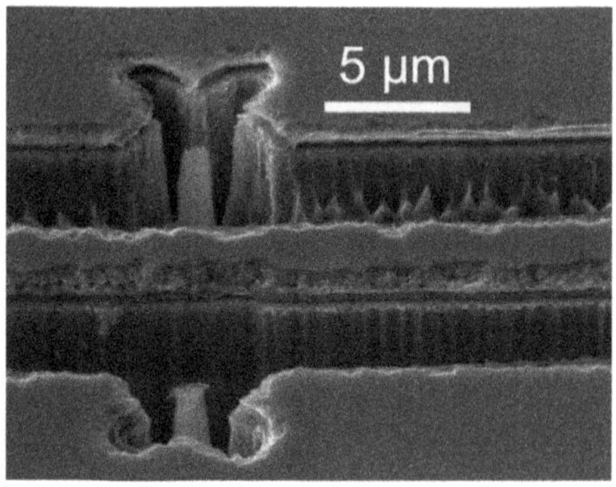

Figure 5.9. SEM micrograph of an I_2-assisted Ga^+ FIB-milled InP-based trench waveguide. Reprinted from [96]. Copyright (2010), with permission from Elsevier.

- Cl_2 and other Cl-based gases have been successfully used for milling rate enhancement in FIB milling of Al, Si, GaAs, InP, Cu, Cr and C.
- XeF_2 has been successfully used for milling rate enhancement in FIB milling of Si, SiO_2, Si_3N_4, C, Ti, Nb, Cr, Al and W.
- I_2 has been successfully used for milling rate enhancement in FIB milling of Si, InP, GaAs, Au, Al, Nb and $CH_3NH_3PbBr_3$ (figure 5.9) [96–98].

The gas-assisted milling has been found to produce more homogeneous milled structures as well as minimization of curtain effects and grain-orientation inhomogeneities [94, 99].

Cryo-FIB milling

It has been shown in devices based on semiconductors and fabricated by means of FIB milling that the performance is enhanced when the devices are patterned at lower temperatures [96, 100]. The reason for this improved behavior is that the use of low temperatures significantly reduces the diffusion of defects created during FIB milling.

Recent results show that the thickness of the amorphization layer created during FIB milling of the topological-insulator Bi_2Se_3 material is much reduced (to be published), showing great potential for the direct FIB patterning of topological circuits.

Cryo-FIB milling in cross-sectional FIB/SEM imaging or during lamellae preparation has been found to be of great interest when soft solid-state materials are present or when organic and inorganic materials are present in the sample of interest [101, 102]. It has also proven extremely useful in life sciences for frozen biological materials (figure 5.10). Adequate work protocols allow these materials to be subsequently imaged by FIB/SEM or TEM [103, 104].

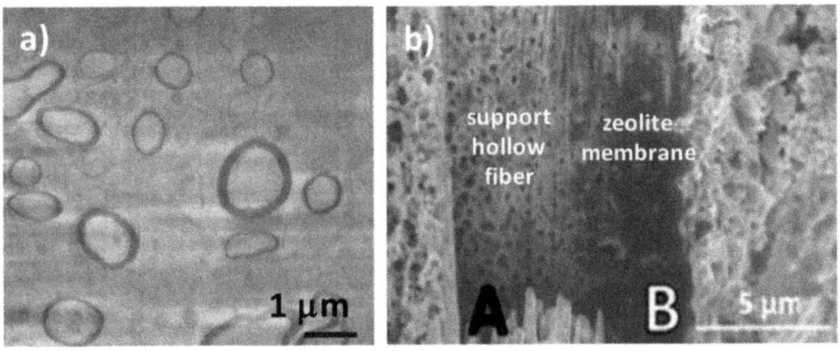

Figure 5.10. (a) SEM image of a hydrated frozen mouse optic nerve after Cryo-FIB milling. Reprinted from [104]. Copyright (2013), with permission from Elsevier. (b) SEM image of a hollow-fiber-supported membrane section after Cryo-FIB milling. Reprinted from [102]. Copyright (2015), with permission from Elsevier.

5.4.2 FIB irradiation for materials modification

Local modification in physical properties induced by FIB irradiation

In this section, a few examples of modifications in physical properties arising from FIB irradiation are listed, in order to show how versatile this technique is, and selected examples are shown in figure 5.11.

- In semiconductor materials, Ga^+ irradiation on porous Si films was found to produce selective photoluminescence on the irradiated areas [105].
- In magnetic materials, several magnetic properties can be tuned by FIB irradiation such as the magnetic-domain-wall propagation field [106, 107], the coercive field and the magnetic anisotropy [108–110], the pinning of magnetic skyrmions [111], etc.
- The mechanical properties of certain materials can be tuned locally by FIB irradiation, as shown for Cu nanopillars [112].
- Nitrogen-vacancy centers can be controllably created in diamond for advanced quantum sensing by N^+ FIB irradiation [113, 114].
- Taking advantage of the great precision of FIB, it is nowadays possible to produce single-ion implantation within a diameter of 100 nm in a deterministic way [115], which can be useful in semiconductor and quantum devices.
- The electrical conductivity of insulating diamond-like carbon material can be tailored through FIB irradiation, with resistivity values as low as 10^3 µΩ cm being achieved [116].
- Spin-filtering regions in non-magnetic InGaAs layers have been created by Ga^+ FIB implantation [117].

Resist exposure for ion beam lithography

Ga^+ FIB was used in the past to expose various types of resists in order to explore the performance for ion lithography. A good result was found using the SAL601 resist with 30-kV Ga^+ FIB irradiation [118]. Line widths with resolution better than 100 nm were found with sensitivity of 8×10^{11} ions/cm^2.

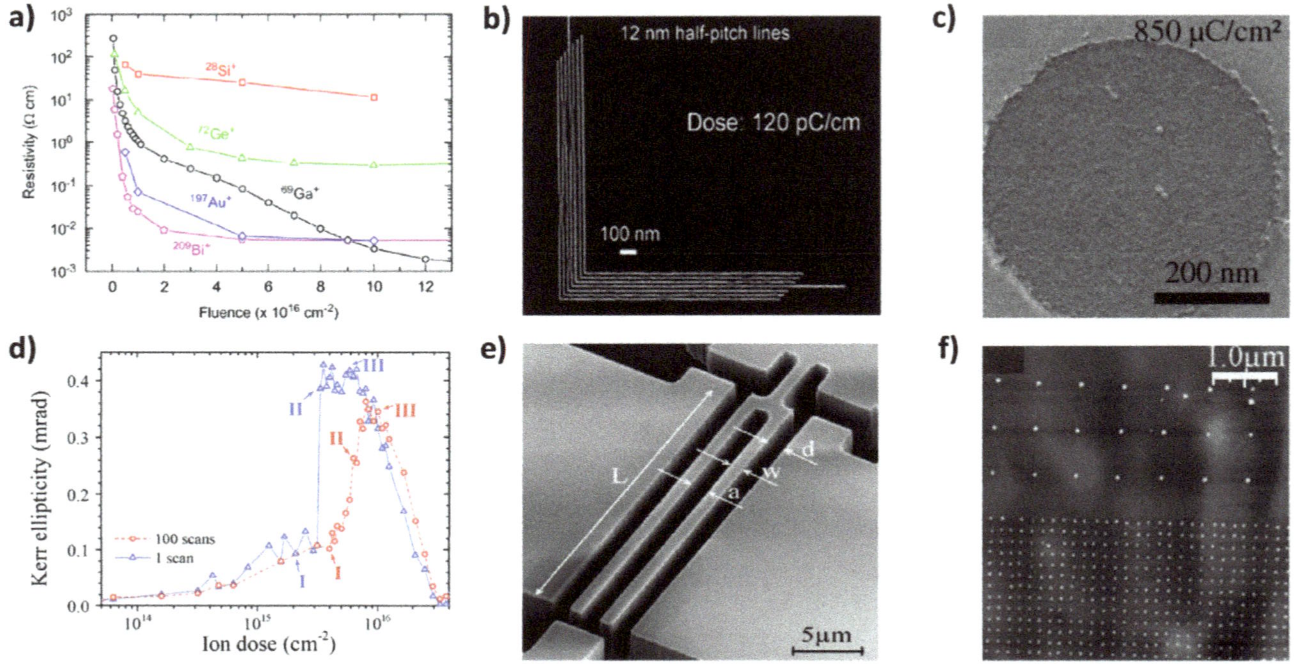

Figure 5.11. (a) FIB irradiation of a 113 nm thick diamond-like-carbon (DLC) film with various ion species accelerated under 30 kV, which allows for decreasing the electrical resistivity of this material. Reprinted from [116]. Copyright (2014), with permission from Elsevier. (b) HSQ resist lines exposed by means of He^+ FIB. Reproduced with permission from [122]. Copyright 2012 American Vacuum Society. (c) Circular-shape carbon nanomembranes fabricated by means of He^+ FIB-irradiated self-assembled monolayers due to cross-linking. Reproduced with permission from [126]. Copyright 2014 Zhang *et al*, licensee Beilstein-Institut. (d) Degree of ferromagnetic behavior (measured through the Kerr ellipticity) of $Fe_{78}Ni_{22}$ thin films submitted to Ga^+ FIB irradiation using 1 scan or 100 scans for a given ion dose. Reprinted from [129]. (e) Microresonator fabricated by Ga^+ FIB irradiation of Al_2O_3 and subsequent selective etching. Reproduced from [141]. Copyright IOP Publishing Ltd. All rights reserved. (f) Selective growth of Ge quantum dots by pre-patterning the Si surface by means of Ga^+ FIB irradiation and subsequent Ge evaporation and sample annealing. Reproduced from [142]. Copyright IOP Publishing Ltd. All rights reserved.

The topic has recently seen a revival given the very good results obtained for resist exposure by He^+ FIB irradiation [119–122]. Using HSQ negative-tone resist and 35 kV He^+ FIB irradiation with 1 pA current and around 100 pC cm^{-1} line dose, resolution values down to 4 nm in half-pitch dense patterns have been obtained [122].

Growth of carbon nanomembranes using FIB irradiation

Previous results on organosilane self-assembled monolayers (SAMs) submitted to FIB irradiation with various ions (Ga^+, Si^+, Au^+, Au^{++}) and subsequent exposure to Pd solutions showed the capability for selective metallization [123]. In addition, in thiol-based SAM-gold systems, the Au–S can be selectively broken under ion irradiation, completely removing the organic molecule [124].

More recently, carbon nanomembranes with thickness in the range of 1 nm were fabricated by the combination of SAMs and He^+ FIB irradiation [125, 126]. The ion irradiation produces the selective cross-linking of the SAM films, producing a carbon nanomembrane (CNM). These CNMs have found applications in TEM supports, diaphragms in pressure sensors, gas and liquid filters, mechanical resonators, biosensing, and as protective layers or to combine with other 2D materials [125]. It has been found that the complete cross-linking of aromatic SAMs requires He^+ FIB irradiation doses around 850 μC cm^{-2}. 1D or 2D cross-linked structures can be generated depending on the FIB scan strategy. The transfer of the CNMs to other substrates can be performed using PMMA resists [126].

Modification of crystallographic phases by FIB irradiation

Using Ga^+ FIB irradiation, it has been shown that sub-100 nm ferromagnetic patterns can be imprinted on a paramagnetic $Fe_{60}Al_{40}$ matrix [127], and then erased by an annealing procedure. This phenomenon correlates with the disorder effect caused by the ion bombardment on $Fe_{60}Al_{40}$, which transforms the original B2 phase into a chemically-disordered A2 phase. This transformed phase can be viewed as an FeAl alloy that harnesses ferromagnetic behavior, becoming stabilized when Al is substituted by Ga. Moreover, the sample surface remains smooth after this phase transformation [127]. This scenario has been confirmed by spectroscopy measurements carried out on the same $Fe_{60}Al_{40}$ material submitted to Ne^+ FIB irradiation [128]. Another similar example is that of paramagnetic fcc $Fe_{0.78}Ni_{0.22}$ films. After Ga^+ FIB irradiation, the irradiated areas transform into a ferromagnetic bcc phase [129]. In this material, it has been reported that the transformation occurs following three regimes. First, the ion irradiation produces stochastically small bcc nuclei, with the number of nuclei being equal to the number of incident ions. As the ion irradiation proceeds, these nuclei grow in size, extending throughout the whole irradiated area. After the maximum saturation magnetization is reached within the irradiated area, further ion bombardment leads to a decrease of the magnetization due to defect creation, atomic intermixing and material sputtering [129]. On the other hand, He^+ FIB irradiation has been used on non-ferromagnetic ordered $L1_2$ fcc Fe_3Pt films, achieving disordered ferromagnetic patterns with resolution down to 10 nm [130].

It has been recently shown that Ga^+ FIB irradiation induces the transformation in $SrFeO_{3-\delta}$ ($\delta \sim 3$) films from the perovskite (PV) phase to the brownmillerite (BM) phase [131]. This phenomenon has been explained by the preferential sputtering produced by the ion bombardment, which leads to an oxygen-deficient phase, $SrFeO_{3-\delta}$ ($\delta \sim 2.5$). The resolution of this patterning has been found to be ~10 nm. The PV and BM phases display different physical properties [132] which could be of interest for the development of useful devices.

Use of FIB irradiation to obtain insulating paths on conductive materials
Due to the defects created introduced by FIB irradiation, it is possible to create insulating paths in films of certain conductive materials in order to create a micro/nano-device. This strategy has been used to generate GaAs quantum nanowires with width down to 20 nm [130]. The non-irradiated central part of the nanowires had a sufficiently low number of defects so that the expected universal quantum fluctuations were observed.

More recently, it has been shown that He^+ FIB irradiation on high-T_C superconducting YBCO films can be used to tune the material conductivity at the local scale [133]. In fact, FIB doses in the range of 10^{16}ions/cm^2 have been used to fabricate Josephson nanojunctions as well as SQUID magnetic nanosensors [134, 135].

Material modification by FIB irradiation for subsequent selective etching
It has been found that Si areas irradiated by Ga^+ FIB behave as hard masks for subsequent wet etching and reactive ion etching processing [136]. The etch selectivity between the irradiated and non-irradiated areas can be very high, which has been used to fabricate nanopillars with an aspect ratio of 15:1 (height of 600 nm and width of 40 nm) [137] and narrow suspended nanowires [138]. Given that the etch selectivity can be precisely controlled by the ion irradiation dose, this technique can be used for gray-scale lithography and has been used to fabricate tailored 3D nanostructures [139].

Other materials have been irradiated by Ga^+ FIB and present similar phenomenology. For example, quartz also presents etching selectivity as a function of the ion irradiation dose, which has been used to fabricate stamps for UV nanoimprint lithography [140]. Similarly, Al_2O_3 presents etching selectivity with Ga^+ FIB irradiation, which has been used to fabricate microresonators [141].

FIB irradiation for subsequent selective material growth
It has been observed that local FIB irradiation of certain substrates produces subsequent preferential growth of semiconductor materials on the irradiated areas. For example, Ge nanocrystals grow preferentially on Ga^+ FIB irradiated areas of Si and SiO_2 substrates [142, 143], which can be used to fabricate ordered arrays. Similarly, it has been found that InAs dots grow preferentially on Ga^+ FIB irradiated areas of GaAs, which is of interest for devices based on semiconductors [144].

Besides, it has been observed that Au clusters grow preferentially on Ga^+ FIB irradiated areas of highly-oriented pyrolytic graphite substrates (figure 5.11). The growth temperature is 400 K, which allows the gold atoms to diffuse on the sample surface to find the irradiated areas acting as pinning centers [145].

5.5 Focused ion beam induced deposition

5.5.1 Fundamentals

As mentioned in previous sections, one of the most relevant applications of FIBs is their capability to locally induce the growth of a material in a single step when used in combination with a precursor material. The ease of use and flexibility of the technique make it appealing for applications that require unique and highly specific designs: custom devices for prototyping, circuit repair, etc.

To perform FIBID, a precursor material in gaseous form must be delivered to the surface of the sample, usually referred to as *substrate*. Si or Si/SiO_2 substrates are the ones most typically found, although it has been shown that is indeed possible to perform the technique using exotic substrates such as insulating, flexible polycarbonate [146]. The amount of precursor that reaches the substrate strongly depends on the angle and distance of the GIS nozzle respective to the sample, on the temperature of the precursor reservoir, and on the controlled pressure of the GIS system, if applicable. In most cases, during deposition the pressure of the process chamber is increased about one order of magnitude above its base value during deposition.

Once they reach the substrate, the precursor molecules adsorb to it and constitute a molecular monolayer in dynamic equilibrium, as the molecules that leave it (according to their specific *residence time*) are swiftly replenished by the continuous precursor delivery flux, which is (conventionally) kept constant during growth. Cryo-FIBID is a notable exception to these operation principles, as will be discussed later in this chapter.

The FIB is then scanned over the sample tracing a pattern defined by the operator, locally inducing the decomposition of the precursor as described in the following section. Once the scanning is complete, with the total time defined by the dimensions of the pattern and on the selected parameters (beam current, dwell time, dose, etc), the precursor delivery is stopped, and the GIS is retracted, with the vacuum system pumping away the remnants of the precursor within the chamber (figure 5.12).

Precursor decomposition

After SEs are emitted from the substrate upon interaction from the beam, they induce a decomposition reaction of the adsorbed molecules of precursor material, dissociating them into volatile and non-volatile parts. The volatile part gets pumped away by the vacuum system, while the non-volatile constituent remains permanently deposited on the substrate.

For instance, for the $W(CO)_6$ precursor material, one can visualize the decomposition reaction as:

$$W(CO)_6 + \text{electrons} \rightarrow W + 6CO \uparrow + \text{electrons}$$

The previous decomposition reaction is, however, an oversimplification, since the actual precursor decomposition is never fully achieved, nor does it occur in a perfectly ideal environment. Firstly, the final deposit always shows 'leftover' constituents from partial precursor separation—in the case of deposition of $W(CO)_6$, as will be discussed in the following sections, the deposits always show a very significant amount of carbon. Secondly, the cleanliness and quality of the vacuum level of the process chamber

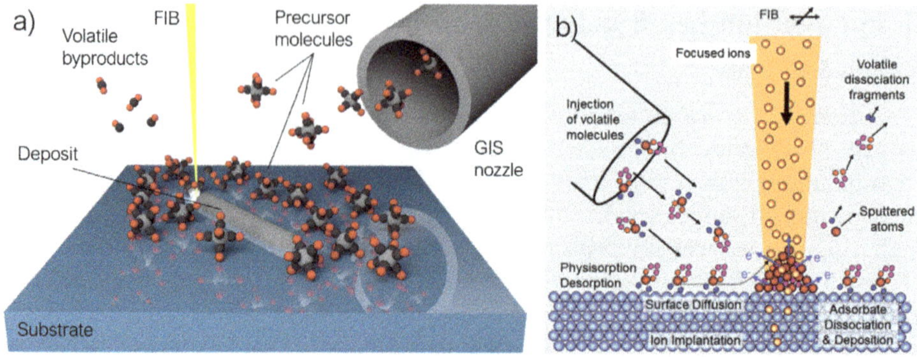

Figure 5.12. FIBID. While the sample stage is shown horizontal for simplicity, in most FIB SEM instruments it will be actually tilted so the FIB incidence is normally into it. (a) General overview, (b) principle of FIB induced deposition using a gas injection system for local deposition of 3D nanostructures by dissociating the physisorbed molecules. Gas related processes include adsorption/desorption and surface diffusion to replenish the FIB dissociated adsorbates. Note that deposition only occurs when the sputter yield is smaller than the deposition yield. Shorubalko *et al* 'Direct-write milling and deposition with noble gases' in [17]. Copyright Springer International Publishing 2016. With permission of Springer.

(usually with a base level of around 10^{-6} mbar and raised to 10^{-5} mbar during deposition) play a significant role on the final composition of the deposit, as the FIB also induces the decomposition of present contaminant species such as water molecules, hydrocarbons, and pump oils [92], if they are present.

Precursor molecules delivered by the GIS can both chemically and physically adsorb to the substrate. The precursor molecular monolayer [147] that forms on the substrate is a dynamical entity whose density depends on the adsorption and spontaneous thermal desorption rates, on the surface molecular diffusion, and on the FIB-induced decomposition rate. The adsorption and desorption rates are, respectively, governed by the GIS flux and the residence time of the molecules, while diffusion depends on the nature of the substrate and of the precursor [92]. An efficient decomposition rate relies on the SEs having an energy close to the binding energy of the precursor molecules, and on the scanning parameters being balanced with the residence time of the molecules. Detailed discussion on the various mechanisms contributing to the precursor dissociation has been carried out in the context of the growth of Pt pillars by He^+ FIBID [148], which illustrates these principles.

The electrical conductivity (or lack thereof) of the substrate also plays a significant role on the outcome of the process. FIB irradiation on substrates with insulating-like behavior results in the appearance of charge imbalance in the substrate, which leads to imaging artifacts, drift, and growth disruption. Charge imbalance can be avoided by performing the irradiation in the vicinity of discharging conducting pads previously grown on the substrate [146]. In systems where the irradiation is expected to leave the sample always positively charged (i.e., irradiating with positively-charged ions and expecting negatively charged SEs to come out of the sample), a *flood gun* that injects electrons into the chamber can be used to neutralize it. The most commonly used substrates in FIBID are Si-based, such as boron-doped Si, SiO_2, or SiN.

Depending on the ratio between the number of generated SEs and the amount of precursor available for decomposition, FIBID takes place in either an *electron-limited regime*, or in a *precursor-limited* regime. In the electron-limited regime, the precursor material is constantly delivered and replenished at a rate much higher than that of the SEs inducing its decomposition. On the other hand, in the precursor-limited regime, the SEs decompose the available precursor faster than it is replenished, resulting in either substrate milling, or milling of the growing material itself. The competition between growth and milling is a fundamental issue that must be addressed when determining the conditions of an FIBID process. For example, the use of low ion beam current is helpful in order to minimize milling effects with respect to deposition.

General properties of FIBID deposits

As with any other nanopatterning technique, the intrinsic nature of the growth procedure sets a series of constraints and features on the final samples.

One paramount limiting factor is the *serial* nature of the technique. Since FIBID is performed with a single beam sequentially scanning one spot after another, spending an amount of time on each point that depends on the required dose (rather than, for instance, exposing a photo-sensitive resist through a mask in optical lithography), the throughput of the technique is rather limited, and in most cases, not scalable for industrial applications. The work field area is, on its most extreme case, limited by the size of the aperture used to define the beam, within the order of millimeters, with the growth being typically performed with a field of view in the order of a few microns. Patterning very large areas requires high processing times and beam drift has to be dealt with. As a consequence, FIBID-grown deposits are usually in the micron-size range.

The attainable resolution strongly depends on the growth conditions, mostly the ion species, the acceleration voltage, the ion beam current, and the substrate thickness and composition. All of these play a fundamental role defining the FIB-substrate interaction volume, which, as discussed before, eventually determines the resolution of the technique. The smallest probe size is attainable at high voltages and low currents, yielding a lateral resolution in the in-plane deposit around 20–30 nm for Ga^+ FIBID, and 10 nm for He^+ FIBID.

As described before, ion-substrate interactions result in *material amorphization* and *ion implantation*. Deposits grown by FIBID are expected to show both of these features, which must be taken into account when assessing whether the procedure is viable for a certain material or application. The magnitude of these events strongly depends on the ion species that is being used, taking high significance when Ga^+ ions are used, and being mostly absent for He^+ ions.

It is not possible to establish a common generalization of the *purity* of the deposits, since it largely depends on multiple growth parameters; as well as on any post-processing methods, if applicable. It will suffice to mention that it is expected for the deposits to show elements present in the precursor material (typically C and O), in variable quantity. In addition, significant amounts of the ionic species used to induce the growth can be present as implanted ions.

FIBID is often compared to its sister technique, focused electron beam induced deposition (FEBID) [149]. The fundamentals and applicability of FEBID largely remain similar—a beam of charged particles is used to induce the decomposition of a gaseous precursor, with all ion-substrate interactions being replaced by electron–substrate interactions. FEBID lacks the growth-milling competition and ion implantation characteristic to most FIBID processes, at the cost of a reduced growth rate due to the FIB-induced SE yield being greater than that induced by a FEB. This generally implies a reduction in the metallic content of the deposit due to the FEB being less efficient than the FIB in dissociating precursor molecules.

Growth under cryogenic conditions

At room temperature, the gaseous precursor adsorbs on the surface of the substrate upon delivery, forming an individual monolayer that is continuously replenished as the FIB causes it to decompose and/or the molecules leave the surface according to their residence time. However, if the temperature of the substrate is lowered below the condensation temperature of each specific precursor material (within the cryogenics range of operation), instead of constituting this dynamical monolayer, a multilayered precursor condensate forms, much thicker than the molecular monolayer occurring in gas phase.

Since the amount of available precursor for decomposition is vastly enhanced when compared to conventional FIBID at room temperature, the volume per dose associated greatly increases. Consequently, the growth speed of a structure increases by two orders of magnitude when grown in cryogenic conditions [147]. Such a significant modification in the required dose also results in differences in the properties of the grown material, both compositional and structural in nature (e.g. absence of implanted Ga, and of substrate and deposit amorphization). This technique has been recently described for the growth of metallic W–C nanostructures, and it has been coined as *Cryo-FIBID* [147] (figure 5.13).

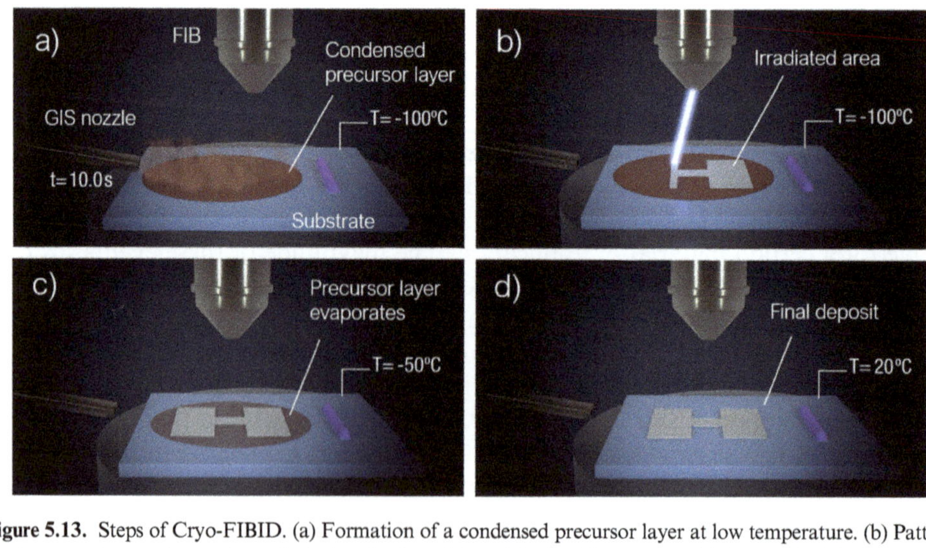

Figure 5.13. Steps of Cryo-FIBID. (a) Formation of a condensed precursor layer at low temperature. (b) Pattern tracing over the condensate layer, inducing the deposition of the irradiated areas. (c) Removal of the unexposed areas by heating the substrate above the condensation temperature. (d) Final deposit after complete evaporation of the precursor layer. Adapted by permission from [147]. Copyright (2019) Macmillan Publishers Ltd.

5.5.2 Applications of FIBID

This section presents a review of the most relevant materials that can be grown via FIBID, summarizing their properties and applications. When no ionic species is explicitly mentioned, references are to be understood as having employed Ga^+ FIBID—the usage of a different ionic species will be clearly indicated.

Platinum

FIBID of platinum is a common technique within the nanofabrication framework, very frequently used as a protective barrier against ion-induced damage in nanoprocessing for TEM sample preparation and cross-sectional FIB/SEM imaging. Pt FIBID nanowires also find application as intermediate contacts for electrical measurements.

The most common gaseous precursor used in the growth of FIBID Pt is trimethyl (methylcyclopentadienyl) platinum (IV) ($MeCpPtMe_3$), which has a melting point of 30 °C. The usage of $CpPtMe_3$ has been widely reported as well.

Characteristic growth conditions and deposit parameters from a selection of reports on Pt FIBID nanostructures are shown in table 5.4.

The presence of impurities, naturally stemming from the FIBID process, raises the resistivity of the deposited materials over its bulk value of 10.4 μΩ cm [150] along a wide range of reported values. Such a dispersion in reported data was accounted

Table 5.4. Resistivity, composition, and microstructure of FIBID Pt deposits obtained from several reports. In all shown references, $MeCpPtMe_3$ was used as a precursor except for those marked with *, for which $CpPtMe_3$ was used instead. AS: Auger spectroscopy; RBS: Rutherford backscattering; EDX: energy dispersive x-ray spectroscopy.

Reference	Resistivity (μΩ cm)	Pt:C:Ga:O (atomic %)	Microstructure
Tao *et al* 1990 [150]	70–700	46:24:28:2 (AS)	Amorphous
Tao *et al* 1991 [151]	70–300	45:25:30:5 (AS)	—
Puretz 1992 [152]	400–2100	37:47:13:- (AS)	—
Telari *et al* 2002 [153]	2 000 000 (?)	20:70:10:- (RBS, AS)	—
Dovidenko *et al* 2002 [154]	1000–5000	30:60:4:2 (AS)	5 nm sized nanocrystals
*Lin *et al* 2003 [155]	62–545	—	—
Langford *et al* 2004 [156]	5000	40:20:20:10	2–4 nm sized nanocrystals
*Tsukatani *et al* 2005 [157]	1300	—	—
Langford *et al* 2007 [158]	1000	45:50:5:5 (EDX)	6–8 nm sized nanocrystals
Peñate-Quesada *et al* 2007 [159]	1000–3500	31:10:50:- (EDX)	5–10 nm sized Pt/Ga grains
*Vaz *et al* 2008 [160]	1600	—	—
Barzola-Quiquia *et al* 2009 [161]	1770	23:64:10:0 (EDX)	Amorphous
De Teresa *et al* 2009 [162]	800	17:72:10:- (EDX)	3 nm sized nanocrystals

for in the systematic study performed by Fernández-Pacheco *et al* [163], in which the thickness of the contacts was found to have a significant influence on the final value of the resistivity, accounted for by changes on the composition of the deposit.

This compositional dependence on thickness is a consequence of the difference in SE yield between the bare substrate and the already-grown deposit on top of which more precursor is deposited. Namely, the presence of an energy gap on SiO_2 substrates [164] results in a lower SE yield than that occurring when deposited Pt–C is further irradiated. The reduced SE yield that drives growth over the bare substrate results in a lower metallic content on the first grown nanometers.

The temperature dependence of the resistivity, $\rho(T)$, is also different between distinct samples. In fact, the actual sign of the thermal coefficient $d\rho/dT$ differs between reports. Fernández-Pacheco *et al* account for this effect by proposing a metal-insulator transition (MIT) between distinct regimes as a function of thickness (figure 5.14). *In situ* resistivity measurements reveal a significant decrease in ρ with increasing thickness up to 50 nm, tending to saturation around 700 μΩ cm when the thickness is further increased. Specifically, the threshold of 3000 μΩ cm for metallic conduction [165] is reached around 50 nm in thickness, setting the ground for the justification of the MIT, also supported by the fact that the metallic content is also low (<33% in atomic percentage).

Thus, below 50 nm in thickness, FIBID Pt–C deposits behave as semiconductors, with conduction governed by variable range hopping processes; whilst above 50 nm, an increased metallic content yields a metallic conduction behavior. Ambient exposure is also found to increase the resistivity of the nanowires, associated with oxidation of the top part of the deposits.

As indicated in table 5.4, most TEM studies show the material exhibiting a microstructure of nanograins with crystalline structure embedded in an amorphous carbonaceous matrix. De Teresa *et al* [162] performed a high-resolution TEM study

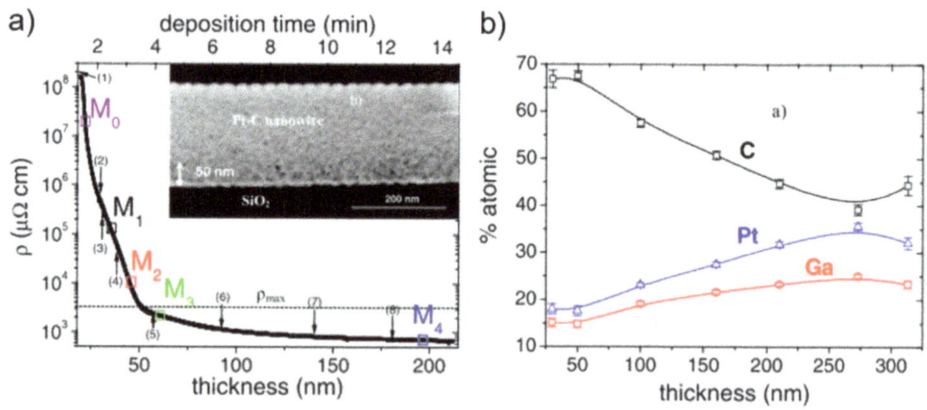

Figure 5.14. (a) Resistivity of FIBID Pt deposits as a function of deposit thickness. The inset shows a TEM image of the cross-section of a deposit, in which a difference in contrast on the first 50 nm can be observed, accounting for the differences in composition. (b) Atomic composition percentages for FIBID-Pt nano-structures as a function of deposit thickness. Reprinted figures with permission from [163]. Copyright (2009) by the American Physical Society.

of thin deposits (20–50 nm) grown on Cu grids, for which fast Fourier transforms of the obtained images were consistent with several orientations of crystalline planes of fcc Pt (lattice parameter of 0.3924 nm). These crystallites are ellipsoidal in shape and their size ranges from 2 up to 10 nm in the reviewed literature.

Several post-processing methods destined to modify the chemical and structural properties of FIBID Pt can be found in the literature. One of the most intuitive, annealing, has been found to have a from moderate to non-existent effect on the resistivity of the nanostructure. However, as previously mentioned, annealing does have a bending effect on freestanding nanowires [62]. This phenomenon is a consequence of the compositional and microstructural anisotropy of the FIBID Pt nanopillars that arises from the fixed position of the GIS with respect to the standing nanostructure, which results in one side of the nanopillar being significantly more exposed to the precursor gas flow than the other. Cui *et al* used this thermal treatment to grow 3D electrical contacts for freestanding nanostructures, and 'nano-cages' to encapsulate nano-objects (figure 5.15).

The injection of a second gas during the growth procedure to favor the formation of new volatile chemical species as a post-processing method has also been investigated. When FIBID Pt is grown with water vapor being injected along the organometallic precursor, an increase in Pt atomic content from 40% to 55% occurs (at the expense of a decrease in C and Ga contents, and an increase in O content), with an associated decrease in resistivity from 5000 to 1000 μΩ cm [156]. If O_2 is injected instead, a change in 0.61–0.27 in the C/Pt ratio is reported, with a decrease in resistivity from 823 to 281 μΩ cm [168], whilst maintaining a linear resistive behavior. Moreover, thickness is found not to have a significant influence in the resistivity when the deposits are grown using the O_2 purification method.

Post-processing based on electron beam irradiation after growth, which is somewhat common for Pt grown via focused electron beams, has no sensitive effect on Pt grown via FIBID [158, 168].

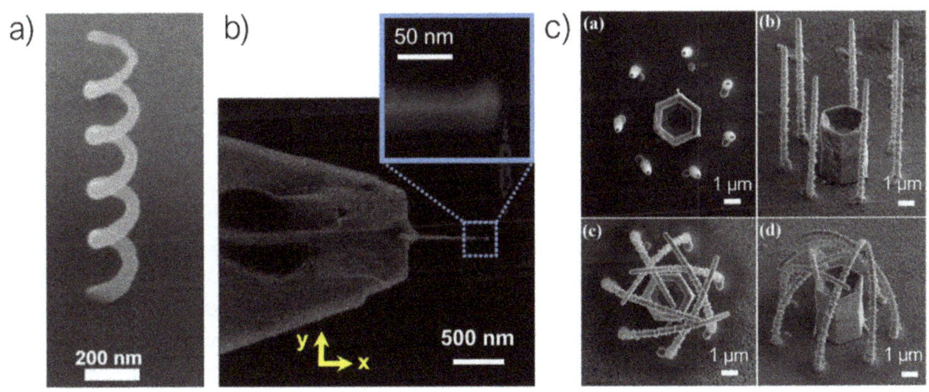

Figure 5.15. (a) 3D Pt nanohelix grown by Ga^+ FIBID. Reprinted with permission from [166]. Copyright (2015) American Chemical Society. (b) Hammerhead Pt nanoneedle grown on top of an AFM tip by He^+ FIBID. Reproduced with permission from [167]. Copyright 2015 American Vacuum Society. (c) Pt nanocage used to encapsulate a ZnO particle, constructed by thermally-induced shape modification of 3D Ga^+ FIBID pillars. Reprinted by permission from [62]. Copyright (2013) Springer.

Despite Ga^+ ions being the most popular option for most FIBID processes, the relatively recent commercialization of FIB columns of He^+ and Ne^+ ions has made possible the investigation of growth of Pt FIBID deposits using different ion types, with associated changes in the final nature of the deposits.

For both Ne^+ and He^+ FIBID processes, the resistivity of Pt deposits strongly decreases with increasing beam current. However, the lowest room-temperature resistivity for these contacts is much higher for He^+ (180 000 μΩ cm) than for Ne^+ (600 μΩ cm) [169]. Their microstructure consists in Pt grains embedded in a carbonaceous matrix, 3.2 and 4.5 nm in average size for He^+ and Ne^+, respectively. The Pt content is roughly the same in both cases, around 16% in atomic percentage.

Tungsten

Along with platinum, tungsten can be considered the flagship of the whole family of metallic materials grown by FIBID. Tungsten–carbon deposits can be grown using tungsten hexacarbonyl, $W(CO)_6$, as a precursor, and present enhanced mechanical properties [170] and relatively-low room-temperature resistivity that enables them to be used as electrical contacts. Most reports agree on room-temperature resistivity values near 200 μΩ cm, which permits the usage of W–C nanodeposits as intermediate contacts for *in situ* and *ex situ* electrical measurements of previously-grown samples.

Undoubtedly, the most prominent feature of FIBID W–C nanodeposits is the presence of *superconductivity* appearing in the vicinity of 5 K, where the resistivity of the material drops from a finite value (normal state) to zero (superconducting state) with decreasing temperature (figure 5.16).

Superconductivity occurs in single-crystal tungsten at much lower and thus harder-to-get-to temperatures (15 mK) [171], and its occurrence in FIBID nanodeposits is, at the writing time of this chapter, a matter of debate. First, an important distinction between the ionic species used to induce the growth needs to be made, for the deposited materials are significantly different in terms of composition and crystallinity—most Ga^+ FIBID deposits show around a 40% W metallic content, whilst He^+ FIBID nanostructures exhibit up to 70%; and while Ga^+ FIBID W–C is either amorphous or has very short-range order only (1 nm sized nanocrystallites), He^+ FIBID is crystalline. On that basis, the origin of superconductivity on both types of deposits is also expected to follow different reasons.

In the case of Ga^+ FIBID, Sadki *et al* [172] relate the occurrence of superconductivity in the deposits to their amorphous microstructure, which is further enhanced by the presence of C and Ga impurities, originating from the precursor and the FIB, respectively. This claim is compatible with the occurrence of superconductivity in amorphous W alloys. On the other hand, for He^+ FIBID, the superconductivity is associated with the presence of a superconducting crystalline phase. The experimental results of Córdoba *et al* [54] point towards the presence of large (20–30 nm) grains whose crystallinity is compatible with a $WC_{1\text{-}x}$ phase (1-x being between 0.13 and 0.36), accounting for the superconducting behavior.

The superconducting phase is bound by the temperature of the material, the magnetic field applied to it, and the current density that flows inside. These

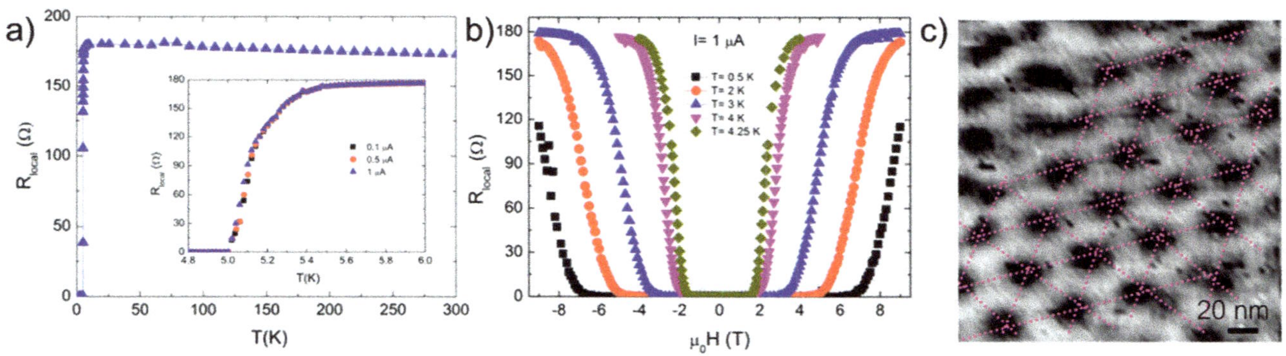

Figure 5.16. (a) Temperature-induced superconducting transition observed in a W–C nanowire grown by Ga^+ FIBID. Inset shows a detailed view of the transition. (b) Magneto-resistance at different temperatures, showing the temperature dependence of the upper critical magnetic field. Reproduced from [173]. Copyright Springer 2014. With permission of Springer. (c) Scanning tunneling spectroscopy image of a Ga+ FIBID W–C film, on which the Abrikosov vortex lattice can be seen. Each dark spot corresponds to a superconducting vortex, where a single quantum of magnetic flux threads the material. The triangular arrangement of the lattice is outlined. Reprinted by permission from [180]. Copyright (2009) Springer.

parameters all have critical values, which are typically used to characterize the superconducting behavior of the material. The most straightforward one is the *critical temperature*, at which the superconducting transition occurs. The reported range of values for this parameters is found from 4.2 to 5.1 K [173] for Ga^+ FIBID, and 6.2–7.1 K for He^+ 3D FIBID [54].

The critical magnetic field, that is, the maximum magnetic field the material can withstand without going back to the dissipative state, is found to lie in the 7 to 8.5 T range [173–175] at a working temperature of 2 K, while it is extrapolated to fall around the 9.5 T value at zero temperature [176]. Values for the critical current density vary between reports, sample geometries, and working conditions. Most of them fall within the 0.01–0.1 mA cm^{-2} range [173, 176–179]. Additionally, two parameters typically used in the Ginzburg–Landau theory of superconductivity, the *coherence length* and *penetration depth*, have been determined for this material to be, respectively, 6.25 and 850 nm for Ga^+ FIBID, and around 5 and 850 nm for He^+ FIBID [54, 180]. Table 5.5 summarizes the superconducting parameters for both Ga^+ and He^+ FIBID W–C.

FIBID W–C deposits exhibit a particular type of superconducting behavior referred to as *type-II superconductivity*. When a magnetic field is applied to a type-II superconductor, it penetrates the material in a quantized and ordered way, forming individual entities named *superconducting vortices*. A superconducting vortex is a small non-superconducting tubular region threaded by the magnetic field flux lines, and screened by a circular supercurrent. Each vortex encloses a finite, quantized amount of magnetic flux (a flux quantum). The supercurrents that surround each of the vortices repel each other and vortices arrange in the (usually) triangular Abrikosov lattice, the density of which depends on the magnitude of the applied magnetic field. An external current injected in the material also interacts with the circular screening supercurrents of the vortices and moves them, resulting in the appearance of electrical dissipation as the non-superconducting cores of the vortices

Table 5.5. Comparison of compositional, microstructural and superconductivity parameters of W–C grown by Ga^+ and He^+ FIBID.

	Ga^+ FIBID W–C [172, 180–182]	3D He^+ FIBID W–C [54]
Metallic W atomic content	Around 40%–50%	Up to 70%
Crystallinity	Amorphous/nanocrystallites	20–30 nm sized grains compatible with a WC_{1-x} phase
Impurities	Significant amounts of implanted Ga	No ion implantation
Critical temperature	4.2–5.1 K	6.2–7.1 K
Upper critical magnetic field (extrapolated to 0 K)	9.5 T	>11.6 T
Critical current density	0.01–0.1 MA cm^{-2}	0.08–0.2 MA cm^{-2}
Coherence length	6.25 nm	4.86–5.05 nm
Penetration depth	850 nm	830–902 nm

drift within the superconducting bulk. This phenomenon requires the splitting of the previously described critical magnetic field into two different values—the *lower* critical magnetic field, upon which vortices start to thread the material, and the *upper* critical magnetic field, upon which superconductivity is destroyed. The latter, which can be easily measured, is the one found in most reports in the literature.

The composition and microstructure studies of Ga^+ FIBID W–C nanodeposits are also mostly in agreement across the literature, with a W atomic content in the 40%–50% range, as well as with significant fractions of C (30%–40%) and implanted Ga (10%–20%). One notable exception to this trend is that of nanowires grown as suspended bridges between two supporting features, on which the W atomic content can be as high as 70% [170].

On table 5.6, relevant parameters from a selection of reports on Ga^+ FIBID W–C nanodeposits are shown.

The patterning flexibility of FIBID makes W–C nanostructures an appealing field of study of several superconductivity-related effects, particularly those related to a finite size or one particular sample geometry. Moving vortices cause measurable resistivity shifts, such as a magnetic field-driven re-entrance in the superconducting state caused by vortex confinement, observed in narrow nanowires (50 nm) [182]; or the appearance of finite voltages in regions of the structure through which the driving current does not flow, solely generated by passing vortices that are being pushed by exerting force on them in another region of the nanostructure [185].

He^+ FIBID of W–C has also proven to be of particular significance in the growth of complex 3D nanostructures with outstanding aspect ratio, such as hollow nanowires [54] and nanohelices [53]. The critical temperature of the deposited nanostructures, in

Table 5.6. Resistivity, composition, and nanostructure of Ga^+ FIBID W–C nanostructures in several reports. EPMA: electron probe microanalysis, EDX: energy dispersive x-ray spectroscopy, EELS: electron energy loss spectroscopy.

Reference	Resistivity (μΩ cm)	Critical temperature (K)	W:C:Ga:O (atomic %)	Microstructure
Sadki *et al* 2004 [172]	200	5.2 ± 0.5	40:40:20:0 (EPMA)	Amorphous (XRD)
Luxmoore *et al* 2007 [177]	200	5.5 ± 1.0	51:37:12:0 (EDX/EELS)	Nanocrystalline order (HRTEM)
Spoddig *et al* 2007 [183]	—	5.2 ± 0.5	17:35:13:4 (EDX)	—
Li *et al* 2009 [184]	100–350	5.0–5.5 ± 0.3	53:34:11:2 (EDX)	Nanocrystallite clusters
Li *et al* 2011 [181]	330	5.0 ± 0.3–0.6	48:30:16:6 (EDX)	Nanocrystallites
Córdoba *et al* 2013 [182]	275	4.3 ± 0.3	40:43:10:7 (EDX)	Nanocrystallites
[a]Córdoba *et al* 2019 [147]	800	—	20:70:30:0 (EDX)	

[a] Cryo-FIBID.

the 6.2–7.1 K range, is up to 1.5 times higher than that of Ga^+ FIBID nanostructures of similar dimensions. TEM-based characterization studies indicate a higher amount of W in the 3D nanostructures than those grown by Ga^+ FIBID (up to 70% of W in atomic content, compensated by a decrease in the amount of C); and the presence of 20–30 nm sized crystalline grains, that it is close to the value of the nanowire diameter. As anticipated, the composition range is compatible with the occurrence of the fcc WC_{1-x} phase on these grains [54].

The aforementioned fabrication of superconducting crystalline hollow nanopillars [54], which can be also thought of as standing nanotubes, is a noteworthy example of the capabilities of 3D He^+ FIBID. The presence of a longitudinal hole piercing the pillar is a consequence of a fine balance between the milling and decomposition mechanisms: to grow a pillar, the highly focused beam is steered at a single point during the whole irradiation time. In the particular conditions of the experiment, the precursor gas is decomposed and deposited over the surface of the sample and newly grown material in the vicinity of said spot, while simultaneously being removed by precise FIB-induced milling at the center of the spot, where the beam is focalized. A 3D hollow nanostructure as small as 32 nm in diameter and aspect ratio of 200 such as this one proves the outstanding capabilities of FIBID.

Anther remarkable application example of WC He^+ FIBID is the fabrication of 3D nanohelices with on-demand geometries by tuning some key deposition parameters [53]. Córdoba *et al* report the smallest and more-densely-packed superconducting nanohelix ever grown so far, as small as 100 nm in diameter, and with an aspect ratio of 65 (figure 5.17).

Lastly, the very recent demonstration of the capabilities of FIBID in cryogenic conditions previously mentioned was assessed by Córdoba *et al* on the growth of W–C

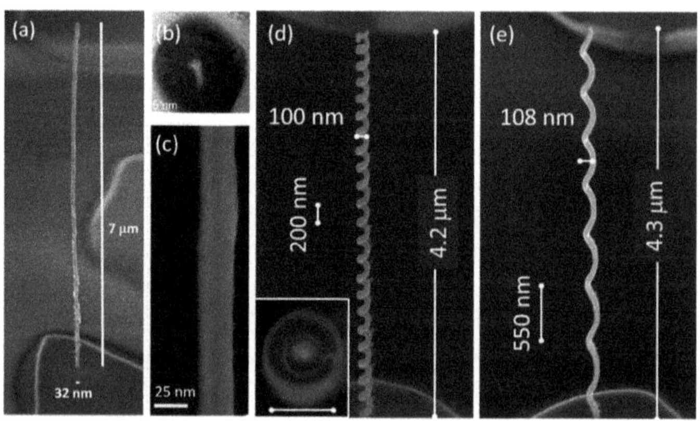

Figure 5.17. (a) SEM images of the standing hollow crystalline W–C 3D superconducting nanopillar grown by He^+ FIBID. (b) Cross-sectional TEM image of the hollow nanowire. The bright central area corresponds to the hole that pierces the nanostructure. (c) High-magnification SEM image of the nanotube. The internal conduct can be detected by the slight change in contrast it produces along the axis. Reprinted with permission from [54]. Copyright (2018) American Chemical Society. (d) and (e) Superconducting W–C 3D nano-helices grown by He^+ with different geometry parameters. Inset in (d) shows the top view. Reprinted with permission from [53]. Copyright (2019) American Chemical Society. Further permissions to the material excerpted should be directed to the ACS.

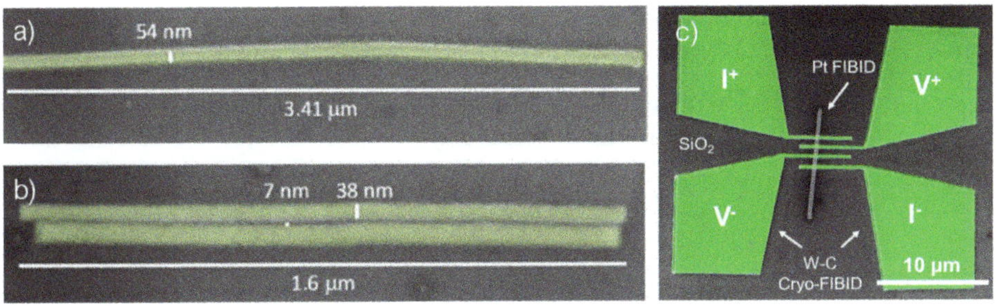

Figure 5.18. (a) Colored SEM micrograph of a high aspect ratio W–C nanowire grown by Cryo-FIBID. (b) Two adjacent W–C nanowires deposited by Cryo-FIBID, showing an enhancement in lateral resolution when compared to room-temperature FIBID. (c) W–C contacts grown by Cryo-FIBID (in green) placed on a Pt nanowire, intended for a four-microprobe measurement. Despite their relatively large size, the growth time of the full set is 50 s, roughly 700 times faster than if the process had been performed at room temperature. Reprinted by permission from [147]. Copyright (2019) Springer.

nanostructures as well [147] (figure 5.18). As aforementioned, the most remarkable feature of the technique is its extreme increase in processing speed, with a reported decrease in growth time from 14 h down to 85 s for a large array of micron-sized W–C squares. TEM-based analyses of the nanostructures show homogeneity in deposit thickness, as well as a composition gradient, with the atomic percentage of W increasing from 10% on the lower part of the deposit up to 22% on the top part.

Concerning their electrical properties, the deposits show higher room-temperature resistivity than those grown at room temperature, but are, however, suitable to be used as electrical contacts, as evidenced by a proof-of-concept contacting test of a Pt nanowire. In addition, and contrary to the deposits grown at room temperature, W–C nanostructures grown by Cryo-FIBID do not present a superconducting transition down to 2 K. The precise reason for this behavior is still unresolved, although the severe differences in deposit composition between the two techniques are likely to play a role on it.

Notably enough, a lateral patterning resolution of 7 nm, significantly higher than that achieved for Ga^+ FIBID at room temperature, is also reported for Cryo-FIBID. The proposed justification for this finding is related to the fact that the precursor decomposition mechanism at cryogenic temperatures could potentially differ from the conventional SE-induced decomposition at room-temperatures. Specifically, the deposition is suggested to be directly induced by the ions in the primary beam, allowing for much higher localization of the growth. Cryo-FIBID is, however, a newborn technique, and further research is required to better understand the underlying mechanisms that govern it.

Copper

In the electronics field, the inclination towards nanostructuring copper stems from the excellent room-temperature properties of the bulk material, including its resistivity (1.7 µΩ cm). Deposition of copper films is achieved by electrochemical deposition, physical and chemical vapor deposition, and sputtering [186].

Table 5.7. Resistivity, composition, and microstructure of FIBID Cu in several growth reports. In all shown references Cu(hfac)TMVS was used as a precursor.

Reference	Substrate temperature	Resistivity (μΩ cm)	Cu:C:Ga:O (atomic %)	Microstructure
Della Rata *et al* 1993 [187]	RT	50–500 [a]	55:45:0:0 (AS)	20 nm sized islands embedded on a C matrix
	100°C	2	100:0:0:0 (AS)	Polycrystalline
Funatsu *et al* 1995 [56] [b]	120°C (local)	3	—	—
Gannon *et al* 2004 [57]	20°C	800–1200	45:50:5:0 (AS)	—
	50°C	300–500	—	—
	75°C	200–300	50:40:10:0	—
	100°C	19	60:30:10:0	—

[a] Dependent on the ion current.
[b] Laser assisted. RT: room temperature; AS: Auger spectroscopy.

FIBID plays a modest role in this scenario, as the resistivity of the deposits is sensitively increased due to the impurities that are intrinsically introduced by the used precursor material (copper (I) hexafluoroacetylacetonate trimethylvinylsilane ($C_{10}H_{13}CuF_6O_2Si$), or Cu(hfac)TMVS for short). Table 5.7 summarizes the reported findings.

The temperature of the substrate plays a fundamental role on the final composition of the deposit, with early reports of the process finding a Cu purity close to 100% when the process is performed at 100 °C [187]. Laser-induced heating for this purpose has been proposed and successfully tested [56]. Later investigations, however, have not been able to fully reach this purity under similar growing conditions [57].

Gold

In a similar fashion, the targeted deposition of gold, which has a bulk resistivity of 2.4 μΩ cm, is highly appealing, with its potential applications encompassing circuit repair, lithography mask repair, and the growth of plasmonics nanostructures, among others.

The precursor material most commonly used to deposit gold via FIBID is dimethyl gold hexafluoro acetylacetonate ($C_7H_7F_6O_2Au$). Table 5.8 presents several reports on the procedure. The work of Blauner *et al* [188] experimentally shows several relevant dependences of the features of the deposit with the growth parameters, including: the thickness of the deposit increasing with the ion dose, but being largely independent of the gas pressure; the carbon content of the deposit increasing with precursor pressure; and a strong decrease of deposition yield with increasing current density.

Ro *et al* [58, 59] report the compositional and microstructural dependences of Au FIBID. The ionic species is found to significantly influence the purity of the deposit, with atomic percentages of Au of 40%, 45%, 80%, and 80% for Ne, Ar, Kr, and Xe ions, respectively. As for the microstructure, three distinct ranges are differentiated —using 70 kV Ar^+ ions, the microstructure strongly depends on the current density:

Table 5.8. Resistivity, composition, and nanostructure of FIBID Au in several growth reports. In all shown references $C_7H_7F_6O_2Au$ was used as a precursor. AS: Auger spectroscopy.

Reference	Ion species	Resistivity (μΩ cm)	Au:C:Ga:O (atomic %)	Microstructure
Shedd *et al* 1986 [189]	15 kV Ga^+	500–1300	75:<5:20:<5 (AS)	—
Blauner 1989 [188]	40 kV Ga^+	500–1500	50:35:15:0 (AS)	—
Dubner and Wagner 1989 [190]	5 kV Ar^+	—	—	—
Ro *et al* 1994 [58]	50–100 kV Ne^+/Xe^+	—	Variable, see text (AS)	—
Ro *et al* 1995 [59]	70 kV Ar^+	—	—	Variable, see text (TEM)

at 7 μA cm^{-2} it is granular, with the deposit exhibiting an atomic percentage of C higher than 50%; whilst the microstructure becomes columnar at 0.7 μA cm^{-2}, with the carbon content decreasing to 30%–50%. Moreover, maintaining the low current density conditions and increasing the substrate temperature at around 120 °C yields a polycrystalline microstructure, and results in the carbon content being reduced to less than 30%. In the first two cases, the grains or columns of gold are embedded in a carbonaceous matrix, and in all three the crystallinity corresponds to that of pure gold.

In the case of both gold and copper, the outstanding conductive properties of the bulk material do not carry to its FIBID-grown counterpart, with the resistivity increasing by up to four orders of magnitude. As a consequence, considering their conductivity only, the utility of FIBID-Pt and W surpasses that of FIBID-Cu and Au. Hence, the actual applicability of the latter is rather limited.

Other materials

The applicability of FIBID is of course limited to the existence of a viable precursor material. These are unfortunately not available to every single material one can think of, restricting the spectrum of typically grown materials to mostly those discussed in the previous section. Other precursors used to induce the growth of different materials, however, do exist, and although their usage and study are found less frequently in the literature, a couple of them are briefly reviewed here as well.

One notable example of such materials is *cobalt*. The usage of dicobalt octacarbonyl, $Co_2(CO)_8$, as a gaseous precursor, is notably absent in the literature concerning its deposition induced by an FIB. However, the usage of this precursor in combination with an FEB is widely reported and studied, mostly thanks to the magnetic properties the deposits exhibit [191]. Lapicki *et al*, however, do report the fabrication of arrays of cobalt dots using this precursor on various substrates, achieving results of micron-sized [192] and 150 nm sized dots [193]. The dots exhibit ferromagnetic properties in both cases, giving the arrays the potential to improve thermal stability in magnetic recordings [192]. Table 5.9 summarizes the reported findings.

Table 5.9. Some additional materials grown by FIBID. EDX: Energy dispersive x-ray spectroscopy, EELS: electron energy loss spectroscopy.

Reference	Precursor material	Ion species	Deposit purity	Comments
Lapicki *et al* 2002 [192]	$Co_2(CO)_8$	30 kV Ga^+	—	Ferromagnetic behavior
Lapicki and Suzuki 2002 [193]	$Co_2(CO)_8$	30 kV Ga^+	—	Ferromagnetic behavior
Kageyama *et al* 2004 [194]	$Co_2(CO)_8$	30 kV Ga^+	—	—
Gazzadi *et al* 2011 [195]	$Co(CO)_3NO$	30 kV Ga^+	54% Co (EDX)	Ferromagnetic after annealing only
Sanz-Martín *et al* 2019 [196]	$Co_2(CO)_8$	30 kV Ga^+	50% Co (EDX-EELS)	Thickness limited to 65 nm
Wu *et al* 2014 [197]	$Co_2(CO)_8$	35 kV He^+	High (EELS)	Width down to 10 nm
Porrati *et al* 2019 [198]	$Nb(Nme_2)_3(N\text{-}t\text{-}Bu)$	30 kV Ga^+	29% Nb (EDX)	Superconducting below 5.0 K

Sanz-Martín *et al* [196] report the milling-deposition competition that takes place when inducing the decomposition of $Co_2(CO)_8$, which limits the thickness of the deposits to around 60 nm within their available growth conditions (contrary to Co FEBID, where 3D growth can be achieved). A Co content of around 50% is found on most samples, and metallic conductivity with $\rho \sim$ 20–40 μΩ cm is observed, partially promoted by Ga doping ($\rho \sim$ 6 μΩ cm for bulk Co). Wu *et al* [197] report He^+ FIBID of Co using the same precursor material, achieving higher Co purity, and ohmic conductivity with $\rho \sim$ 65–75 μΩ cm.

In parallel, Gazzadi *et al* [195] also report the usage of a different precursor material, cobalt tricarbonyl nitrosyl, $Co(CO)_3NO$. In combination with a Ga^+ FIB, the deposition of micron-sized lines with metallic behavior is achieved (with a Co:O:Ga:N:C atomic content ratio of 54:17:9:13:7). Contrary to deposits grown using $Co_2(CO)_8$, these lines are non-magnetic as deposited. However, they do exhibit ferromagnetic behavior if they are annealed at 400 °C after deposition.

One other less common material is *niobium*. Porrati *et al* [198] report 2D and 3D growth of niobium nanostructures, making use of $Nb(Nme_2)_3(N\text{-}t\text{-}Bu)$ as a precursor. The samples exhibit a 29:43:15:13 Nb:C:Ga:N atomic composition ratio, with a polycrystalline microstructure consisting of 15–20 nm sized nanocrystallites. With room-temperature resistivities of 550 and 380 μΩ cm for the 2D and 3D nanostructures, respectively, the nanowires show superconducting behavior similar to that of the extensively studied W–C FIBID nanostructures. A critical temperature of 5 K is detected for the 2D nanowires, while the freestanding 3D nanowires exhibit wide superconducting transitions in the 4–11 K range.

5.6 Other applications of focused ion beams

Deposition, milling and modification of materials are the most conventional application of FIBs. The applicability of them, however, is not restricted to those three at all, for the focused nature of the beam provides with an invaluable tool that can be exploited for other purposes, such as imaging or compositional analysis. In this section some of these somewhat less extended applications are briefly presented.

5.6.1 FIB-based imaging

In both scientific and industrial frameworks, SEM is usually the technique of choice for micro- and nanostructure imaging, thanks to its ease of use, its ample range of application, and its relatively wide equipment availability and affordability. In FIB/SEM instruments with a Ga^+ FIB, FIB imaging is usually performed only incidentally, mostly due to the relatively high material damage the FIB can induce on the sample. On this framework, examples of FIB usage include taking fast images to precisely position a pattern over the sample, or carrying a fine adjustment of beam parameters by focusing and adjusting it over a test sample specifically used for this purpose.

Notwithstanding, the usage of focused beams of charged ions as a probe for image retrieving is indeed a solid and established technique, which, by comparison with the SEM counterpart, goes by the name of *scanning ion microscopy* (SIM) (figure 5.19) [199].

The basis of imaging with charged particles is grounded on the beam-material interactions that occur upon irradiation. In SEM, the focused electron beam induces

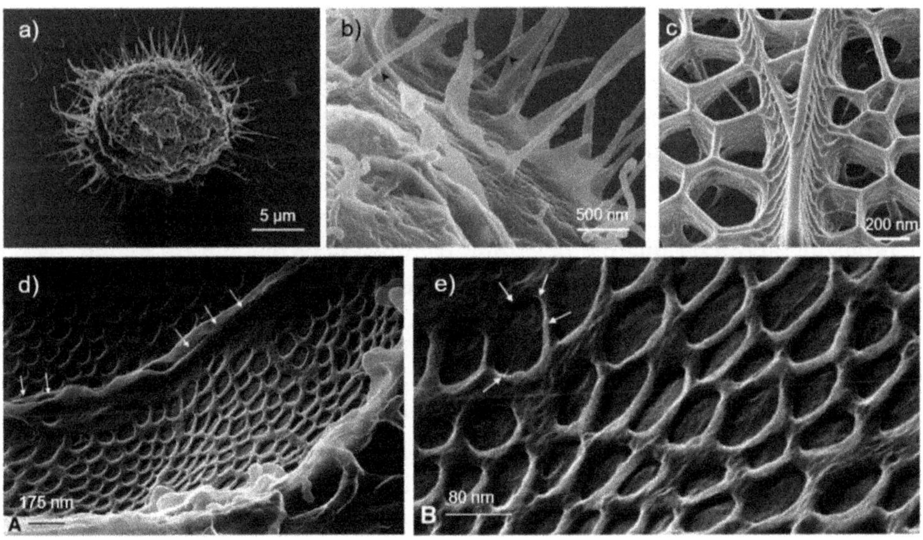

Figure 5.19. Examples of images obtained by HIM imaging. (a) and (b) Images in successive magnification of a mitotic HeLa cell. Notice the high depth of field in (a). Reprinted by permission from [37]. Copyright (2013) Springer. (c) Butterfly wing (*Papilio ulysses)* Reproduced with permission from [202]. Copyright Wiley Periodicals, Inc. (d) and (e) Images in successive magnification of endothelial cells of a rat kidney. Reproduced with permission from [203]. Copyright 2013 Rice *et al.*

the emission of both secondary and backscattered electrons, whose associated signal can be used to generate images, each providing different contrast (SEs are typically used to study the surface topography, whilst the contrast of BSEs relates to the composition of the sample). In SIM, the beam that induces the emission of these electrons is a FIB, but the working principle remains the same.

The key difference between the two approaches is the fact that the de Broglie wavelength associated with the ions in an FIB is smaller than that of the electrons in SEM. This fact, coupled with a smaller interaction volume on the surface, results in a significant resolution enhancement, down to 0.25 nm for He^+ ions [199]. An important drawback is that contrary to electrons, the ions in an FIB can induce surface milling, material amorphization, and ion implantation.

In principle, any ionic species can be used for the purposes of SIM imaging, but lighter ions are preferred, since their associated sputtering rates are reduced. For this reason, whilst Ga^+ SIM is indeed feasible [200], the low sputtering rate and small probe size achievable with He^+ FIBs make them the preferred choice for SIM [17, 199]. In fact, it has just been relatively recently that HIMs have entered the field of FIBIP, for they had been mostly used for imaging purposes so far [17]. The usage of other novel sources, based on more exotic species such as Li^+, has also been reported [201].

Some notable functional differences that arise from the usage of ions rather than electrons, are a higher SE yield, the emission of SIs, and the occurrence of ion channeling. All of them can be exploited for imaging purposes—high resolution imaging, compositional contrast, and crystallographic contrast, respectively [17].

A remarkable feature of SIM is its relatively simple capability to overcome charge imbalance, which is an important issue when imaging electrically insulating samples in SEM. In HIM, since the incident ions are positively charged and the vast majority of the particles emitted by the samples are negatively charged SEs, in all cases the net charge accumulated on the sample is positive (this is not the case in SEM, where the incident particles are negatively charged, rendering a less evident charge balance). The aforementioned *flood gun*, commonly found in HIM equipment, is used to diffuse low energy electrons over the surface of the sample as it is being irradiated with the FIB. By synchronizing and adjusting the operating parameters of the gun to 'match' those of the FIB, it is possible to completely compensate the charge imbalance, eliminating all associated artifacts and improving the quality of the image [17].

One other remarkable feature is the fact that, compared to SEM, HIM imaging allows for a much larger depth of field. Since the He^+ FIB ends as a very sharp probe, with the beam having a very small divergence angle, several planes of the sample away from the focusing point can be still kept 'focused' at the same time, without appearing blurry on the images. This feature is particularly appealing when imaging large 3D structures, such as biological samples.

5.6.2 FIB-based tomography

In general terms, *tomography* is an imaging technique on which the three-dimensional structure of an object is extracted by sequentially obtaining images of successive cross-sectional views of it. Depending on the nature of the probe used to obtain the image, tomography can either be performed by physically slicing the sample and imaging each cut, or by retrieving the signal yielded by penetrating radiation that bounces on it (e.g., a medical CT scan).

Commonplace in medical, life, geological, and material sciences (among others), the applicability of this technique is large, as is the number of available probes and approaches that exist for it. The milling/slicing capabilities of a dual FIB/SEM instrument make a fine addition to that list of probes, for the equipment is capable of achieving notably high resolution (with a size of reconstructed 'volumetric pixel' of tens of nm, down to 5 nm [31]).

FIB-based tomography is performed in FIB/SEM instruments by positioning the sample at the eucentric height, tilted 52°. The working volume is defined by using the FIB to mill trenches on its sides, defining a cube-like volume with edges in the 10 μm range [31]. Then, the FIB is used to mill away a fine slice of the material along the direction of the cube that is parallel to it, i.e., its lateral face that faces the electron beam. An SEM image is taken, and the process is repeated along the whole volume under study. Due to the imaging plane effectively changing height as material slices are milled away during the process, in order to keep the sample in focus, the focus is dynamically tracked and corrected. In addition, due to the extended time the process consumes, it is also necessary to account for beam drift, particularly if the procedure is performed in combination with EDX analysis (which further extends the processing time).

The operating parameters strongly depend on the material under study and its expected microstructure—the thickness of each slice imposes a strong constraint on the quality of the study, potentially obscuring any features smaller than each slice thickness. Moreover, the operating voltage of the SEM column must be adjusted so the interaction volume of the electron beam does not exceed the volume of the features and/or the thickness of the slice. An interaction volume greater than that thickness would include information about the layers below the one under study, also shadowing the validity of the study.

The final step is the 3D reconstruction and data extraction from the retrieved image stack, which is a complex procedure whose nature specifically depends on the material under study and involves corrections for contrast and brightness, image drift, and others. In addition, it can involve statistical and compositional analysis of the revealed features. Currently, some manufacturers provide automated software that handles the procedure and subsequent analyses.

FIB-based tomography finds a strong field of application in the life sciences, for it provides the capability to gain insight into the internal structure of a cell [73] (figure 5.20). This includes, for example, assessing the way magnetic nanoparticles are distributed within loaded cells [204, 205] or 3D reconstruction of various types of animal tissues [206]. Other applications are more closely related to materials science, including the study of porous and granular materials [31].

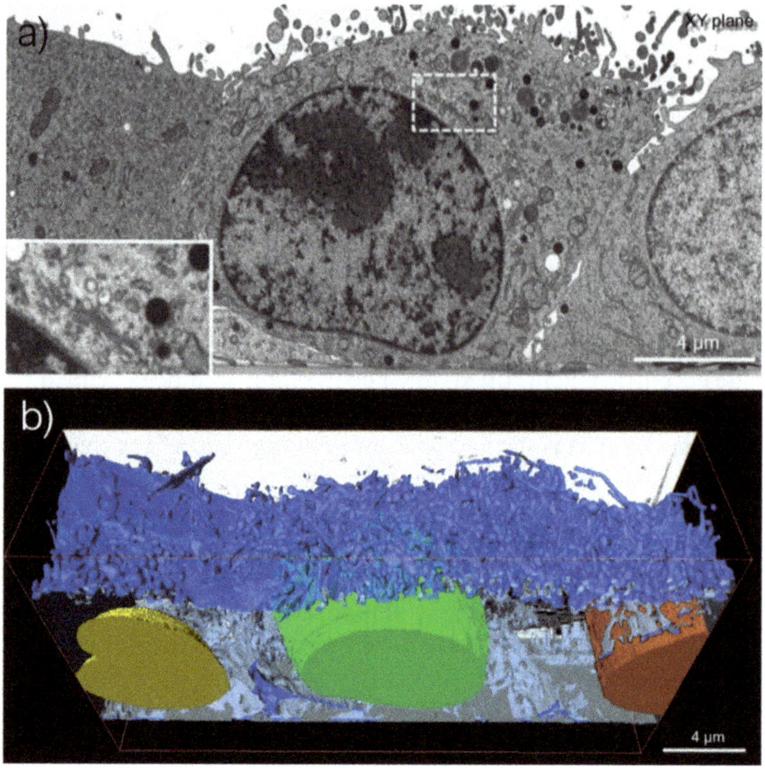

Figure 5.20. A pancreatic carcinoid human cell imaged by FIB/SEM tomography. (a) Tomogram of BON cells. Inset shows nuclear membranes, a part of the Golgi apparatus, and chromogranin A-containing vesicles. (b) Segmentation of the plasma membrane and of three cell nuclei. Reproduced from [207] with permission of Springer. Copyright Springer-Verlag 2012.

5.6.3 Secondary-ion mass spectrometry

The secondary ions (SIs) that are sputtered away from a material when it is irradiated with an FIB (recall that the term 'secondary' is used to differentiate them from those of the FIB itself, which would be considered the 'primary' beam) contain compositional and structural information on the sample. Adequately obtaining these SIs, retrieving them, and performing the subsequent analyses are the fundamentals of SIMS [17].

The procedure begins by irradiating the sample of interest with an FIB, which induces sputtering and ejection of SIs. By applying a potential difference in the chamber, the electrically charged SIs are redirected towards a *mass spectrometer*, on which they follow separate trajectories depending on their mass, and eventually reach detectors that count and track all contributions.

Since the ejection of sputtered atoms primarily occurs at the surface of the material that is being irradiated, the extent to which the information retrieved from the SIs is applicable is limited to the surface of the sample. Depending on the magnitude of the delivered ion dose, SIMS is performed in either *static* mode, with a moderate dose that allows for the study of a 'fixed' surface; or in *dynamic* mode, in which a much higher ion dose results in constant removal of the irradiated surface, which allows for depth profiling.

Mass spectra retrieved by SIMS provide information on the surface composition, both statically and dynamically (*depth profiling*), achieving a depth resolution that ranges from a few nm [208] down to sub-nm resolution when low-energy FIBs are used [17]. In addition, it is possible to reconstruct images of the surface based on the mass filtering. This SIMS imaging is achieved by synchronizing the counting and binning of the data with the sequential scanning of the FIB on the surface, so that for each irradiation spot, a corresponding mass spectrum is stored.

While SIMS is typically performed in dedicated instruments, which include both the FIB and the mass spectrometer [209], a SIMS module can also be incorporated into HIMs.

5.6.4 Current developments and future challenges

The well-established status of FIBIP and the pressing needs from science and manufacturing settle the grounds for future evolution and improvement of the instruments and techniques that surround it.

Concerning instrumentation advancement, the one obvious effervescent field is the development of different ion sources, capable of delivering high enough current densities with small probe sizes. Apart from the previously mentioned ones (LMAISs and PFIBs), *laser-cooled low temperature ion sources* (LOTISs) can be highlighted [34]. These make use of a beam of cold (microkelvin) gaseous beam of neutral Cs atoms that are photoionized using a laser [210].

Within the imaging framework, a very recent development is that of *transmission helium ion microscopy* (THIM), a technology that makes use of FIB-like beams (with acceleration voltages below 50 kV) as imaging probes analogously to TEM, using transmitted ions to retrieve information on the nature of the sample. Mousley *et al* [211] report the usage of the technique in a custom instrument, used for imaging several powders in TEM grids. THIM images of uncoated samples feature a series of bright spots, associated with ion deflection arising from electrostatic charging of the sample. Although these spots are artifacts in nature, they could prove useful in imaging of larger samples, for the ions from the beam are deflected according to the surface topography, eventually yielding the pattern.

The growth of increasingly complex nanostructures using FIBIP is also a strong trend in the field, with 3D nanopatterning being at the forefront. Some noteworthy achievements include the fabrication of custom nano-tools such as manipulators or rotors [95].

5.7 End-of-chapter summary

Focused ion beam processing is an umbrella term used to refer to the wide variety of techniques that make use of a focused beam of charged ions as the main driving engine. FIBIP encompasses the usage of FIBs for material removal and modification (milling, trimming, ion implantation), material deposition via FIBID, imaging, and compositional and structural analysis (SIMS, FIB tomography).

FIBIP is performed in dedicated instruments that hold an ion source, whose nature depends on the ionic species (typically, LMIS for Ga^+ beams and GFIS for He^+/Ne^+), an ion column with several apertures and lenses used to construct and focalize the beam, and a processing chamber that holds the specimen as well as nanofabrication modules (gas injection system, nanomanipulator) and electron and ion detectors.

The fundamentals of each specific technique are related to the nature of the physical interactions between the ions from the beam and the atoms of the material they impinge on. On the one hand, for milling and trimming purposes, mostly nuclear energy losses come into play, with the energetic ions from the beam transferring energy to the atoms in the substrate and displacing them.

On the other hand, growth of material via FIBID is grounded on electronic energy losses, which manifest themselves by the emission of secondary electrons from the atoms in the substrate. These electrons induce the decomposition of gaseous precursors previously adsorbed on the surface of the material, which allows for patterning at will by tracing the required design with the FIB.

Providing both great flexibility, relative ease of use, and a large range of processable materials, FIBIP has established itself as a solid and popular technique in nanofabrication, and it is not unreasonable to expect for it further developments and innovations in the foreseeable future.

5.8 Students' corner

- A video animation on FIB milling and FIBID growth can be imaged at https://iopscience.iop.org/book/978-0-7503-2608-7.
- A video animation on Cryo-FIBID growth can be imaged in reference [147]. We encourage the reader to go deeper into Cryo-FIBID technique, where exciting developments are quickly underway [212, 213].
- Compare the advantages and disadvantages of FEBID and FIBID in these cases: growth of electrical contacts to a nano-object, growth of a 3D magnetic nanostructure, and growth of a protective layer on a given material.
- Compare the advantages and disadvantages of FIB and FEBIE for optical mask repair.
- Compare the advantages and disadvantages for compositional analysis of EDX using a SEM and SIMS using a HIM.

Acknowledgements

Image credit for figures 5.2–5.4 goes to Rafael Marqués. P Orús acknowledges Gobierno de Aragón for funding. R Córdoba acknowledges financial support from 'La Caixa' Foundation (ID 100010434). The fellowship code is LCF/BQ/PR19/11700008. J M De Teresa acknowledges financial support from the Spanish Ministry of Economy and Competitiveness through Projects MAT2018–102627-T and MAT2017–82970-C2, from CSIC through the project PIE 202060E187 and from the Aragon Regional Government (Construyendo Europa desde Aragón) through Project E13_20R, with European Social Fund funding.

List of acronyms

Table 5.10. List of acronyms used throughout this chapter.

AS	Auger spectroscopy
BSE	Backscattered electron
CDEM	Channel detection electron multiplier
CNM	Carbon nanomembrane
EDX/EDS	Energy-dispersive x-ray spectroscopy
EELS	Electron energy loss spectroscopy
EPMA	Electron probe microanalysis
ETD	Everhart–Thornley detector
FEB	Focused electron beam
FEBID	Focused electron beam induced deposition
FIB	Focused ion beam
FIBID	Focused ion beam induced deposition
FIBIP	Focused ion beam induced processing
GFIS	Gas field ion source
GIS	Gas injection system
HIM	Helium ion microscopy/microscope
HRTEM	High-resolution transmission electron microscopy
LMAIS	Liquid metal-alloy ion source
LMIS	Liquid metal ion source
LOTIS	Laser-cooled low temperature ion source
MIT	Metal-insulator transition
PFIB	Plasma focused ion beam
RBS	Rutherford backscattering
SAM	Self-assembled monolayer
SE	Secondary electron
SEM	Scanning electron microscopy/microscope
SI	Secondary ion
SIM	Scanning ion microscopy
SIMS	Secondary ion mass spectrometry
SRIM	Stopping and range of ions in matter
TEM	Transmission electron microscopy
THIM	Transmission helium ion microscopy
TLD	Through-the-lens detector
TMVS	Trimethylvinylsilane
TRIM	Transport of ions in matter
XRD	X-ray diffraction

References

[1] Feynman R P 1960 There's plenty of room at the bottom *Calif. Inst. Technol. Eng. Sci. Mag.* **23** 22–36

[2] Seliger R L and Fleming W P 1974 Focused ion beams in microfabrication *J. Appl. Phys.* **45** 1416–22

[3] Orloff J H and Swanson L W 1975 Study of a field-ionization source for microprobe applications *J. Vac. Sci. Technol.* **12** 1209–13

[4] Orloff J and Swanson L W 1978 Fine-focus ion beams with field ionization *J. Vac. Sci. Technol.* **15** 845–8

[5] Seliger R L, Ward J W, Wang V and Kubena R L 1979 A high-intensity scanning ion probe with submicrometer spot size *Appl. Phys. Lett.* **34** 310–2

[6] Giannuzzi L A (ed) 2013 *Introduction to Focused Ion Beams: Instrumentation, Theory, Techniques and Practice* (Berlin: Springer) 303–6

[7] Gilbert W 1600 *De magnete, magneticisque corporibus, et de magno magnete tellure (On the Magnet and Magnetic Bodies, and on That Great Magnet the Earth)*. (Petrus Short)

[8] Krohn V E and Ringo G R 1975 Ion source of high brightness using liquid metal *Appl. Phys. Lett.* **27** 479–81

[9] Van Es J J *et al* 2004 An improved gallium liquid metal ion source geometry for nanotechnology *Microelctron. Eng.* **73-74** 132–8

[10] Economou N P, Notte J A and Thompson W B 2012 The history and development of the helium ion microscope *Scanning* **34** 83–9

[11] Hill R, Notte J A and Scipioni L 2012 Scanning helium ion microscopy *Advances in Imaging and Electron Physics* vol. 170 (New York: Elsevier) pp 65–148

[12] Müller E W 1951 Das feldionenmikroskop (The field ion microscope) *Z. Phys.* **131** 136–42

[13] Wilbertz C, Maisch T, Hüttner D, Böhringer K, Jousten K and Kalbitzer S 1992 A focused gas-ion beam system for submicron application *Nucl. Instrum. Methods Phys. Res.* B **63** 120–4

[14] Thompson W, Armstrong A, Etchin S, Percival R and Saxonis A 1996 Gas field ion source for finely focused ion beam systems *Mater. Res. Soc. Symp. Proc.* **396** 687–93

[15] Ward B W, Notte J A and Economou N P 2006 Helium ion microscope: a new tool for nanoscale microscopy and metrology *J. Vac. Sci. Technol.* B **24** 2871–4

[16] Scipioni L, Stern L and Notte J 2007 Applications of the helium ion microscope *Microsc. Today* **15** 12–5

[17] Hlawacek G and Gölzhäuser A 2016 *Helium Ion Microscopy* (Cham: Springer) pp 395–414

[18] Shi X and Boden S A 2016 Scanning helium ion beam lithography *Front. Nanosci.* **11** 563–94

[19] Gierak J 2009 Focused ion beam technology and ultimate applications *Semicond. Sci. Technol.* **24** 043001

[20] Swanson L W 1983 Liquid metal ion sources: mechanism and applications *Nucl. Instrum. Methods Phys. Res.* **218** 347–53

[21] Bischoff L 2005 Alloy liquid metal ion sources and their application in mass separated focused ion beams *Ultramicroscopy* **103** 59–66

[22] Bischoff L, Mazarov P, Bruchhaus L and Gierak J 2016 Liquid metal alloy ion sources - An alternative for focussed ion beam technology *Appl. Phys. Rev.* **3** 021101

[23] Hesse E, Bischoff L and Teichert J 1994 Development of a cobalt liquid alloy ion source *J. Phys. D: Appl. Phys.* **27** 427–8

[24] Gila B *et al* 2011 First results from a multi-ion beam lithography and processing system at the University of Florida *AIP Conf. Proc.* **1336** 243–7

[25] Hlawacek G *et al* 2014 Helium ion microscopy *J. Vac. Sci. Technol.* B **32** 020801

[26] Tondare V N 2005 Quest for high brightness, monochromatic noble gas ion sources *J. Vac. Sci. Technol.* A **23** 1498–508

[27] Smith N S *et al* 2006 High brightness inductively coupled plasma source for high current focused ion beam applications *J. Vac. Sci. Technol.* B **24** 2902–6

[28] Freinkman B G, Eletskii A V and Zaitsev S I 2004 A proposed laser source of ions for nanotechnology *Microelectron. Eng.* **73–74** 139–43
[29] Barth J E *et al* 1985 Gas ion source for focused beams *Microelectron. Eng.* **3** 147–52
[30] Hanssen J L, Hill S B, Orloff J and McClelland J J 2008 Magneto-optical-trap-based, high brightness ion source for use as a nanoscale probe *Nano Lett.* **8** 2844–50
[31] Utke I, Moshkalev S and Russell P 2012 *Nanofabrication Using Focused Ion and Electron Beams: Principles and Applications* (Oxford: Oxford University Press)
[32] Mulders J J L 2014 Practical precursor aspects for electron beam induced deposition *Nanofabrication* **1** 74–9
[33] Della Ratta A D, Melngailis J and Thompson C V 2014 Focused ion beam induced deposition of copper *J. Vac. Sci. Technol.* B **11** 2195
[34] Bassim N, Scott K and Giannuzzi L A 2014 Recent advances in focused ion beam technology and applications *MRS Bull.* **39** 317–25
[35] Ul-Hamid A 2018 *A Beginners' Guide to Scanning Electron Microscopy* (New York: Springer)
[36] Goldstein J I, Newbury D E, Michael J R, Ritchie N W M, Scott J H J and Joy D C 2018 *Scanning Electron Microscopy and X-Ray Microanalysis.* (Berlin: Springer)
[37] Joens M S *et al* 2013 Helium ion microscopy (HIM) for the imaging of biological samples at sub-nanometer resolution *Sci. Rep.* **3** 3514
[38] Kinter M 1995 Mass spectrometry *Anal. Chem.* **67** 493–7
[39] Kempshall B W, Schwarz S M, Prenitzer B I, Giannuzzi L A, Irwin R B and Stevie F A 2001 Ion channeling effects on the focused ion beam milling of Cu *J. Vac. Sci. Technol.* B **19** 749–54
[40] Ramachandra R, Griffin B and Joy D 2009 A model of secondary electron imaging in the helium ion scanning microscope *Ultramicroscopy* **109** 748–57
[41] Ohya K and Ishitani T 2004 Monte Carlo study of secondary electron emission from SiO_2 induced by focused gallium ion beams *Appl. Surf. Sci.* **237** 602–6
[42] Biersack J P and Haggmark L G 1980 A Monte Carlo computer program for the transport of energetic ions in amorphous targets *Nucl. Instruments Methods* **174** 257–69
[43] Ziegler J F and Biersack J P 1985 The stopping and range of ions in matter *Treatise on Heavy-Ion Science* (Berlin: Springer) pp 93–129
[44] Smith D A, Joy D C and Rack P D 2010 Monte Carlo simulation of focused helium ion beam induced deposition *Nanotechnology* **21** 175302
[45] Timilsina R, Smith D A and Rack P D 2013 A comparison of neon versus helium ion beam induced deposition via Monte Carlo simulations *Nanotechnology* **24** 115302
[46] Chen P *et al* 2010 Nanopillar growth by focused helium ion-beam-induced deposition *Nanotechnology* **21** 455302
[47] Timilsina R, Tan S, Livengood R and Rack P D 2014 Monte Carlo simulations of nanoscale focused neon ion beam sputtering of copper: elucidating resolution limits and sub-surface damage *Nanotechnology* **25** 485704
[48] Mahady K T, Tan S, Greenzweig Y, Raveh A and Rack P D 2018 Simulating advanced focused ion beam nanomachining: aquantitative comparison of simulation and experimental results *Nanotechnology* **29** 495301
[49] Möller W and Eckstein W 1984 Tridyn—a TRIM simulation code including dynamic composition changes *Nucl. Instrum. Methods Phys. Res.* B **2** 814–8
[50] Möller W 2014 TRI3DYN—Collisional computer simulation of the dynamic evolution of 3-dimensional nanostructures under ion irradiation *Nucl. Instruments Methods Phys. Res. Sect.* B **322** 23–33

[51] Möller W, Johannes A and Ronning C 2016 Shaping and compositional modification of zinc oxide nanowires under energetic manganese ion irradiation *Nanotechnology* **27** 175301
[52] Harriott L R 1993 Digital scan model for focused ion beam induced gas etching *J. Vac. Sci. Technol.* B **11** 2012
[53] Córdoba R *et al* 2019 Three-dimensional superconducting nano-helices grown by He +-focused-ion-beam direct writing *Nano Lett.* **19** 8597–604
[54] Córdoba R, Ibarra A, Mailly D and De Teresa J M 2018 Vertical growth of superconducting crystalline hollow nanowires by He+ focused ion beam induced deposition *Nano Lett.* **18** 1379–86
[55] Oboňa J V, Hrnčíř T, Sharang S, Šikula M and Denisyuk A 2016 Xe plasma FIB Delayering of IC based on 14 nm node technology *Microsc. Microanal.* **22** 56–7
[56] Funatsu J, Thompson C V, Melngailis J and Walpole J N 1996 Laser assisted focused-ion-beam-induced deposition of copper *J. Vac. Sci. Technol.* B **14** 179–80
[57] Gannon T J, Gu G, Casey J D, Huynh C, Bassom N and Antoniou N 2004 Focused ion beam induced deposition of low-resistivity copper material *J. Vac. Sci. Technol.* B **22** 3000–3
[58] Ro J, Thompson C and Melngailis J 1994 Mechanism of ion beam induced deposition of gold *J. Vac. Sci. Technol.* B **12** 73–7
[59] Ro J S, Thompson C V and Melngailis J 1995 Microstructure of gold grown by ion-induced deposition *Thin Solid Films* **258** 333–5
[60] Fang J Y, Qin S Q, Zhang X A, Liu D Q and Chang S L 2014 Annealing effect of platinum-incorporated nanowires created by focused ion/electron-beam-induced deposition *Chinese Phys.* B **23** 088111
[61] Batra N M, Patole S P, Abdelkader A, Anjum D H, Deepak F L and Costa P M F J 2015 Structural changes of electron and ion beam-deposited contacts in annealed carbon-based electrical devices *Nanotechnology* **26** 445301
[62] Cui A *et al* 2013 Thermally induced shape modification of free-standing nanostructures for advanced functionalities *Sci. Rep.* **3** 2429
[63] Zalm P C 1983 Energy dependence of the sputtering yield of silicon bombarded with neon, argon, krypton, and xenon ions *J. Appl. Phys.* **54** 2660–6
[64] Adams D P and Vasile M J 2006 Accurate focused ion beam sculpting of silicon using a variable pixel dwell time approach *J. Vac. Sci. Technol.* B **24** 836–44
[65] Frey L, Lehrer C and Ryssel H 2003 Nanoscale effects in focused ion beam processing *Appl. Phys. A Mater. Sci. Process.* **76** 1017–23
[66] Gierak J 2014 Focused Ion Beam nano-patterning from traditional applications to single ion implantation perspectives *Nanofabrication* **1** 35–52
[67] Burnett T L *et al* 2016 Large volume serial section tomography by Xe Plasma FIB dual beam microscopy *Ultramicroscopy* **161** 119–29
[68] Moll P J W 2018 Focused ion beam microstructuring of quantum matter *Annu. Rev. Condens. Matter Phys.* **9** 147–62
[69] Mohiuddin T 2014 Focused ion beam (FIB) circuit edit *Electron. Device Fail. Anal.* **16** 20–3
[70] Drezner Y, Fishman D, Greenzweig Y and Raveh A 2011 Characterization of damage induced by FIB etch and tungsten deposition in high aspect ratio vias *J. Vac. Sci. Technol.* B **29** 011026
[71] Prenitzer B I *et al* 2003 The correlation between ion beam/material interactions and practical FIB specimen preparation *Microsc. Microanal.* **9** 216–36

[72] Mayer J, Giannuzzi L a, Kamino T and Michael J 2007 TEM sample preparation and damage *MRS Bull.* **32** 400–7

[73] Márquez Neila P, Baumela L, González-Soriano J, Rodríguez J R, DeFelipe J and Merchán-Pérez Á 2016 A fast method for the segmentation of synaptic junctions and mitochondria in serial electron microscopic images of the brain *Neuroinformatics* **14** 235–50

[74] Lanyon Y H *et al* 2007 Fabrication of nanopore array electrodes by focused ion beam milling *Anal. Chem.* **79** 3048–55

[75] Tang Z, Zhang D, Cui W, Zhang H, Pang W and Duan X 2016 Fabrications, applications and challenges of solid-state nanopores: a mini review *Nanomater. Nanotechnol.* **6** 35

[76] Ali M Y and Lim B H 2006 Characterisation of surface texture using AFM with trimmed probe tip *Surf. Eng.* **22** 443–6

[77] Saxey D W, Cairney J M, McGrouther D, Honma T and Ringer S P 2007 Atom probe specimen fabrication methods using a dual FIB/SEM *Ultramicroscopy* **107** 756–60

[78] Cryan M J *et al* 2005 Focused ion beam-based fabrication of nanostructured photonic devices *IEEE J. Sel. Top. Quantum Electron.* **11** 1266–76

[79] Yasaka A, Aramaki F, Kozakai T and Matsuda O 2016 Nanoscale imaging, material removal and deposition for fabrication of cutting-edge semiconductor devices ion-beam-based photomask defect repair technology *Hitachi Rev.* **65** 233

[80] Gonzalez C M *et al* 2014 Evaluation of mask repair strategies via focused electron, helium, and neon beam induced processing for EUV applications *Proc. SPIE* **9048** 90480M

[81] Bret T, Hofmann T and Edinger K 2014 Industrial perspective on focused electron beam-induced processes *Appl. Phys. A Mater. Sci. Process.* **117** 1607–14

[82] Mayer M *et al* 2011 Multilayer Fresnel zone plate for soft x-ray microscopy resolves sub-39nm structures *Ultramicroscopy* **111** 1706–11

[83] Keskinbora K, Sanli U T, Baluktsian M, Grévent C, Weigand M and Schütz G 2018 High-throughput synthesis of modified Fresnel zone plate arrays via ion beam lithography *Beilstein J. Nanotechnol.* **9** 2049–56

[84] Kollmann H *et al* 2014 Toward plasmonics with nanometer precision: nonlinear optics of helium-ion milled gold nanoantennas *Nano Lett.* **14** 4778–84

[85] Abbas A N *et al* 2014 Patterning, characterization, and chemical sensing applications of graphene nanoribbon arrays down to 5 nm using helium ion beam lithography *ACS Nano.* **8** 1538–46

[86] Wu H, Ferranti D and Stern L 2014 Precise nanofabrication with multiple ion beams for advanced circuit edit *Microelectron. Reliab.* **54** 1779–84

[87] Mouchet S R *et al* 2016 Liquid-induced colour change in a beetle: the concept of a photonic cell *Sci. Rep.* **6** 19322

[88] Gierak J *et al* 2009 Exploration of the ultimate patterning potential achievable with focused ion beams *Ultramicroscopy* **109** 457–62

[89] Kollmann H *et al* 2014 Toward plasmonics with nanometer precision: nonlinear optics of helium-ion milled gold nanoantennas *Nano Lett.* **14** 4778–84

[90] Casey J D, Doyle A F, Lee R G, Stewart D K and Zimmermann H 1994 Gas-assisted etching with focused ion beam technology *Microelectron. Eng.* **24** 43–50

[91] Xia D, Notte J, Stern L and Goetze B 2015 Enhancement of XeF_2-assisted gallium ion beam etching of silicon layer and endpoint detection from backside in circuit editing *J. Vac. Sci. Technol.* B **33** 06F501

[92] Utke I, Hoffmann P and Melngailis J 2008 Gas-assisted focused electron beam and ion beam processing and fabrication *J. Vac. Sci. Technol.* B **26** 1197
[93] Khan Malek C, Hartley F T and Neogi J 2003 Fast prototyping of high-aspect ratio, high-resolution x-ray masks by gas-assisted focused ion beam *Microsyst. Technol.* **9** 409–12
[94] Gonzalez J C, Griffis D P, Miau T T and Russell P E 2001 Chemically enhanced focused ion beam micromachining of copper *J. Vac. Sci. Technol.* B **19** 2539–42
[95] Kim C S, Ahn S H and Jang D Y 2012 Review: developments in micro/nanoscale fabrication by focused ion beams *Vacuum* **86** 1014–35
[96] Callegari V, Sennhauser U and Jäckel H 2010 Analysis of the propagation losses of InP/InGaAsP trench waveguides fabricated by focused ion beam *Microelectron. Eng.* **87** 2061–64
[97] Alias M S *et al* 2016 Enhanced etching, surface damage recovery, and submicron patterning of hybrid perovskites using a chemically gas-assisted focused-ion beam for subwavelength grating photonic applications *J. Phys. Chem. Lett.* **7** 137–42
[98] Gazzadi G C, Angeli E, Facci P and Frabboni S 2006 Electrical characterization and Auger depth profiling of nanogap electrodes fabricated by I-2-assisted focused ion beam *Appl. Phys. Lett.* **89** 173112
[99] Santschi C, Jenke M, Hoffmann P and Brugger J 2006 Interdigitated 50 nm Ti electrode arrays fabricated using XeF_2 enhanced focused ion beam etching *Nanotechnology* **17** 2722–9
[100] Schneider M, Gierak J, Marzin J Y, Gayral B and Gérard J M 2000 Focused ion beam patterning of III-V crystals at low temperature: a method for improving the ion-induced defect localization *J. Vac. Sci. Technol.* B **18** 3162–7
[101] Kim S, Jeong Park M, Balsara N P, Liu G and Minor A M 2011 Minimization of focused ion beam damage in nanostructured polymer thin films *Ultramicroscopy* **111** 191–9
[102] Cacho-Bailo F *et al* 2015 Metal-organic framework membranes on the inner-side of a polymeric hollow fiber by microfluidic synthesis *J. Membr. Sci.* **476** 277–85
[103] Gierak J *et al* 1996 3D defect distribution induced by focused ion beam irradiation at variable temperatures in a GaAs/GaAlAs multi quantum well structure *Microelectr. Eng.* **30** 253–6
[104] Schertel A *et al* 2013 Cryo FIB-SEM: volume imaging of cellular ultrastructure in native frozen specimens *J. Struct. Biol.* **184** 355–60
[105] Steckl A J, Xu J, Mogul H C and Mogren S 1993 Doping-induced selective area photoluminescence in porous silicon *Appl. Phys. Lett.* **62** 1982–4
[106] Serrano-Ramón L *et al* 2013 Improvement of domain wall conduit properties in cobalt nanowires by global gallium irradiation *Nanotechnology* **24** 345703
[107] Serrano-Ramón L *et al* 2013 Modification of domain-wall propagation in Co nanowires via Ga+ irradiation *Eur. Phys. J.* B **86** 2–7
[108] Chappert C *et al* 1998 Planar patterned magnetic media obtained by ion irradiation *Science (80-.)* **280** 1919–22
[109] Ozkaya L D, Langford R M, Chan W L and Petford-Long A K 2002 Effect of Ga implantation on the magnetic properties of permalloy thin films *J. Appl. Phys.* **91** 9937–42
[110] Vieu C *et al* 2002 Modifications of magnetic properties of Pt/Co/Pt thin layers by focused gallium ion beam irradiation *J. Appl. Phys.* **91** 3103–10
[111] Fallon K *et al* 2020 Controlled individual skyrmion nucleation at artificial defects formed by ion irradiation *Small* **16** 1–1907458

[112] Guo Q, Landau P, Hosemann P, Wang Y and Greer J R 2013 Helium implantation effects on the compressive response of Cu nanopillars *Small* **9** 691–6

[113] Lesik M *et al* 2013 Maskless and targeted creation of arrays of colour centres in diamond using focused ion beam technology *Phys. Status Solidi Appl. Mater. Sci.* **210** 2055–9

[114] Pezzagna S, Naydenov B, Jelezko F, Wrachtrup J and Meijer J 2010 Creation efficiency of nitrogen-vacancy centres in diamond *New J. Phys.* **12** 065017

[115] Shinada T, Koyama H, Hinoshita C, Imamura K and Ohdomari I 2002 Improvement of focused ion-beam optics in single-ion implantation for higher aiming precision of one-by-one doping of impurity atoms into nano-scale semiconductor devices *Jpn. J. Appl. Physics,* **41** L287

[116] Philipp P *et al* 2014 The origin of conductivity in ion-irradiated diamond-like carbon - Phase transformation and atomic ordering *Carbon* **80** 677–90

[117] Nguyen C T *et al* 2013 Fabrication of an InGaAs spin filter by implantation of paramagnetic centers *Appl. Phys. Lett.* **103** 052403

[118] Hartney M A, Shaver D C, Shepard M I, Melngailis I, Medvedev J and Robinson W P 1991 Surface imaging of focused ion-beam exposed resists *J. Vac. Sci. Technol.* B **9** 3432–5

[119] Winston D *et al* 2009 Scanning-helium-ion-beam lithography with hydrogen silsesquioxane resist *J. Vac. Sci. Technol.* **B 27** 2702–6

[120] Sidorkin V, van Veldhoven E, van der Drift E, Alkemade P, Salemink H and Maas D 2009 Sub-10-nm nanolithography with a scanning helium beam *J. Vac. Sci. Technol.* **B 27** L18

[121] Shi X, Prewett P, Huq E, Bagnall D M, Robinson A P G and Boden S A 2016 Helium ion beam lithography on fullerene molecular resists for sub-10 nm patterning *Microelectron. Eng.* **155** 74–8

[122] Li W-D, Wu W and Stanley Williams R 2012 Combined helium ion beam and nanoimprint lithography attains 4 nm half-pitch dense patterns *J. Vac. Sci. Technol.* B **30** 06F304

[123] Ada E T 1995 Ion beam modification and patterning of organosilane self-assembled monolayers *J. Vac. Sci. Technol.* **B 13** 2189

[124] Cyganik P *et al* 2005 Modification and stability of aromatic self-assembled monolayers upon irradiation with energetic particles *J. Phys. Chem.* B **109** 5085–94

[125] Turchanin A and Gölzhäuser A 2016 Carbon nanomembranes *Adv. Mater.* **28** 6075–103

[126] Zhang X, Vieker H, Beyer A and Gölzhäuser A 2014 Fabrication of carbon nano-membranes by helium ion beam lithography *Beilstein J. Nanotechnol.* **5** 188–94

[127] Menéndez E *et al* 2009 Direct magnetic patterning due to the generation of ferromagnetism by selective ion irradiation of paramagnetic FeAl alloys *Small* **5** 229–34

[128] La Torre E *et al* 2018 Local probe of irradiation-induced structural changes and orbital magnetism in $Fe_{60}Al_{40}$ thin films via an order-disorder phase transition *Phys. Rev.* B **98** 024101

[129] Urbánek M *et al* 2018 Research update: focused ion beam direct writing of magnetic patterns with controlled structural and magnetic properties *APL Mater.* **6** 060701

[130] Causer G L *et al* 2018 Direct measurement of the intrinsic sharpness of magnetic interfaces formed by chemical disorder using a He^+ Beam *ACS Appl. Mater. Interfaces* **10** 16216–24

[131] Ferreiro-Vila E, Bugallo D, Magén C, Rivadulla F and De Teresa J M 2020 Topotactic transformation in $SrFeO_{3-\delta}$ triggered by low-dose Ga^+ focused ion irradiation *Appl. Phys. Lett.* **116** 163103

[132] Enriquez E *et al* 2016 Oxygen vacancy-driven evolution of structural and electrical properties in $SrFeO_{3-\delta}$ thin films and a method of stabilization *Appl. Phys. Lett.* **109** 3–8

[133] Cybart S A *et al* 2015 Nano Josephson superconducting tunnel junctions in $YBa_2Cu_3O_{7-\delta}$ directly patterned with a focused helium ion beam *Nat. Nanotechnol.* **10** 598–602

[134] Cho E Y, Li H, Lefebvre J C, Zhou Y W, Dynes R C and Cybart S A 2018 Direct-coupled micro-magnetometer with Y-Ba-Cu-O nano-slit SQUID fabricated with a focused helium ion beam *Appl. Phys. Lett.* **113** 162602

[135] Müller B *et al* 2019 Josephson junctions and SQUIDs created by focused helium-ion-beam irradiation of $YBa_2Cu_3O_7$ *Phys. Rev. Appl.* **11** 044082

[136] Schmidt B, Oswald S and Bischoff L 2005 Etch rate retardation of Ga^+-ion beam-irradiated silicon *J. Electrochem. Soc.* **152** G875

[137] Chekurov N, Grigoras K, Peltonen A, Franssila S and Tittonen I 2009 The fabrication of silicon nanostructures by local gallium implantation and cryogenic deep reactive ion etching *Nanotechnology* **20** 065307

[138] Böttger R, Bischoff L, Schmidt B and Krause M 2011 Characterization of Si nanowires fabricated by Ga^+ FIB implantation and subsequent selective wet etching *J. Micromech. Microeng.* **21** 095025

[139] Erdmanis M and Tittonen I 2014 Focused ion beam high resolution grayscale lithography for silicon-based nanostructures *Appl. Phys. Lett.* **104** 10–5

[140] Rumler M, Fader R, Haas A, Rommel M, Bauer A J and Frey L 2013 Evaluation of resistless Ga^+ beam lithography for UV NIL stamp fabrication *Nanotechnology* **24** 365302

[141] Liu Z, Iltanen K, Chekurov N, Grigoras K and Tittonen I 2013 Aluminum oxide mask fabrication by focused ion beam implantation combined with wet etching *Nanotechnology* **24** 175304

[142] Portavoce A, Kammler M, Hull R, Reuter M C and Ross F M 2006 Mechanism of the nanoscale localization of Ge quantum dot nucleation on focused ion beam templated Si (001) surfaces *Nanotechnology* **17** 4451–55

[143] Karmous A, Berbezier I and Ronda A 2006 Formation and ordering of Ge nanocrystals on SiO_2 *Phys. Rev. B Condens. Matter Mater. Phys.* **73** 075323

[144] Mehta M, Reuter D, Melnikov A, Wieck A D and Remhof A 2007 Focused ion beam implantation induced site-selective growth of InAs quantum dots *Appl. Phys. Lett.* **91** 3–6

[145] Gierak J *et al* 2005 Exploration of the ultimate patterning potential achievable with high resolution focused ion beams *Appl. Phys. A Mater. Sci. Process* **80** 187–94

[146] Peinado P, Sangiao S and De Teresa J M 2015 Focused electron and ion beam induced deposition on flexible and transparent polycarbonate substrates *ACS Nano* **9** 6139–46

[147] Córdoba R, Orús P, Strohauer S, Torres T E and De Teresa J M 2019 Ultra-fast direct growth of metallic micro- and nano-structures by focused ion beam irradiation *Sci. Rep.* **9** 14076

[148] Alkemade P F A and Miro H 2014 Focused helium-ion-beam-induced deposition *Appl. Phys.* A **117** 1727–47

[149] Van W F, Dorp and Hagen C W 2008 A critical literature review of focused electron beam induced deposition *J. Appl. Phys.* **104** 0081301

[150] Tao T, Ro J, Melngailis J, Xue Z and Kaesz H D 1990 Focused ion beam induced deposition of platinum *J. Vac. Sci. Technol.* B **8** 1826–9

[151] Tao T, Wilkinson W and Melngailis J 1991 Focused ion-beam induced deposition of platinum for repair processes *J. Vac. Sci. Technol.* B **9** 162–4

[152] Puretz J 1992 Focused ion beam deposition of Pt containing films *J. Vac. Sci. Technol.* B **10** 2695

[153] Telari K A, Rogers B R, Fang H, Shen L, Weller R A and Braski D N 2002 Characterization of platinum films deposited by focused ion beam-assisted chemical vapor deposition *J. Vac. Sci. Technol.* B **20** 590

[154] Dovidenko K, Rullan J, Moore R, Dunn K A, Geer R E and Heuchling F 2002 FIB-assisted Pt deposition for carbon nanotube integration and 3-D nanoengineering *Material Research Society Symposium Proceedings* **vol 739** (Warrendale, PA: Materials Research Society) pp 193–8

[155] Lin J-F F, Bird J P, Rotkina L and Bennett P A 2003 Classical and quantum transport in focused-ion-beam-deposited Pt nanointerconnects *Appl. Phys. Lett.* **82** 802–4

[156] Langford R M, Ozkaya D, Sheridan J and Chater R 2004 Effects of water vapour on electron and ion beam deposited platinum *Microsc. Microanal.* **10** 1122–3

[157] Tsukatani Y, Yamasaki N, Murakami K, Wakaya F and Takai M 2005 Transport properties of Pt nanowires fabricated by beam-induced deposition *Jpn. J. Appl. Phys.* **44** 5683–6

[158] Langford R M, Wang T X and Ozkaya D 2007 Reducing the resistivity of electron and ion beam assisted deposited Pt *Microelectron. Eng.* **84** 784–8

[159] Peñate-Quesada L, Mitra J and Dawson P 2007 Non-linear electronic transport in Pt nanowires deposited by focused ion beam *Nanotechnology* **18** 8–13

[160] Vaz A R, Da Silva M M, Leon J, Moshkalev S A and Swart J W 2008 Platinum thin films deposited on silicon oxide by focused ion beam: characterization and application *J. Mater. Sci.* **43** 3429–34

[161] Barzola-Quiquia J *et al* 2009 Transport properties and atomic structure of ion-beam-deposited W, Pd and Pt nanostructures *Nanotechnology* **20** 165704

[162] De Teresa J M, Córdoba R, Fernández-Pacheco A, Montero O, Strichovanec P and Ibarra M R 2009 Origin of the difference in the resistivity of as-grown focused-ion- and focused-electron-beam-induced Pt nanodeposits *J. Nanomater.* **2009** 936863

[163] Fernández-Pacheco A, De Teresa J M, Córdoba R and Ibarra M R 2009 Metal-insulator transition in Pt-C nanowires grown by focused-ion-beam- induced deposition *Phys. Rev. B Condens. Matter Mater. Phys.* B **79** 174204

[164] Ohya K and Ishitani T 2004 Monte Carlo study of secondary electron emission from SiO_2 induced by focused gallium ion beams *Appl. Surf. Sci.* **237** 606–10

[165] Mott N F and Davis E A 2012 *Electronic Processes in Non-Crystalline Materials* (Oxford: Oxford University Press)

[166] Esposito M, Tasco V, Cuscunà M, Todisco F, Benedetti A, Tarantini I, De Giorgi M, Sanvitto D and Passaseo A 2014 Nanoscale 3D chiral plasmonic helices with circular dichroism at visible frequencies *ACS Photon.* **2** 105–14

[167] Nanda G, van Veldhoven E, Maas D, Sadeghian H and Alkemade P F A 2015 Helium ion beam induced growth of hammerhead AFM probes *J. Vacuum Sci. Technol.* B **33** 06F503

[168] Perez-Roldan M J, Mulders J J L and Trompenaars P H F 2017 Oxygen-assisted purification of platinum structures deposited by ion and electron beam induced processes *J. Phys. D: Appl. Phys.* **50** 205307

[169] Wu H M *et al* 2013 Synthesis of nanowires via helium and neon focused ion beam induced deposition with the gas field ion microscope *Nanotechnology* **24** 175302

[170] Córdoba R, Lorenzoni M, Pablo-Navarro J, Magén C, Pérez-Murano F and De Teresa J M 2017 Suspended tungsten-based nanowires with enhanced mechanical properties grown by focused ion beam induced deposition *Nanotechnology* **28** 445301

[171] Li W, Fenton J C, Wang Y, McComb D W and Warburton P A 2008 Tunability of the superconductivity of tungsten films grown by focused-ion-beam direct writing *J. Appl. Phys.* **104** 093913
[172] Sadki E S, Ooi S and Hirata K 2005 Focused ion beam induced deposition of superconducting thin films *Phys. C Supercond. Appl.* **426–431** 1547–51
[173] Córdoba Castillo R 2014 *Functional Nanostructures Fabricated by Focused Electron/Ion Beam Induced Deposition* (Berlin: Springer)
[174] Dai J *et al* 2013 Superconductivity in tungsten-carbide nanowires deposited from the mixtures of $W(CO)_6$ and C_{14} H_{10} *Jpn. J. Appl. Phys.* **52** 075001
[175] Sun Y, Wang J, Zhao W, Tian M, Singh M and Chan M H W 2013 Voltage-current properties of superconducting amorphous tungsten nanostrips *Sci. Rep.* **3** 2307
[176] Sadki E S, Ooi S and Hirata K 2004 Focused-ion-beam-induced deposition of superconducting nanowires *Appl. Phys. Lett.* **85** 6206–08
[177] Luxmoore I J *et al* 2007 Low temperature electrical characterisation of tungsten nano-wires fabricated by electron and ion beam induced chemical vapour deposition *Thin Solid Films* **515** 6791–97
[178] Spoddig D *et al* 2007 Transport properties and growth parameters of PdC and WC nanowires prepared in a dual-beam microscope *Nanotechnology* **18** 495202
[179] Guillamón I *et al* 2008 Nanoscale superconducting properties of amorphous W-based deposits grown with a focused-ion-beam *New J. Phys.* **10** 093005
[180] Guillamón I *et al* 2009 Direct observation of melting in a two-dimensional superconducting vortex lattice *Nat. Phys.* **5** 651–5
[181] Li W, Fenton J C, Gu C and Warburton P A 2011 Superconductivity of ultra-fine tungsten nanowires grown by focused-ion-beam direct-writing *Microelectron. Eng.* **88** 2636–8
[182] Córdoba R *et al* 2013 Magnetic field-induced dissipation-free state in superconducting nanostructures *Nat. Commun.* **4** 1437
[183] Spoddig D *et al* 2007 Transport properties and growth parameters of PdC and WC nanowires prepared in a dual-beam microscope *Nanotechnology* **18** 495202
[184] Li W, Fenton J C and Warburton P A 2009 Focused-ion-beam direct-writing of ultra-thin superconducting tungsten composite films *IEEE Trans. Appl. Supercond.* **19** 2819–22
[185] Córdoba R *et al* 2019 Long-range vortex transfer in superconducting nanowires *Sci. Rep.* **9** 12386
[186] Ghosh S 2019 Electroless copper deposition: a critical review *Thin Solid Films* **669** 641–58
[187] Della Ratta A D, Melngailis J and Thompson C V 1993 Focused-ion-beam-induced deposition of copper *J. Vac. Sci. Technol.* B **11** 2195–9
[188] Blauner P G 1989 Focused ion beam fabrication of submicron gold structures *J. Vac. Sci. Technol.* B **7** 609
[189] Shedd G M, Lezec H, Dubner A D and Melngailis J 1986 Focused ion beam induced deposition of gold *Appl. Phys. Lett.* **49** 1584–6
[190] Dubner A D and Wagner A 1989 Mechanism of ion beam induced deposition of gold *J. Vac. Sci. Technol.* B **12** 73–7
[191] De Teresa J M, Fernández-Pacheco A, Córdoba R, Serrano-Ramón L, Sangiao S and Ibarra M R 2016 Review of magnetic nanostructures grown by focused electron beam induced deposition (FEBID) *J. Phys. D: Appl. Phys.* **49** 243003
[192] Lapicki A, Ahmad E and Suzuki T 2002 Ion beam induced chemical vapor deposition (IBICVD) of cobalt particles *J. Magn. Magn. Mater.* **240** 47–9

[193] Lapicki A and Suzuki T 2002 Fabrication of magnetic dot arrays by ion beam induced chemical vapor deposition (IBICVD) *INTERMAG Eur. 2002 - IEEE Int. Magn. Conf.* **38** 2589–91

[194] Kageyama Y, Lapicki A and Suzuki T 2004 Fabrication of cobalt particles by ion beam-induced chemical vapor deposition (IBICVD) *J. Magn. Magn. Mater.* **272–276** 2003–4

[195] Gazzadi G C *et al* 2011 Characterization of a new cobalt precursor for focused beam deposition of magnetic nanostructures *Microelectron. Eng.* **88** 1955–8

[196] Sanz-Martín C, Magén C and De Teresa J M 2019 High volume-per-dose and low resistivity of cobalt nanowires grown by Ga^+ focused ion beam induced deposition *Nanomaterials* **9** 1715

[197] Wu H *et al* 2014 Focused helium ion beam deposited low resistivity cobalt metal lines with 10 nm resolution: implications for advanced circuit editing *J. Mater. Sci., Mater. Electron.* **25** 587–95

[198] Porrati F *et al* 2019 Crystalline niobium carbide superconducting nanowires prepared by focused ion beam direct writing *ACS Nano* **13** 6287–96

[199] Hill R, Notte J and Ward B 2008 The ALIS He ion source and its application to high resolution microscopy *Phys. Procedia* **1** 135–41

[200] Ishitani T and Tsuboi H 1997 Objective comparison of scanning ion and scanning electron microscope images *Scanning* **19** 489–97

[201] Twedt K A, Chen L and McClelland J J 2014 Scanning ion microscopy with low energy lithium ions *Ultramicroscopy* **142** 24–31

[202] Boden S A, Asadollahbaik A, Rutt H N and Bagnall D M 2011 Helium ion microscopy of Lepidoptera scales *Scanning* **34** 107–20

[203] Rice W L, Van Hoek A N, Păunescu T G, Huynh C, Goetze B, Singh B, Scipioni L, Stern L A and Brown D 2013 High resolution helium ion scanning microscopy of the rat kidney *PLoS ONE* **8** e57051

[204] Calatayud M P *et al* 2013 Neuronal cells loaded with PEI-coated Fe_3O_4 nanoparticles for magnetically guided nerve regeneration *J. Mater. Chem.* B **1** 3607–16

[205] Calatayud M P *et al* 2017 Cell damage produced by magnetic fluid hyperthermia on microglial BV2 cells *Sci. Rep.* **7** 8627

[206] Vidavsky N *et al* 2016 Cryo-FIB-SEM serial milling and block face imaging : large volume structural analysis of biological tissues preserved close to their native state *J. Struct. Biol.* **196** 487–95

[207] Villinger C, Gregorius H, Kranz C, Höhn K, Münzberg C, Wichert G, Mizaikoff B, Wanner G and Walther P 2012 FIB/SEM tomography with TEM-like resolution for 3D imaging of high-pressure frozen cells *Histochem. Cell Biol.* **138** 549–56

[208] Schwarz S A 2001 Secondary ion mass spectroscopy *Encycl. Mater. Sci. Technol.* 8283–90

[209] Boxer S G, Kraft M L and Weber P K 2009 Advances in imaging secondary ion mass spectrometry for biological samples *Annu. Rev. Biophys.* **38** 53–74

[210] Smith N S, Notte J A and Steele A V 2014 Advances in source technology for focused ion beam instruments *MRS Bull.* **39** 329–35

[211] Mousley M *et al* 2019 Stationary beam full-field transmission helium ion microscopy using sub-50 keV He^+: projected images and intensity patterns *Beilstein J. Nanotechnol.* **10** 1648–57

[212] De Teresa J M *et al* 2019 Comparison between focused electron/ion beam-induced deposition at room-temperature and under cryogenic conditions *Micromachines* **10** 799

[213] Salvador-Porroche A *et al* 2020 Optimization of Pt-C deposits by Cryo-FIBID: substantial growth rate increase and quasi-metallic behaviour *Nanomaterials* **10** 1906

Chapter 6

Scanning probe lithography

Ricardo Garcia

The atomic force microscope enabled a variety of approaches to modify surfaces with nano or atomic-scale features. Some of those approaches had the robustness and nanoscale accuracy to evolve into advanced nanopatterning and nanolithography methods. This chapter describes the principles of scanning probe lithography (SPL) and some of its applications in materials science and nanotechnology. The focus is placed on selected SPL methods based on oxidation, thermal and deposition processes, which offer high-resolution nanolithographic applications and are applied by several groups.

6.1 Introduction: from atomic-scale modifications to scanning probe lithography

The inventions of scanning tunneling microscope (STM) [1] and atomic force microscope (AFM) [2] brought about a rich variety of methods to modify and/or manipulate surfaces at atomic and nanometer scales [3–10]. The local character of the tip–surface interaction opened several routes to pattern surfaces and materials which eventually led to the establishment of scanning probe lithography (SPL) [6]. This process was the result of a combination of factors, among them, a technological context of the late 1990s that favoured the development of alternative lithographies, the curiosity of some scientists and the appeal of combining the powerful metrology provided by probe microscopy with nanoscale patterning.

Several STM and AFM methods demonstrated the capability to modify or manipulate surfaces with atomic-scale resolution [5, 11–13]. Those approaches were inspiring due to their atomic-scale features but they lacked the required throughput for large scale nanopatterning or device applications. There were some remarkable exceptions such as the fabrication of atomic-scale transistors [14, 15] or very high-density memories [16] by manipulating a few atoms in each processing step.

Both STM and AFM were used to perform nanolithographic experiments. However, the majority of SPL applications are performed with a force microscope.

doi:10.1088/978-0-7503-2608-7ch6

The AFM is very convenient for nanolithography because the feedback parameter that controls the imaging process [17] does not depend on the conductivity of the sample surface. In addition, the AFM might be operated at its highest spatial resolution in air, liquid and vacuum.

This chapter aims to provide an introduction to scanning probe lithography (SPL) by focusing on the probe-based methods that have a very high degree of repeatability by patterning a large number of nanometer-scale features and/or enabled the fabrication of electronic or optical nanoscale devices. The described methods are oxidation SPL [3, 6, 18], thermal SPL [6, 19, 20] and deposition SPL, which includes dip-pen nanolithography [4] and some variations [21, 22]. At the end of the chapter, some considerations will be provided for other SPL methods. A general scheme to classify SPL methods is shown in figure 6.1, which also serves the purpose for the description of the physical and chemical interactions relevant in SPL.

A variety of physical and chemical interactions were applied to pattern surfaces by using an AFM, as shown in figure 6.1: electrochemical reactions [3, 8], mechanical forces [9, 23–25], bias-induced processes [26–29], thermal-induced processes [19–21, 30, 31], chemical reactions including catalytic reactions [32, 33] or the controlled deposition of molecules [4, 22, 34]. SPL methods either add or remove material from the surface. The local character of the interactions in a probe microscope interface gives rise to the formation of nanoscale features.

6.2 Oxidation SPL

Oxidation SPL is based on the formation of nanoscale oxides on a surface. The oxidation reaction is driven by the application of a voltage bias between the AFM tip (negative) and the sample surface (positive) in the presence of oxygen species (oxyanions). The oxygen species are usually provided by the presence of a

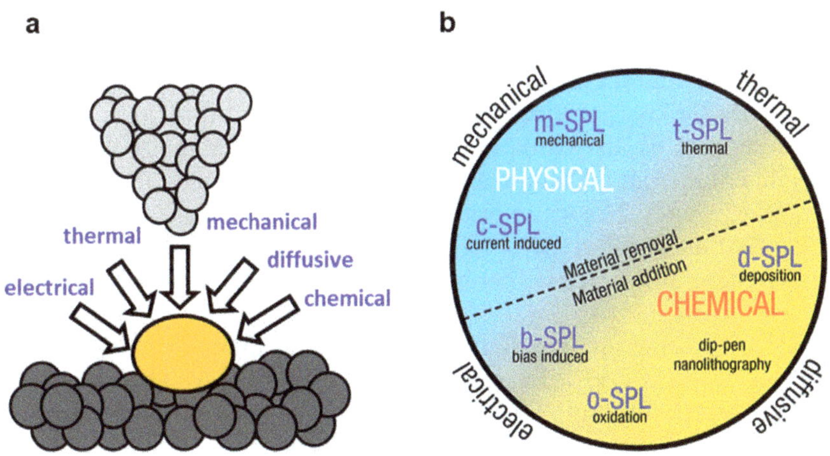

Figure 6.1. (a) Types of tip–surface interactions in SPL. (b) Classification of SPL methods according to the dominant interaction/process used for nanopatterning [6].

spontaneous water bridge joining tip and surface. The polarity of the tip indicates that the oxidation happens in the positive electrode (anode).

There are some similarities between the local oxidation process in o-SPL and the macroscopic anodic oxidation happening in a conventional electrochemical cell. The tip (cathode), the sample (anode) and the water bridge (solution) can be considered the elements of a nanoscale electrochemical cell. The absence of a reference electrode is a key difference between o-SPL and an anodic oxidation in a conventional electrochemical cell.

The first evidence of a probe-based oxidation process was reported by John Dagata and co-workers in 1990 [35]. They observed that a hydrogen-terminated silicon surface was modified while imaging it with an STM operated in ambient air. At about the same time, Thomas Thundat and co-workers used an STM to smooth tantalum surfaces [36]. In both cases a voltage of a few volts was applied between the STM tip and sample surface. High-spatial resolution spectroscopic measurements showed that the modifications were due to the oxidation of the surfaces [37, 38].

A relevant feature of o-SPL is that the patterning and the imaging processes operate independently. This means that the patterning processes could be interrupted at any time to examine its metrology aspects. Long range van-der-Waals forces and/or short-range repulsive forces are used to image the surface while an external voltage is applied to induce the local oxidation of the surface [17]. High resolution images of the sample can be obtained and used for precise alignment or assessment of the lithography without introducing further modifications on the fabricated nanostructures.

6.2.1 Key aspects of o-SPL

The main factors that participate in the nanopatterning process by o-SPL are the water bridge, the electric field and the oxidation kinetics (figure 6.2(a)). A brief description of those factors is given below. Full details of o-SPL are provided in comprehensive reviews [3, 6, 18] and/or in original publications.

The water bridge. The formation of a nanoscale water bridge (meniscus) connecting the tip and the sample is a key feature in o-SPL (figure 6.2). The water bridge provides the oxygen species (mostly OH^-) needed to oxidize the surface. It also confines spatially the oxidation reaction. The water molecules forming the bridge might come from the humid air or from water molecules previously adsorbed on the surface [39, 40]. The water bridge acts as an effective electrostatic lens which focuses the field lines within the water bridge limits. The focusing effect is enhanced by the differences in the electric permittivities inside and outside the liquid bridge, respectively, 81 and 1.

The bias voltage and the electric field. To drive the anodic oxidation process, the sample is positively biased with respect to the tip. In an AFM interface, the tip–surface separation is very small (~1 nm). Those tip–surface distances imply that the application of a voltage in the 10 V range might generate extremely high electric fields of ~10 GV m^{-1}. The electric field in o-SPL has three main roles. First, it induces and stabilizes the formation of a water bridge (or meniscus). Second, it

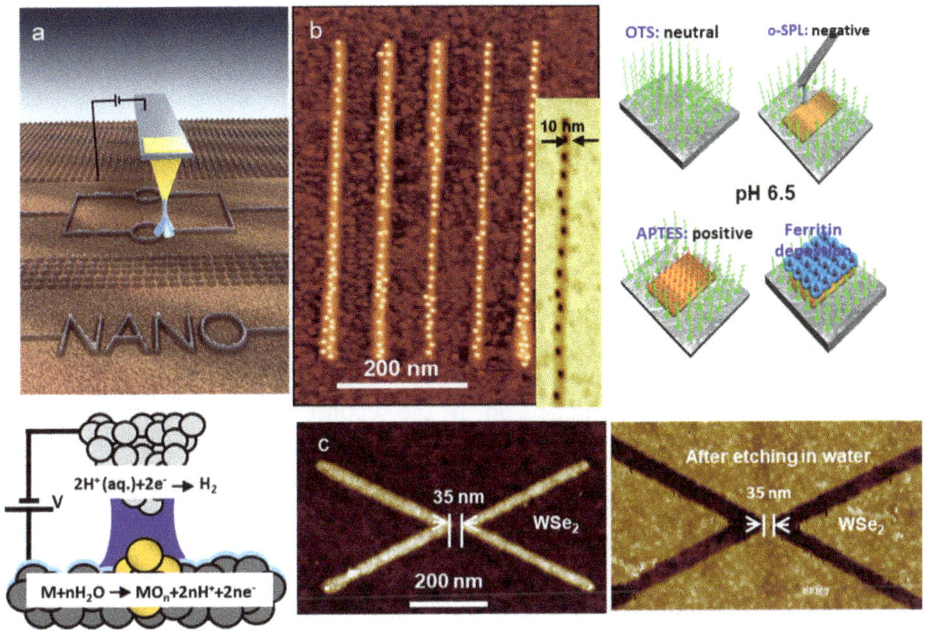

Figure 6.2. (a) Top, scheme of o-SPL. Bottom, a generic electrochemical reaction in o-SPL. (b) Left, nanopatterning of ferritin proteins by combining o-SPL and bottom–up electrostatic interactions. Right, main steps of the patterning process. At pH = 6.5 a ferritin protein is negatively charged. Adapted with permission from [60]. Copyright 2010 John Wiley and Sons. (c) Left, a 35 nm constriction fabricated on a few layer WSe_2 flake. Right, the oxide is removed by etching the nanostructure in water. This process leaves a dielectric barrier separating the WSe_2 constriction from the main WSe_2 flake. Adapted with permission from [57]. Copyright 2016 American Institute of Physics.

produces the hydrolysis of the water molecules. This process generates the oxyanions that participate in the oxidation process. Third, it drives the oxyanions through the growing oxide.

Kinetics. In o-SPL, the oxide height shows a linear dependence on the applied voltage and a logarithmic dependence on the oxidation time (power-of-time law) [42, 43]

$$h(t,\ V) = Vx[h_0(t/t_0)^\gamma] \tag{6.1}$$

where h, V, t are the total thickness of the oxide, applied voltage and oxidation time; h_0, t_0 and γ are fitting parameters; γ depends on the material. For Si(100) and MoS_2, the γ values are, respectively, ~0.2 and ~0.5. The power-of-time law dependence was theoretically explained by assuming that the rate-controlling step of the oxidation reaction involves the interaction of oxyanions with holes, electrons and defects at a material–oxide interface. To deduce the above equation, it is necessary to consider that the oxide tends to acquire an excess of negative charge. The production of charged defects leads to a buildup of space charge within the oxide that causes a reduction of the oxidation growth rate for longer voltage pulse times.

In general, for a metallic surface the half-cell reaction (anode) can be described as

$$M + nH_2O \rightarrow MO_n + 2n\,H^+ + 2ne^- \qquad (6.2)$$

Charge and mass conservation require another half-cell reaction at the tip (cathode), in this case, hydrogen generation

$$2nH^+(aq) + 2ne^- \rightarrow nH_2 \qquad (6.3)$$

6.2.2 Nanolithography

The electric field that drives the local anodic oxidation process is generated by applying a voltage bias between the tip and the sample. The voltage can be applied either continuously or as a pulse. In the first case, the tip is displaced along the sample while the voltage is switched on. In the second case, the tip rests on a fixed position of the surface during the application of the voltage pulse, then the voltage is switched off and the tip is displaced to another position.

The value of the bias and the oxidation time depend on the material and tip's size. Typical ranges for voltage and oxidation times are, respectively, 5–30 V and 0.0005–0.1 s. It is worth mentioning that most commercial AFMs might be easily transformed into an o-SPL instrument [3, 6, 18].

The direct result of an o-SPL process is the generation of a thin and narrow oxide nanostructure on a surface. Those structures might have three different functionalities: dielectric barriers, masks for an etching process or templates. Those elements were used either independently or in combination to generate of a large variety of nanopatterns and nanodevices. Some examples are shown in figures 6.2 and 6.3.

Nanopatterning. The minimum feature size achieved by o-SPL was of ~6 nm. Some examples were provided by defining a tunnel junction on a Ti film [44] and by patterning a periodic array of silicon oxide lines with a periodicity of 15 nm [18]. The minimum feature size seems to be limited by the tip's size, the lateral extension of the liquid bridge and the local oxidation kinetics [41].

A large variety of materials were patterned by o-SPL, among them, semiconductors (Si, GaAs, InAs, SiC), metals (Ta, Ti, Nb, or Hf), dielectric materials, self-assembled monolayers, polymers, proteins, complex oxides, correlated oxide materials, and two-dimensional layered materials [3, 6, 7, 18, 45–52]. The direct writing capabilities of o-SPL found a niche of applications in the nanopatterning single and a few layer materials such as graphene [53, 54], few layer MoS_2 [55, 56], WSe_2 [57] and black phosphorus [58].

Several examples illustrated how the physical and chemical properties of the nanopatterns were used as templates to promote the growth or ordered alignment of materials such as organic semiconductors [59], proteins [60], nanoparticles [61–63], block-copolymers [50] or quantum dots [64].

Nanodevices. Oxidation SPL was used to fabricate a metal-oxide transistor where the dielectric barrier was a local oxide. This was considered the first transistor with truly nanometer-size features [65]. Similarly, one of the first single-electron transistors that operated at room temperature was fabricated by using an STM tip to

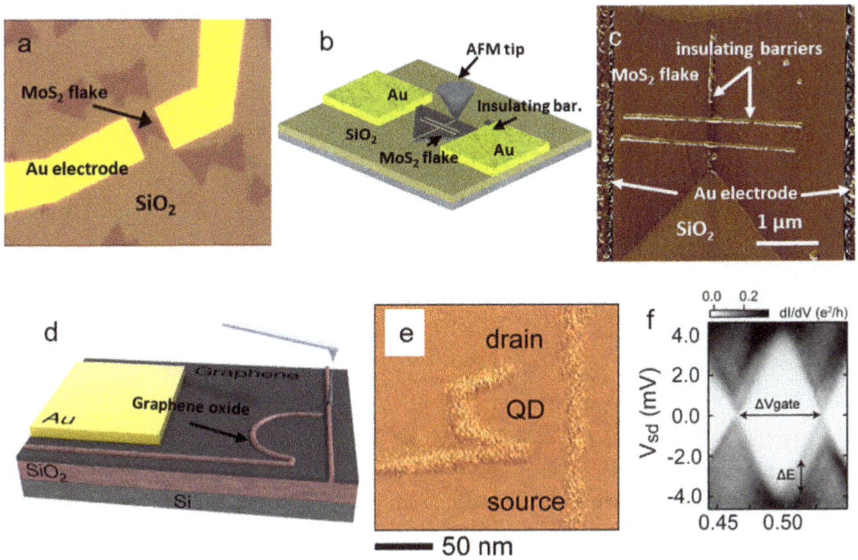

Figure 6.3. (a) Optical image of the metallic contacts on a MoS_2 flake. (b) Scheme of the o-SPL fabrication of a MoS_2 FET transistor [55]. (c) AFM phase image of a MoS_2 thin layer FET. The narrowest section of the channel is 200 nm wide. The dielectric barriers prevent the current flowing outside the channel. Reprinted with permission from [55]. Copyright 2015 American Institute of Physics. (d) Scheme of the fabrication of a graphene quantum dot; o-SPL generates very narrow graphene oxide lines that enclose a graphene quantum dot [6]. (e) AFM image of a single quantum dot. (f) Coulomb blockade diamond of the quantum dot measured at $T = 50$ mK. Panels (e) and (f) reprinted with permission from [69]. Copyright 2013 American Institute of Physics.

produce the dielectric barriers (local Ti oxides) that separated a Ti nanoscale dot from the metal electrodes [66]. Since then, a large variety of devices such as sub-10 nm silicon nanowire FET transistors [67], optical cavities or quantum devices were produced [3, 7, 18].

The direct and resistless features of o-SPL drew the interest of the 2D materials community. A variety of quantum and nanoelectronic devices were fabricated based on 2D and few-layered materials [7, 10, 18, 68–72]. Figure 6.3 illustrates the procedure to fabricate a nanoFET channel transistor on a monolayer MoS_2 [55] and a quantum dot on graphene [69].

6.2.3 Beyond oxidation processes

The experimental set-up used in o-SPL might be easily reconfigured to produce other chemical reactions or to modify the local electronic properties of a material. For example, the chemical composition of the nanopatterns was transformed from oxides to carbon-based materials by changing the solvent in the liquid bridge [73]. It was also shown that the electrical properties of complex oxides heterostructures can be reversibly switched by changing the polarity of the AFM tip during the nanolithographic process [51, 74, 75]. Similarly, the AFM was used to control the phase transitions at the nanoscale in strongly correlated materials such as VO_2 [76].

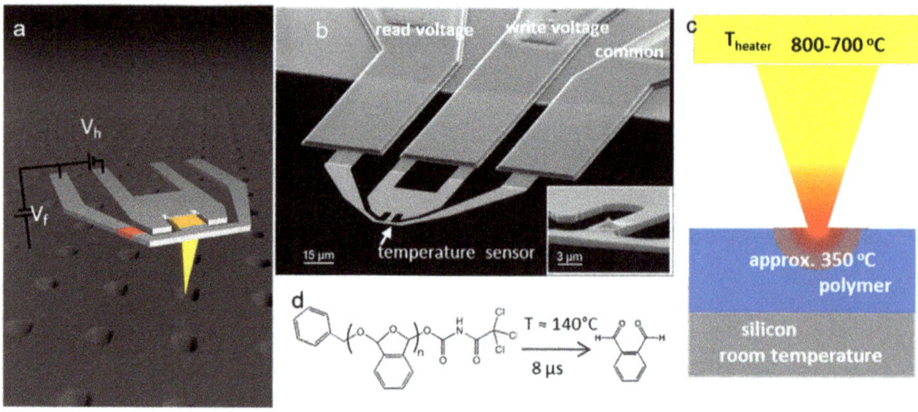

Figure 6.4. (a) Scheme of t-SPL [31]. V_h is the voltage applied to the heater to control the temperature at the tip's base; V_f is the voltage applied between the substrate and the cantilever to control the tip–polymer distance. The region in red marks the position of the temperature sensor. (b) SEM image of a silicon cantilever designed for t-SPL. Reproduced with permission from [78]. Copyright 2011 IOP Publishing. (c) Scheme of heat flow between the heater and the surface of the resist that is in contact with the tip. (d) Molecular structure of PPA. The minimum temperature and the time needed to decompose PPA are indicated [78].

The above results might be important in fields such as energy conversion and storage, catalysis or to fabricate reprogrammable electronic devices.

6.3 Thermal SPL

Thermal scanning probe lithography (t-SPL) is based on the use of heat to generate local modifications on a polymer layer (resist). The heat applied by the tip produces the melting and removal (evaporation) of the polymer molecules under the tip. The resist is modified by the combined effects of mechanical forces and temperature [6, 19, 20]. This method was proposed by John Mamin and Dan Rugar en the mid-1990s for data storage applications [19]. Since then, t-SPL experienced a number of evolutions [20, 30, 31, 77–79]. In some cases, the resulting patterns were made of a material with structure and chemistry different from the original one. This t-SPL process was called thermochemical SPL [80].

6.3.1 Key aspects of t-SPL

The patterning process is controlled by three elements, the heater (cantilever-tip system), the heat flow and the resist.

Heater. In the first t-SPL experiments, a laser was used to heat a transparent polymer under the AFM tip [19]. Later, the heater was integrated into the AFM cantilever system [20, 30, 31, 77–79].

In integrated systems, the tip is resistively heated by a current flowing in the cantilever legs. The cantilever legs are highly doped except for the region where the tip is positioned. In one design, the microcantilever (figure 6.4(a) and (b)) consists of three highly-doped sections (legs) and a lower-doped heater region [78]. Three voltage inputs at the base of the legs selectively power the resistors. The central leg comprises a large area platform in order to provide an additional capacitive force. The capacitive force was applied to control the tip–surface distance and, thus, the force exerted on the resist layer (cantilever force constant of ~1 N m^{-1}).

Heat flow. For silicon-based cantilevers, the maximum sustainable temperature at the heater position is in the 800 °C–1000 °C range. Electro-migration of the dopants across the legs and the type of dopants are the factors that limit the maximum temperature. The thermal time constant of the integrated heaters was in the 5–100 μs range [79]. Those values enabled for fast switching of the thermal stimulus. The effective temperature at the polymer surface depends on the ratio of thermal resistance of the substrate and of the combined resistance of the tip and tip–sample interface [20]. For sharp tips, say with a radius of 5 nm, and a polymer layer thicker than the lateral size of the contact, the temperature at the polymer surface might be reduced by a factor of two with respect to the heater's temperature (figure 6.4(c)).

Materials. The first step in t-SPL is to pattern the resist layer. This requires the use of highly temperature sensitive materials. The etching properties of the resist determine the throughput and the minimum feature size of the patterning process. Several polymer materials were studied, among them, polymethylmethacrylate (PMMA), phenolic compounds, poly(p-phenylene) vinylene, and polyphthalaldehyde (PPA) [20, 81–83]. Thermally responsive PPA is the most common resist in t-SPL because it quickly decomposes into volatile monomers when a hot tip touches it. The process leaves a void in the resist layer whenever the tip's temperature is higher than the decomposition temperature of PPA ~140 °C [82]. In particular, PPA was used for high-throughput nanopatterning because the fission of a single bond was amplified by the spontaneous decomposition of the remaining polymer chains. By using PPA, a 4×10^4 μm^2 s^{-1} throughput was reached at a patterning rate of 2 μs/pixel [78].

Imaging. In some instruments optimized for high-speed nanopatterning, the thermal heater acted as a sensor to track the topography. Imaging was achieved by using only electrical control without the need for an optical beam deflection scheme [17]. This scheme made the t-SPL instrument more compact and faster. However, this method is unsuitable for tracking the topography of rough surfaces.

6.3.2 Nanolithography

Thermal SPL operates by modifying a resist layer to generate a local void in the resist. A succession of etching steps enables transferring the pattern written on the resist to a material of interest (figure 6.5(a)). In one of the fastest t-SPL protocols [78], for each programmed pixel, two different voltages V_h and V_f are applied for a few microseconds to write the patterns; V_h produces the current that heats the tip and V_f provides the force exerted on the resist by pulling the tip into contact with the surface. The heater might sustain temperatures of about 1000 °C for several days. During patterning, the tip is displaced across the substrate at a defined distance from its surface.

Resolution. In t-SPL it is necessary to distinguish the minimum feature written on the polymer resist and the minimum feature size transferred to the material of

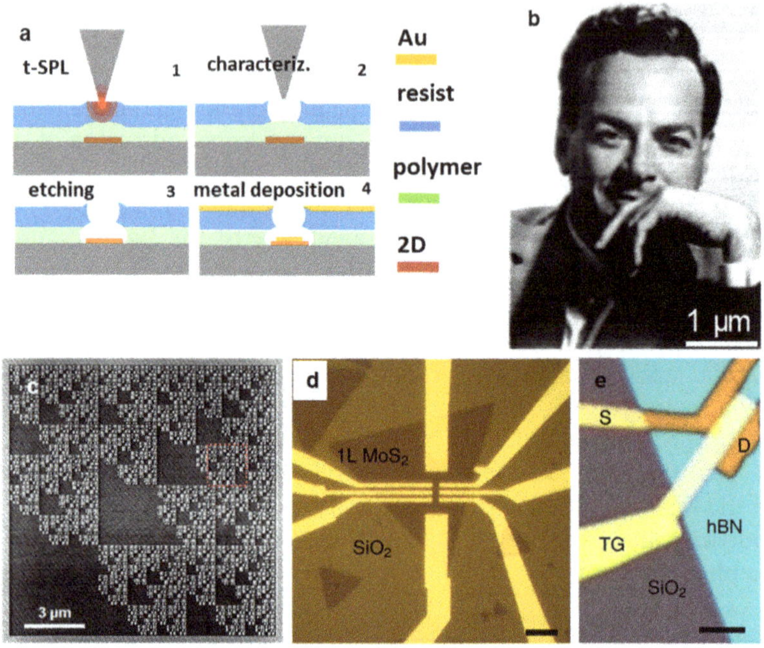

Figure 6.5. t-SPL nanopatterns and devices. (a) Main steps in the fabrication process of metal contacts to a 2D materials flake [95]. (b) Greyscale patterning of a photograph of Richard Feynman. Adapted with permission from [6]. Copyright 2014 Springer Nature. The original portrait (not shown) belongs to the Archives, California Institute of Technology. (c) Topographical image of a fractal pattern made of 880 × 880 pixels written in 12.8 s. Adapted with permission from [78]. Copyright IOP Publishing. (d) Optical image of a 2D material FET device. Scale bar, 5 µm. (e) Optical image of a top-gate contact fabricated on an exfoliated h-BN flake. Scale bar, 4 µm. Panels (d) and (e) adapted with permission from [95]. Copyright 2019 Springer Nature.

interest. An array of parallel lines with a periodicity of 20 nm was generated in PPA while Si nanowires were patterned with a width of 10 nm [84]. The minimum feature size transferred to a material is limited by the thickness of the polymer layer which for PPA is about 6–7 nm. This value is a consequence of the low etch resistance of the PPA.

Greyscale nanopatterning. A singular feature of t-SPL is the capability to generate nanopatterns of different depths in the resist. Those changes might be transferred to a silicon substrate by using an etching protocol. The control of height comes from the reproducible motion of the cantilever, the applied force and the stability of the heater temperature. Those features enabled the generation of three-dimensional greyscale height nanopatterns [6, 31, 82]. The patterning depth in a PPA polymer layer was controlled to about 1 nm, this is, less than the linear size of a single resist molecule. Figure 6.5(b) shows a greyscale pattern of a portrait of R Feynman who in 1959 wrote a visionary essay about nanotechnology. The same method was applied to fabricate nanofluidic ratchet-shaped Brownian motors [85].

Nanopatterning. Detailed descriptions of the nanopatterning capabilities of t-SPL might be found elsewhere [20, 86]. Let's underline some representative results.

Thermal SPL achieved the fastest nanopatterning application of SPL. The pattern shown in figure 6.5(c) contains 880 × 880 pixels. It was written in less than 12.8 s [78].

A combination of greyscale nanopatterning and template growth was applied to control the density of amine groups on a polymer surface. By attaching fluorescence labels on the amines, the resulting pattern could be imaged by fluorescence microscopy [87]. Thermal SPL was used to deprotect active groups such as carboxylic acid and amine groups which can subsequently be used for biochemical conjugation of nano-objects. Multifunctional patterns of proteins, DNA and C_{60} were obtained [88, 89] as well as guided assembly of block copolymers [90] and gold nanorods [91]. PPA surfaces patterned by t-SPL were used to build artificial models to mimic the tissue environment [92].

Nanodevices. Hot probes were applied to promote the thermal reduction of graphene oxide and graphene fluoride layers. Sub-20 nm wide nanoribbons of reduced graphene oxide were fabricated as elements of a FET [93]. The process enabled tuning the conductivity of the nanoribbons over four orders of magnitude. Similarly, the locality of the thermal interaction was used to transform pentacene precursors into a semiconducting pentacene nanostructured transistor [94] or to pattern metal contacts on monolayer 2D materials [95]. Other devices fabricated by t-SPL include magnonic structures [96] or supramolecular polymer-based color-switchable QR codes [97]. It is worth mentioning that in some of the above examples [93], the direct thermal interaction of the hot probe with a material generates functionality. In other words, an etching process is not always needed to reach device functionality.

6.4 Deposition SPL

Deposition SPL is based on the controlled delivery of a material from the tip to the surface [4, 21, 34, 98, 99] (figure 6.6). Dip-pen nanolithography and nanoscale dispensing were the first approaches meeting the above definition. Nanoscale dispensing uses hollow cantilevers integrated with fluidic channels to deliver small liquid drops onto a surface [99–101]. This method has yet to reach the sub-100 nm patterning resolution. For that reason, nanoscale dispensing methods will not be discussed in the chapter.

Deposition SPL (d-SPL) provides a straightforward approach to transform the chemical and biological functionality of surfaces at the 100 nm range. The first deposition of organic molecules (octadecanethiols) from an AFM tip onto a surface (mica) dates back to 1995 [102]. This process was comprehensively developed for nanopatterning purposes under the name of dip-pen nanolithography (DPN) [4, 34]. The use of liquid inks and polymer-based materials [21, 103, 104] expanded the capabilities initially envisioned by DPN [4]. For that reason, we prefer to use the term d-SPL to include all the probe-based methods based on the controlled transfer of material previously deposited on a tip to the sample surface.

6.4.1 Key aspects of deposition SPL

Molecular and liquid inks. Deposition SPL was performed by using a variety of inks, including organic and biological molecules, polymers, colloidal particles, and metal

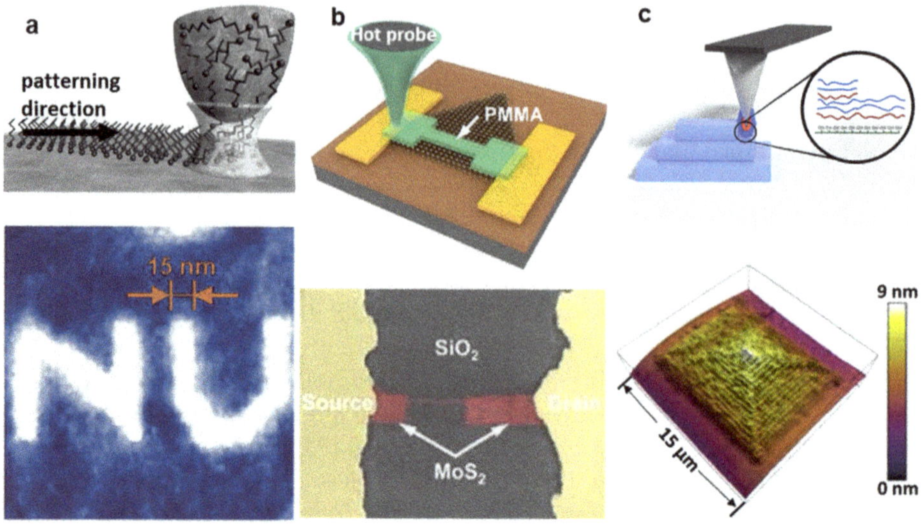

Figure 6.6. Deposition SPL methods. (a) Top, dip-pen nanolithography scheme [4]. Bottom, self-assembled monolayers written on a Au(111) surface. Bottom panel reproduced with permission from [111]. Copyright 1999 AAAS. (b) Top, d-SPL scheme by using a heated tip to deposit a solid ink onto a 2D materials surface. Bottom, SEM image of a 36 nm wide monolayer MoS_2 nanoribbon device. Adapted with permission form [112]. Copyright 2019 American Chemical Society. (c) Top, scheme for delivering materials layer-by-layer in 3D nanoprinting. Bottom, AFM image of a square pyramid made of 15 layers. Bottom panel reproduced with permission from [22]. Copyright 2016 American Chemical Society.

ions [4, 21, 34, 98, 104–106]. The inks might be divided into diffusive and liquid inks. Diffusive inks are soluble in water. Liquid inks are polymer-based such as molten polyethylene, polysterne, PMMA or polyelectrolyte complexes. Solid inks include a sub-category of liquid inks. The deposition of liquid and, in particular of solid inks, is performed with a heated tip to promote the flow of the polymer from the tip to the surface [21].

Ink transport. The transport of the ink molecules from the tip to the surface depends on the type of ink. In the case of diffusive inks, the transport process might be divided into three steps: dissolution of the ink molecules in the water, motion of the ink molecules in the water meniscus formed between the tip and surface and assembly of the molecules on the surface through chemical reactions [107, 108].

The transport of liquid inks from the tip to the surface is rather complex. It depends on the ink composition, viscosity, surface tension, temperature, volume of the ink on the tip and interfacial energy [109]. The tip temperature can be used to control the ink deposition. The rise of the temperature of the tip causes solid ink to melt and wet the tip [110]. The advantage of this approach is two-fold. The ink flow can be turned on and off by changing the temperature. This is a major difference with respect to other d-SPL methods where the ink deposition happens continuously as long as the tip stays in contact with it. In addition, the temperature provides a mechanism to control the rate of the ink diffusion.

Resolution. The feature size depends on the balance between the flow from the tip, deposition time and surface diffusion. For diffusive inks sub-20 nm feature sizes were reported [111]. For liquid and solid inks, the feature size might be influenced by several factors among them the size of the liquid meniscus, temperature and deposition time [109]. Feature sizes in the sub-50 nm range were claimed [112].

6.4.2 Nanolithography

Deposition SPL functions by facilitating the direct transport of molecules to surfaces, much like the transfer of ink from a macroscopic dip-pen to paper. With independence of the type of ink, d-SPL follows three major steps: first, the tip is loaded with the ink; secondly, the ink is transferred to a pre-selected position of the substrate. For liquid and, especially solid inks, this step is facilitated by using hot probes; in a third step, the deposited ink is processed to reach a determined functionality. This last step might be divided into several sub-steps.

Nanopatterning. Deposition SPL enables generating a range of chemical functionalities on the same substrate by depositing different kinds of molecules. Some examples of molecules deposited by d-SPL include oligonucleotides [105], nanoparticles [113] or block-copolymers [114]. The use of polymer-based inks has enhanced the patterning and device applications. It has been shown that poly (N-isopropylacrylamide), a type of protein adhesion molecules, can be reproducibly written from the melt [110]. These nanostructures reversibly bind and release proteins when actuated through a hydrophilic–hydrophobic phase transition. A layer-by-layer process led to a 3D nanoprinting approach [22]. The local deposition of resists led to the fabrication of different devices such as nanomechanical resonators [115], microfluidic channels and 2D materials nanoribbon transistors [112, 116].

6.5 Other SPL methods

Voltage, mechanical or chemical interactions led to other methods of patterning surfaces [3–10, 24, 32]. Those approaches have yet to become widely used or to reach a common set of protocols. For those reasons, they have not been explicitly described in this chapter. It is worth realizing that some SPL-based approaches might combine two or more interactions. For example, in t-SPL a mechanical force is always applied to deform the resist while in o-SPL electric fields and chemical reactions co-exist. Let's provide a brief overview of bias, mechanical and catalytic SPL.

Mechanical SPL or nanomachining (m-SPL) uses the mechanical force exerted by the tip to produce the selective removal of material from a surface. Almost any AFM could be fitted to perform nanolithographic applications by controlling the force. For this reason, m-SPL was widely applied to modify materials [24, 32] such as self-assembled monolayers [23, 32], polymers [25], proteins [32, 117], supramolecular films [118] or even metallic films [119]. The tip's lifetime is a major issue in m-SPL. The mechanical force applied to modify the material might eventually wear or contaminate the tip.

Bias SPL. A great variety of probe-based nanopatterning methods beyond o-SPL were based on the application of a voltage bias between tip and surface. The tip's proximity and the application of a voltage generates very high electric fields (high electric fields ~10 V nm^{-1} or 10 GV m^{-1}). Those fields were used to induce the desorption of atoms from the tip and to deposit them on the sample [26, 27, 120] or to decompose the gas molecules surrounding the tip [28, 121]. In other examples, the electronic current produced by the voltage bias applied between the tip and a conductive substrate was used to break molecular bonds of a polymer layer or soft material deposited on the substrate [29, 122–124]. The electric field at the tip–surface interface was controlled to invert the polarization of a small region in a ferroelectric film. This generated a nonvolatile ferroelectric domain. This mechanism was proposed for data storage [125].

Oxidation SPL might be the best example of using electrochemical reactions for nanopatterning but other electrochemical processes were also explored by probe-based patterning methods [3, 8, 73, 126–128].

Catalytic SPL. The deposition of catalysts in the AFM tip enabled performing and controlling the spatial extension of catalytic reactions [129]. Examples of catalytic reactions using AFM tips included transition metal coatings, metal complexes, organic catalysts, photocatalysts or enzymes. A nanodevice application of c-SPL was the reduction of insulating graphene oxide in the presence of hydrogen. Nanoribbons with widths ranging from 20 to 80 nm were fabricated [130].

The potential of bias SPL goes beyond the field of nanolithography. The method has been applied to understand new processes to decompose very stable chemical species such as carbon dioxide [129], to decompose organic solvents [131, 132] or to study the diffusion of metal ions [126].

6.6 End-of-chapter summary

A scanning probe microscope interface provides a wealth of physical and chemical approaches to modify a surface with nanometer-scale precision. Scanning probe lithography methods have drawn the attention of the nanoscience and nanotechnology communities for two main reasons. First, they provide relatively inexpensive routes to pattern surfaces or devices with nanoscale precision. Secondly, SPL enables studying materials and processes at the nanoscale with an ease that is unparalleled.

Several SPL approaches have evolved to become widely applied for nanopatterning and nanodevice fabrication. Thus, oxidation, thermal and deposition processes are commonly applied in scientific laboratories for designing novel surfaces and prototyping nanodevices. Table 6.1 summarizes the figures of merit of these techniques in terms of minimum feature size and single-dot patterning time. The three methods demonstrated a very high degree of repeatability by generating well-defined structures and patterns with feature sizes in the sub-50 nm range. Oxidation SPL has generated arrays of silicon oxide lines with a periodicity of 15 nm [18]. Almost the same periodicities were obtained by t-SPL while patterning arrays of silicon nanowires and slightly higher values by d-SPL while patterning SAM monolayers [111]. Larger differences are observed in the patterning time of a single

Table 6.1. Figures of merit of SPL methods (clearly resolved patterns).

SPL method	Minimum size (nm)	Smallest periodicity (nm)	Minimum time for writing a single dot (s)
Deposition SPL	~15	32 [111]	0.1–1 [103, 107]
Oxidation SPL	~7	15 [18]	5×10^{-4} [133]
Thermal	~10	20 [84]	5×10^{-6} [78]

dot. Thermal SPL can operate in the μs range while o-SPL operates in the submillisecond range and d-SPL might operate in the subsecond range.

The development of user-friendly AFMs and powerful software tools to control the tip's displacement together with the availability of AFMs will ensure the use of SPL by the scientific community for years to come. The transformation of SPL from a scientific tool to a technology for large-scale nanomanufacturing seems far at best. The sequential character of scanning probe microscopy and the relatively short lifetime of an AFM tip imposes limits to the size of the patterned areas and the throughput. Critical breakthroughs are needed to bring the SPL technology closer to large scale (~cm^2) nanolithography applications.

6.7 Student's corner

- Which of the following SPL methods do pattern a resist in the first nanolithographic step?
  - A. oxidation SPL;
  - B. thermal SPL;
  - C. deposition SPL;
  - D. dip-pen nanolithography.
- Select from the list the factors that influence the feature size in oxidation SPL.
  - A. the tip's radius;
  - B. the temperature of the tip;
  - C. the applied voltage;
  - D. the thickness of the resist.
- What are the functionalities of the oxide structures fabricated by o-SPL?
- (a) Describe thermal scanning probe lithography. (b) Mention the dominant factors that control the feature size and (c) its order of magnitude (1, 10, 100 or 1000 nm?).
- Dip-pen nanolithography is based on the deposition of diffusive inks on a surface. Describe the main steps of the transport process of diffusive inks.
- Provide an estimation of the electric field generated by applying 10 V for a tip that is placed 1 nm above a gold surface? Is this field larger or smaller than the average electric fields applied in electron beam lithography?

- Design a micro-optical tag made of multi-layered MoS_2 that alternates blue, green and yellow regions. The color of MoS_2 flakes is related to the number of monolayers (see scheme below). Each optical region of the tag is a square of 800 nm × 800 nm. A triangular MoS_2 flake of 10 μm (side length) and 30 monolayers in thickness is deposited on a Si substrate (see below). It is the raw material for the tag. Hint: the number of MoS_2 monolayers oxidized per unit of time in o-SPL is known. The MoS_2 oxides generated in o-SPL are soluble in water (see figure 6.2(c)).

Acknowledgments

Financial support from the Ministerio de Ciencia e Innovación is acknowledged through grants MAT2016-76507-R, RED2018-102627-T and PID2019-106801GB-I00.

References

[1] Binnig G, Rohrer H and Gerber C 1982 Surface studies by scanning tunneling microscopy *Phys. Rev. Lett.* **49** 57

[2] Binnig G, Quate C F and Gerber C 1986 Atomic force microscope *Phys. Rev. Lett.* **56** 930

[3] Garcia R, Martinez R V and Martinez J 2006 Nano-chemistry and scanning probe nanolithographies *Chem. Soc. Rev.* **35** 29–38

[4] Ginger D S, Zhang H and Mirkin C A 2004 The evolution of dip-pen nanolithography *Angew. Chem. Int. Ed* **43** 30–45

[5] Custance O, Perez R and Morita S 2009 Atomic force microscopy as a tool for atom manipulation *Nat. Nanotechnol.* **4** 803–10

[6] Garcia R, Knoll A W and Riedo E 2014 Advanced scanning probe lithography *Nat. Nanotechnol.* **9** 577–87

[7] Kurra N, Reifenberger R G and Kulkarni G U 2014 Nanocarbon-Scanning probe microscopy synergy: fundamental aspects to nanoscale devices *ACS Appl. Mater. Interfaces* **6** 6147–63

[8] Liu H, Hoeppener S and Schubert U S 2016 Nanoscale materials patterning by local electrochemical lithography *Adv. Eng. Mater.* **18** 890–902

[9] Xie X N, Chung H J, Sow C H and Wee A T S 2006 Nanoscale materials patterning and engineering by atomic force microscopy nanolithography *Mater. Sci. Eng.* R **54** 1–48

[10] Rani E and Wong L S 2019 High-resolution scanning probe lithography of 2D materials: novel nanostructures *Adv. Mater. Technol.* **4** 1900181

[11] Lyo I W and Avouris P 1991 Field-induced nanometer-scale to atomic-scale manipulation of silicon surfaces with the STM *Science* **253** 173–6

[12] Eigler D M and Schweizer E K 1990 Positioning single atoms with a scanning tunneling microscope *Nature* **344** 524–6

[13] Garcia R G 1992 Atomic-scale manipulation in air with the scanning tunneling microscope *Appl. Phys. Lett.* **60** 1960–2

[14] Fuechsle M, Miwa J A, Mahapatra S, Ryu H, Lee S, Warschkow O, Hollenberg L C L, Klimeck G and Simmons M Y 2012 A single-atom transistor *Nat. Nanotechnol.* **7** 242–6

[15] Koch M, Keizer J G, Pakkiam P, Keith D, House M G, Peretz E and Simmons M Y 2019 Spin read-out in atomic qubits in an all-epitaxial three-dimensional transistor *Nat. Nanotechnol.* **14** 132–40

[16] Kalff F E, Rebergen M P, Fahrenfort E, Girovsky J, Toskovic R, Lado J L, Fernández-Rossier J and Otte A F 2016 A kilobyte rewritable atomic memory *Nat. Nanotechnol.* **11** 926–9

[17] Garcia R 2010 *Amplitude Modulation Atomic Force Microscopy* (Weinheim: Wiley-VCH)

[18] Ryu Y K and Garcia R 2017 Advanced oxidation scanning probe lithography *Nanotechnology* **28** 142003

[19] Mamin H and Rugar D 1992 Thermomechanical writing with an atomic force microscope tip *Appl. Phys. Lett.* **61** 1003–5

[20] Howell S T, Grushina A, Holzner F and Brugger J 2020 Thermal scanning probe lithography—a review *Microsyst. Microeng.* **6** 21

[21] Sheehan P E, Whitman L J, King W P and Nelson B A 2004 Nanoscale deposition of solid inks via thermal dip pen nanolithography *Appl. Phys. Lett.* **85** 1589–91

[22] Zhao J, Swartz L A, Lin W F, Schlenoff P S, Frommer J, Schlenoff J B and Liu G Y 2016 Three-dimensional nanoprinting via scanning probe lithography-delivered layer-by-layer deposition *ACS Nano* **6** 5656–62

[23] Xu S and Liu G Y 1997 Nanometer-scale fabrication by simultaneous nanoshaving and molecular assembly *Langmuir* **13** 127–9

[24] Liu M, Amro N A and Liu G Y 2008 Nanografting for surface physical chemistry *Annu. Rev. Phys. Chem.* **59** 367–86

[25] Liu G, Chen L, Liu J, Qiu M, Xie Z, Chang J, Zhang Y, Li P, Lei D Y and Zheng Z 2018 Scanning nanowelding lithography for rewritable one-step patterning of sub-50 nm high-aspect ratio metal nanostructures *Adv. Mater.* **30** 1801772

[26] Mamin H J, Guethner P H and Rugar D 1990 Atomic emission from a gold scanning-tunneling-microscope tip *Phys. Rev. Lett.* **65** 2418

[27] Calleja M, Tello M, Anguita J, García F and Garcia R 2001 Fabrication of gold nanowires on insulating substrates by field-induced mass transport *Appl. Phys. Lett.* **79** 2471–3

[28] Vasko S E, Kapetanovic A, Talla V, Brasino M D, Zhu Z, Scholl A, Torrey J D and Rolandi. M 2011 Serial and parallel Si, Ge, and SiGe direct-write with scanning probes and conducting stamps *Nano Lett.* **11** 2386–9

[29] Kaestner M, Hofer M and Rangelow I W 2013 Nanolithography by scanning probes on calixarene molecular glass resist using mix-and-match lithography *J. Micro-Nanolithogr. MEM* **12** 031111

[30] Binnig G, Despont M, Drechsler U, Haberle W, Lutwyche M, Vettiger P, Mamin H J, Chui B W and Kenny T W 1999 Ultrahigh-density atomic force microscopy data storage with erase capability *Appl. Phys. Lett.* **74** 1329–31

[31] Pires D, Hedrick J L, De Silva A, Frommer J, Gotsmann B, Wolf H, Despont M, Duerig U and Knoll A W 2010 Nanoscale three-dimensional patterning of molecular resists by scanning probes *Science* **328** 732–5

[32] Krämer S, Fuierer R R and Gorman C B 2003 Scanning probe lithography using self-assembled monolayers *Chem. Rev.* **103** 4367–418

[33] Carnally S A M and Wong L S 2014 Harnessing catalysis to enhance scanning probe lithography *Nanoscale* **6** 4998–5007

[34] Piner R D, Zhu J, Xu F, Hong S and Mirkin C A 1999 'Dip-pen' nanolithography *Science* **283** 661–3

[35] Dagata J A, Schneir J, Harary H H, Evans C J, Postek M T and Bennet J 1990 Modification of hydrogen-passivated silicon by a scanning tunneling microscope operating in air *Appl. Phys. Lett.* **56** 2001–3
[36] Thundat T, Nagahara L A, Oden P I, Lindsay S M, George M A and Glaunsinger W S 1990 , Modification of tantalum surfaces by scanning tunneling microscopy in an electrochemical cell *J. Vac. Sci. Technol.* A **8** 3537–41
[37] Mori G, Lazzarino M, Ercolani D, Biasiol G, Sorba L, Heun S and Locatelli A 2005 Desorption dynamics of oxide nanostructures fabricated by local anodic oxidation nanolithography *J. Appl. Phys.* **98** 114303
[38] Dago A I *et al* 2018 Chemical and structural analysis of sub-20 nm graphene patterns generated by scanning probe lithography *Carbon* **129** 281–5
[39] Gomez-Moñivas S, Saenz J J, Calleja M and Garcia R 2003 Field-induced formation of nanometer-sized water bridges *Phys. Rev. Lett.* **91** 056101
[40] Cramer T, Zerbetto F and Garcia R 2008 Molecular mechanism of water bridge buildup: field-induced formation of nanoscale menisci *Langmuir* **24** 6116–20
[41] Calleja M, Tello M and Garcia R 2002 Size determination of field-induced water menisci in noncontact atomic force microscopy *J. Appl. Phys.* **92** 5539–42
[42] Dagata J, Pérez-Murano F, Martin C, Kuramochi H and Yokoyama H 2004 Current, charge, and capacitance during scanning probe oxidation of silicon I: maximum charge density and lateral diffusion *J. Appl. Phys.* **96** 2386–92
[43] Kinser C R, Schmitz M J and Hersam M C 2006 Kinetics and mechanism of atomic force microscope local oxidation on hydrogen-passivated silicon in inert organic solvents *Adv. Mater.* **18** 1377–80
[44] Gotoh Y *et al* 2000 Experimental and theoretical results of room-temperature single-electron transistor formed by atomic force microscope nano-oxidation process *J. Vac. Sci. Technol.* A **18** 1321–5
[45] Snow E S and Campbell P M 1994 Fabrication of Si nanostructures with an atomic force microscope *Appl. Phys. Lett.* **64** 1932–4
[46] Garcia R, Calleja M and Pérez-Murano F 1998 Local oxidation of silicon surfaces by dynamic force microscopy: nanofabrication and water bridge formation *Appl. Phys. Lett.* **72** 2295–7
[47] Chien F S S, Chou Y C, Chen T T, Hsieh W F, Chao T S and Gwo S 2001 Nano-oxidation of silicon nitride films with an atomic force microscope: chemical mapping, kinetics, and applications *J. Appl. Phys.* **89** 2465–72
[48] Maoz R, Cohen S R and Sagiv J 1999 Nanoelectrochemical patterning of monolayer surfaces: toward spatially defined self-assembly of nanostructures *Adv. Mater.* **11** 55–61
[49] Liu H, Hoeppener S and Schubert U S 2016 Reversible nanopatterning on polypyrrole films by atomic force microscope electrochemical lithography *Adv. Funct. Mater.* **26** 614–9
[50] Oria L, Ruiz de Luzuriaga A, Alduncin J A and Perez-Murano F 2013 Polystyrene as a brush layer for directed self-assembly of block co-polymers *Microelectron. Eng.* **110** 234
[51] Cen C, Thiel S, Hammerl G, Schneider C W, Andersen K E, Hellberg C S, Mannhart J and Levy J 2008 Nanoscale control of an interfacial metal-insulator transition at room temperature *Nat. Mater.* **7** 298–302
[52] Schrecongost D *et al* 2019 On-demand nanoscale manipulations of correlated oxide phases *Adv. Funct. Mater.* **29** 1905585

[53] Weng L, Zhang L, Chen Y P and Rokhinson P L 2008 Atomic force microscope local oxidation lithography of graphene *Appl. Phys. Lett.* **93** 093107
[54] Byun I S, Yoon D, Choi J S, Hwang I and Lee D H *et al* 2011 Nanoscale lithography on monolayer graphene using hydrogenation and oxidation *ACS Nano* **5** 6417–24
[55] Espinosa F M, Ryu Y K, Marinov K, Dumcenco D, Kis A and Garcia R 2015 Direct fabrication of thin layer MoS_2 field-effect nanoscale transistors by oxidation scanning probe lithography *Appl. Phys. Lett.* **106** 103503
[56] Zhao P, Wang R X, Lien D H, Zhao Y B, Kim H, Cho J, Ahn G H and Javey A 2019 Scanning probe lithography patterning of monolayer semiconductors and application in quantifying edge recombination *Adv. Mater.* **31** 1900136
[57] Dago A I, Ryu Y K and Garcia R 2016 Sub-20 nm patterning of thin layer WSe_2 by scanning probe lithography *Appl. Phys. Lett.* **109** 163103
[58] Liu X, Chen K S, Wells S A, Balla I, Zhu J, Wood J D and Hersam M C 2017 Scanning probe nanopatterning and layer-by-layer thinning of black phosphorus *Adv. Mater.* **29** 1604121
[59] Garcia R, Tello M, Moulin J F and Biscarini F 2004 . Size and shape controlled growth of molecular nanostructures on silicon oxide templates *Nano Lett.* **4** 1115–9
[60] Martinez R V, Martinez J, Chiesa M, Garcia R, Coronado E, Pinilla-Cienfuegos E and Tatay S 2010 Large-scale nanopatterning of single proteins used as carriers of magnetic nanoparticles *Adv. Mater.* **22** 588–91
[61] Liu S, Maoz R and Sagiv J 2004 Planned nanostructures of colloidal gold via self-assembly on hierarchically assembled organic bilayer template patterns with *in situ* generated terminal amino functional *Nano Lett.* **4** 845–51
[62] Germain J, Rolandi M, Backer S A and Fréchet M 2008 Sulfur as a novel nanopatterning material: an ultrathin resist and a chemically addressable template for nanocrystal self-assembly *Adv. Mater.* **20** 4526–9
[63] Fabre B and Herrier C 2012 Automated sub-100 nm local anodic oxidation-directed nanopatterning of organic monolayer-modified silicon surfaces *RSC Adv.* **2** 168–75
[64] Mukai K, Hirota A and Nakashima S 2015 Position control of PbS quantum dot using nanohole on silicon substrate processed by scanning probe lithography *Japan. J. Appl. Phys.* **54** 04DJ02
[65] Snow E S and Campbell P M 1995 AFM fabrication of sub-10-nanometer metal-oxide devices with *in situ* control of electric properties *Science* **270** 1639–41
[66] Matsumoto K, Ishii M, Segawa K, Oka Y, Vartanian B J and Harris J S 1996 Room temperature operation of a single electron transistor made by the scanning tunneling microscope nanooxidation process for the TiO_x/Ti system *Appl. Phys. Lett.* **68** 34–6
[67] Martinez J, Martinez R V and Garcia R 2008 Silicon nanowire transistors with a channel width of 4 nm fabricated by atomic force microscope nanolithography *Nano Lett.* **8** 3636–9
[68] Neubeck S, Ponomarenko L A, Freitag F, Giesbers A J M, Zeitler U, Morozov S V, Blake P, Geim A K and Novoselov K S 2010 From one electron to one hole: quasiparticle counting in graphene quantum dots determined by electrochemical and plasma etching *Small* **6** 1469–73
[69] Puddy R K, Chua C J and Buitelaar M R 2013 Transport spectroscopy of a graphene quantum dot fabricated by atomic force microscope nanolithography *Appl. Phys. Lett.* **103** 183117

[70] Lee D H, Kim C K, Lee J H, Chung H J and Park B H 2016 Fabricating in-plane transistor and memory using atomic force microscope lithography towards graphene system on chip *Carbon* **96** 223–8
[71] Masubuchi S, Arai M and Machida T 2011 Atomic force microscopy based tunable local anodic oxidation of graphene *Nano Lett.* **11** 4542–6
[72] Lipatov A, Li T, Vorobeva N S, Sinitski A and Gruverman A 2019 Nanodomain engineering for programammable ferroelectric devices *Nano Lett.* **19** 3194–8
[73] Tello M, Garcia R, Martín-Gago J A, Martínez N F, Martín-González M S, Aballe L, Baranov A and Gregoratti L 2005 Bottom–up fabrication of carbon-rich silicon carbide nanowires by manipulation of nanometer-sized ethanol menisci *Adv. Mater.* **17** 1480–3
[74] Irvin P, Ma Y, Bogorin D F, Cen C, Bark C W, Folkman C M, Eom C B and Levy J 2010 Rewritable nanoscale oxide photodetector *Nat. Photonics* **4** 849
[75] Chen L, Li J, Tang Y, Pi Y Y, Chen Y, Pryds N, Irwin P and Levy J 2018 Extreme reconfigurable nanoelectronics at the $CaZrO_3/SrTiO_3$ interface *Adv. Funct. Mater.* **30** 1801794
[76] Scherecongost D *et al* 2019 On-demand nanoscale manipulations of correlated oxides *Adv. Funct. Mater.* **29** 190585
[77] Vettiger P, Despont M, Drechsler U, Durig U, Haberle W, Lutwyche M I, Rothuizen H, Stutz R, Widmer R and Binnig G K 2000 The 'Millipede'—more than thousand tips for future AFM storage *IBM J. Res. Dev.* **44** 323–40
[78] Paul P C, Knoll A W, Holzner F, Despont M and Durig U 2011 Rapid turnaround scanning probe nanolithography *Nanotechnology* **22** 275306
[79] King W P, Bhatia B, Felts J R, Kim H J, Kwon B, Lee B, Somnath S and Rosenberger M 2013 Heated atomic force microscope cantilevers and their applications *Annu. Rev. Heat Transfer* **16** 287–326
[80] Szoszkiewicz R, Okada T J, Simon C, Li T D, King W P, Marder S R and Riedo E 2007 High-speed, sub-15 nm feature size thermochemical nanolithography *Nano Lett.* **7** 1064–9
[81] Gotsmann B, Duerig U, Frommer J and Hawker C J 2006 Exploiting chemical switching in a Diels-Alder polymer for nanoscale probe lithography and data storage *Adv. Funct. Mater.* **16** 1499
[82] Knoll A W, Pires D, Coulembier O, Dubois P, Hedrick J L, Frommer J and Duerig U 2010 Probe-based 3-D nanolithography using self-amplified depolymerization polymers *Adv. Mater.* **22** 3361–5
[83] Fenwick O, Bozec L, Credgington D, Hammiche A, Lazzerini G M, Silberberg Y R and Cacialli F 2009 *Nat. Nanotechnol.* **4** 664–8
[84] Ryu Cho Y K, Rawlings C, Wolf H, Spieser M, Bisig S, Reidt S, Sousa M, Khanal S, Jacobs T and Knoll A W 2017 Sub-10 nanometer feature size in silicon using thermal scanning probe lithography *ACS Nano* **11** 11890–7
[85] Skaug M J, Schwemmer C, Fringes S, Rawlings C D and Knoll A W 2018 Nanofluidic rocking Brownian motors *Science* **359** 1505–8
[86] Ryu Y K and Knoll A W 2019 *Electrical Atomic Force Microscopy for Nanoelectronics* (New York: Springer) pp 143–72
[87] Carroll K M, Giordano A J, Wang D, Kodali V K, Scrimgeour J, King W P, Marder S R, Riedo E and Curtis J E 2013 Fabricating nanoscale gradients with thermochemical nanolithography *Langmuir* **29** 8675–82

[88] Wang D *et al* 2009 Thermochemical nanolithography of multifunctional nanotemplates for assembling nano-objects *Adv. Funct. Mater.* **19** 3696–702
[89] Duvigneau J, Schoenherr H and Vancso G J 2008 Atomic force microscopy based thermal lithography of poly (tert-butyl acrylate) block copolymer films for bioconjugation *Langmuir* **24** 10825–32
[90] Gottlieb S, Lorenzoni M, Evangelio L, Fernández-Regúlez M, Ryu Y K, Rawlings C, Spieser M, Knoll A W and Perez-Murano F 2017 Thermal scanning probe lithography for the directed self-assembly of block copolymers *Nanotechnology* **28** 175301
[91] Holzner F, Kuemin C, Paul P, Hedrick J L, Wolf H, Spencer N D, Duerig U and Knoll A W 2011 Directed placement of gold nanorods using a removable for guided assembly *Nano Lett.* **11** 3957–62
[92] Tang S W *et al* 2019 Replication of a tissue microenvironment by thermal scanning probe lithography *ACS Appl. Mater. Interfaces* **11** 18988–94
[93] Wei Z *et al* 2010 Nanoscale tunable reduction of graphene oxide for graphene electronics *Science* **328** 1373–6
[94] Shaw J E, Stavrinou P N and Anthopoulos T D 2013 On-demand patterning of nano-structured pentacene transistors by scanning thermal lithography *Adv. Mater.* **25** 552–8
[95] Zheng X, Calo A and Albisetti E *et al* 2019 Patterning metal contacts on monolayer MoS_2 with vanishing Schottky barriers using thermal nanolithography *Nat. Electron.* **2** 17–25
[96] Albisetti E *et al* 2016 Nanopatterning reconfigurable magnetic landscapes via thermally assisted scanning probe lithography *Nat. Nanotechnol.* **11** 545–51
[97] Zimmermann S T, Balkenende D W R, Lavrenova A, Weder C and Brugger J 2017 Nanopatterning of a stimuli-responsive fluorescent supramolecular polymer by thermal scanning probe lithography *ACS Appl. Mater. interfaces* **9** 41454–61
[98] Su M and Dravid V P 2002 Colored ink dip-pen nanolithography *Appl. Phys. Lett.* **80** 434
[99] Lewis A, Kheifetz Y, Shambrodt E and Radko A 1999 Fountain pen nanochemistry: atomic force control of chrome etching *Appl. Phys. Lett.* **75** 2689
[100] Hong M H, Kim K H, Bae J and Jhe W 2000 Scanning nanolithography using a material-filled nanopipette *Appl. Phys. Lett.* **77** 2604
[101] Meister A, Liley M, Brugger J, Pugin R and Heinzelmann H 2004 Nanodispenser for attoliter volume deposition using atomic force microscopy probes modified by focused-ion-beam milling *Appl. Phys. Lett.* **85** 6260–2
[102] Jaschke M and Butt H J 1995 Deposition of organic material by the tip of a scanning force microscope *Langmuir* **11** 1061–4
[103] Nakashima H, Higgins M J, O'Connell C, Torimitsu K and Wallace G G 2012 Liquid deposition patterning of conducting polymer ink onto hard and soft flexible substrates via dip-pen nanolithography *Langmuir* **28** 804–11
[104] Felts J R, Somnath S, Ewoldt R H and King W P 2012 Nanometer-scale flow of molten polyethylene from a heated atomic force microscope tip *Nanotechnology* **23** 215301
[105] Demers L M, Ginger D S, Park S J, Li Z, Chung S W and Mirkin C A 2002 Direct patterning of modified oligonucleotides on metals and insulators by dip-pen nanolithography *Science* **296** 1836–8
[106] Bellido E, de Miguel R, Ruiz-Molina D, Lostao A and Maspoch D 2010 Controlling the number of proteins with dip-pen nanolithography *Adv. Mater.* **22** 352–5
[107] Sheehan P E and Whitman L J 2002 Thiol diffusion and the role of humidity in 'dip pen nanolithography' *Phys. Rev. Lett.* **88** 156104

[108] Saha S K and Culpepper M L 2010 An ink transport model for prediction of feature size in dip pen nanolithography *J. Phys. Chem.* C **114** 15364–9

[109] O'Connell C D, Higgins M J, Sullivan R P, Moulton S E and Wallace G G 2014 Ink-on-probe hydrodynamics in atomic force microscope deposition of liquid inks *Small* **10** 3717–28

[110] Lee W K, Whitman L J, Lee J, King W P and Sheehan P E 2008 The nanopatterning of a stimulus-responsive polymer by thermal dip-pen nanolithography *Soft Matter* **4** 1844–7

[111] Hong S, Zhu J and Mirkin C A 1999 Multiple ink nanolithography: toward a multiple-pen nano-plotter *Science* **286** 523–5

[112] Chen S, Kim S P, Chen W, Yuan J, Bashir R, Lou J, van der Zande A M and King W P 2019 Monolayer MoS_2 nanoribbon transistors fabricated by SPL *Nano Lett.* **19** 2092–8

[113] Zhang H and Mirkin C A 2004 DPN-generated nanostructures made of gold, silver, and palladium *Chem. Mater.* **16** 1480–4

[114] Felts J R, Onses M S, Rogers J A and King W P 2014 Nanometer scale alignment of block-copolymer domains by means of a scanning probe tip *Adv. Mater.* **26** 2999–3002

[115] Hu H, Cho H, Somnath S, Vakakis A F and King W P 2014 Silicon nano-mechanical resonators fabricated by using tip-based nanofabrication *Nanotechnology* **25** 275301

[116] Lee W K *et al* 2011 Chemically isolated graphene nanoribbons reversibly formed in fluorographene using polymer nanowire masks *Nano Lett.* **11** 5461–4

[117] Ngunjiri J and Garno J C 2008 AFM-based lithography for nanoscale protein assays *Anal. Chem.* **80** 1361–9

[118] Cavallini M, Biscarini F, Leon S, Zerbetto F, Bottari G and Leigh D A 2003 Information storage using supramolecular surface patterns *Science* **299** 531–3

[119] Chen H A, Lin H Y and Lin H N 2010 Localized surface plasmon resonance in lithographically fabricated single gold nanowires *J. Phys. Chem.* C **114** 10359–64

[120] Liu J F and Miller G P 2009 Field-assisted nanopatterning of metals, metal oxides and metal salts *Nanotechnology* **20** 055303

[121] Garcia R, Losilla N S, Martinez J, Martinez R V, Palomares F J, Huttel Y, Calvaresi M and Zerbetto F 2010 Nanopatterning of carbonaceous structures by field-induced carbon dioxide splitting with a force microscope *Appl. Phys. Lett.* **96** 143110

[122] Garcia R, Tamayo J, Soler J M and Bustamante C 1995 Physical parameters that control the imaging of purple membranes with the scanning tunneling microscope *Langmuir* **11** 2109–14

[123] Lyuksyutov S F, Vaia R A, Paramonov P B, Juhl S, Waterhouse L, Ralich R M, Sigalov G and Sancaktar E 2003 Electrostatic nanolithography in polymers using atomic force microscopy *Nat. Mater.* **2** 468–72

[124] Lenk C *et al* 2019 High-throughput process chain for single electron transistor devices based on field-emission scanning probe lithography and smart nanoimprint lithography technology *J. Vac. Sci. Technol.* B **37** 021603

[125] Tayebi N *et al* 2012 Tuning the built-in electric field in ferroelectric $Pb(Zr_{0.2}Ti_{0.8})O_3$ films for long-term stability of single-digit nanometer inverted domains *Nano Lett.* **12** 5455–63

[126] Berson J, Burshtain D, Zeira A, Yoffe A, Maoz R and Sagiv J 2015 Single-layer ionic conduction on carboxyl-terminated silane monolayers patterned by constructive lithography *Nat. Mater.* **14** 613–21

[127] Yang S M, Paranthaman M P, Noh T W, Kalinin S V and Stelcov E 2016 Nanoparticle shape evolution and proximity effects during tip-induced electrochemical processes *ACS Nano* **10** 663–71

[128] Obermair C, Kress M, Wagner A and Schimmel T 2012 Reversible mechano-electrochemical writing of metallic nanostructures with the tip of an atomic force microscope *Beilstein J. Nanotechnol.* **3** 824–30

[129] Carnally S A M and Wong. L S 2014 Harnessing catalysis to enhance scanning probe nanolithography *Nanoscale* **6** 4998–5007

[130] Zhang K, Fu Q, Pan N, Yu X, Liu J, Luo Y, Wang X, Yang J and Hou J 2012 Direct writing of electronic devices on graphene oxide by catalytic scanning probe lithography *Nat. Commun.* **3** 1194

[131] Martinez R V, Losilla N S, Martinez J, Huttel Y and Garcia R 2007 Patterning polymeric structures with 2 nm resolution at 3 nm half pitch in ambient conditions *Nano Lett.* **7** 1846–50

[132] Suez I, Rolandi M, Backer S A, Scholl A, Doran A, Okawa D, Zettl A and Fréchet J M J 2007 High-field scanning probe lithography in hexadecane: transitioning from field induced oxidation to solvent decomposition through surface modification *Adv. Mater.* **19** 3570–3

[133] Calleja M and Garcia R 2000 Nano-oxidation of silicon surfaces by noncontact atomic-force microscopy: size dependence on voltage and pulse duration *Appl. Phys. Lett.* **76** 3427–9

IOP Publishing

Nanofabrication
Nanolithography techniques and their applications
José María De Teresa

Chapter 7

Soft thermal nanoimprint and hybrid processes to produce complex structures

Isabel Rodríguez and Jaime Hernández

This chapter is particularly dedicated to illustrating soft thermal nanoimprint lithography (T-NIL) and hybrid processes as enabling technology to create complex surface nanoarchitectures such as high aspect ratio and hierarchical micro and nanostructures with high fidelity and resolution in a cost effective manner. The ability to produce complex surface nanoarchitectures has opened up new possibilities to create a large variety of functional and multifunctional nanostructured surfaces that can provide materials with exceptional properties beyond those expected from the bulk material itself, and very often, they are inspired from surface structures found in nature. In the first part of the chapter, the basic NIL processes are introduced. Then, a comprehensive description of the basic tools and processes including the preparation of soft molds is given. In the second part of the chapter, the main processing challenges concerning the production of high aspect ratio and multilevel hierarchical structures using soft T-NIL are described. Finally, some of the most efficient methodologies to produce multiscale hierarchical surface structures with high precision using soft T-NIL in combination with other technologies are reviewed.

7.1 Fundamentals of nanoimprint processes

7.1.1 Introduction

Nanoimprint lithography (NIL) is a high-resolution patterning technology which, after intense development in the last 25 years, has evolved to be one of the most resourceful and versatile technologies for nanofabrication and nanomanufacturing of products integrating nanoscale features [1, 2]. The fundamental principles underlying the technology, the materials and processing methodology of the current state of the art of NIL processes have been well documented by a vast number of publications and books [3–6]. The main applications enabled by NIL have also been

doi:10.1088/978-0-7503-2608-7ch7

reviewed [7–9]. NIL is indeed a replication technology for mechanically molding or embossing onto a polymeric material by means of pressure exerted through a mold whose patterns are then transferred to the polymer material and fixed either by heating and cooling or UV light crosslinking.

The practical worth of NIL derives from several distinct advantages over other lithography methods. First and foremost is its inherent simplicity, only involving a mold patterned with the features to transfer and a suitable flowable resist material. It only requires a simple instrumentation that does not imply expensive optical or high vacuum components. Due to its purely mechanical nature, NIL offers a high resolution not limited by light diffraction or beam scattering like traditional photolithographic processes. In fact, NIL offers almost unlimited pattern resolution that is basically decided by the availability of the mold. A pattern resolution below 10 nm was realized since the early developments of NIL [10] and, recently, atomic scale resolution has also been demonstrated [11].

Nanoimprint is a parallel processing technology that can provide a high throughput at nanoscale resolution. Additionally, through roll-to-roll (R2R) or roll-to-plate (R2P) techniques, the processing capabilities of NIL can be up-scaled to produce nano-patterns in a continuous fashion. The technology can be applied to a large variety of materials, generating functional patterns in almost any kind of substrate, extending the applicability of nanoimprint technology enormously. Because NIL is a cost effective and versatile lithographic tool offering high performance in terms of both resolution and throughput, it has emerged as one of the most promising technologies for the fabrication and commercial production of nano-enabled products.

Nanoimprinting was born as a lithographic technique and was proposed originally as a high resolution technique for pattern transfer from a mold to a substrate for subsequent post-processing. In the early times, a thermoplastic polymer was employed as a resist material. Thus, the technology was referred to as thermal nanoimprint lithography (T-NIL) [12, 13]. Subsequently, UV-curable resins were introduced as resist material, giving rise to UV-NIL whereby the structures are fixed via photopolymerization or crosslinking.

Due to the material patterning versatility of NIL, the technology soon evolved from being a lithographic pattern transfer technology to being a technology to create functional nanostructured surfaces where the imprinted structures themselves remain as active components providing for functionality. Hence, because of simple processing and large choice of material options, NIL has become a nanofabrication technology ideally suited for applications beyond microelectronics. NIL has demonstrated to be particularly useful to produce functional nanopatterns in areas such as polymer nano-optics and photonics [14, 15], biomimetic functional surfaces [16–18] and biological and tissue culture applications [19].

7.1.2 Basic nanoimprint processes and variants

The NIL processes have evolved into different approaches and these can be broadly classified into two technologies: T-NIL or UV-NIL. Figure 7.1 shows schematics of the general processing steps involved.

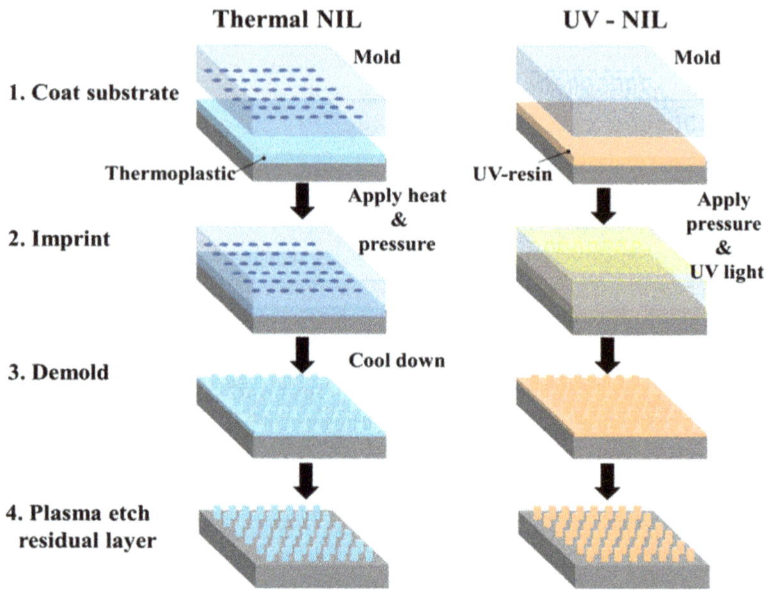

Figure 7.1. Schematic illustration of standard thermal nanoimprint lithography (T-NIL) and UV nanoimprint lithography (UV-NIL).

In T-NIL, a thermoplastic thin polymer film is first produced by spin casting a solution of the polymer. Alternatively, a free standing polymeric film can be used if required. The film is pressed and imprinted with a mold under elevated temperature. This process has been also referred to as nano hot embossing whereby the thermoplastic polymer is mechanically deformed by the mold under pressure at an elevated temperature. In practice, this temperature is set at about 50 °C–90 °C above the polymer's glass transition temperature (T_g) so that the polymer reaches the viscous flow state. The typical pressure employed is within the range of 10–50 bar depending primarily on the pattern resolution and geometry of the cavities to fill in the mold [3]. After the process, the mold-substrate assembly is cooled down below the T_g, where the polymer recovers its hard elastic, glassy properties and then separated revealing the replica of the imprinted mold nanopattern on the thermoplastic polymer film.

In UV-NIL, a liquid UV sensitive resin (formulation of monomers and photo initiators) is deposited by spin coating on a suitable substrate. Then the mold is brought into contact with the substrate and a low to moderate pressure (0–5 bar) is applied to fill the mold cavities. In this case, the mold or the substrate needs to be UV transparent so that by illumination through, it is possible to induce crosslinking and hardening of the resin. Since UV-NIL is performed using low-viscosity resins, it can be realized at room temperature and at low pressures allowing for a much faster processing and as a result, the imprint throughput can be significantly larger than that of the more traditional T-NIL.

In both processes, a so-called thin residual polymer layer is left underneath the mold relief features. For lithographic applications where post-processing is required to transfer the pattern into the underlying substrate, this residual layer needs to be removed to have the substrate exposed. This is typically accomplished by a short anisotropic oxygen plasma treatment to dry etch the polymeric material [3]. The thickness of this residual layer depends upon the initial film thickness of the resist and the size of the features to pattern. To minimize this layer it is necessary to apply the right amount of material since too thick initial layers lead to large residual layers [20]. However, too thin applied layers are more difficult to pattern, particularly when using hard mold and or substrates because these are more prone to wear or damage. The residual layer etching step is obviated when NIL is directed to create functional topographical patterns. In this case, NIL is indeed ideal to pattern directly the surface of polymeric materials in the form of free-standing films.

Initially, hard molds made in silicon, quartz or nickel were used for the NIL process. However, hard molds have several drawbacks such that they do not adapt well to the surface of another substrate and often make a poor conformal contact [21]. This has a number of implications on the fidelity of the pattern transfer and on the number of imprint defects on the patterned area, including poor uniformity and defects due to particles that propagate to a large extent within the substrate. Additionally, nonconforming molds undergo a significant mechanical stress under the applied pressure which translates into a high mold wearing and breaking rate with the resulting associated high cost. An additional source of defects arises from air trapped in the gaps between resist and mold cavities due to poor gas dissolution within the imprint materials causing defects such as air bubbles on the imprinted features [22, 23].

In addressing these challenges, new methodologies have appeared allowing the expansion of NIL as a versatile nanopatterning tool. These can be categorized broadly in two groups including those that rely on step-and-repeat imprint approaches and those that make use of a soft mold allowing for conformal contact over large substrate areas.

The first approach relies on the use of a small area mold that is repeatedly imprinted by the relative displacement of the substrate over the mold to finally pattern a large area substrate. This technology was termed initially step and flash imprint lithography (S-FIL™) and is a form of UV-NIL process where low-viscosity UV-curable resist is dispensed sequentially field by field using an inkjet process and cured by a UV light while a transparent template is then pressed into the resist, and this process is repeated in a step-and-repeat fashion. The technology has been termed more recently jet and flash imprint lithography (J-FIL) as it uses inkjet techniques to dispense variable picoliter volumes of low-viscosity UV curable resists, matching pattern size variations on the template [24]. J-FIL steppers have made noteworthy improvements in reducing defectivity in NIL by making use of rectangular (~26 mm × 33 mm) bendable molds that initiate contact to the substrate at the middle avoiding trapping gas bubbles. Furthermore, J-FIL steppers dispense well distributed and with high precision a resin volume according to the volume of the mold cavities to imprint. This technology was devised at the University of Texas and developed for semiconductor fabrication by Molecular Imprints Inc. Now, the technology is part of Canon Corporation and it is likely to lead the commercial fabrication of cutting-edge semiconductor integrated circuits (ICs) [25].

A second approach to improve upon the conformal contact relies on the use of soft molds made of plastics or elastomers ensuring conformal contact with the surface of the substrate even on curved or rough substrates. The concept of soft lithography using flexible poly(dimethylsiloxanes) (PDMS) molds was implemented into NIL processes through the use of molds made of an elastomeric material like PDMS or other polymers like fluoropolymers, polyurethanes or inorganic–organic hybrid polymers. However, PDMS is by far the most widely utilized soft material. Polymeric soft molds are typically permeable, particularly at high temperature, which advantageously helps to eliminate the trapped air bubbles in the resist layer and in addition, facilitate diffusion and elimination of the resin solvent through the mold. Due to the flexibility of the elastomeric stamp, it ensures a perfect conformal assembly onto the substrate even with curved or flexible substrates, thus allowing high fidelity replication on areas up to several tens of cm^2. Moreover, polymeric soft molds allow reducing the cost of the mold substantially as the expensive master can be replicated easily by soft lithography while it is preserved.

Soft UV-NIL is considered today the most promising NIL variant to replicate patterns in the sub-50 nm range for mass production at low cost. In industry settings, some variations of soft UV-NIL have been developed, these include the substrate conformal imprint lithography (SCIL) developed by Philips and SUSS [26] and the Soft SmartNIL® developed by EVG [27]. In SCIL, a flexible composite mold (made of two layers of PDMS of different modulus, soft and hard) is mounted on a bendable glass carrier and kept in place by vacuum applied through an array of grooves on the carrier. Wafer-scale imprints are made using the SCIL tool to execute a sequential imprinting process by the stepwise release of the vacuum in the grooves, causing the progressive contact of the flexible mold with a wafer substrate which is coated with a thin resist layer. After the imprinted resist is cured by UV exposure, separation of the mold from the substrate is performed by switching on the vacuum in the grooves in reverse sequence to the conformal imprint process.

The Soft SmartNIL technology uses a conformal bilayer polymer stamp on a transparent polyethylene (PET) carrier that is pressed by a roll onto a UV-resin coated substrate, allowing for automated sequential molding and demolding [28].

These new developments in UV-NIL have allowed the technology to meet the industry manufacturing requirements and therefore, are gradually being adopted by industry. The J-FIL is the process primarily implemented in semiconductor IC manufacturing [24, 29]. The Soft UV-NIL modalities are currently being applied as patterning tool in application fields like flexible electronics [30], photovoltaics [28, 31], photonics [15, 32], photonic crystals in LEDs [33] and flexible displays [34]. The T-NIL is being used primarily in the fabrication of metallic nanoplasmonic devices [35–37] and in emerging application such as bioinspired surfaces [16] or biotechnology applications [38, 39].

This chapter is particularly dedicated to illustrate soft T-NIL processes, discussing the critical process issues to produce complex architectures such as high aspect ratio and hierarchical nano and micro patterns as functional topographical surfaces.

7.2 Soft thermal nanoimprint lithography (soft T-NIL)

7.2.1 Thermoplastic polymers for soft T-NIL

The proper selection of the thermoplastic polymer material for an imprinting task is important as its properties determine to a large extent the selection of process parameters such as processing time, temperature, pressure and also influence the outcome of the process in terms of pattern fidelity and stability.

In table 7.1, the main thermoplastic polymers and their properties with relevance in soft T-NIL processing are collected. These properties will be referred to throughout the chapter as the optimization of the T-NIL processing parameters is discussed.

7.2.2 Molds for soft T-NIL

Molds are an essential component in NIL and they are the decisive element for resolution, uniformity and reproducibility. Since NIL is a contact technique, the quality of the imprint directly depends on the quality of the mold. The fabrication of the original master molds with well-defined features at the micro- and sub-micrometric scales is carried out by direct writing lithographic techniques, including the established silicon based clean room technologies such as UV and electron beam lithography (EBL) and more recently direct laser writing by two-photon polymerization.

7.2.2.1 Mold fabrication techniques

Traditional master molds have been originated in silicon using deep UV lithography and electron beam lithography (EBL). Indeed, EBL is already an established technology for patterning at the nanometer range with high precision (see chapter 3 in this book). However, EBL has limited applicability when large area or high throughput patterning is required. In this case, other technologies such as deep UV lithography (at 248 nm light wavelength), immersion optical lithography (at 193 nm) and an extreme UV lithography (light at a wavelength of 13.5 nm) have been utilized [47]. However, these technologies require advanced and costly equipment and their utilization is for the most part limited to the semiconductor industry (see chapters 1 and 2 in this book).

After the lithographic process step, the pattern produced is normally transferred to the silicon wafer substrate by a reactive ion etching process. Traditionally, in order to create deep nanostructures with vertical side walls, the Bosch process has been employed. This is a two-step process of etching and passivation using sulfur hexafluoride (SF_6) and octafluorocyclobutane (C_4F_8) gases, respectively. The process enables the generation of structures with aspect ratio up to 100:1. However, this process usually produces scalloping of the side walls that increases the side wall friction hampering the process of demolding in NIL [48]. This problem can be improved using the silicon cryo-etch process, (see chapter 3 in this book), which today has emerged as an alternative process for high aspect ratio silicon etching [49]. The cryo-etch is a continuous process that uses SF_6 and O_2 gases at low

Table 7.1. Commonly used thermoplastic polymers and their soft thermal nanoimprint processing conditions [40].

Thermoplastic polymer	Young's modulus (MPa) 25 °C	Surface free energy [41] 20 °C (mJ m^{-2})	Tg[a] T imprint[a] (°C)	Viscosity (Pa s)[a] E^b (MPa[a])
PMMA (poly(methyl metacrylate)) [2, 3]	2500	41.1	Tg: 100–120 T imprint:140–190	10^7–10^4 380–540 @ 80 °C
PS (polystyrene) [3]	2350	40.7	Tg: 110 T imprint:140–190	10^7–10^5 10^2 @ 110 °C
PC (polycarbonate) [42]	2270	34.2	Tg: 150 T imprint: 160–200	10^6–10^4 438 @ 160 °C 2.18 @ 170 °C
COC (cyclic olefin copolymer) [43]	2600	30	Tg 75–180 T imprint 170	10^4
PP (polypropylene) [44, 45]	1300	30.1	Tg: −20 to −10 T imprint: 200	10^6–10^3 200 @ 125 °C
PVDF (polyvinylidene fluoride) [46]	1500	30.3	Tg : −40 to −35 T imprint: 200	10^4–10^3

[a] Molecular weight dependent.
[b] Young Modulus of the viscous polymer at T imprint.

temperatures (typically below −100 °C) and allows for the formation of a passivation layer of SiO_xF_y and fluorine radicals for the etching action. The technique produces somewhat lower aspect ratio structures with typical aspect ratio about 50:1, depending on the process.

Direct laser writing by two-photon polymerization (2PP-DLW), with resolution down to 100 nm (see chapter 2 in this book), is emerging as a new and exceptionally versatile technique that allows for preparing high aspect ratio 3D structures with high complexity to serve as master stamps [50, 51]. Compared to EBL, 2PP-DLW can achieve higher writing speed and a larger working area of up to several centimetres [52]. This combined with the ability to directly form 3D structures, makes 2PP-DLW a remarkable technique for rapid mastering of complex structures. The master structures can then be replicated via soft lithography by PDMS into working molds allowing for repeatable nanoimprint processing. Hence, the combination of 2PP DWL and NIL is particularly advantageous as it combines great flexibility in terms of structural design together with the scalability of NIL.

A concern to take into account when using molds replicated from masters created with 2PP-DWL is the lateral wall roughness due to the laser scanning trajectories and layering build up. This issue has been solved or minimized by H Schift and coworkers by the process termed TASTE (temperature activated selective thermal topography equilibration) [53]. TASTE is a process suitable for thermoplastic imprinted materials and it involves the application of post-processing steps to smoothen the surface [54, 55]. This includes the radiation of the surface with vacuum UV (VUV) light at a wavelength of 172 nm emitted from an excimer lamp which causes polymer chain scission reactions and consequently the reduction of the molecular weight of the polymer at the outermost surface only (up to 400 nm). This allows for decreasing the glass transition temperature compared to the bulk region, and by thermal heating, a selective polymer reflow of the top exposed surface can be achieved with the consequent smoothening effect while the bulk and overall geometry is unaffected. The process allows reducing the roughness from about 400 nm to sub-10 nm values while preserving the sub-micrometric structural details [53]. The smoothening effect obtained on micro-prism PMMA structures can be observed in figure 7.2, before and after post-processing by TASTE.

7.2.2.2 Coating technologies for molds and anti-sticking treatments

An essential requirement for molds employed in NIL is to have a low surface energy to allow for easy demolding of the imprinted polymer structures after the process. This is critical to accomplish a good imprint quality without structural defects and also to protect the mold from failure while increasing its life time. During the demolding stage, the imprinted polymer is subjected to several forces as it is detached from the mold including the friction force, the adhesion force, and the thermal contraction force caused by the different shrinkage coefficients between the imprinted polymers and the mold material. As a result of these forces acting on both the polymer and mold during demolding, fracture, elongation, or structural deformation of imprinted structures may take place.

These issues can be readily improved by anti-adhesive coatings that lead to the reduction of surface energy and the associated interfacial friction arising when the mold and polymer replica are being separated. Fluoroalkyl trichlorosilanes with different carbon chain lengths are the most commonly used anti-adhesive layers because of their low surface energy, high surface reactivity, and high resistance

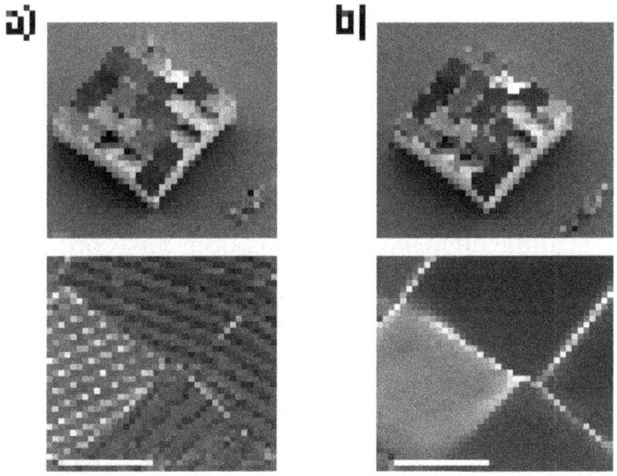

Figure 7.2. SEM images of a 5 × 5 micro-prism array imprinted in PMMA and obtained from a 2PP-DLW original master mold (a) before post-processing and (b) after VUV exposure and thermal reflow of the surface selectively modified (overview and close-up: scale bar 10 μm). Reproduced from [53] with the permission of SPIE.

against temperature and pressure [56]. The most popular fluorosilanes employed in NIL are 1H,1H,2H,2H-perfluorodecyltrichlorosilane (FDTS) and 1H,1H,2H, 2H-perfluorooctyl-trichlorosilane (FOCTS).

The anti-adhesive functionalization is achieved through the process known as silanization whereby silane functional compounds form a robust Si–O covalent linkage with a surface that is rich in hydroxyl groups. This process results in the formation of a monolayer that acts as a coupling agent for the fluorocarbon chains turning the surface into highly hydrophobic. Trichlorosilanes have been commonly utilized to modify glass, silica, alumina, titania, and other metal oxide surfaces which are rich in hydroxyl groups. For polymer surfaces, the silanization process can be enhanced via oxygen plasma treatment for the introduction of hydroxyl groups on the polymer surface which are the anchor points for covalent grafting of trichlorosilane molecules.

The silanization process is best carried out by chemical vapour deposition (CVD) because vapors are more efficient than liquids in penetrating into the nanoscale gaps of a mold, producing a more effective coverage. CVD with silanes can be carried out at ambient temperature under a moderate vacuum. In a typical process, prior to coating, molds are exposed to gentle low power oxygen plasma (20 W, 30 s) and immediately after, they are placed into a vacuum chamber (100 Pa) whereby the liquid silane evaporates and reacts with the mold surface. Typically, the deposition takes about 2 h, after which the molds are heated upto 120 °C for in-plane silane crosslinking.

Schift and co-workers [57] developed a silane coating process using mixtures of mono- and trichlorosilanes that allowed for higher molecular packing of the alkylfluorochains self-assembled on the surface while the roughness caused by cross polymerization between silane molecules was reduced. This surface coating producing higher density of fluoroalkyl molecules, showed improved anti-adhesive properties.

7.2.2.3 Flexible soft molds based on PDMS

The use of PDMS has been widely popularized through the development of the soft-lithography family of patterning techniques [58]. PDMS, typically referred to as the commercial thermocured siloxane polymer Sylgard 184 (Dow), is today the most widely used polymer as soft-mold material in NIL due to a number of reasons, but particularly for providing high fidelity replication, easy processing and low cost. PDMS, also called s-PDMS, is an elastomeric material with low Young's modulus (~2.0 MPa) and a low surface energy of 25 mJ m^{-2}. This allows for conformal contact to non-planar surfaces and easy release from both the master mold or from the imprinted patterns, generally without the need for anti-sticking agents as required for rigid molds. PDMS holds a fairly high toughness with a high elongation at break (>150%) which allows for significant deformation before failure during demolding. It is also highly transparent in the visible and in the UV spectrum being suitable for UV-NIL photocuring based processes. A critical advantage for NIL is its high gas permeability, which allows for the efficient elimination of the air trapped within the soft mold and the imprinted substrate through the soft mold itself.

The common steps to produce PDMS soft molds by soft lithography or replica molding to form a master are shown in figure 7.3. Conventional PDMS molds are produced by first mixing the siloxane oligomers and crosslinking agent in a proportion of 10:1. The mixture is then degassed in a low pressure vacuum chamber and casted over a master template. PDMS can be cured over the master at 80 °C for 2 h. However, a longer curing time of 24 h at a temperature of 60 °C is recommended in order to allow for spontaneous degassing and to avoid tension build-up due to thermal shrinkage. Cured PDMS can be thermally aged to reduce the low molecular weight chains through a higher temperature treatment at 120 °C for 60 min. Both the curing thermal treatment and the crosslinker to pre-polymer ratio influence the mechanical properties of a PDMS. Thus, they should be controlled precisely for reproducible and repeatable imprint results. The stiffness of PDMS can be slightly modulated by varying the crosslinker-to-PDMS base ratio over the range 19:1 to 2:1, and by varying the curing temperature. Within two extreme curing processes, i.e., 120 min at 100 °C and 2 d at 165 °C, it was possible to vary the stiffness from 800 kPa to 10 MPa [59].

However, conventional soft PDMS (sPDMS) has two important shortcomings when it comes to replicating nanofeatures of small size and/or high aspect ratio.

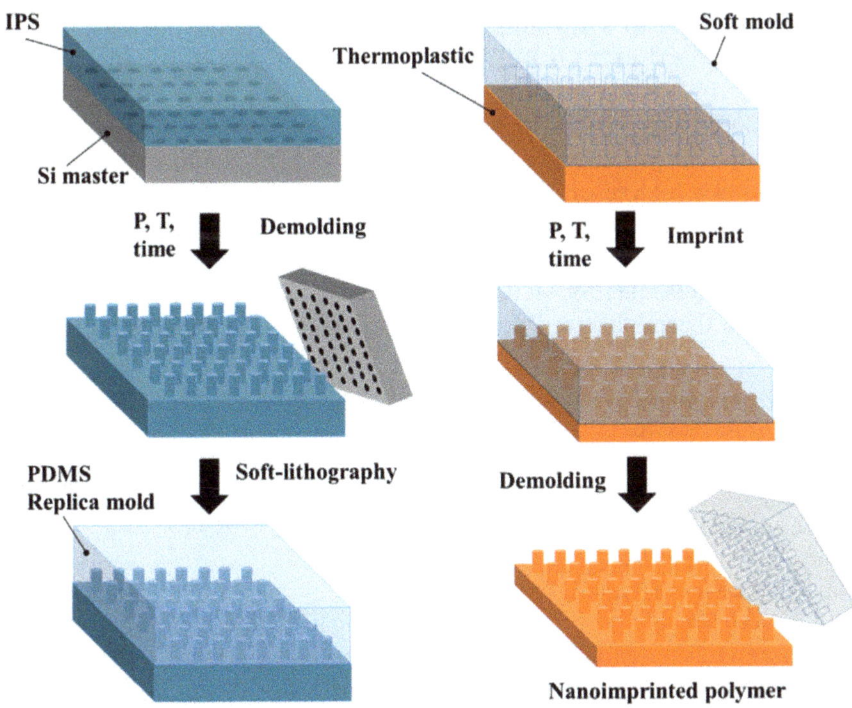

Figure 7.3. Schematic illustration of the process to make a soft mold of PDMS from an initial master mold involving the steps of imprinting the pattern features on an intermediate polymer substrate (IPS) that serves as template to cast PDMS and produce a replica mold of the original master. On the right, the PDMS replica mold is used for T-NIL process on a free standing thermoplastic polymer.

The first drawback is its high viscosity of 3900 mPa s. With such high viscous fluid, it is not possible to fill nanocavities below 200 nm and consequently, it will not be suitable for mold replication below these dimensions. Secondly, sPDMS has a very low Young's modulus which often limits the replication of features with high aspect ratio below 500 nm since these sorts of features will be very prone to deformation or collapse. Both these two issues have been tackled by using crosslinking formulations with variable proportions of vinyl and hydrosilane pre-polymers in the polymerization mixture. In the pioneer work by Schmid and Michel at IBM [60], they showed that hydrosilane-to-vinyl ratio in the polymerization mixture determines the degree of crosslinking of the polymer and as a result, the hardness. They also provided some guidelines and formulation examples on how to manipulate the pre-polymer ratios to tune the hardness of the resulting material as desired. These formulations have a low viscosity in their precursor state, which facilitates an accurate replication of sub-500 nm master structures. In addition, the formulation can be tuned to provide for different hardness according to the dimensions of the feature to replicate. Formulation examples employed in our lab are shown in table 7.2 with high (hPDMS) and mid-range hardness (mPDMS).

One needs to bear in mind that as the elastic modulus and hardness of the siloxane polymer material increases, it also becomes increasingly more brittle and it will not be able to withstand the bending forces applied during demolding, nor would it be adaptable to conform to a substrate. The surface energy also increases slightly to reach a value of 35 mJ m^{-2} for hPDMS. Hence, to have both beneficial properties within a mold, bi-layer composite stamps of soft and hard PDMS have been developed whereby a thin hPDMS layer bearing the patterns is supported by a thick sPDMS backing that provides for flexibility [61]. As such, composite molds are able to provide for both, high resolution patterning and a good compliance to a substrate. The fabrication procedure involves first preparing an appropriate siloxane mixture for hPDMS which after degassing, is spin-coated over a nanostructured master mold and soft cured. Subsequently, a degassed liquid precursor mixture of sPDMS is poured on it and thermally cured forming the flexible mold backing. It is important

Table 7.2. PDMS formulations with two different proportions of vinyl and hydrosiloxane pre-polymers and stiffness offered by the material.

	mPDMS	hPDMS
(7.0%–8.0%) Vinylmethylsiloxane-dimethylsiloxane copolymer (trimethylsiloxy terminated)—PREPOLYMER	1 g	1.7 g
Platinum-divinyltetramethyldisiloxane complex in xylene (1:2)—CATALYST	18 μL	9 μL
2,4,6,8-Tetramethyl-2,4,6,8-tetravinylcyclotetrasiloxane MODULATOR	100 μL	50 μL
(25%–35%) Methylhydrosiloxane-dimethylsiloxane copolymer, trimethylsiloxane terminated—HYDROSILANE PREPOLYMER CROSSLINKER	1 g	0.5 g
Young's modulus (N mm^{-2})[a]	9 ± 2	14 ± 4

[a] Modulus obtained from the nanoindentation testing.

to control the thickness of the hard layer to be as thin as possible and restrict it to filling just the mold features. This is because hPDMS is highly brittle, and small bending or flexing forces on the mold will lead to cracking of this layer. As a general rule, sPDMS would be appropriate only to pattern structures above 500 nm, below this dimension, hPDMS composite molds would be more suitable to pattern features down to 15 nm [62].

The main drawback of PDMS-based soft molds is the short life time giving out only a limited number of nanoimprinting cycles depending on the particular process. It is a consequence of the porous nature of PDMS whereby organic solvents easily penetrate, causing swelling and deformation of the mold features. In addition, monomers from resist mixtures are also able to penetrate and due to inter cross-linking, cause hardening and embrittlement to the material over time. These events cause the deterioration of the molds and have a negative effect on reproducibility in the course of repeated patterning.

7.2.2.4 Soft molds based on fluoro-polymers

Some of the difficulties with PDMS molds are precluded by fluorinated polymers. Fluorinated polymers are formed by linear chains of multiple carbon–fluorine bonds. These bonds are intrinsically much stronger than carbon–hydrogen bonds and as such, fluorinated polymers offer high temperature and solvent resistance. Particularly employed in NIL are the perfluoropolyether (PFPE) acrylates based elastomers because they have a number of advantages as intermediate molds including a combination of higher stiffness and a lower surface energy which allows for sustaining higher imprint pressures reducing the friction and adhesion forces during demolding, thus facilitating the final detachment of the imprinted polymers. These characteristics allow having a significantly improved mold life time compared to that for PDMS [63].

With a molecular weight ~2000 g mol^{-1}, a dynamic viscosity at 20 °C of 430 cP and a low surface energy, PFPE based materials can be casted directly onto the master mold filling nanocavities down to 50 nm [64]. Due to their high thermal stability (up to 288 °C) [65] PFPE based molds are suitable for T-NIL processing. The mechanical properties of the PFPE films depend upon the processing conditions and precursor formulation including the concentration of the radical initiator. The Young's modulus of the PFPE materials ranges from several MPa for the acryloxy PFPE to several 10 s MPa for the dimethacryloxy form (DMA-PFPE), and over 150 MPa for the trimethacryloxy form (TMA-PFPE) [66].

Commonly employed in NIL is the acryloxy PFPE (denoted as α-PFPE) commercialized as Fluorolink® MD700 perfluoropolyether (Solvay Speciality Polymers) with a modulus of 10.5 MPa and a lower surface energy (12.7 mJ m^{-2}). For replica molding, it is typically mixed with a 3%–4% w/w alpha-hydroxyketone type of photoinitiator (2-Hydroxy-2-methylpropiophenone or Darocur 1173) and crosslinked by UV-light (365 nm, 25 mW cm^{-2}) for 2–3 min depending on the film thickness. The exposure needs to be performed in the absence of oxygen to avoid the inhibition of the free radical photocuring reactions, and it can be done by covering the photo resin with a film of polyethylene terephthalate (PET) while layered on the mold to replicate, or else by carrying out the process in N_2 atmosphere. The PFPE replicated substrates can be used as free-standing molds. However, for large area molds, the PFPE structured material can be supported on thin, flexible PET sheets as backing film allowing for more practical handling. Sylgard 184 PDMS backing layers have also been utilized by curing the PDMS on the PFPE cured material before peeling it from the master mold [67].

7.2.2.5 Soft molds based on hybrid organic–inorganic materials

OrmoStamp is the most common organic–inorganic hybrid material and it has been formulated by Microresist Technology GmbH specifically for the fabrication of NIL transparent working stamps. The material is a UV curable liquid with viscosity of 0.75 Pa; this relatively low viscosity allows for efficient filling of the master template cavities. After curing and thermal post treatment, the polymer becomes durometric, which means that no glass transition takes place when heating, exhibiting a modulus of elasticity of 0.650 GPa and hardness of 0.036 GPa. These values of elasticity and hardness make OrmoStamp useful for imprinting dense nanostructures as well as high-aspect-ratio features.

OrmoStamp working molds have a high UV-transparency at 350 nm that is preserved even after thermal exposure at 270 °C. As such, OrmoStamp working molds can be used for both thermal and UV-NIL processes. However, the material has a high surface energy and therefore after UV-curing and separation from the master mold, the replicated OrmoStamp molds must be post-baked to harden the polymer and an antistiction layer must be deposited before the stamp can be used. OrmoStamp working molds can be used for up to 50 imprints with almost no pattern degradation, and features sizes down to 50 nm have been successfully replicated [68].

OrmoStamp working molds are commonly prepared as flexible composite molds on a backing support. On the choice of this backing material, compatibility aspects need to be considered. Firstly, it is the large shrinkage due to crosslinking by UV light exposure and hard baking of approximately 6% that will cause bending of the backing film. This can be minimized by decreasing the OrmoStamp layer thickness or/and using a thick backplane [69]. A second important aspect to consider is the interfacial adhesion between the OrmoStamp and the backing film material. Sylgard 184 PDMS roughened by UV exposure and treated with adhesion primer was found to produce a suitable and stable interface, and at the same time it allows for good conformation to the substrate [70, 71].

Table 7.3 summarizes the key physical properties for common polymers employed in the construction of flexible soft NIL molds.

7.2.2.6 Sequential soft mold replication

Because of the low surface energy characteristic of most polymers, and in particular for the group described above, it is possible to apply them in sequential replication processes of molding, soft lithography and nanoimprinting. The combination of materials and processes allows for producing a second generation of replica molds in soft materials. It also provides for an intermediate step to invert the relief features to create the negative or the positive contrast profile from the original master on the

Table 7.3. Physical properties and resolutions reported for common polymers employed in the construction of flexible soft NIL molds.

Material	Young's modulus (MPa)	Surface energy (mJ m^{-2})	Viscosity (mPa s)	Resolution (nm)	Reference
s-PDMS	1.8–2	25	3900	250	[58]
h-PDMS	10–14	35	<490 tunable	15	[62]
PFPE	315	12.7	430	15	[65]
Ormostamp	650	41	750	Sub-10 nm	[72]

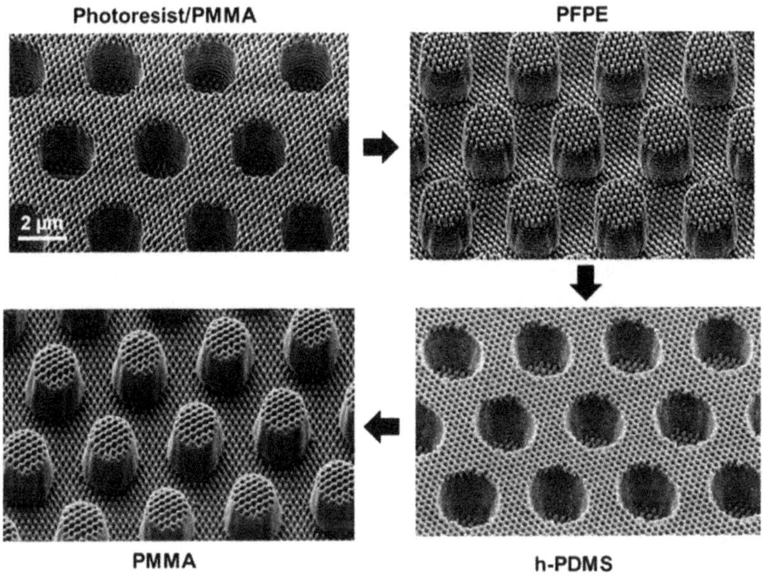

Figure 7.4. Sequential polymer replications of topographic features from the original master template produced in a photo resin and PMMA. The original template is replicated into PFPE and subsequently in h-PDMS/s-PDMS composite mold by soft lithography. The PDMS composite mold is then employed as second generation working mold to imprint by T-NIL the features into PMMA. Reprinted with permission from reference [75]. Copyright (2019) American Chemical Society.

desired functional materials. Figure 7.4 illustrates a multiple replication sequence to transfer the hierarchical structures generated on a photoresin/PMMA template onto a PMMA substrate in three replication steps involving soft lithography and T-NIL. However, it should be taken into consideration that at each polymer replication step, shrinkage of the replication material takes place due to curing or thermal contraction [73]. The level of shrinkage will vary depending on the particular material, molecular weight and curing conditions. Typically it ranges from shrinkage ratios about 2% (PDMS) [74] to 6% (PFPE, OrmoStamp) [70]. Hence, when accuracy is required, appropriated early corrections onto the master mold should be implemented.

7.3 Complex structures by soft thermal nanoimprint and hybrid processes

7.3.1 Introduction

The ability to carry out nanoimprinting with soft molds has opened up new possibilities to create a large variety of complex functional surface structures with high fidelity and resolution in a cost effective manner. These surfaces may exhibit unique properties and have often been inspired by surface structures found in nature. Indeed, the complex topographies found in nature reveal important design principles to impart advanced functionalities into materials and commonly these involve high aspect ratio structures often organized within a hierarchical multiscale arrangement of micro- and nanostructures.

Multilevel and or hierarchical structures including micro- and nanoscale features are certainly ubiquitous in nature and provide biological materials with exceptional properties beyond those expected from the bulk material itself. Some of the well-known examples include the lotus leaf covered with hierarchical micro- and nanostructures that provide for anti-wetting or superhydrophobicity [44, 76, 77], the branched nanopillars found in gecko feet that provide for dry adhesive properties [78, 79], butterfly wings with a multiscale arrangement of microridges and crossing nanoribs on the wing´s surface that provide for structural coloration [80–82] or moth eyes with subwavelength multiscale ommatidial arrays enclosing conical structures conferring antireflection and bactericidal properties [83, 84].

A close look into many of these natural surfaces shows that they often comprise a multi-scale/hierarchical arrangement of nano- and microscale features to achieve optimal properties for species adaptation to their living environment. Multi-functionality in these natural systems is generally accomplished due to convergence of both the structural elements and the material chemical composition. Hence, to follow nature design principles, it is essential to be able to produce patterned structures in a broad variety of materials, adequate to the target application.

Inspired by nature to achieve functionality, a number of artificial materials with advanced physical or mechanical properties have been engineered. A summary of the different technologies available to create bioinspired or biomimetic multi-functional surfaces has been published in several reviews [16, 85, 86]. The interest in engineering high aspect ratio and multilevel hierarchical topographies to impart advanced functionalities into manufactured consumer products has been recognized in different fields of application [87], including energy harvesting [82], membrane technology [88], anti-icing [89], photonics and optical devices [14, 15], robotics [90], tissue engineering [91], micro/nano-fluidics [92] and fog harvesting [93] among others.

T-NIL is one of the most promising approaches to produce artificially complex multifunctional surfaces for its high resolution and large choice of thermoplastic materials that can be imprinted. Some examples of natural high aspect ratio and hierarchical micro- and nanostructures and their bio-inspired replicas created by T-NIL technology are shown in figure 7.5. The next sections will describe in detail

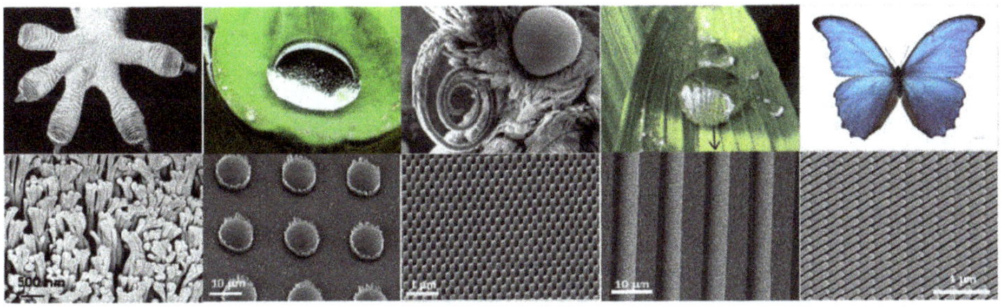

Figure 7.5. Some bioinspired hierarchical micro and nano structures created by T-NIL and hybrid methods. Reproduced with permission from: gecko [78] copyright American Chemical Society, lotus leaf [94] copyright American Chemical Society, butterfly eye [95], rice leaf [96] Royal Society of Chemistry, butterfly wings [80] John Wiley & Sons.

the basic processing issues to produce complex high aspect ratio and multilevel hierarchical structures by soft T-NIL and combined lithographic process.

7.3.2 High-aspect ratio imprinting

Nanopatterned features with high aspect ratio (HAR) are interesting to bring out the true effect of the topography without interfering effects from the underlying substrate. This is especially true for micro-optical elements and photonic devices [14, 15, 97], SERS substrates [98, 99] or substrates for cell biology studies [91]. HAR imprinting entails additional processing challenges related to cavity filling and demolding due to the huge increase in mold–substrate contact area [100]. During the imprint process of HAR structures, the polymer needs to flow over long distances to fill the cavities of the mold. This would require very high applied pressure and long imprinting times sufficient to drive the flow along the nanocavities. Nonetheless, the filling process is often enabled by capillary forces as the heat-softened viscous polymer is able to wet the surface of the mold and, as a result, the pressure driven viscous flow is assisted by capillary action.

On the other hand, a low surface energy is required to reduce the adhesive forces and wall friction to ease the separation of the mold from the imprinted structures during demolding, avoiding possible stretching or brakeage of the small imprinted structures. Accordingly, a HAR imprinting process demands reaching a balance between mold wetting properties for capillary filling action and good mold release properties for demolding. Considerations on these two process steps during HAR imprinting are given below.

7.3.2.1 HAR mold filling

The imprint process of HAR filling relies on the external imprint pressure as well as on capillary driven flow within the cylindrical nanocavities. Capillary pressure arises due to the ability of the viscous heated polymer to wet the surface of the mold cavities. The capillary pressure across the interface polymer—air is the Laplace pressure given by the difference between the pressure of the viscous polymer (P_l) and the gas phase (P_g) at the interface,

$$\Delta P_c = P_g - P_l = \frac{2\gamma \cos\theta}{r} \tag{7.1}$$

where γ is surface tension of the viscous polymer, θ is the solid–liquid contact angle and r is the capillary radius.

The Lucas–Washburn equation describes the rate dh/dt of fluid flow through a cylindrical capillary of radius r as a function of the driving capillary pressure. Making the assumptions that flow is laminar, viscous and incompressible and that the capillary is much longer than it is wide, Washburn applied Poiseuille's Law to describe the pressure drop in a fluid flowing through a cylinder, deriving the following equation:

$$\frac{dh}{dt} = \frac{r\,\gamma\cos\theta}{4\eta h} \tag{7.2}$$

where η is the viscosity of the heated polymer and h is the height of polymer inside the nanocavity.

Integrating from $h = 0$ @ $t = 0$ gives the Lucas–Washburn equation where the height of the polymer scales with the $t^{1/2}$ and gives:

$$h^2 = \frac{r\,\gamma\,\cos\theta}{2\eta}t,\; t = \frac{2\eta h^2}{r\gamma\cos\theta} \tag{7.3}$$

The Lucas–Washburn equation predicts that when the contact angle is 90°, there is no Laplace pressure across the interface and, as such, there is no driving force to generate fluid flow. For a contact angle above 90°, the sign changes and there will be a negative Laplace pressure opposing the filling of the capillary cavities. In these cases, the external imprint pressure applied has to be increased substantially to overcome the Laplace pressure. In the situation when the contact angle of the viscous polymer increases due to poor mold wetting, the pressure or the imprint time to fill the cavities may be excessively long. Figure 7.6 illustrates these situations.

Predicted by equation (7.2), the filling of cylindrical nanocavities with viscous polymers via capillary action is very slow due to the small value of r and the high viscosity of polymer typically in the range of 10^4–10^7 Pa s. To give an estimate, the time needed for complete filling of 0.5×2 μm pore cavities in a mold with a polymer with viscosity about 10^6 (Pa s) and surface tension ~30 mN m^{-1} and wetting contact angle of 85° calculated with equation (7.2), it will take about 102 min. Reducing the viscosity to 10^5 Pa s will decrease drastically the filling time to 10 min, or by reducing the contact angle to 60°, it will take 17 min.

The process of filling HAR structures is highly favoured in T-NIL working at high temperature since viscosity and surface tension decrease dramatically with temperature at the same time that wetting increases reducing the contact angle. These circumstances facilitate the viscous polymer flow entry into the mold cavities and filling by Laplace pressure. High temperature is also beneficial to facilitate gas diffusion, particularly in the case of soft polymer molds where the gas enclosed in the cavities can be eliminated through the mold permeable material [23].

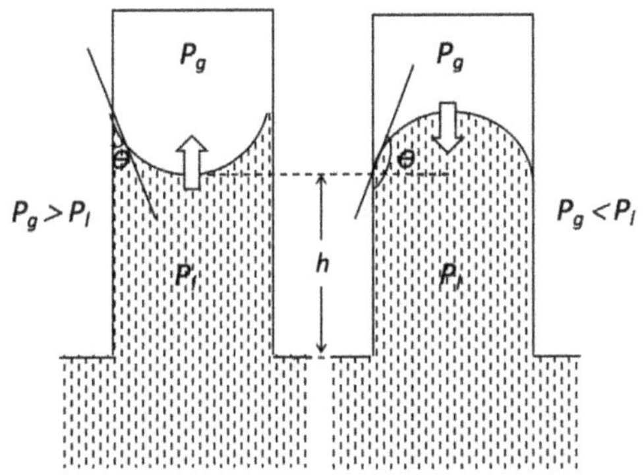

$$P_c = \frac{2\,\gamma\cos\theta}{r} \qquad \frac{dh}{dt} = \frac{r\,\gamma\cos\theta}{4\eta h}$$

Figure 7.6. Schematic of capillary filling situations when (left) the polymer wets the capillary ($\theta < 90°$) and (right) when the polymer does not wet the capillary ($\theta > 90°$). The Laplace pressure across the interface is indicated by the arrows.

High energy mold materials that provide lower contact angle and higher wettability will be also advantageous to increase Laplace pressure, thus allowing to significantly reduce the imprinting time or the imprinting pressure. However, high energy materials will increase adhesion and will preclude the separation of the stamp from the imprinted polymer during demolding. As such, there is a need to find a suitable surface energy to balance these two processes of molding and demolding for each particular imprinting task.

7.3.2.2 HAR mold releasing

The process of demolding involves the detachment or separation of the mold, releasing the imprinted structures. The required demolding force is related to the adhesion force between the mold and the polymer, the friction force of the structures on the side walls as they are being released and the deformation force due to the presence of undercuts or roughness at the sidewalls of the mold structures [101]. In addition, there may be a thermally-induced residual stress at the polymer–mold interface arising during the cooling stage due to the difference in thermal shrinkage between the polymer within the cavities and the mold, which results in high contact stress and, in turn, contributes to the demolding force.

Demolding forces are directly determined by the geometrical parameters of the mold, the aspect ratio and density of the structures. Due to the large increase in contact area, structures with HAR are susceptible to ripping at the base where the stress is concentrated. When the density of structures is high, the resulting force will be extremely high such that whole areas of the polymer may break or detach from the supporting substrate. Likewise mold-filling, surface energy and temperature are two parameters critical in demolding as both directly influence the friction and adhesion forces involved. Furthermore, the thermal induced stress and the mechanical strength of the polymer-imprinted features are directly related to temperature

[102, 103]. Hence, the choice of demolding temperature needs to be made wisely. The recommended demolding temperatures were predicted by measuring the variation of demolding force with temperature [102, 104]. It was found that there is an optimum demolding temperature when the adhesive force is decreased and the friction force is at its lowest. This is a direct consequence of the inverse contribution of temperature on the friction or adhesion forces. This optimum demolding temperature was found to be somewhere slightly below the Tg and for PMMA, it was around 70 °C [101, 103]. In practice, this implies that demolding at lower temperature may cause the imprinted patterns to break off due to high adhesion and brittleness arising from thermal induced stress. On the other hand, higher temperatures may allow easy demolding due to softness and higher elasticity but it may result in losing the replication fidelity due to stretching and collapse of the structures as a result of low mechanical strength. These issues will be particularly more severe as the aspect ratio increases.

The mold surface energy is a determinant factor on adhesion, and as such, low values of mold surface energy and of polymer surface tension will lead to low adhesion. Conversely, low surface energy values will reduce wetting and the capillary force during mold filling which is not helpful in HAR imprinting. Consequently, all parameters need to be taken into account and balanced for a particular imprinting task. It should be borne in mind that silane-based anti-sticking agents allow adjusting the surface energy within the range of 10–20 mJ m^{-2}.

Soft molds with high elasticity and inherent low surface energy can provide a number of advantages when it comes to demolding HAR structures. Soft polymeric molds, having similar thermal expansion coefficient as the polymer imprint material, will generate less thermally induced residual stress during the cooling stage, thus inducing less damage during demolding. Due to their elastic nature, soft molds will be more tolerant towards deformation forces and demolding shear due to inhomogeneous mold walls. A determinant beneficial factor of soft molds is that demolding is realized by a sequential action of peeling the mold from the imprinted substrate. This effectively results in demolding a smaller area at the time, and as a result, a lower detachment force for demolding needs to be applied, reducing the probability of damage to the imprinted structures or the mold, improving the uniformity and reproducibility and, in general, the yield of the process.

7.3.2.3 Stability of soft molds with mechanical and thermal loading

Since T-NIL is a mechanical process, the mechanical properties of the mold and dependence with temperature are decisive to the nanoimprint replication result. Constraints on thermal nanoprinting with soft molds are imposed primarily by the stamp deformation at high temperature and high pressure.

While choosing the mold material, the first consideration to make is the geometrical characteristics of the features to imprint, as this will determine the degree of deformation. Molds with features protruding will bear a considerable compressive stress on the small feature areas when imprint pressure is applied and consequently, these small areas will experience a large compressive strain and generally a reduction of the mold features' dimensions with the force. Hence, in the case of protruding mold features, there will be a need for materials with high modulus to resist the compressive stress and deformation. Typically, this situation calls upon the use of flexible composite molds consisting of a structured thin rigid layer of a high modulus material bonded to a flexible supporting backing such as those described earlier (section 7.2.2) including h-PDMS, OrmoStamp or PFPE composite molds.

In the situation of molds having structural intruding features like pores, the elastic modulus of the material is not as critical since the compressive stress will be basically borne by the large area comprising the mold cavities. In this case, it is possible to employ molds made homogenously of the standard s-PDMS elastomer; imprints of pillar features with high fidelity have been demonstrated in several works [105, 106].

While choosing the material for producing soft molds, it is imperative to consider its mechanical and wetting characteristics and adjust the process parameters accordingly. It is the interplay between pressure, temperature and process time that will determine the compressive strain on the mold features and, as a result, the replication fidelity. While working with soft molds, two premises should be followed: the imprint materials should have a Young's modulus lower than that of the mold, and the imprint pressure applied should be lower than the modulus of the polymer mold. Since thermoplastic materials generally have an elastic modulus higher than that of the elastomeric mold, pressing the mold against these materials will cause a severe compressive strain onto the mold features. Hence, in this case, the initial step of an imprint process should be to heat up the thermoplastic materials to increase the mobility of the polymer chains reducing its elastic modulus (the modulus decreases by a factor of about 10^3 above the glass transition or melting temperature [107]) and, in a subsequent step, pressing the soft mold into the viscous material will cause the flow of the viscous polymer into the mold cavities, minimizing the mold compression.

Working with soft molds, one would typically require reducing the imprint pressure in order to decrease the compressive strain. To this end, materials like h-PDMS with surface energy in the range of 35 mJ m^{-2} that can sustain capillary forces, would be helpful in assisting the filling process and, as a result, reduce the level of the required external pressure. The elastic recovery of the mold material under compression is an additional factor to consider during the imprint process in order to achieve high imprinting fidelity. Schift and co-workers experimentally observed that even after complete filling of the mold, in some cases, a small deformation on the imprinted structures could still be observed [108]. To explain this observation they put forward the hypothesis that during the initial stage of the embossing process, the features on a soft mold would be partially compressed. Following this, as the mold features are filled with viscous polymer, the compression is normalized and with sufficient time, the mold is allowed to elastically relax and recover its initial shape. This relaxation time for elastic recovery was found to have a direct relationship with temperature, that is, at higher temperature, the viscosity of the imprint polymer is lower producing less compressive stress and at the same time facilitating the compressive recovery. In addition, it was found that the elastic recovery was also time dependent. The analysis of the results indicated that depending on the imprinting temperature and related polymer viscosity, there was a time span required for compressive strain recovery. Thus, the findings in this work signify the need to take into account within the imprint processing time, the time for the compressive recovery of the mold material in connection with other process parameters such as pressure and temperature, and this should be customized for a particular mold structural design.

7.3.2.4 Material limitation and pattern collapse in high aspect ratio imprinting

During the replication process of HAR features, it is important to consider the mechanical stability of the features in relation to their effective stiffness to predict the upper limit in terms of the feature aspect ratio to possibly imprint. HAR imprinted patterns are prone to feature collapse primarily driven by adhesion forces between the pillars and between pillars and the substrate [109]. In this process, besides stiffness, other factors such as pattern density and surface energy of the imprinted materials have a direct influence on the stability or otherwise adhesive failure of the imprinted features. In general terms, two failure mechanisms can take place. If the spacing between patterns is larger than their height, the imprinted features can undergo ground collapse. On the other hand, if the pattern density is high, that is, if the spacing between the patterns is smaller than their height, with only a small bending, the imprinted features in close proximity may experience adhesion forces toward the neighbouring patterns causing a lateral collapse failure.

The critical aspect ratio of ground collapse, $(h/d)_c$, under which the pillars are stable against ground collapse, was derived by Roca-Cusachs and co-workers [109] by balancing the bending energy for ground collapse and the energy of the pillar adhesion to the substrate, arriving at the following expression:

$$\left(\frac{h}{d}\right)_c = \frac{\pi^{\frac{5}{3}}}{2^{\frac{11}{3}}3^{\frac{1}{2}}}(1 - v^2)^{-\frac{1}{6}}\left(\frac{Ed}{2\gamma_s}\right)^{\frac{2}{3}} \tag{7.4}$$

where h is the feature height, d is the feature width, E is the Young's modulus of the material, v is the Poisson's ratio and γ_s is the surface energy.

The limiting aspect ratio for lateral collapse was derived by Glassmaker and co-workers [110] for a square array of posts by balancing the bending energy and energy of deformation of laterally adhered pillars with their energy of adhesion and it was shown to be for arrays of cylindrical and square shape pillars, respectively:

$$\left(\frac{h}{d}\right)_c = \frac{0.57 \times E^{\frac{1}{3}}w^{\frac{1}{2}}}{\gamma_s^{\frac{1}{3}}d^{\frac{1}{6}}(1 - \nu^2)^{\frac{1}{12}}} \tag{7.5}$$

$$\left(\frac{h}{d}\right)_c = \left(\frac{3Ew^2}{8d\gamma_s}\right)^{\frac{1}{4}} \tag{7.6}$$

where w is the lateral separation between adjacent pillars.

These equations allow making predictions on the mechanical stability of structures that can be replicated via nanoimprinting and show that the polymer elastic modulus plays a critical role on the stability of polymer nanostructures. The stability can also be enhanced by reducing the surface energy of the material. As a general guideline, the typical aspect ratio range for sub-micrometer pillars will vary from 2 to 10 as we move from elastomeric materials such as PDMS (modulus on the MPa range) to stiffer materials such as PMMA with elastic modulus on the GPa range [111].

7.3.3 Multilevel hierarchical complex nanoimprinting

Extensive efforts have been undertaken to develop methodologies leading to the creation of multiscale hierarchical surface structures, particularly those inspired by nature. The fabrication processes to create these complex structures in a controlled manner often require several steps and generally involve a high level of complexity compared to simple structures. Among the methodologies employed, NIL is one of the most practical to produce multilevel and hierarchical micro- and nanostructures with high precision. Top–down approaches involving lithographic processes typically confer a higher degree of controllability and repeatability over the geometrical dimensions of the generated structures. Appropriate combinations of NIL and other lithographic approaches, particularly photolithography, have permitted the fabrication of hierarchical topographies with well-defined geometry and feature control. Some of the most efficient methodologies enabling the scalable and high-throughput fabrication of multiscale hierarchical surface structures with high precision using T-NIL in combination with other technologies are described below.

7.3.3.1 Sacrificial layer mediated nanoimprint (SLAN)

The basis of the sacrificial layer mediated NIL technology relies on the use of a sacrificial layer to protect the structures produced on a first NIL step before a second NIL step is performed. After the final step, the sacrificial layer is removed typically by solvent dissolution. The function of the sacrificial layer is to encapsulate and fix the structural patterns produced in a first step to prevent deformation during additional thermal imprint process steps. The main requirement for the sacrificial material is that it should remain thermally stable during the second thermal NIL processing conditions. In addition, it should allow for selective dissolution using an orthogonal solvent to the imprinted material. Another consideration that should be kept in mind is the chemical compatibility and adhesion energies between the different polymer materials involved.

Raut and co-workers demonstrated the fabrication of moth's eye inspired multiscale hierarchical ommatidial arrays on a PC substrate [83]. In a first step, the moth's eye nanocone structures were fabricated by imprinting the suitable mold onto a PC film. Subsequently, these structures were covered with a thin (500 nm) sacrificial layer of poly(sodium 4-styrenesulfonate) (PSS) by spin coating a 20 wt. % solution in water. Then, a second thermal imprint step of a microlens array was performed at a temperature higher than the glass transition temperature of PC. Because the PSS has a higher melting temperature (T_m = 450 °C) than that of the PC (T_g = 155 °C), the temperature during the second nanoimprinting did not soften the PSS and the nanocone patterns being encapsulated into the PSS sacrificial layer remain fixed and undeformed. After demolding the micro structures, the imprinted substrate was immersed in water for the dissolution of the PSS sacrificial layer releasing the nanocone imprinted structures on the microlenses.

On a second approach, the sacrificial method has also been employed to fabricate hierarchical fibrillar gecko inspired dry adhesive structures [112]. The geometrical characteristics of this particular design includes micropillars of 3 μm in diameter, with an aspect ratio of 2, and fibrils structures of about 200 nm in diameter with an aspect ratio of 10, which was below the critical aspect ratio for lateral collapse for the PC fibrillar arrays estimated to be 13. The processing steps are schematically depicted in figure 7.7. Initially, micropillar arrays are imprinted onto a PC sheet. Subsequently, the arrays are immersed in a sacrificial matrix of PSS such that only the pillar tips remain exposed. This is achieved by controlling the spin coating conditions. Next, the fibrillar structures are imprinted on top of the micropillar arrays using a porous anodic aluminium oxide template (AAO). PSS exhibited excellent thermal stability at elevated temperatures and the underlying micropillars enclosed within the PSS matrix did not show any appreciable deformation. After demolding the AAO template, the imprinted sheet was immersed in water for the selective dissolution of the PSS sacrificial layer to expose the hierarchical HAR array structures. In this step, the evaporation of the water after the dissolution of PSS caused capillary-driven coalescence of the flexible fibrils. This issue was solved by freeze-drying the substrate in liquid nitrogen to freeze and fix the free-standing fibrillar nanostructure, followed by freeze-dry elimination of the water.

The SLAN approach has revealed itself as a process capable of efficiently generating nanoimprinted hierarchical structures with high resolution, high uniformity and reproducibility. It appears particularly beneficial for the fabrication of HAR structures and hierarchies where demolding of HAR nanostructures can lead to damage of the imprinted features. It provides for relative high freedom in

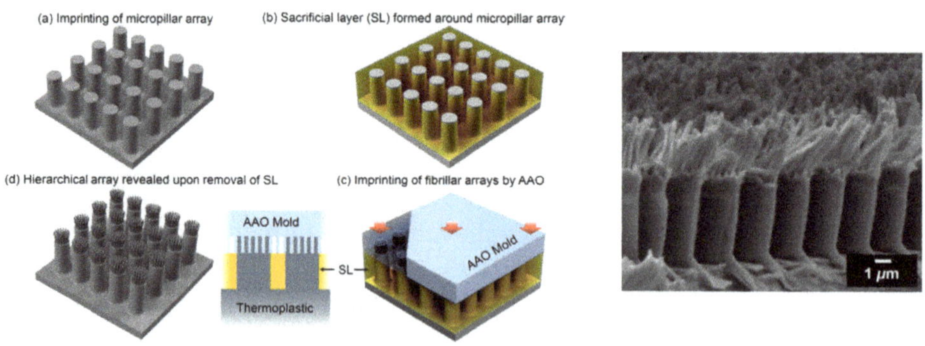

Figure 7.7. Schematic of the sacrificial layer mediated nanoimprint process for the fabrication of hierarchical fibrillar gecko inspired structures. Reprinted with permission from reference [112]. Copyright (2018) American Chemical Society.

choosing the geometrical characteristics of the features and it can be scaled-up over large areas. Dissolving the sacrificial polymer materials appears to be a critical step as it may cause elastocapillary coalescence or structural collapse due to solvent evaporation, or the dissolved material may leave a residue on the imprinted structures if it is not rinsed sufficiently.

7.3.3.2 Nanoimprint hybrid techniques

7.3.3.2.1 Combination of nanoimprint and other lithographic techniques

The combination of several nanofabrication techniques, particularly lithographic technologies with T-NIL and or UV-NIL, has permitted the generation of a variety of hierarchical topographies with well-defined geometry.

The combination of nanoimprint and photolithographic (CNP) processes was first described by Guo and co-workers [113] using a hybrid mask-mold consisting of a transparent quartz substrate with metal patterns acting as a photomask together with nanoscale relief features. The hybrid mold is utilized to imprint by T-NIL onto a photoresist that is subsequently exposed to UV light. After demolding, the substrate is developed to eliminate the non-crosslinked resist. This methodology can be carried out with many of the exiting photoresists exhibiting a thermoplastic behaviour. This methodology produces self-aligned nano and micro patterns of high precision. However, the preparation of hybrid mask-molds requires a challenging clean room fabrication process. An alternative for more practical methodology to produce these mask-molds was proposed by Schift *et al*, by producing the mold nanofeatures using OrmoStamp material over a metal pre-patterned glass substrate [114]. This approach allowed for more flexibility and simpler fabrication of the mold design. The downside is that due to the finite residual thickness of the OrmoStamp layer on top of the masking metal layer, there are some diffraction effects at the edges causing some edge shadowing effect on the imprinted patterns.

T-NIL has also been combined with e-beam lithography to produce multilevel patterned topographies. An exemplary process is outlined in figure 7.8 [115]. The process involves initially performing T-NIL on an electron sensitive thermoplastic polymer such as PMMA with the appropriated molecular weight. Subsequently, grey-scale e-beam exposure is carried out on the patterned polymer film creating sloped reliefs. Finally, a TASTE process of polymer reflow is performed to eliminate the sidewall roughness caused by the dose dependent e-beam exposure to produce smooth inclined slopes, while the T-NIL surface patterns stay unaffected.

7.3.3.2.2 Combination of nanoimprint and UV flood exposure

The fabrication of hierarchical topographies with nanopatterns entirely covering the micron-scale surface was demonstrated by Steinberg and co-workers [116]. The methodology involves two T-NIL steps with an intermediate step of surface hardening. Initially, nanofeatures on an SU-8 precursor-film are produced by T-NIL and subsequently, the surface is subjected to VUV short treatment (3–10 s) which exposes the SU-8 surface-near only, triggering the polymer crosslinking of a

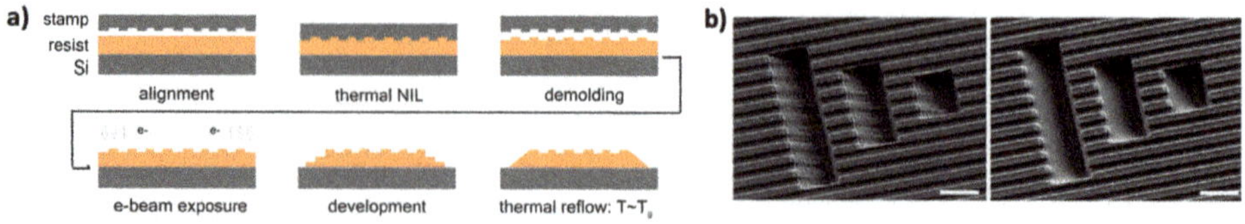

Figure 7.8. (a) Process sequence to create multilevel PMMA resist structures based on the combined sequential processes of T-NIL, grey-scale e-beam exposure and thermal reflow by TASTE process. (b) The process allows for the selective smoothening of the roughness created by grey scale e-beam patterning of prism structures onto NIL prepatterned gratings, which remains unchanged (scale bar = 2 μm). Reprinted from [115], with the permission of AIP Publishing.

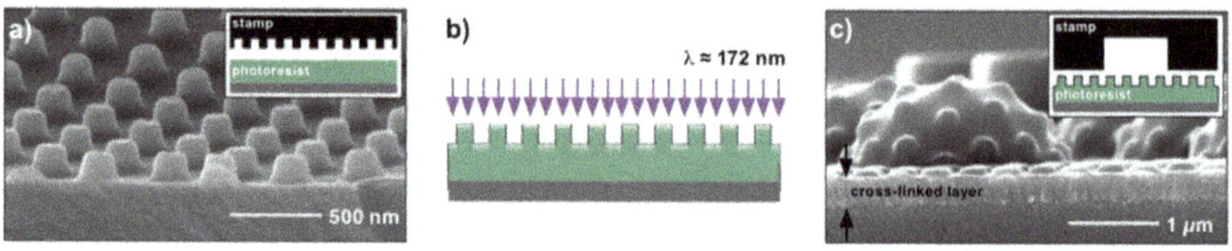

Figure 7.9. Process steps for the fabrication of hierarchical nano–micro structures on a negative tone photoresist material (SU8). (a) T-NIL of the nano-structures; (b) VUV flood exposure of the surface-near only; (c) second T-NIL of the microstructures. Reproduced with permission from [116], Copyright (2017), with permission from Elsevier.

limited film depth. Finally, a second T-NIL is performed, using a mold containing the micrometric structures, generating the hierarchical nano- and microstructures. Figure 7.9 illustrates this process sequence.

The procedure applies only to negative tone photoresist materials like SU-8 (when exposed to UV light, the exposed areas become crosslinked). SU-8 in the non-crosslinked state displays a thermoplastic behaviour, that is, it becomes viscoelastic above the T_g and almost liquid at high temperatures. The T_g of unexposed SU-8 is in the range of 35 °C–50 °C. After UV treatment, the post-exposure bake promotes and accelerates the crosslinking and increases the elastic modulus of the material. The VUV exposure dose determines the thickness of the crosslinked top layer; if the exposure is too low, the layer may be too thin and the stability insufficient. On the other hand, long treatment times may produce an excessively thick layer reducing the flowability and imprinting ability during the second imprint step. Therefore, in this approach, the VUV-exposure dose and process temperature are critical processing parameters to adjust in conjunction with the geometrical parameters of the features in order to find the adequate process window for producing the hierarchical structures.

7.3.3.3 Multilevel hierarchical topographies

The combination of consecutive T-NIL and direct laser writing processes allows for the fabrication of complex multiscale hierarchical structures with high precision in a practical and reproducible manner [75]. This process comprises three steps involving first a T-NIL step on a first layer of a thermoplastic material to create a first level of nanopatterns. In a second step, a photoresist second layer is spin coated on top of it and texturized by a second T-NIL process. The approach is enabled by the thermoplastic behavior of most photoresists before they are photo-crosslinked. In a final step, to define the micron size features, optical lithography is carried out by direct laser writing on the nanopatterned photoresist second layer. Alternatively, a photolithography step through a photomask can also be performed. After developing the micropatterns by eliminating the exposed photoresist, a hierarchical topography is obtained. Figure 7.10(a) illustrates schematically the fabrication steps.

The noteworthy advantage of this fabrication approach is that it allows customizing the nanostructure geometry independently at each level of the hierarchy by simply using the molds with the desired features for the first and the second T-NIL steps. Thus, the process allows for an enormous versatility in terms of topographical designs that can be produced. Some examples are shown in figures 7.11(a) and (b) which show SEM images of two-level, second order and two-level, third order topographies, where the order refers to the number of different topographical

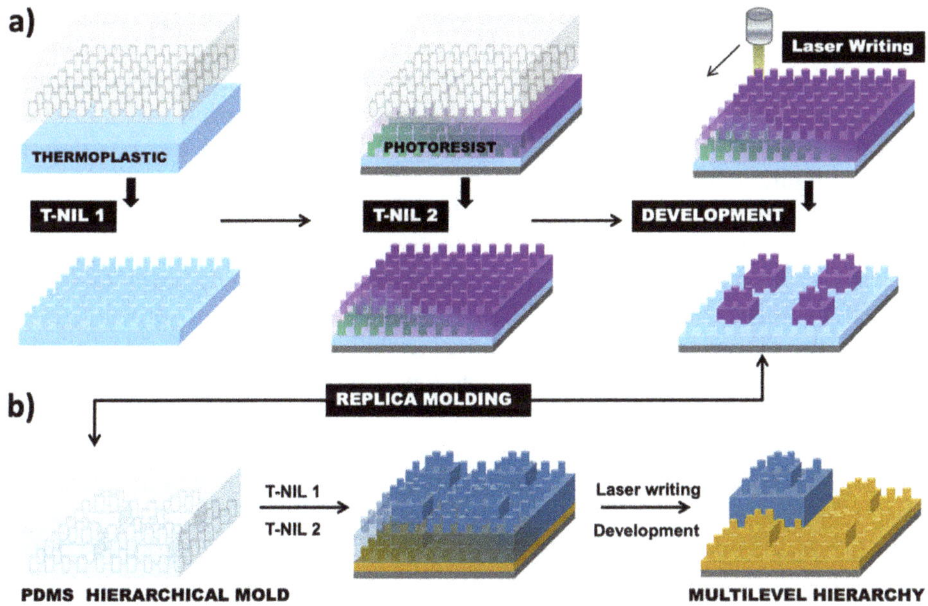

Figure 7.10. (a) Outline of the fabrication process to produce multilevel hierarchical topographies. First, a polymer film of PMMA is spin coated on a silicon wafer and imprinted with nanofeatures via T-NIL. Second, a photoresist film is spin coated over the imprinted polymer and a process of T-NIL is carried out with the same or different geometrical nanofeatures. Thirdly, direct laser writing of micron size features is performed. Finally the exposed resist is eliminated. (b) Fabrication of multilevel fractal-like hierarchical structures. This process uses the same sequential fabrication steps but makes use of a two-level hierarchical mold obtained by soft-lithographic replication of a previously produced hierarchical substrate.

Figure 7.11. SEM images of (a) two-level, second order; (b) two-level, third order and (c) four-level third order hierarchical structures.

features. In these examples, the hierarchical topographies include combinations of a moth-eye topography or 500 nm periodic gratings at two different levels of micron-size features (pillars, ridges or cubes).

When using thermoplastic polymers, it is necessary to select among those with sufficient differences in glass transition temperature (T_g) to avoid softening of the material of the first layer during the second imprinting step. Nonetheless, since the photoresist is imprinted at a low temperature of about 100 °C most thermoplastics would be suitable. Alternatively, UV sensitive thermosetting resins can be used for the first layer while they will be imprinted via UV-NIL. This approach can be combined with replica molding to produce multilevel hierarchical topographies. In this case, a two-level hierarchical mold obtained by soft-lithographic replication of a previously produced hierarchical substrate is employed to imprint the first layer or even the second layer following the same sequential fabrication steps as described above (see figure 7.10(b)).

When the process is repeated sequentially, complex 3D fractal-like topographical arrangements of self-similar structures at reducing length-scales can be generated. An example of such, showing a third-order, four-level multi-scale topography produced by this process, it is displayed in figure 7.11(c). The process requires increasing the size of the microfeatures during the successive laser writing steps to accommodate the pre-exiting features on the substrate. In the case of isotropic structural designs, the enclosed features are self-aligned into the previous level, and alignment is not required. The maximum number of structural levels to possibly accommodate will depend on the material and geometrical characteristics of the different levels and importantly, the feature density and aspect ratio as these parameters will lead to the increase in contact area, and as such, the friction and adhesion forces involved in demolding. In general, as the architectural complexity increases, the possibility to cause damage or ripping of the structures due to demolding also increases, and this will be the main constraint. This strategy is an easy-to-implement fabrication process to produce multilevel hierarchical structures with very high feature control over large areas and a wide range of materials. As such it appears as very attractive strategy to enable many advanced nanofabrication possibilities.

7.4 End-of-chapter summary

NIL performed with soft polymeric molds has emerged as the most viable technology for the realistic production of nanoenabled products. Particularly, soft T-NIL, due to its versatility in materials processing, has become especially useful to produce functional nanopatterns where the imprinted structures themselves are the active components to provide for functionality. To fulfil this purpose, these nanopatterns require specific geometrical designs that are often inspired by nature and involve complex architectures such as high aspect ratio and hierarchical nano- and micropatterns.

This chapter has reviewed the most important processing issues including the properties and requirements on materials for the fabrication of soft polymeric molds and imprint process to produce these functional complex structures by soft T-NIL. The fabrication of the required soft molds is realized by replica molding from masters created using direct writing techniques. In addition to traditional silicon micromachining technologies, we can now take advantage of the huge flexibility in terms of structural design of direct laser writing by two-photon polymerization tools for rapid mastering of complex structures.

The most popular materials employed to produce flexible molds were discussed. One important consideration to take into account when selecting a polymer material is the viscosity of the liquid precursors since it limits the minimum feature size to possibly fill. On the other hand, the Young's modulus and surface energy of the crosslinked material impact the imprinting and demolding processes. In HAR imprinting, the surface energy is one of the most important parameters to take into account as it conditions the mold filling through the generation of capillary filling action and also the demolding through the adhesion forces between the mold and the polymer. The mechanical properties of the imprint material will determine the maximum aspect ratio that can be achieved before patterns collapse takes place. The composite mold approach, whereby the nanostructures are formed by a stiffer material on a flexible backing should be implemented when the mold contains protruding high-resolution features.

Generally, the production of hierarchical topographies with well defined geometry and feature control involves a significant engineering challenge. These have mostly been created by appropriate combinations of NIL and other lithographic approaches, principally photolithography. In particular, it has been shown that the combination of consecutive T-NIL and direct laser writing processes allows for the fabrication of complex multiscale hierarchical structures with high precision in a practical and reproducible manner.

7.5 Student's corner

- How is it possible to use s-PDMS as a mold for T-NIL of nanostructures on PC if the elastic modulus of the PC is three orders of magnitude higher?
- During T-NIL of HAR nanostructures on PMMA it is observed that a large number of nanopillars are affected by ripping. What could be done to improve the replica quality in this case?

- During the fabrication of nanopillars by T-NIL on a PS film, it is observed that upon demolding, a large number of pillars are taller than expected according to the mold dimensions, what could be the reason? How would you select the demolding temperature?
- How does the aspect ratio influence the detachment force during demolding of nanoimprinted structures? What is the advantage of using flexible molds during the demolding step of HAR nanostructures?
- Which are the main factors related to the pattern collapse of HAR?
- Given a polymer with viscosity $=10^6$ (Pa s), surface tension ~20 mN m^{-1} and wetting contact angle of 80°:
  a. What is the time needed to fill 0.5 × 2 μm pore size cavities by capillary action?
  b. What viscosity should the polymer have to reduce that time to about 7 min?
  c. If mold silanization is accomplished in order to facilitate the demolding process, how is the polymer wetting contact angle expected to be affected?
- What provides the gecko with its amazing climbing capability?

References

[1] Fang F Z, Zhang X D, Gao W, Guo Y B, Byrne G and Hansen H N 2017 *CIRP Ann.* **66** 683–705

[2] Schift H and Kristensen A 2017 *Nanoimprint Lithography* (Berlin: Springer)

[3] Schift H 2008 *J. Vac. Sci. Technol.* B **26** 458–80

[4] Guo L J 2007 *Adv. Mater.* **19** 495–513

[5] Zhou W 2013 *Nanoimprint Lithography: An Enabling Process for Nanofabrication* (Berlin: Springer)

[6] Miyauchi A 2019 *Nanoimprinting and Its Applications: Jenny Stanford Publishing* (Boca Raton, FL: CRC Press)

[7] Kam A P and Torres C M S 2019 *Nanoimprint lithography: A competitive fabrication technique towards nanodevices* (Boca Raton, FL: CRC Press)

[8] Hongbo L 2013 *Soft UV Nanoimprint Lithography and Its Applications* (Rijeka: IntechOpen)

[9] Yu C-C and Chen H-L 2015 *Microelectron. Eng.* **132** 98–119

[10] Austin M D, Ge H, Wu W, Li M, Yu Z and Wasserman D *et al* 2004 *Appl. Phys. Lett.* **84** 5299–301

[11] Li R, Chen Z, Datye A, Simon G H, Ketkaew J and Kinser E *et al* 2018 *Commun. Phys.* **1** 75

[12] Chou S Y, Krauss P R and Renstrom P J 1995 *Appl. Phys. Lett.* **67** 3114–6

[13] Chou S Y, Krauss P R and Renstrom P J 1996 *Science* **272** 85–7

[14] Bar-On O, Brenner P, Siegle T, Gvishi R, Kalt H and Lemmer U *et al* 2018 *Sci. Rep.* **8** 7833

[15] Zhang C, Subbaraman H, Li Q, Pan Z, Ok J G and Ling T *et al* 2016 *J. Mater. Chem.* C **4** 5133–53

[16] Sun J and Bhushan B 2019 *Tribol. Int.* **129** 67–74

[17] Zhang F, Chan J and Low H Y 2008 *Appl. Surf. Sci.* **254** 2975–9

[18] Li Y, John J, Kolewe K W, Schiffman J D and Carter K R 2015 *ACS Appl. Mater. Interfaces* **7** 23439–44
[19] Chen W, Shao Y, Li X, Zhao G and Fu J 2014 *Nano Today* **9** 759–84
[20] Scheer H-C 2012 Pattern definition by nanoimprint *Proc. Micro-Optics 2012* **8428** 842802
[21] Kwon B and Kim J H 2016 *J. Nanosci.* **2016** 6571297
[22] Liang X, Tan H, Fu Z and Chou S Y 2006 *Nanotechnology* **18** 025303
[23] Scheer H-C, Papenheim M, Dhima K, Wang S and Steinberg C 2015 *Microsyst. Technol.* **21** 1595–605
[24] Sreenivasan S V 2017 *Microsyst. Nanoeng.* **3** 17075
[25] Higashiki T, Nakasugi T and Yoneda I 2011 *J. Micro/Nanolithogr. MEMS MOEMS* **10** 043008
[26] Verschuuren M A, Megens M, Ni Y, Hv S and Polman A 2017 *Adv. Opt. Technol.* **6** 243
[27] https://evgroup.com/technologies/nanoimprint-lithography-smartnil/
[28] Hauser H, Tucher N, Tokai K, Schneider P, Wellens C and Volk A *et al* 2015 *J. Micro/Nanolithogr. MEMS MOEMS* **14** 031210
[29] Zhang W, Fletcher B, Thompson E, Liu W, Stachowiak T and Khusnatdinov N *et al* 2016 High throughput jet and flash imprint lithography for semiconductor memory applications *Proc. Alternative Lithographic Technologies VIII* SPIE 9777 97770A
[30] Shao J, Chen X, Li X, Tian H, Wang C and Lu B 2019 *Sci. China Technol. Sci.* **62** 175–98
[31] Haslinger M, Moharana A and Mühlberger M 2019 Antireflective moth-eye structures on curved surfaces fabricated by nanoimprint lithography *Proc. 35th European Mask and Lithography Conf. (EMLC 2019)* SPIE **11177** 111770K
[32] Shen Y, Istock A, Zaman A, Woidt C and Hillmer H 2018 *Appl. Nanosci.* **8** 1415–25
[33] Ji R, Hornung M, Verschuuren M A, Laar R, Eekelen J and Plachetka U *et al* 2010 *Microelectron. Eng.* **87** 963–7
[34] Jeans A, Almanza-Workman M, Cobene R, Elder R, Garcia R and Gomez-Pancorbo F *et al* 2010 Advances in roll-to-roll imprint lithography for display applications *Proc. Alternative Lithographic Technologies II (San Jose, CA)* SPIE **7367** 763719
[35] Espinha A, Dore C, Matricardi C, Alonso M I, Goñi A R and Mihi A 2018 *Nat. Photonics* **12** 343–8
[36] Nagel R D, Filser S, Zhang T, Manzi A, Schönleber K and Lindsly J *et al* 2017 *J. Appl. Phys.* **121** 084305
[37] Mohapatra S, Pant U and Moirangthem R S 2020 *Mater. Today Proc.* **28** 215–7
[38] Li K, Morton K, Veres T and Cui B 2011 *5.11—Nanoimprint Lithography and Its Application in Tissue Engineering and Biosensing* (Burlington, MA: Academic)
[39] Ventre M, Coppola V, Iannone M, Netti P A, Tekko I and Larrañeta E *et al* 2018 *Nanotechnologies for Tissue Engineering and Regeneration* (Amsterdam: Elsevier) ch 2
[40] Van Krevelen D W and Te Nijenhuis K 2009 *Chapter 13—Mechanical Properties of Solid Polymers* (Amsterdam: Elsevier)
[41] Gooch J W 2011 *Encyclopedic Dictionary of Polymers* (New York: Springer)
[42] Liang C-C, Lin C-H, Cheng T-C, Shieh J and Lin H-H 2015 *Adv. Mater. Interfaces* **2** 1500030
[43] Nielsen T, Nilsson D, Bundgaard F, Shi P, Szabo P and Geschke O *et al* 2004 *J. Vac. Sci. Technol.* B **22** 1770–5
[44] Kehagias N, Francone A, Guttmann M, Winkler F, Fernández A and Torres C M S 2018 *J. Vac. Sci. Technol.* B **36** 06JF1

[45] Li J, Zhu Z, Li T, Peng X, Jiang S and Turng L-S 2020 *J. Appl. Polym. Sci.* **137** 48581
[46] https://espanol.kynar.com/export/sites/kynar-latam/.content/medias/downloads/literature/kynar-kynar-flex-pvdf-performance.pdf
[47] van de Kerkhof M A, Benschop J P H and Banine V Y 2019 *Solid State Electron* **155** 20–6
[48] Miller K, Li M, Walsh K M and Fu X-A 2013 *J. Micromech. Microeng.* **23** 035039
[49] Dussart R, Tillocher T, Lefaucheux P and Boufnichel M 2014 *J. Phys. D Appl. Phys.* **47** 123001
[50] Saive R, Bukowsky C R and Atwater H 2017 arXiv:1702.04012v1
[51] Cadarso V J, Chidambaram N, Jacot-Descombes L and Schift H 2017 *Microsyst. Nanoeng.* **3** 17017
[52] Ohrt C, Acar Y, Seidel A, Cheng W, Kiyan R and Chichkov B N 2012 *Int. J. Adv. Manuf. Technol.* **63** 103–8
[53] Kirchner R and Schift H 2019 *Mater. Sci. Semicond. Process.* **92** 58–72
[54] Chidambaram N, Kirchner R, Fallica R, Yu L, Altana M and Schift H 2017 *Adv. Mater. Technol.* **2** 1700018
[55] Schift H, Chidambaram N, Altana M and Kirchner R 2017 Selective surface smoothening of 3D micro-optical elements *Proc. Emerging Patterning Technologies (San Jose, CA)* SPIE **10144** 101440B
[56] Jung G-Y, Li Z, Wu W, Chen Y, Olynick D L and Wang S-Y *et al* 2005 *Langmuir* **21** 1158–61
[57] Schift H, Saxer S, Park S, Padeste C, Pieles U and Gobrecht J 2005 *Nanotechnology* **16** S171–5
[58] Qin D, Xia Y and Whitesides G M 2010 *Nat. Protoc.* **5** 491–502
[59] Seghir R and Arscott S 2015 *Sens. Actuators* A **230** 33–9
[60] Schmid H and Michel B 2000 *Macromolecules* **33** 3042–9
[61] Odom T W, Thalladi V R, Love J C and Whitesides G M 2002 *J. Am. Chem. Soc.* **124** 12112–3
[62] Li Z, Gu Y, Wang L, Ge H, Wu W and Xia Q *et al* 2009 *Nano Lett.* **9** 2306–10
[63] John J, Tang Y, Rothstein J P, Watkins J J and Carter K R 2013 *Nanotechnology* **24** 505307
[64] Masciullo C, Sonato A, Romanato F and Cecchini M 2018 *Nanomaterials* **8** 609
[65] Con C, Zhang J, Jahed Z, Tsui T Y, Yavuz M and Cui B 2012 *Microelectron. Eng.* **98** 246–9
[66] Williams S S, Retterer S, Lopez R, Ruiz R, Samulski E T and DeSimone J M 2010 *Nano Lett.* **10** 1421–8
[67] Truong T T, Lin R, Jeon S, Lee H H, Maria J and Gaur A *et al* 2007 *Langmuir* **23** 2898–905
[68] Schift H, Spreu C, Saidani M, Bednarzik M, Gobrecht J and Klukowska A *et al* 2009 *J. Vac. Sci. Technol.* B **27** 2846–9
[69] Papenheim M, Steinberg C, Dhima K, Wang S and Scheer H-C 2015 *J. Vac. Sci. Technol.* B **33** 06F601
[70] Papenheim M, Mayer A, Wang S, Steinberg C and Scheer H-C 2016 *J. Vac. Sci. Technol.* B **34** 06K406
[71] Papenheim M, Eidemüller W, Steinberg C, Mayer A and Scheer H-C 2017 *J. Vac. Sci. Technol.* B **35** 06G305
[72] Klukowska A, Kolander A, Bergmair I, Mühlberger M, Leichtfried H and Reuther F *et al* 2009 *Microelectron. Eng.* **86** 697–9

[73] Mayer A, Rond J, Staabs J, Leifels M, Zajadacz J and Ehrhardt M *et al* 2019 *J. Vac. Sci. Technol.* B **37** 061601
[74] Madsen M H, Feidenhans'l N A, Hansen P-E, Garnæs J and Dirscherl K 2014 *J. Micromech. Microeng.* **24** 127002
[75] Alameda M T, Osorio M R, Hernández J J and Rodríguez I 2019 *ACS Appl. Nano Mater.* **2** 4727–33
[76] Nosonovsky M and Bhushan B 2007 *Nano Lett.* **7** 2633–7
[77] Yao X, Song Y and Jiang L 2011 *Adv. Mater.* **23** 719–34
[78] Autumn K, Liang Y A, Hsieh S T, Zesch W, Chan W P and Kenny T W *et al* 2000 *Nature* **405** 681–5
[79] Ho A Y Y, Yeo L P, Lam Y C and Rodríguez I 2011 *ACS Nano* **5** 1897–906
[80] Kustandi T S, Low H Y, Teng J H, Rodriguez I and Yin R 2009 *Small* **5** 574–8
[81] Zhang S and Chen Y 2015 *Sci. Rep.* **5** 16637
[82] Dumanli A G and Savin T 2016 *Chem. Soc. Rev.* **45** 6698–724
[83] Raut H K, Dinachali S S, Loke Y C, Ganesan R, Ansah-Antwi K K and Góra A *et al* 2015 *ACS Nano* **9** 1305–14
[84] Navarro-Baena I, Jacobo-Martín A, Hernández J J, Castro Smirnov J R, Viela F and Monclús M A *et al* 2018 *Nanoscale* **10** 15496–504
[85] Bae W-G, Kim H N, Kim D, Park S-H, Jeong H E and Suh K-Y 2014 *Adv. Mater.* **26** 675–700
[86] Kong T, Luo G, Zhao Y and Liu Z 2019 *Adv. Funct. Mater.* **29** 1808012
[87] Malshe A P, Bapat S, Rajurkar K P and Haitjema H 2018 *CIRP Ann.* **67** 627–50
[88] Wong H C, Grenci G, Wu J, Viasnoff V and Low H Y 2018 *Ind. Eng. Chem. Res.* **57** 13759–68
[89] Durret J, Frolet N and Gourgon C 2016 *Microelectron. Eng.* **155** 1–6
[90] Li Y, Krahn J and Menon C 2016 *J. Bionic Eng.* **13** 181–99
[91] Higgins S G, Becce M, Belessiotis-Richards A, Seong H, Sero J E and Stevens M M 2020 *Adv. Mater.* **32** 1903862
[92] Chen X and Zhang L 2018 *Sens. Actuators* B **254** 648–59
[93] Raut H K, Ranganath A S, Baji A and Wood K L 2019 *Appl. Surf. Sci.* **465** 362–8
[94] AYY H, Luong Van E, Lim C T, Natarajan S, Elmouelhi N and Low H Y *et al* 2014 *J. Polym. Sci. B Polym. Phys.* **52** 603–9
[95] Source picture: public domain, https://commons.wikimedia.org/wiki/File:Butterfly_tongue.jpg. Author: Dartmouth College
[96] Bixler G D and Bhushan B 2014 *Nanoscale* **6** 76–96
[97] Cadarso V J, Llobera A, Puyol M and Schift H 2016 *ACS Nano* **10** 778–85
[98] Suresh V, Ding L, Chew A B and Yap F L 2018 *ACS Appl. Nano Mater.* **1** 886–93
[99] Chen J, Li Y, Huang K, Wang P, He L and Carter K R *et al* 2015 *ACS Appl. Mater. Interfaces* **7** 22106–13
[100] Scheer H-C, Mayer A, Dhima K, Wang S and Steinberg C 2014 *Microsyst. Technol.* **20** 1891–8
[101] Omar F, Brousseau E, Elkaseer A, Kolew A, Prokopovich P and Dimov S 2014 *J. Micromech. Microeng.* **24** 055007
[102] Park S, Song Z, Brumfield L, Amirsadeghi A and Lee J 2009 *Appl. Phys.* A **97** 395–402
[103] Dirckx M E and Hardt D E 2011 *J. Micromech. Microeng.* **21** 085024

[104] Trabadelo V, Schift H, Merino S, Bellini S and Gobrecht J 2008 *Microelectron. Eng.* **85** 907–9

[105] Viela F, Granados D, Ayuso-Sacido A and Rodríguez I 2016 *Adv. Funct. Mater.* **26** 5599–609

[106] Hernández J J, Monclús M A, Navarro-Baena I, Viela F, Molina-Aldareguia J M and Rodríguez I 2017 *Sci. Rep.* **7** 43450

[107] Rudin A and Choi P 2013 *Mechanical Properties of Polymer Solids and Liquids* (Boston, MA : Academic) ch 4

[108] Chidambaram N, Kirchner R, Altana M and Schift H 2016 *J. Vac. Sci. Technol.* B **34** 06K401

[109] Roca-Cusachs P, Rico F, Martínez E, Toset J, Farré R and Navajas D 2005 *Langmuir* **21** 5542–8

[110] Glassmaker N J, Jagota A, Hui C-Y and Kim J 2004 *J. R. Soc. Interface* **1** 23–33

[111] Finn A, Lu B, Kirchner R, Thrun X, Richter K and Fischer W-J 2013 *Microelectron. Eng.* **110** 112–8

[112] Raut H K, Baji A, Hariri H H, Parveen H, Soh G S and Low H Y *et al* 2018 *ACS Appl. Mater. Interfaces* **10** 1288–96

[113] Cheng X, Jay and Guo L 2004 *Microelectron. Eng.* **71** 277–82

[114] Schift H, Spreu C, Schleunitz A, Gobrecht J, Klukowska A and Reuther F *et al* 2009 *J. Vac. Sci. Technol.* B **27** 2850–3

[115] Schleunitz A, Spreu C, Vogler M, Atasoy H and Schift H 2011 *J. Vac. Sci. Technol.* B **29** 06FC1

[116] Steinberg C, Rumler M, Runkel M, Papenheim M, Wang S and Mayer A *et al* 2017 *Microelectron. Eng.* **176** 22–7

IOP Publishing

Nanofabrication
Nanolithography techniques and their applications
José María De Teresa

Chapter 8

Stencil lithography

Oscar Vazquez-Mena and Luis Guillermo Villanueva

Stencil lithography is an ancient method that has been used for centuries to transfer patterns on different substrates. In recent decades, it has been applied to micro- and nano-fabrication. It is a resist-less method that allows wafer-scale transfer of nano-structures using deposition, etching and implantation. In this chapter we describe the technique, highlighting its advantages and challenges, and we provide a detailed guide on how to achieve an optimized result. We finalize the chapter by briefly enumerating some of the most interesting applications that stencil lithography has been used for.

8.1 Introduction

Stencil lithography, hereinafter also referred to as 'SL', is a lithographic technique that is based on the use of a solid mask to locally block/allow the flux of certain atoms/molecules/particles onto our substrate resulting in pattern generation. The idea is similar to the classical analog, where a shape is cut-out in a solid material, which we call *stencil* or *mask*, and we use it to locally stop/allow the flow of, e.g. spray paint (see figure 8.1).

This concept, which macroscopically has been used for centuries, can also be used to transfer patterns in the micro-/nano-scale [1]. In this latter case, the masks must be made separately on a solid material with apertures in the shape of the pattern that we want to transfer. To perform these apertures, we must use different lithographic techniques, as will be explained later in the chapter. The solid mask will stop/allow the flux of be atoms, ions, molecules, or even particles (see video 8.1 available at https://iopscience.iop.org/book/978-0-7503-2608-7). The available processes to perform the pattern transfer are deposition, etching or ion implantation. We will see later in the chapter how the choice of patterning method determines the type of mask to be used and has different implications in the shape of the transferred pattern.

In contrast to more conventional types of lithography, based on resist processing, stencil lithography typically presents a distance (gap) between the stencil and the

Figure 8.1. Macroscopic stencils. (Left) a paper stencil with the symbol for *wifi* cut out. Material could go through the holes in the paper to transfer the pattern on a surface. (Right) street graffiti made using another paper stencil where the letters have been cut out.

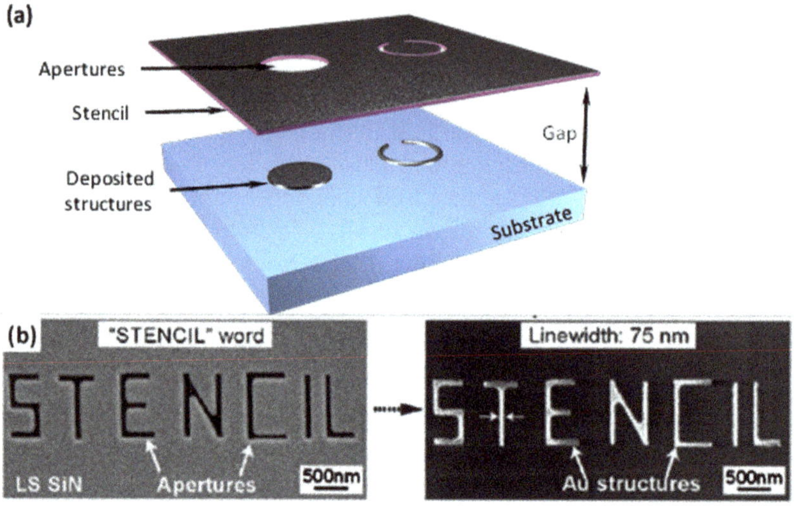

Figure 8.2. Canonical example for SL–metal deposition. (a) Schematic of the metal deposition setup, where a membrane with micro-/nano-apertures is placed between the metal source and the substrate to stop metal from passing except from the regions where the apertures are opened. (b) Scanning electron microscope (SEM) micrographs showing on the left a membrane with nanoapertures and on the right the same pattern replicated on the surface with 75 nm thick Au layer. Image with permission from [2] copyright 2012 American Chemical Society.

substrate (see figure 8.2(a)). The existence of this gap has implications on both the advantages and disadvantages of the technique. To give two quick examples: on the one hand, it is possible to pattern transfer without touching the substrate, which is useful for fragile substrates. On the other hand, however, there are no physical walls limiting the landing and spreading of the material on the substrate, so the patterns will not be transferred 1:1 but they will observe some change of the lateral dimensions (this is what we call *blurring*).

Figure 8.2 shows the canonical use of SL–metal deposition through a stencil membrane. We can see the membrane with micro-/nano-apertures that stands between the metal source and the substrate, masking locally the flow of atoms. This membrane is in most cases fabricated on silicon or silicon nitride, which are durable materials to reduce fracture and allow re-utilization of the stencils.

8.1.1 Brief history of stencil in micropatterning

The use of stencils in micromachining has a long history. We can go back to 1969, where a stencil was proposed to etch glass by hydrofluoric acid [3]. In the 1970s, SL was used to mask the deposition of metal by sputtering, and achieving lateral resolution in the millimeter range [4]. Later, SL was demonstrated for the fabrication of thin film Josephson devices [5], and to deposit contacts on solar cells, thus avoiding photoresist processing [6]. The minimum attained lateral dimensions have been improving significantly since then, with several milestones on the way, such as: 10 μm in 1982 [7], 3 μm in 1996 [8], 500 nm in 2003 [9]; and a record of 50 nm in 2008 [10]. This reduction in the lateral resolution has indeed been possible thanks to the advances in the other lithographic processes used for the fabrication of the stencil membranes: electron beam lithography (EBL) [11], deep-UV lithography [12] and focused ion beam (FIB) [10], nanosphere lithography [13] and nano-imprint lithography (NIL) [14].

8.1.2 Why is stencil lithography interesting?

SL has the following advantages that make it attractive for nanopatterning:

Resistless processing: SL does not involve the coating of the substrate with any resist, avoiding contact with solvents, organic materials, or energy radiation. This enables the use of the technique on chemically fragile surfaces such as polymers or self-assembled monolayers. At the same time, it also allows the patterning on mechanically fragile substrates and structures, such as bad adhesion materials or released micromechanical structures (membranes, cantilevers, etc). In addition, SL can also be performed on substrates that show a 3D topography, i.e. substrates where resist casting (via spin or spray coating) is challenging or even impossible.

Easy manipulation and implementation: SL can be used by anyone that has access to a deposition, etching or implantation machine. Once the stencil (mask) is made, one can fix it on a substrate even with practical methods, like adhesive tape or mechanical clamps, and insert the coupled system into the processing machine. The stencil will mask the entities and thus allow the transfer of the pattern into the substrate.

Stencil reusability: Following the previous point, the stencil (mask) is the fundamental part of the process. Even though the fabrication might be complicated (see next section), the stencils can be reused many times. This makes the pattern transfer via SL quite cost-effective. For deposition of nanostructures (through nanoscale apertures), reusability is limited by the so-called clogging, i.e. the progressive closing of the apertures due to lateral deposition. For etching and ion-implantation, there is no lateral deposition, but the membranes might be damaged during the process if proper protection is not implemented.

Parallel patterning: As opposed to other nanopatterning techniques, SL can be used to pattern in parallel a full wafer. The results, as we will cover later below, might not be optimum when using large stencils, since the gap between the stencil and the substrate might be larger, but it is still possible to do the pattern transfer and use blurring corrective methods.

Dynamic stencil lithography: Since (typically) stencils do not stick to the substrate, there exists the possibility of moving them with respect to the substrate during pattern transfer. This has been tested with deposition and it opens up a rainbow of different options: variable thickness structures, multi-material structures with different patterns using a single mask, complex designs, etc.

8.1.3 Why is stencil lithography challenging?

Not everything is positive regarding SL. There are several challenges to overcome for SL to become a mainstream option, namely blurring, clogging, shadowing and membrane stability. These three points are very much referring to the use of SL for nanopatterning. When apertures are in the micron-size, these points are much less important.

Blurring: As we can see in figure 8.2, there is always a gap between the stencil and the substrate. This means that when species travel through the apertures they are not laterally confined, and thus the lateral resolution of the technique is compromised. This can be caused by surface diffusion of the deposited material and by the divergence of the material flux, which in turn depends on the process pressure and the collimation of the flux. Blurring becomes more relevant with decreasing structure size, since the absolute value of the dimension loss remains constant, thus in relative terms grows inversely proportional to the structure size. At the nanoscale, the dimensions of the *blurred* zone can be as large as the structures themselves. A deeper analysis and strategies for blurring reduction are discussed later in the chapter.

Shadowing: In the case that the flux of material is not collimated, but divergent, depending on the aspect ratio of the membrane apertures, we could also have a partial masking of the line of deposition into the substrate. This generates patterns that do not have a uniform thickness over the width of the aperture.

Clogging: In the case of deposition through a stencil mask, there is an issue that arises when the deposited material thickness is of the same order or greater than the lateral size of the stencil aperture. In that case, the deposition of material on the aperture sidewall significantly reduces the size of the aperture. This implies, essentially, that the aperture width dynamically changes from the beginning till the end of the deposition. Indeed, if the deposited thickness is larger than the stencil aperture width, it could happen that the apertures close during a given deposition. Clogging also affects the reusability of the stencil membrane since the subsequent deposition will start at a narrower aperture width. However, clogging can also be used to fabricate conical or pyramidal patterns [15, 16], which is a distinct feature compared to other types of lithography.

Membrane stability (and low fill factor): To perform pattern transfer in the nano (sub-micron) scale, we need to etch apertures in the stencil membrane with lateral dimensions in the nanoscale. This is only possible if the membrane thickness is also in the nanoscale, since otherwise the aspect ratio of the aperture makes it extremely hard to be opened. The previous effect, shadowing, is another reason to have small aspect ratio in the stencil apertures. Having thin membranes means that they are more fragile and that they are more prone to mechanical deformation arising from material deposition, or damage during etching or implantation. The first logical solution to this is to make the membranes smaller in lateral dimensions, so that they are sturdier mechanically. However, this also means that the area to be patterned is smaller, and the overall surface of the wafer cannot be used within a single deposition, reducing the fill-factor available with SL.

8.2 Fabrication of stencil membranes

In SL everything starts with the fabrication of the membranes. It is a process very much analogous to the fabrication of masks/reticles for optical lithography and the fabrication of stamps for nanoimprint lithography.

8.2.1 Membrane material

Unlike optical lithography, where the typical mask material is quartz, stencil membranes can be made of different materials, depending on the design requirements.

First, a flat membrane is favored over a wrinkled one. This is something that happens in certain structures and is called *buckling* [17]. In order to avoid this buckling, membranes made out of materials with compressive stress should be avoided.

Second, as we just mentioned in the previous section, if one needs nano-apertures, the thickness of the membrane must also be in the sub-micron range. If one needs micro-apertures, then the thickness can be much larger. This has a direct impact in the choice of material. For small apertures, in order to gain stability of the membrane with a small thickness, a material with tensile stress should be used, and the most used one is silicon nitride [18]. For larger apertures, the material choice opens up, but people have mostly used either silicon [19–21] or certain polymers [22].

8.2.2 Fabrication steps

The steps to fabricate a stencil membrane are relatively straightforward. They will depend, of course, on the actual material chosen for the membrane itself and the dimensions of the apertures. Here we give what we consider to be the canonical example of stencil membrane fabrication: a flat silicon nitride membrane with nano-apertures (see figure 8.3).

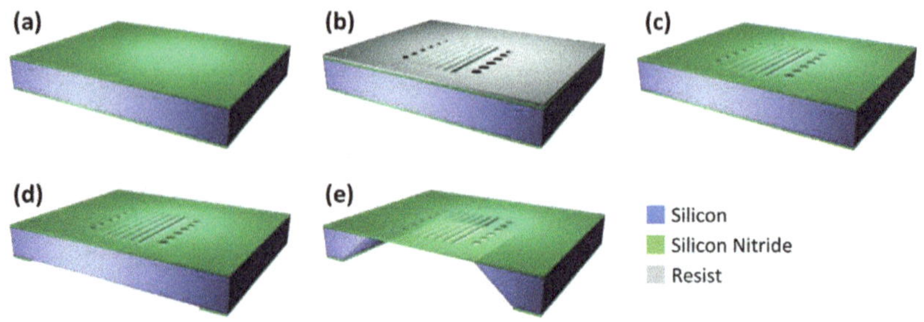

Figure 8.3. Canonical example for stencil membrane fabrication. (a) Si wafer with a thin layer of SiN coated on both sides. (b) A lithography is performed on the frontside using EBL, NIL, DUV... or any nano-lithography technique we include in this book. (c) Frontside SiN is patterned, and resist is removed. (d) SiN on the backside is etched to define the windows for a subsequent etching step. (e) KOH etching that removes the Si and releases the membrane.

A silicon wafer is coated on both sides with silicon nitride (figure 8.3(a)). Here, one typically chooses low-stress silicon nitride (made by biasing the stoichiometry and enriching the silicon content). Then a lithography is performed on the frontside (figure 8.3(b)) to define the apertures on the frontside SiN layer via dry etching (figure 8.3(c)). Subsequently, windows are opened on the backside SiN layer (figure 8.3(d)) that will allow KOH to etch the bulk Si and thus release the membrane (figure 8.3(e)). These steps have been described thoroughly in the literature [23].

Variations of this process flow can be found for example on the bulk Si etching, which in some cases has been made combining deep reactive ion etching (DRIE) and KOH, or even only DRIE. The main reason to do this is to be able to place membranes closer together, since KOH defines an angle of 54.7° in the sidewalls and that limits the number of membranes that can be placed per area.

8.2.3 Membrane reinforcement

As we have described previously, membrane stability is an important concern for SL. We have mentioned that it is possible to improve the membrane stability by increasing the thickness (if the aperture size allows it) and by reducing the lateral dimensions of the membrane (if the required fill-factor allows it).

There might be cases where none of these options is a possibility, meaning that nano-apertures over a large area are required. This would imply that the membranes must be thin and large and thus very fragile mechanically. In order to improve membrane yield and endurance, it is possible to add mechanical reinforcements to the membrane. This is typically done (but there are other ways) using corrugations in the membrane, as we can see in figure 8.4.

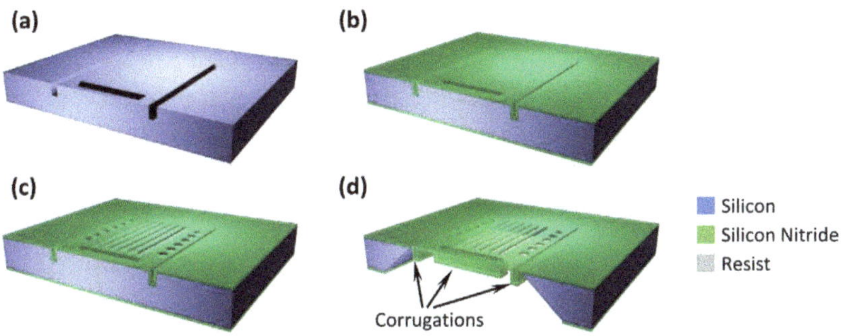

Figure 8.4. Canonical example for stencil membrane fabrication with corrugations. (a) Si wafer is patterned with trenches where the reinforcements are required. (b) SiN is deposited on the wafer, conformally covering the trenches. (c) SiN on the frontside is patterned using lithography and dry etching. (d) SiN on the backside is etched to define the windows for a subsequent KOH etching that removes the Si and releases the membrane, revealing the corrugations that add mechanical stability to the membrane.

8.3 Challenges of stencil lithography

We have already listed the main challenges or this technique, but now we are going to perform a deeper analysis on each of them.

8.3.1 Blurring

Blurring is the increase in lateral dimensions of the desired pattern when being transferred from the membrane to the substrate. Depending on the fabrication step where we use the stencil, the behavior of the species travelling through the apertures is different. A schematic for each case can be seen in figure 8.5.

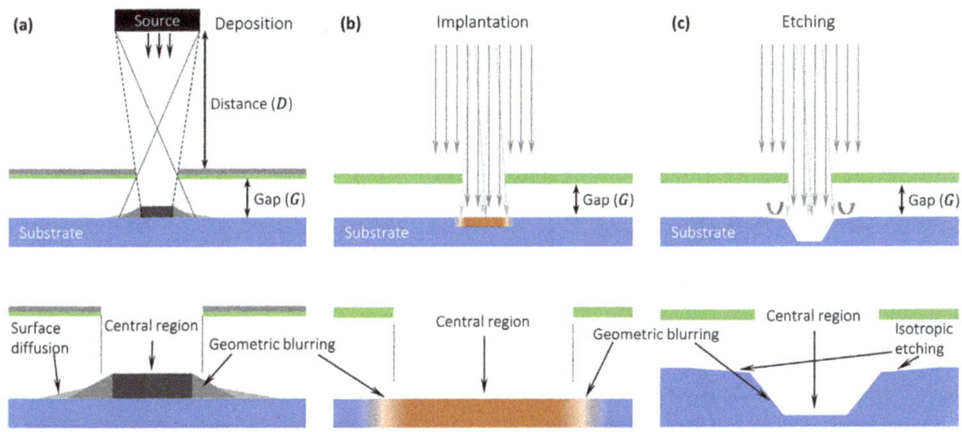

Figure 8.5. Schematic for the different blurring effects on stencil lithography depending on the fabrication technique. (a) Deposition. Since the beam of material is not collimated, the deposition yields a zone of *penumbra* that corresponds to the so-called *geometric blurring*, which is continued by a *surface diffusion* zone. (b) Implantation. Even though in this case the beam is quite collimated, the *geometric blurring* always exists. (c) Etching. On top of the *geometric blurring* due to the material not being collimated, we can also find an isotropic etching component that is caused by free radicals which are not directed by the bias voltage on the substrate.

8.3.1.1 Deposition

The most common case, based on the typical use of stencil lithography, is for the deposition of material and, more in particular, evaporation. In this case (figure 8.5(a)), the material source is heated around an area of diameter (d) and placed at a distance (D) of the membrane/substrate, which are separated by a gap (G). Via a simple geometrical analysis, it is possible to see that the aperture of width (w) transfers onto the substrate forming a structure which does not have constant thickness across the width. On the final structure, we can see three distinct regions. The darker part, that we can call the *central region*, corresponds to the zone with the desired thickness (dark grey in figure 8.5(a)). Moving laterally, we observe a region where not all the points from the source material reach the substrate. This zone is the one that corresponds to the *geometric blurring*. It happens because the material emerging from the source is not perfectly collimated. Finally, the last portion shown in the schematic (in light gray) is a very thin layer of material due to *surface diffusion*.

Geometric blurring is relatively easy to calculate. Considering that the gap is much smaller than the distance to the source ($G \ll D$, please note that the drawing in figure 8.5 is not made to scale) and that the thickness of the membrane is smaller than the aperture width ($t \ll w$), we can see that the deposited structure will have a part where the thickness is maximum, and then a zone where the thickness decreases linearly. The dimensions can be calculated by equation (8.1):

$$\Delta w_{\text{geom}} \approx G \cdot \frac{d}{D};\; w_{\max} \approx w_{\text{aper}} - \Delta w_{\text{geom}}; \tag{8.1}$$

where $w_{\max}$ corresponds to the width of the region with maximum thickness. From equation (8.1) it is clear that this effect disappears when $G = 0$, as we mentioned before. The fact that the effect reduces with the ratio $\frac{d}{D}$ is intuitive if we think of the collimation of the material emanating from the source. In an analogy to a light source, rays arrive collimated when the source is distant and punctual, i.e. $\frac{d}{D} \to 0$.

In practical terms, we can see that to have better results when depositing material through a stencil, we should seek large deposition chambers, where the source and the substrate are as far away as possible. This typically happens in evaporators with only one chamber. At the same time, we would like to reduce the size of the source. This implies that the best technique is e-beam evaporation (since it is possible to minimize the heated area); then thermal evaporation (with a small heating crucible); then multi-target sputtering with a small target size; and finally single-target sputtering. The difference in d between evaporation and sputtering is at least of one order of magnitude, and this is the main reason why a stencil is rarely used with sputtering. In addition to this purely geometrical effect, the fact that sputtering takes place at a much higher chamber pressure than evaporation makes the collimation even harder, even though in some cases a substrate bias could be applied to convey directionality to the deposition.

The second contribution to *blurring* during deposition is *surface diffusion*. The size of this contribution, which has also been called *halo*, is difficult to calculate *a priori* since it depends on many factors, for example the state of the substrate surface [24], the material being deposited [25], the energy of the material when reaching the surface, etc. The only realistic way to estimate the width passes through experimenting and measuring its size. However, this can be proven to be quite cumbersome since this *halo* is only a few nanometers thick (typically < 5 nm), and it is not visible under the scanning electron microscope (SEM). To facilitate the visualization, a technique called *structures development* is used, where a mild dry etching is performed on the underlying substrate, therefore revealing the size of the halo. A clear example of this can be seen in figure 8.6(a) [25], where we show respective SEM micrographs of an aperture in a stencil membrane, the deposited aluminum film on Si, and the *developed* structure after a mild dry etching, revealing the size of the *halo*.

By using this technique, it has been possible to analyze the dependence of *surface diffusion* on different parameters. Figure 8.6(b) shows how the gap makes larger the *geometric blurring*, but also how it increases the size of the *halo*. Figure 8.6(c) shows that for larger deposition rates we observe a larger *surface diffusion*. This is probably because material arrives at the surface more abundantly and thus there is more

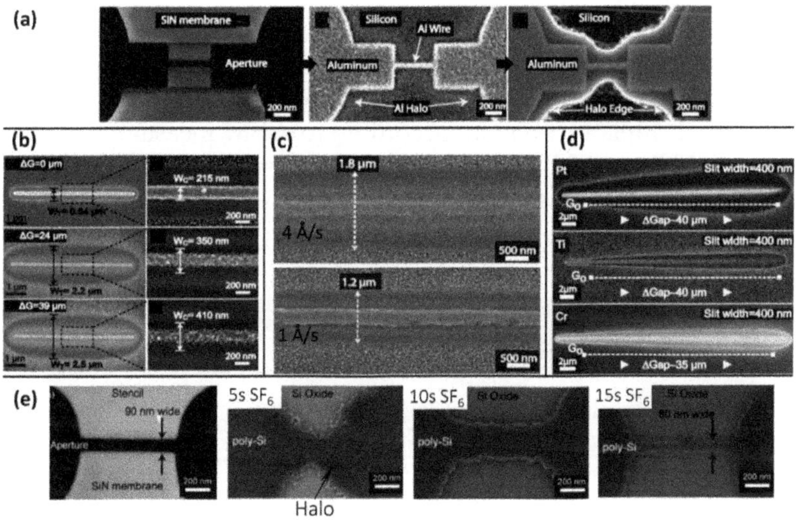

Figure 8.6. Examples of blurring. (a) Method to reveal the size of the *halo* created by *surface diffusion*. From left-to-right, SEM micrographs of an aperture in the SiN; the metal deposited on the substrate making it difficult to observe the thin *halo*; the same substrate after a mild dry-etching of the substrate material, where the *halo* acts as a mask, therefore making it visible. (b) Dependence of blurring on the gap. SEM micrographs showing deposited structures from the same aperture width (w_{aper}) but different gaps. An increase in the width of the central structure ($w_c = w_{aper} + w_{geom}$) with the gap is evident, as it is an increase of the *halo*. (c) SEM micrographs showing the dependence of the halo with the deposition rate. Higher rate means larger *surface diffusion*, probably because atoms arrive with more energy to the substrate. (d) SEM micrographs showing the effect of material and gap on the size of the *halo*. Different materials, depending on their adhesion (surface diffusion coefficient), expand more on the surface than others. (e) Technique to reduce the effect of blurring, specially the one coming from surface diffusion. SEM micrographs showing a stencil aperture and the result of the deposition of 50 nm-thick amorphous-Si layer. A mild dry-etching is performed for different times, reducing the size of the halo until it is completely gone after 15 s etching. Images adapted from [25] copyright IOP Publishing, and [26] copyright IEEE.

energy to allow the atoms to travel faster. Another example can be seen in figure 8.6(d), that evidences the difference in material and, at the same time, due to the gap. This latter point can also be seen in figure 8.6(b). However, the dependence on the material is very much striking and indicates the importance of studying this issue beforehand.

Even though this *halo* can reach very large dimensions, since its thickness is quite small, it is also possible to bypass the issue by removing it before continuing the processing. This can be done via a mild dry etching very selective to the material we want to remove. An example of this can be seen in figure 8.6(e) [26], where 50 nm of amorphous silicon were deposited on silicon oxide. The visible *halo* is removed by a mild silicon etching. After 15 s of etching, the nanowire-like shape is recovered with no visible *halo*.

8.3.1.2 Implantation and etching

We group implantation and etching because both present rather collimated beams of ions going to the substrate. In the first case (figure 8.5(b)), as we know from the functioning principles for ion implanters, the ions are generated from a source, accelerated, filtered and then collimated with the aid of thin slits along the way. However, this collimation is not perfect and thus the pattern will suffer a slight distortion. We can see this in the examples of figures 8.7(a) and (b), where the stencil apertures (of 140 nm (a) and 310 nm (b)) are used to let ions through and the final dimensions of the implanted region, according to AFM measurements are slightly larger (around 100 nm) [27]. This is right after implantation. Depending on the subsequent annealing process, the overall loss of dimensions can be somewhat larger.

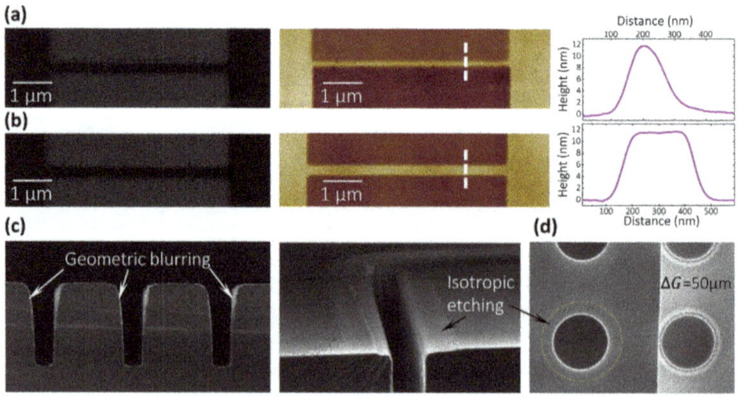

Figure 8.7. Effect of blurring for implantation and etching. (a) SEM micrograph of a 140 nm stencil aperture; followed by an AFM topography image of the substrate after implantation has been performed; and the profile of the implanted region. The 10 nm protrusion might be due to a faster native oxidation rate within the doped region. (b) Same as (a) but with an aperture of 310 nm. (c) SEM micrographs showing trenches in silicon being etched by means of a deep reactive ion etching (Bosch® process). Geometric blurring can be observed, as well as the effect of a shallow isotropic etching emanating from the free radicals. (d) SEM micrograph showing the dependence of blurring on the gap, where the rightmost part has been patterned with a gap that is 50 µm larger than the section on the left.

In the case of dry etching, as schematically shown in figure 8.5(c) and mentioned above, the ionized molecules within the generated plasma will be attracted towards the substrate thanks to the applied bias. This generates quite collimated beams of ions that will etch the substrate providing some *geometrical blurring*. It is possible to observe this in figure 8.7(c), in the cross section, where the corners of the trenches are being rounded up by non-purely directional etching. In addition to this, we also observe the effect of the free-radicals that are not collimated by the bias, since they are charge-neutral. In this case, an isotropic component of the etching appears and damages the surface. This is shown in figures 8.7(c) and (d). Finally, as in the case of the deposition, it is easy to see that blurring for etching and implantation increases with the gap, as is shown in figure 8.7(d).

8.3.2 Shadowing

The second challenge that we must face when using SL is the so-called *shadowing*. This is purely a geometrical effect so, as for the case of the *geometrical blurring*, it is more important when the source material is not collimated. Therefore, we will focus our discussion on the deposition. In contrast to *blurring*, *shadowing* does not depend uniquely on the gap (G), but also on the thickness of the membrane. From that point of view, *shadowing* can be considered an issue for any type of lithography that uses a (thick) resist with a non-collimated process (e.g. lift-off).

Figure 8.8 shows a schematic of the shadowing effect with its two main consequences: reduction in thickness and displacement of the structures with respect to the center of the aperture.

The first effect relates to the part of the source that can send material to the substrate in an effective way. We can also call it the 'line of sight' from the substrate. It is also a purely geometrical effect, as the *geometric blurring*. As we can see in figure 8.8(a), the central part of a structure always receives more material than the borders. It is possible to calculate the width of the portion with maximum thickness, similar to what we did in equation (8.1), only that now we will not neglect the effect of the membrane thickness (t):

$$w_{\max} \approx w_{\text{aper}} - \frac{d}{D}(G + t). \tag{8.2}$$

What equation (8.2) shows is that there is a point where we will not find a region with the maximum nominal thickness. This is given by equation (8.3)

$$(G + t)_{\text{limit}} \approx w_{\text{aper}}\frac{D}{d}. \tag{8.3}$$

Interestingly, even though the gap (G) disappears, if the aspect ratio t/w is large enough, we will reach this limit point. Beyond that, if we continue to increase $(G + t)$, we will see an effective reduction of the source side, given by equation (8.4), and exemplified in figure 8.8(a).

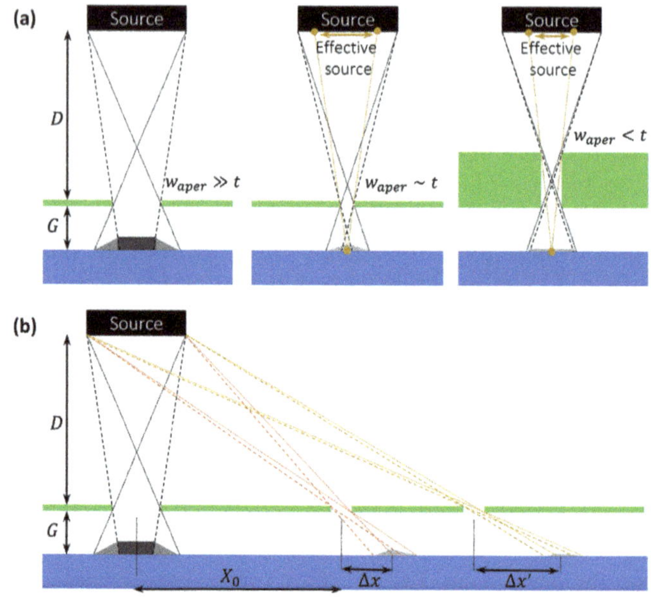

Figure 8.8. Shadowing effect. (a) Schematic showing the reduction in thickness due to the aspect ratio $\frac{w_{\text{aper}}}{G+t}$. It is easy to see that at a certain point the substrate receives material from a subset of the original source, which we call *effective source*. It is also important to note that even though the gap (G) goes to zero, this effect persists. (b) Schematic showing the effect of a deposition on the border of the wafer, far from the center of the material source. The patterned structures are displaced with respect to the center of the aperture. This effect is maximized on the border of the substrate wafer. In addition to this displacement on the center, a reduction of the thickness is observed when moving to the border of the wafer.

$$d_{\text{effective}} \approx \frac{w_{\text{aper}}}{G+t} D. \tag{8.4}$$

The second effect is that the deposited structure is not located at the same position as the aperture in the membrane (figure 8.8(b)). This displacement is approximately equal to

$$\Delta x \approx G \cdot \frac{X_0}{D}; \tag{8.5}$$

where X_0 is the radial distance of the aperture to the point of the substrate that is centered with respect to the material source. Considering typical values for the gap, distance, and wafer diameter (50 µm, 50 cm, and 100 mm, respectively), one can see that Δx can reach up to several microns. This is a big issue if we need to perform alignment within our process, since we would have to take this into account when performing the design.

In addition to this displacement of the center of the structure, we can see in figure 8.8(b) that the thickness reduces when the aperture is not centered with respect to the source. This is essentially the same as the one described before and shown in figure 8.8(a). Except that, when you are far from the center, the distance to the center, X_0 plays an important role. Indeed, we could write the limit value of said distance for which the maximum thickness is reached, in an analogous way to equation (8.3)

$$X_{0,\text{limit}} \approx D \cdot \frac{w_{\text{aper}}}{t} - \left(\frac{G}{t} + \frac{1}{2}\right)d. \tag{8.6}$$

And we can also calculate the effective diameter of the source that determines material being deposited on the substrate, beyond the limit case:

$$d_{\text{eff}} \approx \frac{w_{\text{aper}}D - tX_0}{G + t/2}. \tag{8.7}$$

From equations (8.6) and (8.7) we can see that this effect also exists when the gap (G) is zero and that moving away from the center or having large aspect ratio apertures are not beneficial for a proper pattern transfer.

8.3.3 Clogging

The third challenge for SL is the so-called *clogging*. This is an issue solely affecting *deposition* and it consists of the fact that the material being deposited *through* the membrane will also be deposited *on* the membrane and partially on the sidewalls of the apertures. It is an issue that affects mostly the patterning of nano-structures since it becomes important when the aperture width (w_{aper}) is of the same order as the thickness of the deposited material (t_{dep}) [28, 29].

Two consequences can be seen from the clogging of an SL membrane: the reduction of structure width along the deposition, which can even yield in some cases ($w_{\text{aper}} < t_{\text{dep}}$) that the structure does not reach the intended thickness, as we can see in figure 8.9. The second effect, a direct consequence of this latter point, is that clogging

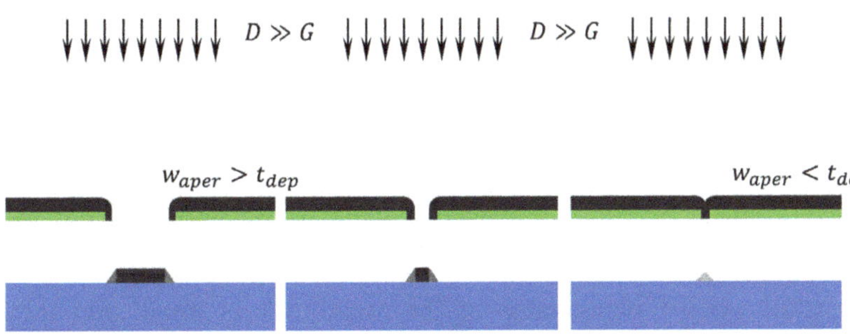

Figure 8.9. Clogging effect. Schematic showing the reduction in width along the thickness of the structure due to a reduction in the aperture width caused by the deposited material on the side walls. When the aperture width is very large compared to the thickness ($w_{\text{aper}} \gg t_{\text{dep}}$), this effect can be neglected. But when this ratio starts being large, we can even reach the point where the aperture closes completely, therefore ending up with structures that do not reach the maximum thickness overall. The different colors represent different regions of the deposited structure with different thickness. For simplicity, and better illustration, we show in this schematic the case where *blurring* and *shadowing* can be neglected, i.e. the case with a collimated beam of material arriving at the stencil membrane.

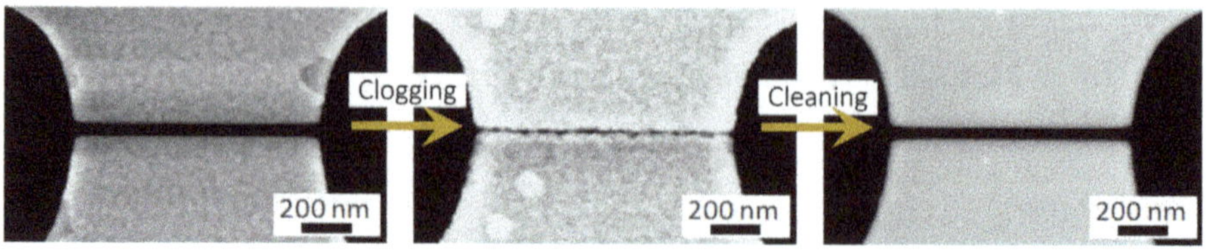

Figure 8.10. Clogging and cleaning. SEM micrographs of (left) a nanoslit in a stencil membrane after fabrication; (center) the same aperture after deposition of metal (100 nm of evaporated Al), showing almost complete clogging; (right) the same aperture after cleaning in a wet chemical etchant for Al. Images reprinted with permission from [32].

might render the stencil unusable, since the apertures are closed, and they cannot be used anymore.

The deposition of material on the sidewalls depends on the material being deposited and the deposition method. It has been seen, for example, that Al shows that the thickness deposited on the sidewalls is half that deposited on the flat substrate [29], but other materials behave differently. It also depends on the material, treatment, and state of the membrane. As two examples, surface treatment on the membrane [15] and high membrane temperature during deposition [30, 31], have both been shown to significantly reduce the membrane clogging.

The second issue arising from *clogging* that we mentioned is that the apertures are completely blocked, thus rendering the stencil unusable. In order to solve this, the easiest option is to perform wet etching of the covered membranes to remove the additional material. This has been proven for some metals [24, 29] as is shown in figure 8.10, and it is a useful and easy method to use whenever a suitable etching solution can be found.

8.3.4 Cracking and bending

The last challenge for SL that we will mention here is the *bending* and *cracking* of the membranes. This is inherent to SL as we use (thin) membranes to perform the patterning. It is important for all the different processes we can do with SL (figure 8.5—deposition, implantation and etching), however, the reasons are quite different.

In the case of *deposition*, the material that is deposited on top of the membrane causes a mechanical stress that can induce bending, buckling or even rupture of the membrane. As we mentioned before, in the fabrication section, these effects are more important when the thickness of the membrane is small, and they can be minimized by including mechanical reinforcements in the membrane, as for example corrugations. A clear example is shown in figure 8.11 [33], where we compare the deformation of a stencil membrane before and after deposition. The intrinsic stress in the metal has a much smaller effect if reinforcements are placed within the membrane.

In the case of *implantation*, the ions are much more energetic when arriving at the membrane than in the case of evaporation and thus, they penetrate the membrane material. This is an important point since the thickness of the stencil needs to be

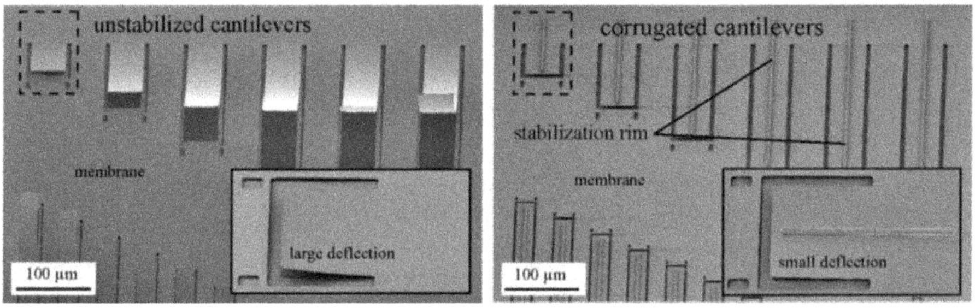

Figure 8.11. Membrane bending. SEM micrographs showing two stencil membranes after deposition of 50 nm of Cr on top. Since Cr is a metal with a large intrinsic stress, we can see the bending of the cantilever-like part of the flat membrane (left), whereas the structures in the corrugated membrane (right) remain flat. Images from [33] reprinted with permission from Elsevier.

large enough to *stop* the ions. The actual value of the thickness depends on (i) the ions being sent (size and mass) and (ii) the energy at which they are accelerated. With these two parameters it is possible to estimate via simulations what is the penetration depth within the membrane, and thus the minimum thickness required for the membrane to be functional as a mask for implantation. In any case, the ions penetrating the membrane cause significantly larger damage than in the case of deposition. This is one of the main reasons why SL has not been used for implantation very often, but only in some cases [27, 34, 35].

When performing *etching* through a stencil membrane, the situation is also much more aggressive than for deposition. In this case, the energy of the ions is not enough to penetrate the membrane material, but the etching process can thin the membrane rendering it more fragile. In fact, it is necessary to use a membrane material that is compatible with the type of etching that is going to be performed. In principle, we want a selectivity with the material on the membrane that is large enough so as to not have any etching. An example is the use of aluminum to coat the membrane as a preparation for its use to etch silicon in the substrate. Si etching (based on SF_6) does not etch aluminum (selectivity Al:Si is larger than $1:10^6$). Thus, with a protection layer of 50 nm Al it is possible to etch through a full Si wafer (525 μm) [36]. As in the case of the implantation, this process has not been used very frequently because it is much more aggressive than deposition [3, 36–39].

8.4 Design considerations

Following the discussion on challenges, we would like now to offer a synthesis and to elaborate in the different design considerations to obtain a successful patterning with SL.

8.4.1 Process as collimated as possible

As we have seen in the previous section, *blurring* and *clogging* are associated with geometrical effects that would minimize if the entities (ions, atoms …) arriving at the stencil membrane come as parallel as possible. To do this one can: (i) collimate the ions via electronic lenses; (ii) use a very small source size; and (iii) at a large distance from the substrate; and (iv) reduce the pressure of the process to increase the mean free path, thus avoiding unnecessary collisions between the atoms.

8.4.2 Thin but robust membranes

We have also shown that *shadowing* is increased by having thick membranes. In addition, fabrication of high-aspect ratio apertures is challenging. Therefore, smaller aspect ratios are preferred ($w \gg t$). However, thin membranes are more fragile, thus one might consider adding mechanical reinforcements and the use of a proper material. In addition, as explained above, in the case of implantation we need a minimum thickness of the membrane that is larger than the penetration depth of the ions into the membrane material. Otherwise the membrane does not stop the ions, i.e. no masking is performed.

8.4.3 Durable material (both mechanically and chemically)

In order to ensure a long lifetime of the stencil membranes, we have to choose a material that is durable/robust mechanically. Materials with high Young's modulus, maybe tensile stress, and non-brittle are ideal from the mechanical point of view. In addition, to ensure reusability of the membranes after different processes, the material should be chemically stable, i.e. not easy to etch via chemical methods. These points are the main reason why silicon nitride (low stress, silicon rich, silicon nitride to be precise) is the most used materials for stencil membranes.

8.4.4 Surface treatment

Chemical stability is beneficial to ensure reusability of stencils after deposition. If the membranes are used for etching, as we mentioned before, it is necessary to coat the membranes with a selective material in order to protect the membrane. The selection of this protective material depends on the type of dry etching to be performed. To reduce the effect of *clogging* we also mentioned that some surface coatings can be used, for example some self-assembled monolayers.

Another way to reduce *clogging* goes through the fabrication of more complex membranes, which include electrodes to enable heating during deposition. This temperature increase reduces *clogging*, but it makes fabrication (extra steps to place metallic heaters on the membranes) and implementation (electrical connection must be ensured during the deposition) more difficult.

8.4.5 Reducing gap

As we mentioned in the introduction, the major difference between patterning through a stencil membrane and through a layer of resist is that the membrane is not conformal to the substrate, but it has a gap (G) in between. We have seen how this is a parameter of fundamental importance for both *blurring* and *shadowing*. Thus, many people have devoted time and effort into solving this issue.

The simplest way to reduce the gap is to reduce the size of the stencil. The overall curvature of the stencil mask (wafer) yields a larger gap the larger the size of the

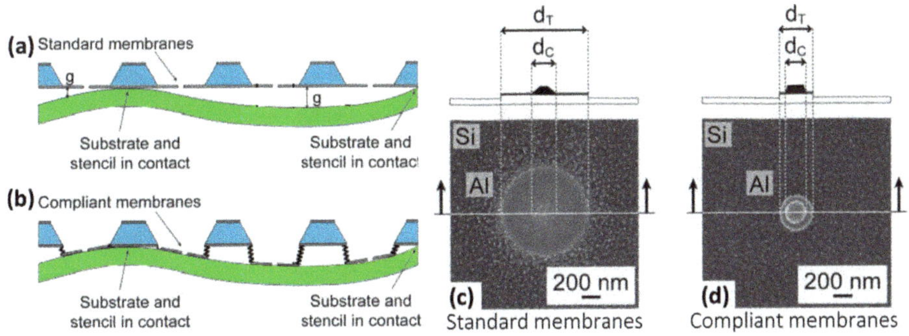

Figure 8.12. Compliant stencil. (a) and (b) Schematic drawings of a stencil mask on a substrate, showing the change in gap across the mask, for standard and compliant membranes, respectively. (c) and (d) SEM micrographs showing the results of depositing 50 nm Al on top of Si through standard or compliant membranes, and after a mild dry etching to reveal the size of the *halo*. The difference in the size of the central structure (d_c) and the total structure (d_T) is significant. Images adapted from [40] with permission of The Royal Society of Chemistry.

mask, and this goes roughly with the square of the mask dimensions. This means that moving from a 10 mm square chip to a 100 mm wafer, the gap might be up to 100 times larger. This is the reason why nanopatterning with SL at the full wafer scale has seldom been accomplished and most of the time it has been done using single chips. Importantly, here we refer to the size of the mask, not the size of the membranes. Clearly, this solution compromises a lot on the interest of the technique since it limits the size of the area to be patterned on the substrate.

If full-wafer patterning is desired, the compromise comes from somewhere else. One explored option follows from the mentioned solution for *clogging*: fabricating membranes with integrated heaters. Not only can a reduction in the clogging rate be observed, but also via heating the gap is reduced [30, 31]. However, as mentioned above, the implementation and fabrication become more complicated, partially losing the advantage of the technique.

Another possibility is to only complicate the fabrication, while remaining with a simple implementation. This solution [40, 41] consists of fabricating membranes that are recessed from the wafer-level, and attached to the wafer via mechanical (MEMS) springs. This way, when clamping together stencil mask and substrate, the recessed membranes will approach the substrate notably minimizing the gap, as shown schematically in figures 8.12(a) and (b). Figures 8.12(c) and (d) show the dramatic consequence of the use of these compliant stencils, with a substantial reduction in the *halo* as well as a much better defined central structure.

8.5 Alignment

Any nano-lithography process must address two major points to become relevant. The first is the resolution, which we have already widely analyzed in the previous sections, and then is the alignment. This is a clear challenge mostly for novel lithography methods that do not have the history of, e.g. optical or electron beam lithography.

Figure 8.13. Alignment in SL. (a) Commercial transfer chuck for a bonding aligner. The chuck is used to clamp both wafers (substrate and stencil) and then transfer them into the processing equipment. (b) Detail of one of the clamps holding both wafers. (c) SEM micrograph showing the alignment of two depositions. First, electrode contacts in Ti/Pt are deposited, and subsequently an aluminum layer is deposited contacting the Pt below. Image from [10] with permission from American Chemical Society.

In the case of SL, alignment needs to be performed between two solid pieces that need to stay together during the processing. The most comfortable solution to perform this alignment is using bonding aligners. These machines are specifically made to align two wafers and to clamp them, allowing the pair then to be transported to the bonding equipment. Figures 8.13(a) and (b) show pictures of an example of a clamping tool used to perform bond alignment of two full 100 mm wafers. Figure 8.13(c) shows an SEM micrograph of an example of alignment, where a nanowire-like structure in aluminum was deposited on some pre-defined platinum contacts. Alignment accuracy is better than 2 µm in this case.

In order to enable this alignment, one needs to prepare membranes that contain the alignment marks, to allow the microscope to see through. The accuracy of the alignment depends on the machine being used and can be better than 500 nm. The main issue with this process is that, after alignment, the full assembly (wafers and clamping device) must be introduced into the processing equipment. This is doable for deposition machines (in particular evaporators), but for implantation and etching it is more difficult. It is rare to find commercial implanters and etchers that allow us to introduce this type of assembly in the processing chamber. This is the reason why alignment at full-wafer scale has been only done with evaporation [10].

Alignment at the chip level can be performed using dye bonder machines. The main point is the clamping mechanism which, in some cases, must be made manually. With lower resolution, one can also perform manual alignment using a microscope. However, achieving a better resolution than 10 µm is rather difficult manually. Still, this has been used to demonstrate complex integration of systems, even fabricating transistors fully by SL (implantation, etching and metallization [42]).

8.6 Dynamic stencil

One of the most unique and attractive features of SL is the fact that the movement of the mask with respect to the substrate is allowed during the deposition. This, in turn, enables the fabrication of any shape *in situ*, and without the need to break vacuum. In this sense, it is similar to what one can do with focused ion/electron beam induced deposition (FIBID/FEBID), but in this case the patterning can happen at full wafer scale and at many places simultaneously. However, in order to perform this, it is required to add the functionality of moving the stencil mask with respect to the substrate in vacuum. This, as is the case for the alignment that we discussed before, is exceedingly difficult to perform for etching and implantation, and it has only been done for deposition. Furthermore, only a handful of groups in the world have been able to do dynamic patterning with SL.

First, some groups have performed it at the chip level, using a combination of SL and scanning probe microscopy (SPM), where the cantilever played the role of solid mask, being patterned with nano-apertures. To do this, they modified setups for SPM in ultra-high vacuum (UHV) to also enable deposition. One of the advantages is that it is possible to image the dimensions of the deposited patterns *in situ* [43–47].

For full-wafer dynamic stencil, one approach can be seen in figure 8.14. Three motors are connected to the stencil mask in order to enable movement in all three axes, and they are placed inside an evaporator. Both dynamic (motion while deposition is happening, see video 8.3 available at https://iopscience.iop.org/book/978-0-7503-2608-7) and quasi-dynamic (deposition happens when the stencil is static, and motion when the deposition is stopped, see video 8.2 available at https://iopscience.iop.org/book/978-0-7503-2608-7) SL can be performed [48]. One of the challenges is to perform alignment with former structures on the wafer or with the axis of the motors. A solution could be to include optical microscopes inside the chamber, or to use nanoelectromechanical sensors (NEMS) to detect where deposition is happening [49].

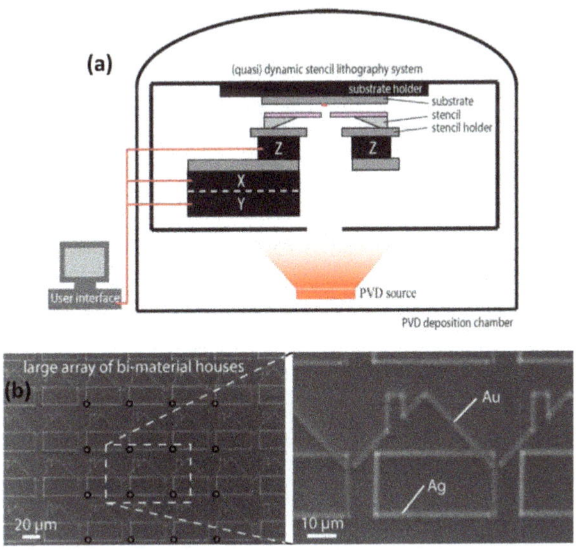

Figure 8.14. Dynamic SL. (a) Schematic of the concept for a full wafer dynamic (and quasi-dynamic) stencil. Three different motors are connected to the stencil mask and are controlled by an external computer. The full setup is inserted in an evaporation chamber to enable patterning of the substrate. (b) SEM micrographs showing the drawing of (i) a rectangle in silver and (ii) the roof of a house in gold, without breaking vacuum. Images adapted from [48] with permission of AVS.

8.7 Applications

SL is a very promising technique. It offers a different way to perform nano- and micropatterning that in some cases enables certain applications that otherwise would not be possible. We do not wish to make here an exhaustive list of the applications for which SL has been used, there are other works where one can follow that [1], but we just aim to give a short idea.

Since SL does not require any resist, people have used it to deposit sensitive materials like organic conductive molecules [50–53], magnetic alloys [54], complex oxides [55, 56], and fullerenes [44]. In addition, SL has been used on delicate and unconventional substrates such as non-planar substrates [36, 42, 57], CMOS devices [58], SAMs [59], and organic layers [2, 60–62].

SL has been used to fabricate periodic structures on the back reflectors of a-Si solar cells: to improve their quantum efficiency [63–65]; to pattern electrical contacts on 2D materials to keep them cleaner [66, 67]; to fabricate transistors on non-planar substrates [42]; to fabricate infrared plasmonic nanostructure antenna arrays such as nanodots [68]; to realize metal nanopatterns to promote binding of biomaterials [69]; to fabricate pentacene thin film transistors (TFT) [53, 70]; to fabricate NEMS resonators integrated with CMOS circuitry [58]; and many more that we will not mention but we refer the interested reader elsewhere [1].

8.8 End of chapter summary

In this chapter we have covered the functioning principles of SL, its major challenges and advantages. We have shown how this ancient method for surface patterning has reemerged and it has proven to be very reliable for nanopatterning a wide variety of materials and on different substrates.

We have covered the different processes that can be performed, enumerated some of the major applications, described how to fabricate a stencil membrane, ... But, above all, we have put special emphasis on the different points that need to be checked in order for a successful implementation of SL, in order to minimize the impact of each of the four main challenges present in SL. As such, this chapter can be taken as an initiation guide to SL.

8.9 Student's corner

- If you want to deposit 100 nm structures, and you can afford ±10 nm, would you use sputtering, e-beam evaporation, or thermal evaporation? Why?
- Considering a distance source–substrate of 75 cm and a typical source size of 2 mm, how much should the gap be to ensure the requirements above?
- If these 100 nm structures are supposed to be 200 nm thick, what are the main issues that we would face, and why?
- How would you pattern with SL a ring of 10 μm in diameter and 200 nm in width, plus 20 nm thick?

Acknowledgements

We thank Professor J Brugger for valuable contributions to this chapter, as well as the staff at CMi (Center for Micro and Nano technology at EPFL) for their assistance with some of the pictures.

References

[1] Vazquez-Mena O, Gross L, Xie S, Villanueva L G and Brugger J 2015 Resistless nanofabrication by stencil lithography: a review *Microelectron. Eng.* **132** 236–54
[2] Vazquez-Mena O, Sannomiya T, Tosun M, Villanueva L G, Savu V and Voros J *et al* 2012 High-resolution resistless nanopatterning on polymer and flexible substrates for plasmonic biosensing using stencil masks *ACS Nano* **6** 5474–81
[3] Alix E C 1969 A convenient stencil for glass etching *J. Chem. Educ.* **46** 863
[4] Ingle F W 1974 A shadow mask for sputtered films *Rev. Sci. Instrum.* **45** 1460–1
[5] Dunkleberger L N 1978 Stencil technique for the preparation of thin-film Josephson devices *J. Vac. Sci. Technol.* **15** 88–90
[6] Nguyen H B 1980 A proposed slotted mask for direct deposition of metal contact pattern on MIS solar cells *IEEE Trans. Electron Devices* **27** 1303–4
[7] Maah S and Van Duzer T 1982 Silicon shadow masks for fine-feature thin-film deposits *IEEE Electron Device Lett.* **3** 21–3
[8] Burger G J, Smulders E J T, Berenschot J W, Lammerink T S J, Fluitman J H J and Imai S 1996 High-resolution shadow-mask patterning in deep holes and its application to an electrical wafer feed-through *Sens. Actuators A Phys.* **54** 669–73
[9] Kim G M, van den Boogaart M A F and Brugger J 2003 Fabrication and application of a full wafer size micro/nanostencil for multiple length-scale surface patterning *Microelectron. Eng.* **67-68** 609–14
[10] Vazquez-Mena O, Villanueva G, Savu V, Sidler K, van den Boogaart M A F and Brugger J 2008 Metallic nanowires by full wafer stencil lithography *Nano Lett.* **8** 3675–82
[11] Vazquez-Mena O, Sannomiya T, Villanueva L G, Voros J and Brugger J 2011 Metallic nanodot arrays by stencil lithography for plasmonic biosensing applications *ACS Nano* **5** 844–53
[12] van den Boogaart M A F, Doeswijk L M and Brugger J 2006 Silicon-supported membranes for improved large-area and high-density micro/nanostencil lithography *J. Microelectromech. Syst.* **15** 1663–70
[13] Hulteen J C and Vanduyne R P 1995 Nanosphere lithography: a materials general fabrication process for periodic particle array surfaces *J. Vac. Sci. Technol.* A **13** 1553–8
[14] Chou S Y, Krauss P R and Renstrom P J 1996 Imprint lithography with 25-nanometer resolution *Science* **272** 85–7
[15] Kolbel M, Tjerkstra R W, Brugger J, van Rijn C J M, Nijdam W and Huskens J *et al* 2002 Shadow-mask evaporation through monolayer-modified nanostencils *Nano Lett.* **2** 1339–43
[16] Kölbel M, Tjerkstra R W, Kim G, Brugger J, van Rijn C J M and Nijdam W *et al* 2003 Self-assembled monolayer coatings on nanostencils for the reduction of materials adhesion *Adv. Funct. Mater.* **13** 219–24
[17] Guan D, Bruccoleri A R, Heilmann R K and Schattenburg M L 2014 Stress control of plasma enhanced chemical vapor deposited silicon oxide film from tetraethoxysilane *J. Micromech. Microeng.* **24** 027001

[18] Iliescu C, Wei J S, Chen B T, Ong P L and Tay F E H 2007 *Proc. SPIE* **6415** 64150L
[19] Butschke J, Ehrmann A, Hofflinger B, Irmscher M, Kasmaier R and Letzkus F *et al* 1999 SOI wafer flow process for stencil mask fabrication *Microelectron. Eng.* **46** 473–6
[20] Mauger P E, Shimkunas A R, Wolfe J C, Sen S, Loschner H and Stengl G 1992 Silicon stencil masks for lithography below 0.25 Mu-M by ion-projection exposure *J. Vac. Sci. Technol.* B **10** 2819–23
[21] Eguchi H, Kurosu T, Yoshii T, Sugimura H, Itoh K and Tamura A 2003 Low stress stencil masks using SOI substrates for EPL and LEEPL *Proc. SPIE* **5130** 879–87
[22] Yun H, Kim S, Kim H, Lee J, McAllister K and Kim J *et al* 2015 Stencil nano lithography based on a nanoscale polymer shadow mask: towards organic nanoelectronics *Sci Rep.* **5** 10220
[23] Xing X, Yang J M, Tai Y C and Ho C M 1999 Micromachined membrane particle filters *Sens. Actuators A Phys.* **73** 184–91
[24] Tun T N, Lwin M, Kim A H H, Chandrasekhar A N and Joachim C 2007 Wetting studies on Au nanowires deposited through nanostencil masks *Nanotechnology* **18** 335301
[25] Vazquez-Mena O, Villanueva L G, Savu V, Sidler K, Langlet P and Brugger J 2009 Analysis of the blurring in stencil lithography *Nanotechnology* **20** 415303
[26] Vazquez-Mena O, Sidler K, Savu V, Park C-W, Villanueva L G and Brugger J 2011 Reliable and improved nanoscale stencil lithography by membrane stabilization, blurring, and clogging corrections *IEEE Trans. Nanotechnol.* **10** 352–7
[27] Villanueva L G, Martin-Olmos C, Vazquez-Mena O, Montserrat J, Langlet P and Bausells J *et al* 2011 Localized ion implantation through micro/nanostencil masks *IEEE Trans. Nanotechnol.* **10** 940–6
[28] Lishchynska M, Bourenkov V, van den Boogaart M A F, Doeswijk L, Brugger J and Greer J C 2007 Predicting mask distortion, clogging and pattern transfer for stencil lithography *Microelectron. Eng.* **84** 42–53
[29] Vazquez-Mena O, Villanueva G, van den Boogaart M A F, Savu V and Brugger J 2008 Reusability of nanostencils for the patterning of aluminum nanostructures by selective wet etching *Microelectron. Eng.* **85** 1237–40
[30] Xie S, Savu A V and Brugger J 2012 In-situ actuated gap reduction and clogging-free apertures for quasi-dynamic stencil lithography *The 7th Annual IEEE Int. Conf. on Nanol Micro Engineered and Molecular Systems (Kyoto, Japan, 5–8 March 2012)*
[31] Xie S Q, Savu V and Brugger J 2011 Heated membranes prevent clogging of apertures in nanostencil lithography *TRANSDUCERS 2011 16th Int. Solid-State Sensors, Actuators and Microsystems Conf.* pp 998–1001
[32] Vazquez Mena O 2010 Development of stencil lithography for nanopatterning and for electronic and biosensing applications *Thesis* EPFL
[33] van den Boogaart M A F, Lishchynska M, Doeswijk L M, Greer J C and Brugger J 2006 Corrugated membranes for improved pattern definition with micro/nanostencil lithography *Sens. Actuators* A **130-131** 568–74
[34] Villanueva G, Martin C, Vazquez-Mena O, Montserrat J, Langlet P and Bausells J *et al* 2008 Resistless ion implantation of sub-micron scale features through nano-stencil *MNE'08 (Athens, Greece)*
[35] Dumas C, Grisolia J, Ressier L, Arbouet A, Paillard V and Ben Assayag G *et al* 2007 Synthesis of localized 2D-layers of silicon nanoparticles embedded in a SiO_2 layer by a stencil-masked ultra-low energy ion implantation process *Phys. Status Solidi (a)* **204** 487–91

[36] Villanueva G, Vazquez-Mena O, Hibert C and Brugger J 2009 Direct etching of high aspect ratio structures through a stencil *2009 IEEE 22nd Int. Conf. Micro Electro Mechanical Systems MEMS* pp 144–7
[37] Viallet B, Grisolia J, Ressier L, Van Den Boogaart M A F, Brugger J and Lebraud T 2008 Stencil-assisted reactive ion etching for micro and nano patterning *Microelectron. Eng.* **85** 1705–8
[38] Villanueva G, Vazquez-Mena O, van den Boogaart M A F, Sidler K, Pataky K and Savu V *et al* 2008 Etching of sub-micrometer structures through stencil *Microelectron. Eng.* **85** 1010–4
[39] Pang S W, Geis M W, Goodhue W D, Efremow N N, Ehrlich D J and Goodman R B *et al* 1988 Pattern transfer by dry etching through stencil masks *J. Vac. Sci. Technol.* B **6** 249–52
[40] Sidler K, Villanueva L G, Vazquez-Mena O, Savu V and Brugger J 2012 Compliant membranes improve resolution in full-wafer micro/nanostencil lithography *Nanoscale* **4** 773–8
[41] Sidler Arnet K 2010 Organic pentacene thin film transistors on flexible substrates fabricated by stencil lithography *Thesis* EPFL
[42] Villanueva L G, Vazquez-Mena O, Martin-Olmos C, Savu V, Sidler K and Montserrat J *et al* 2012 All-stencil transistor fabrication on 3D silicon substrates *J. Micromech. Microeng.* **22** 095022
[43] Luthi R, Schlittler R R, Brugger J, Vettiger P, Welland M E and Gimzewski J K 1999 Parallel nanodevice fabrication using a combination of shadow mask and scanning probe methods *Appl. Phys. Lett.* **75** 1314–6
[44] Zahl P, Bammerlin M, Meyer G and Schlittler R R 2005 All-in-one static and dynamic nanostencil atomic force microscopy/scanning tunneling microscopy system *Rev. Sci. Instr.* **76** 023707
[45] Egger S, Ilie A, Fu Y, Chongsathien J, Kang D-J and Welland M E 2005 Dynamic shadow mask technique: a universal tool for nanoscience *Nano Lett.* **5** 15–20
[46] Guo H, Martrou D, Zambelli T, Dujardin E and Gauthier S 2008 Development of UHV dynamic nanostencil for surface patterning *Rev Sci Instrum.* **79** 103904
[47] Guo H, Martrou D, Zambelli T, Polesel-Maris J, Piednoir A and Dujardin E *et al* 2007 Nanostenciling for fabrication and interconnection of nanopatterns and microelectrodes *Appl. Phys Lett.* **90** 093113
[48] Savu V, van den Boogaart M A F, Brugger J, Arcamone J, Sansa M and Perez-Murano F 2008 Dynamic stencil lithography on full wafer scale *J. Vac. Sci. Technol.* B **26** 2054–8
[49] Arcamone J, Sansa M, Verd J, Uranga A, Abadal G and Barniol N *et al* 2009 Nanomechanical mass sensor for spatially resolved ultrasensitive monitoring of deposition rates in stencil lithography *Small* **5** 176–80
[50] Egger S, Higuchi S and Nakayama T 2006 A method for combinatorial fabrication and characterization of organic/inorganic thin film devices in UHV *J. Comb. Chem.* **8** 275–9
[51] Egger S, Ilie A, Machida S and Nakayama T 2007 Integration of individual nanoscale structures into devices using dynamic nanostenciling *Nano Lett.* **7** 3399–404
[52] Lee J W, Ju B K, Jang J, Yoon Y S and Kim J K 2007 High mobility organic transistor patterned by the shadow-mask with all structure on a plastic substrate *J. Mater. Sci.* **42** 1026–30
[53] Sidler K, Cvetkovic N V, Savu V, Tsamados D, Ionescu A M and Brugger J 2010 Organic thin film transistors on flexible polyimide substrates fabricated by full-wafer stencil lithography *Procedia Chem.* **162** 155–9

[54] Rodriguez A F, Heyderman L J, Nolting F, Hoffmann A, Pearson J E and Doeswijk L M *et al* 2006 Permalloy thin films exchange coupled to arrays of cobalt islands *Appl. Phys Lett.* **89** 142508

[55] Cojocaru C-V, Harnagea C, Pignolet A and Rosei F 2006 Patterning of functional materials by pulsed laser deposition through nanostencils *2006 IEEE Conf. on Emerging Technologies (10–13 January 2006)*

[56] Cojocaru C-V, Harnagea C, Rosei F, Pignolet A, van den Boogaart M A F and Brugger J 2005 Complex oxide nanostructures by pulsed laser deposition through nanostencils *Appl. Phys. Lett.* **86** 183107

[57] Brugger J, Berenschot J W, Kuiper S, Nijdam W, Otter B and Elwenspoek M 2000 Resistless patterning of sub-micron structures by evaporation through nanostencils *Microelectron. Eng.* **53** 403–5

[58] Arcamone J, Boogaart M A F, Serra-Graells F, Fraxedas J, Brugger J and Perez-Murano F 2008 Full-wafer fabrication by nanostencil lithography of micro/nanomechanical mass sensors monolithically integrated with CMOS *Nanotechnology* **19** 305302

[59] Speets E A, Riele P T, van den Boogaart M A F, Doeswijk L M, Ravoo B J and Rijnders G *et al* 2006 Formation of metal nano- and micropatterns on self-assembled monolayers by pulsed laser deposition through nanostencils and electroless deposition *Adv. Funct. Mater.* **16** 1337–42

[60] Zhou Y X, Johnson A T, Hone J and Smith W F 2003 Simple fabrication of molecular circuits by shadow mask evaporation *Nano Lett.* **3** 1371–4

[61] Sidler K, Cvetkovic N V, Savu V, Tsamados D, Ionescu A M and Brugger J 2009 Organic thin film transistors on flexible polyimide substrates fabricated by full wafer stencil lithography *XXIII Eurosensors Conf., Lausanne, Switzerland. Proc. Eurosensors XXIII Conf. 2009*

[62] Sidler K, Vazquez-Mena O, Savu V, Villanueva G, van den Boogaart M A F and Brugger J 2008 Resistivity measurements of gold wires fabricated by stencil lithography on flexible polymer substrates *Microelectron. Eng.* **85** 1108–11

[63] Pahud C, Savu V, Klein M, Vazquez-Mena O, Haug F-J and Brugger J *et al* 2013 Stencil-nanopatterned back reflectors for thin-film amorphous silicon n-i-p solar cells *IEEE J. Photovolt.* **3** 22–6

[64] Pahud C, Savu V, Klein M, Vazquez-Mena O, Soederstroem K and Haug F-J *et al* 2012 Stencil-nanopatterned back reflectors for thin-film amorphous silicon n-i-p solar cells *2012 38th IEEE Photovoltaic Specialists Conf. (PVSC)*

[65] Pahud C, Isabella O, Naqavi A, Haug F J, Zeman M and Herzig H P *et al* 2013 Plasmonic silicon solar cells: impact of material quality and geometry *Opt. Express* **21** A786–97

[66] Novoselov K S, Geim A K, Morozov S V, Jiang D, Zhang Y and Dubonos S V *et al* 2004 Electric field effect in atomically thin carbon films *Science* **306** 666–9

[67] Novoselov K S, Geim A K, Morozov S V, Jiang D, Katsnelson M I and Grigorieva I V *et al* 2005 Two-dimensional gas of massless Dirac fermions in graphene *Nature* **438** 197–200

[68] Aksu S, Yanik A A, Adato R, Artar A, Huang M and Altug H 2010 High-throughput nanofabrication of infrared plasmonic nanoantenna arrays for vibrational nanospectroscopy *Nano Lett.* **10** 2511–8

[69] Pataky K, Vazquez-Mena O and Brugger J 2009 Nanostencil and ink jet printing for bionanotechnology applications *Nano-Net* ed A Schmid, S Goel, W Wang, V Beiu and

S Carrara (Lecture Notes of the Institute for Computer Sciences Social Informatics and Telecommunications Engineering vol 20) (Berlin: Springer) pp 222–8
[70] Sidler K, Cvetkovic N V, Savu V, Tsamados D, Ionescu A M and Brugger J 2009 Organic thin film transistors on flexible polyimide substrates fabricated by full wafer stencil lithography *Procedia Chem.* **1** 762–5

Chapter 9

Ice lithography

Anpan Han, Ding Zhao and Min Qiu

We present ice lithography (IL), which is electron beam lithography (EBL) using ices or condensed gases as resists. Examples of ices are water-, alcohol-, and diesel ice. Water ice can be used as a positive tone resist, while the organic ices serve as negative tone resists. Instead of spin-coating of resist on mirror polished wafers, the ice resists can be condensed onto cold surfaces with complicated geometries. This is demonstrated by patterning on suspended carbon nanotubes. In the future, IL is promising for 3D printing of functional materials such as metals and semiconductors, and entirely new applications might emerge.

9.1 Introduction and historical perspective

This book chapter is organized into five sections. The first section is a brief introduction to the organization of the chapter, and how the different sections are organized to give the reader the best possible experience. In the second section, to understand the mechanisms of ice lithography and how to make nanostructures using ices and electrons, we start with refreshing three important and key fundamental topics; the energetic electron, condensing gases inside vacuum, and electron–matter interactions. We will introduce the reader to concepts and terminology that are necessary and facilitate the following chapters. In the third section, we present ice lithography, which was invented by Gavin King, Gregor Schurman, Jene Golovchenko and Dan Branton in 2005 [1]. Here, we present instrumentation, materials and processes. Section 9.4 focuses on published applications of ice lithography, which contains the main results obtained by this emerging technology. Finally, we discuss future research challenges and opportunities, which include fundamental research on electron–matter interactions, applications in lithography and additive manufacturing (3D printing), and we introduce the new concept of Icetronics.

For historic perspectives, it is up to debate when to start; nevertheless, we think it should start in 1956 when Dr Steere at UC Berkeley built an apparatus to freeze and fracture biological specimens for electron microscopy studies [2]. After a post-doc at

doi:10.1088/978-0-7503-2608-7ch9

ETH Zurich, Dan Branton, returned to the US and used the refined Swiss freeze fracturing method to conclude that cell membranes were lipid bilayers, which later led to the discovery of membrane proteins, and the development of the 'fluid mosaic model' of the cell membrane in 1972 [3]. In 1990, the Human Genome Project was launched, and it had the aim and goal to determine human DNA. Dan Branton and Jene Golovchenko joined forces to research and develop nanopore sequencing, which is a commercial success today [4]. During their push for nanopore sequencing in 2005, they needed a pristine nanolithography method for building nanopore sequencing devices. Because Jene was involved in proton–ice interactions earlier at Bell labs, and Dan worked with freeze fracturing which involved freezing samples at liquid nitrogen temperatures and electron microscopy, they had the enabling knowledge, and agreed to push for ice lithography. Anpan Han joined the Nanopore group to build the first dedicated ice lithography instrument [5]. Min Qiu and Ding Zhao became involved in ice lithography in 2012. The first organic ice work was reported in 2017 [6], and 3D metallic structures were reported in 2018 [7]. Our research is still in its infancy, and we see great potential that we share in this book chapter.

9.2 Fundamentals

9.2.1 Energetic electrons

The first ingredient in ice lithography is the electrons. The electron is a fundamental particle, and evidently the literature concerning the electron is colossal. In this section, we will present its importance in two areas that are relevant for ice lithography. First, its role in chemistry will be presented. Second, its role in scanning electron microscopy (SEM) for nanometer imaging and electron beam lithography will be summarized.

First, in the Bohr model of the hydrogen atom, the positive charged proton nucleus is in the center of the atom, and the negatively charged electron forms a shell around the nucleus. If the electron is removed from the hydrogen atom, the remaining positively charged nucleus is referred to as an ion, which is chemically unstable, and it will react with other atoms or molecules to reduce its chemical energy. Two hydrogen atoms can form the more energetically favorable hydrogen molecule. Here, the covalent bond between the two atoms consists of the two electrons that form an electron pair (Lewis pair). The Lewis pair form a common shell around the two nuclei that is separated by a distance defined as the bond length. If one electron is removed from the hydrogen molecule, there is an unpaired valence electron, and a dihydrogen cation is formed. In chemistry, these energetic dihydrogen cations with unpaired valence electron are also referred to as radicals. The same picture can be translated to most molecules.

In the next sections, free electrons will be interacting with matter, which will cause chemical reactions. In other words, the free electrons will interact with the electrons bound to the nucleus. We choose to approach chemical reactions by considering and focusing on the energy exchange. Breaking a chemical covalent bond, i.e. removing an atom from a molecule requires energy. The bond-dissociation energy (BDE) is the energy needed for breaking chemical bonds, i.e. removing electrons from the Lewis pair, which is also known as radicalization. The energy is between 1 and 12 eV per bond. The BDE for C–C and C–H bond in alkanes is about 4 eV. Nitrogen bonds are 9.79 eV, and fluorine bonds are 1.63 eV.

While breaking chemical bonds happens in physical processes and provides the upper limits of required energy, one central subset of chemistry is the science of electron exchange and sharing between atoms and molecules. Examples are reduction–oxidation (redox) reactions in which the oxidation states of atoms are changed. In a redox reaction, the molecule that loses an electron is oxidized, and the molecule that gains an electron is reduced. Note that oxygen does not have to be present in a redox reaction, but electrons are essential. The energy exchange in redox reactions is in the order of 100 kJ mol^{-1} or 1 eV per molecule. Some redox reactions are spontaneous, and others require the addition of energy. For reduction to take place, electrons do not have to come from another molecule, and they can be delivered by an external electron source. This leads to electro-chemistry, and here the electrons required for redox reactions are directly injected or removed by an external source. The energies that are involved in electrochemical cells are also in the order of 1 eV. For galvanic Daniell cells, the produced energy is 213 kJ mol^{-1} or 2.1 eV per molecule. In conclusion, the energies required for many chemical reactions are in the order of 1–10 eV.

Second, electrons play a central role in transmission electron microscopy (TEM) and SEM. For imaging applications, please refer to Williams and Carter [8], Goldstein *et al* [9], and other excellent textbooks in electron microscopy. In brief, inside the SEM vacuum chamber, free electrons are extracted from point electron sources and an external electric field is applied. Provided by the applied field, the electron energy can be adjusted from 1 keV to 30 keV. Using electrostatic or electromagnetic lenses, the electrons are tightly focused into an electron beam (e-beam). Modern SEMs create beams with diameters less than 1 nm. The e-beam current ranges between a few pA to mA. The SEM image is formed by scanning the e-beam across the sample. When the e-beam interacts with the sample, secondary electrons (SE) are ejected from the sample and collected by a detector. The signal is transduced into an electrical signal, digitized, and displayed on a computer screen. The displayed signal is correlated with the number of ejected SE, such that a brighter signal corresponds to a larger number of SE. Each pixel on the screen corresponds to the scanning e-beam hitting a spot on the sample.

As in optical lithography, in EBL the electron microscope is adapted for lithography applications (see chapter 3 in the present book). For a book on EBL please refer to [10], chapter 3. Regarding instrumentation, for most research applications, Gaussian EBL systems are used. The electron optics are very similar to SEMs, and many EBL systems are adapted SEMs. Some SEM companies offer EBL as an optional module. The EBL option consists of an electrostatic e-beam blanker which switches the beam on-off quickly, a pattern generator which takes control of the scanning coils in the SEM to position the e-beam and draws user-designed patterns, a picoampere meter that measures the beam current, and lithography software.

9.2.2 Ices, condensed gases in vacuum

The second ingredient for ice lithography is ice. We define ices as solids of condensed gases. Water ice is an example of ice. Other notable ices are CO_2 ice, which is also known as dry ice, ammonia and methane ices, which are found on cold planets and asteroids. A large group of matter that are gases or liquids at ambient conditions are solids in vacuum and low temperatures. Many organic molecules, as we normally know them, are gases or liquids. Examples are octane which is petrol and isopropanol which is used for, e.g. cleaning. Theoretically, most organic gases and liquids condense to solids or ices at low temperature in vacuum. The condensation condition of matter is given by its phase diagram (figure 9.1). Under closed equilibrium conditions, depending on the temperature and pressure, matter can be in four different phases; solid, gas, liquid or supercritical fluid. To change phases the materials often have to pass the 'lines of equilibrium'. At certain conditions two or three phases coexist. For example, at the triple point the material takes the forms of gas, liquid and solid, and all these phases are in equilibrium. At the lines of equilibrium, the matter contains two phases. An example is the liquid–air interface of water; here the water vapor (gas) is in equilibrium with its liquid phase. At the surface of solids, the solid phase is in equilibrium with the gas phase. The vapor pressure (VP) is given by the semi-empirical Antoine equation, which is derived from the Clausius–Clapeyron relation.

$$\log_{10}p(T) = A - \frac{B}{C + T} \tag{9.1}$$

p is VP, A, B, C are material specific constants, T is temperature.

Moving away from the closed system, and into a dynamic open system, the rate of phase change becomes important. The sublimation rate from a planar surface of water is derived by Estermann [11].

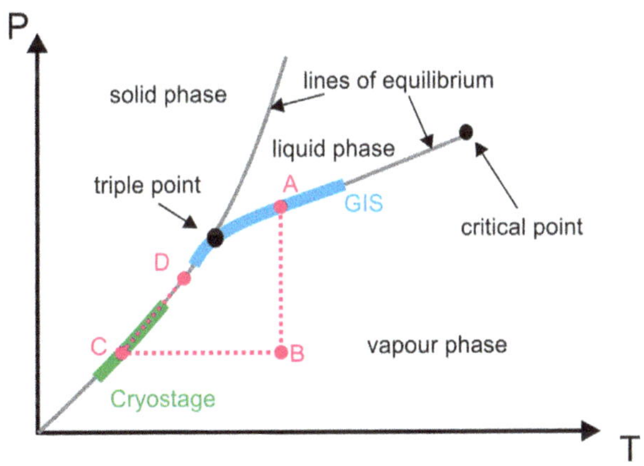

Figure 9.1. Phase diagram illustration of a selected material for ice lithography. Depending on the pressure (P) and temperature (T) the material can exist in three different phases. At the lines of equilibrium, the material exists in two different phases. Inside the GIS, the vapor phase of the material is in equilibrium with the solid or liquid phase. The GIS temperature can be adjusted such that the vapor pressure can be adapted, and the blue line highlights the material phases in the GIS environment. The magenta dashed line and the four letters illustrate the ice condensation process. The material vapor in the GIS (A) is injected to the low-pressure vacuum chamber (B), and the vapor is condensed on the cold sample on the cryostage (C). The material vapor can be sublimated by increasing the cryostage temperature (D). The green line shows the material phases on the cryostage.

$$S = p(T)\left(\frac{M_w}{2\pi RT}\right)^{1/2} \tag{9.2}$$

S is the sublimation mass flux and the unit is given in kg m^{-2} s^{-1}, M_w is the molecular weight, R is the universal gas constant. Equation (9.2) states the sublimation rate is proportional to the VP at equilibrium, and less dependent on the molecular weight. For water ice, at 70 K, the sublimation rate is 1 molecule cm^{-2} h^{-1} [11], and hence under the appropriate conditions, ice is very stable.

The phase diagram, and equations (9.1) and (9.2) guide our selection of materials for ice lithography applications. We follow the phases and phase changes of the selected material during the ice deposition process (figure 9.1); first the selected material is loaded into the gas injection system (GIS, see section 9.3 for process and instrumentation), and inside the GIS the vapor phase is in equilibrium with the solid or liquid phase of the material. Second, the vapor is injected into the SEM vacuum, and for a fraction of a second, the material is only in the vapor phase. Third, the material is then condensed into a solid that is in equilibrium with its vapor phase at low temperature. Finally, we can heat the cryostage to sublimate the ice. Following this process, a suitable ice must have two desired properties. First, at room temperature (298 K), the material must have a suitable VP which could be practically controlled by the GIS. The GIS temperature could be adjusted to control the VP of the selected material. Second, the ice must be stable at cryogenic temperature in vacuum (10^{-6} Torr in the SEM chamber), which requires its VP must be sufficiently low such that the sublimation is small. The low VP is also required to maintain a stable SEM vacuum for subsequent e-beam processing. The cryostage temperature could be adjusted for freezing ice.

In table 9.1, we present selected materials that have and have not the two desired properties. We give the extrapolated VP values at 60, 150 and 298 K. 298 K is room temperature, 150 K is easily obtained by liquid nitrogen (LN2) cooled cryostages, and the more expensive helium based cryotechnology is needed to cool samples below 60 K. A selection of the materials in table 9.1 has already been used in our publications, while others are being evaluated or considered for future studies. Specific Antoine constants of materials are from www.webbook.nist.gov, however, the constants are not measured at low temperatures, and hence the extrapolated VPs are rough estimates and there might even be large deviations. We use water as a benchmark material. The water VP at room temperature is 23.5 Torr, and VP of hydrated salts are a few Torr, which work nicely with GIS. Water VP at 150 K is 5×10^{-8} Torr, which is much smaller than the SEM vacuum pressure. Water VP at 60 K is extrapolated to 2×10^{-57} Torr, which is extremely low, and it cannot be measured yet (UHV pressures are in the range of 10^{-9} Torr). To condense Xe, CO_2, the cryostage must be cooled by He based cryogenic technology. Methane, hydrogen and oxygen are volatile at 60 K. Linear alkanes with more than seven carbons, and small alcohol molecules can be used for ice lithography using LN2 cooled cryostages. The styrene and naphthalene aromatics can be used with LN2 based cryostages, while benzene needs He based cryotechnologies. Importantly, the aromatics are the basic building blocks of graphene, carbon nanotubes, and organic semiconductors. Many

Table 9.1. (Potential) ices, vapor pressure estimates, and chemistry estimates of ice lithography products.

Chemical	Formula	A	B	C	p(60 K) Torr	p(150 K) Torr	p(298 K) Torr	Ionic and volatile products	Cross-linked deposit	References
Water	H_2O	5.4	1839	−32	2×10^{-57}	5×10^{-8}	23.5	OH	None	[1, 7, 17, 18]
Xenon	Xe	2.8	327	−50	5×10^{-27}	3×10^{2}	3×10^{4}	Xe	None	
Carbon Dioxide	CO_2	6.8	1302	−3	4×10^{-14}	6	2×10^{5}	O	CO	
Oxygen	O_2	4.0	340	−4	5	3×10^{4}	5×10^{5}	O	None	
Hydrogen	H_2	3.5	99	8	9×10^{4}	6×10^{5}	1×10^{6}	H	None	
Methane	CH_4	4.0	443.03	0	0.3	8×10^{3}	2×10^{5}	H	CH	
Heptane (l)	C_7H_{16}	4.8	1635.4	−27	4×10^{-43}	2×10^{-6}	44.8	CH	CH	
Octane	C_8H_{18}	5.2	1936.3	−20	3×10^{-41}	1×10^{-7}	12.8	CH	CH	[19]
Nonane	C_9H_{20}	3.8	1492.9	−56	NA	7×10^{-10}	3.4	CH	CH	[6, 19, 20]
Tetradecane	$C_{14}H_{30}$	4.1	1739.6	−106	NA	7×10^{-33}	0.01	CH	CH	[19]
Benzene	C_6H_6	4.7	1660.7	−1	2×10^{-21}	3×10^{-4}	100.2	CH	CH	
Styrene	C_8H_8	4.1	1459.9	−60	NA	6×10^{-10}	6.5	CH	CH	
Naphthalene	$C_{10}H_8$	4.3	1831.6	−61	NA	3×10^{-14}	0.3	CH	CH	
1-propanol	C_3H_8O	5.3	1690.9	−52	NA	9×10^{-10}	20.9	CHO	CHO	[6]
1-pentanol	$C_5H_{12}O$	4.0	1106.1	−135	NA	1×10^{-65}	1.2	CHO	CHO	[6]
Anisole	C_7H_8O	4.2	1489.8	−70	NA	3×10^{-12}	3.4	CHO	CHO	[6]
Dimethylsulfoxide	C_2H_6OS	4.5	1807	−61	NA	1×10^{-13}	0.6	CHOS	CHOS	
Xenon difluoride	F_2Xe	2.8	326.6	−50	5×10^{-27}	3×10^{2}	3×10^{4}	XeF	None	
Iodine	I_2	3.4	1039.2	−147	NA	4×10^{-29}	0.2	I	None	
Octafluorocyclobutane	C_4F_8	4.3	1007.4	−30	2×10^{-27}	5×10^{-2}	2×10^{3}	CF	CF	
Nitropropane	$C_3H_7NO_2$	4.2	1466.4	−58	NA	1×10^{-9}	10.1	CHN	CHN	
Trimethylaluminum	C_3H_9Al	4.6	1724	−31	1×10^{-52}	1×10^{-7}	10.4	H	AlCH	
Diethylzinc	$C_4H_{10}Zn$	4.4	1571	−35	3×10^{-56}	4×10^{-7}	20.0	H	ZnCH	
Tungsten hexafluoride	WF_6	4.6	1021	−65	NA	3×10^{-5}	1×10^{3}	F	WF	
HMDSO	$C_6H_{18}OSi_2$	3.9	1203.9	−65	NA	5×10^{-8}	41.4	CHO	SiCHO	[21]

different halogens containing organic chemicals can be used for LN2 based cryo-stage. Another interesting group of organic materials is organometallics, which is often applied for chemical vapor deposition (CVD) and atomic layer deposition (ALD). Most elements can be incorporated into organic chemicals. Many organometallics are suitable for ice lithography applications. It is important to mention the safety concerns; some organic chemicals are very reactive, even at room temperature. In table 9.1, while most chemicals have a sufficiently low VP at 150 K, a few chemicals have low VP at 298 K too (e.g. tetradecane, naphthalene), which means that the gas injection process and hence ice formation can be very slow. Fortunately, higher VP pressures and ice deposition rates can be obtained by increasing the GIS temperature.

The condensed gases, solids, can be crystalline or amorphous (glassy) or a mixture of both. Crystals have well organized lattice structures, while atoms in glasses are packed in a random fashion. Glasses are formed when the phase change is extremely rapid such that crystal formation is not possible. Because of the difference in structure, the material properties are very different, which we will not discuss in detail. One important difference is the density of glasses can be significantly smaller than crystals. Because the atoms in glasses are not organized in the positions with the least energy as in the case of crystals, if the temperature is finite, the atoms in the glass will slowly reorganize to form stable crystals. At high temperatures, the lattice vibrations enhance the glass-to-crystal transition process. The critical temperature is referred to as the glass transition temperature (T_g). For ice lithography, the gases are condensed on a cold surface, and a glass thin-film is often formed. If the cold surface temperature is near or below the T_g, nanocrystalline thin-films are formed in the condensation process. If a glass thin-film is heated to the T_g, the glass thin-film reorganizes into nanocrystalline thin-films (unpublished results). If the sample temperature approaches the glass transition temperature, crystals are formed, and voids in the original film can be observed because glasses have a smaller density than crystals. The transition of glasses to crystals takes place at high temperatures, and it is often associated with increased sublimation.

As for the fundamental research of organic glasses, an extremely relevant research thrust led by Edigar and colleagues, studies organic glass thin-films made by physical vapor deposition (PVD). Unlike with our work, they started to work with organic molecules with T_g higher than room temperature. In 2007, they discovered that the glasses prepared by PVD were exceptionally thermodynamically stable [12]. The stability is manifested by a higher density. Furthermore, during heating organic glasses, here, the thin-film sublimates at an 'onset temperature' (T_{onset}) which can be 20 K higher than the T_g [13]. For unstable organic glasses T_{onset} is equal to T_g. Following this discovery, an ongoing world-wide effort has the aim to understand the fundamental mechanisms behind this stability. The theoretical foundation was laid down by de Pablo and team [14]. In 2017, Edigar and team published condensation of organic molecules such as isopropanol at cryogenic temperatures [13]. Their fundamental findings are nicely presented and discussed in a 2017 review by Edigar [15]. For ice lithography application, the three most relevant findings are the following. First, some organic molecules can form very stable glasses, while others cannot. Tylinski *et al*, showed that for alcohol glasses, despite their chemical similarities, there was a wide range of kinetic stability [13]. Second, there is a sweet spot for the stability of organic glasses, i.e. during PVD, the substrate temperature (T_S) should be slightly lower than T_g, and the deposition should be slow such the molecules have sufficient time to densely pack. Third, depending on the molecular structure, the organic glass film can be anisotropic [16].

9.2.3 Electron–ice interactions

For the overview of electron–matter interactions (EMI), we refer to several excellent textbook chapters [8, 9]. There is also a very relevant EMI for electron-beam lithography application (chapter 3 in [10]). In what follows, we will summarize from the above references and then apply them for electron–ice interactions.

Figure 9.2 illustrates different processes when energetic electrons interact with a solid. The energetic electron is called the primary electron (PE). The electron–matter interaction can be inelastic or elastic. In elastic interactions, the incoming electron does not lose any energy and it is scattered. Backscattered electrons are ejected out of the sample by elastic interactions, and their energy is slightly smaller than the PE. In inelastic interactions, the incoming PE energy is reduced, and depending on the energy

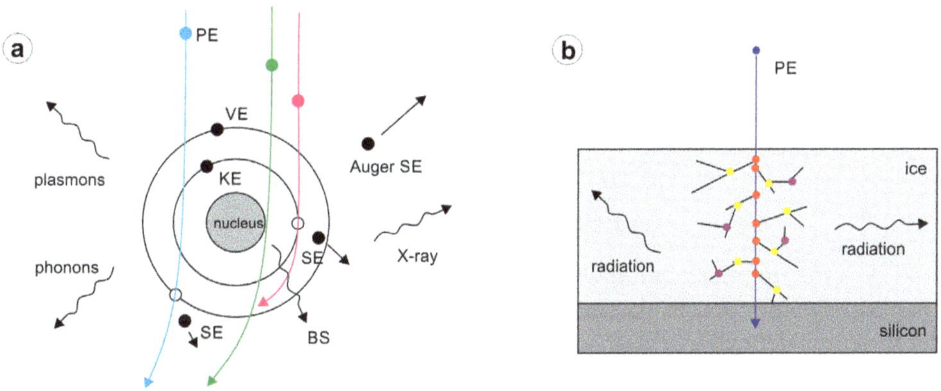

Figure 9.2. (a) Three energetic primary electrons (PE) interacting with an atom in solid matter. Note that the change in trajectory is highly exaggerated. The first PE (blue) interacts with a valence electron (VE), and a secondary electron is ejected (SE). The second PE (green) interacts with the nucleus and its trajectory changes and radiates bremsstrahlung (BS). The third PE (red) interacts with a strongly bound K-shell electron, and the electron is ejected as an SE. The atom is ionized in the first and third PE–matter interaction. The ionized atom can relax to lower energies by chemical reaction using the VE, or emitting Auger SE or characteristic x-rays. PE can also interact with the material lattice and generate plasmons and phonons. (b) The PE interacts with ice thin-film on a silicon substrate. A cascade of SE is generated in close proximity to the PE path. The first cascade of SE is illustrated in red, the second cascade of SE is in yellow and the third in purple. Later in this section, we will see that the range of the cascade of reactions is few nm for a 100-nm-thick organic ice thin-film. The long-range radiation will also cause cascading interactions, however different from electrons, it is spread over much larger distances.

magnitude, different processes could take place. It is important to point out that in SEM a high energy PE hitting a sample is likely to cause a cascade of different processes. The likelihood of an electron–matter interaction is given by the interaction cross section. The most likely interaction is generation of plasmons (10^{-22} m^2), phonons, bremsstrahlung, followed by ionization (10^{-23} m^2), and finally SE (10^{-25} m^2). Plasmons and phonons are, respectively, electron and lattice oscillations in the sample, which ultimately generates heat. The energy of bremsstrahlung ranges from a few eV to near PE energy, and the high energy radiation can cause ionization. Energetic electrons or radiation cause ionization, which could involve removal of electrons in the inner electron shells, or by removing valence electrons (VE) from the Lewis pair, i.e. breaking bonds (section 9.2.1). While the energy needed to break C–C bonds is 3.6 eV, the energy needed to remove K-shell electrons (KE) ranges from 100 to 10^5 eV. The energetic molecules and atoms can release their energies in three different processes. First, they can react with other molecules or recombine as described in section 9.2.1. At lower electron doses, bonds are broken and larger molecules will be broken into smaller molecules. At electron doses above a critical threshold, cross-linking to large molecule networks takes place. We will discuss this chemistry in detail later. Second, the energetic molecule can emit characteristic x-rays. Third, characteristic Auger electrons can be ejected to reduce excitation state. High energy characteristic x-rays and Auger electrons can cause further cascading ionization. While the free mean path of x-rays is long, the short ranged electrons will cause ionization a few nm next to its creation.[1]

A quantitative view of the electron–matter interaction is to consider the energy loss of the PE as it interacts with the sample. The energy loss ($-dE$) along the path (ds) is given by the Bethe equation

$$-\frac{dE}{ds} = \frac{NZe^4}{8\pi E_0 \epsilon_0^2} \, ln\left(\frac{4E_0}{I}\right)^{½} \tag{9.3}$$

Here, N is the number of atoms per unit volume, which is proportional to density, Z is the atomic number, E_0 is the primary electron energy, I is the mean ionization energy that is defined as $I = (9.76 + 58.8\,Z^{-1.9})Z$. As a comparison, for a 20 keV PE interacting with carbon, the energy loss is 2.24 eV nm^{-1}, while it is 6.94 eV nm^{-1} for silver. Important for organic ices that consist mainly of hydrogen, carbon and oxygen, the electrons will mainly interact with the carbon and oxygen. For 100-nm-thick thin-films deposited on a silicon substrate the energy loss is about 200 eV, and hence sufficient for about 100 chemical reactions. Most of the energy will be released in the silicon substrate.

Based on the Bethe equations, the PE trajectory and its energy loss can be visualized with Monte Carlo simulations. Casino freeware was used for the

[1] The definition of SE is blurred in the literature. Seen from an electron–matter interaction perspective, SE are all electrons that are generated by the cascade of PE and matter interaction. While for electron microscopist, SE are low energy electrons that can be detected, i.e. only the electrons that escape the material surface are considered as SE. In the following discussions, we take the electron–matter interaction perspective, and the energy of SE ranges between a few eV to near PE electron energy.

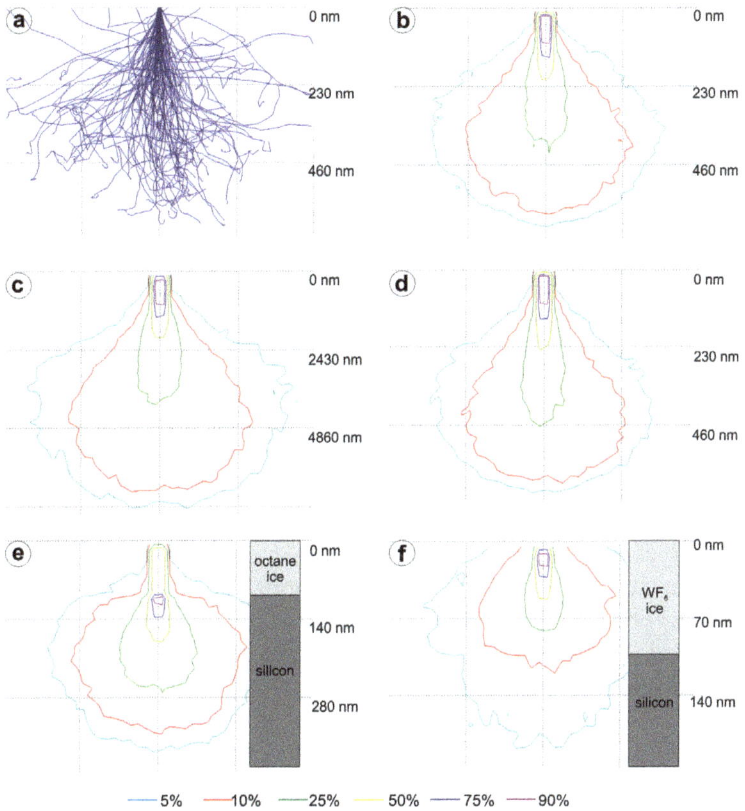

Figure 9.3. Monte Carlo simulations using Casino freeware. PE trajectories of 5 keV electrons that are interacting with water ice (a) and energy by position distribution (b). 90% of the energy is outside the purple line, 50% of the energy is outside the yellow line, and 5% of the energy is outside the cyan line. 20 keV electrons interacting with water ice (c). 5 keV electrons interacting with solid octane (d) and 100-nm-thick octane thin-film on silicon substrate (e). 5 keV electrons interacting with 100-nm-thick tungsten hexafluoride thin-film on silicon (f).

simulations (figure 9.3(a)). In the simulation, we set the e-beam diameter to 20 nm, and PE energy to 5 keV. For water ice, the interaction volume has an onion shape. The initial PE beam and its energy spread out. 50% of the energy is deposited in a cylinder that is 200 nm tall and 50 nm in diameter (figure 9.3(b)). 95% of the energy is contained in the top 600 nm of the ice. Note the energy extends 15 nm in the lateral direction. For water ice interacting with 20 keV PE, 50% is deposited in a cylinder that is 2000 nm tall and 50 nm in diameter (figure 9.3(c)), and the interaction volume is much larger than for the 5 keV case. At 5 keV, the interaction volume for octane is very similar to water (figure 9.3(d)), which can be anticipated from the Bethe equation. If the octane ice is deposited as a 100-nm-thick thin-film on a silicon substrate (figure 9.3(e)), we see that because Si has a larger Z than C, there is an energy concentration at the silicon surface, and the interaction volume is smaller than for pure octane ice. For tungsten hexafluoride ice condensed on silicon substrate, 90% of the energy is deposited in the ice, while very little energy enters the underlying silicon substrate.

The direct consequence of the above analysis for ice lithography, using water and carbon containing organic ices, the interaction volume is almost independent on the chemistry. This makes the method very robust and predictable. In contrast, if organometallic ices are used, the interaction depth will depend on the density and Z of the metal atom in the organic molecule. The interaction volume can be adjusted by modulating the PE energy, and 20 keV can be used for processing at least 2 μm of hydrocarbon ice. For octane thin-films deposited on silicon substrates (figure 9.3(e)), the energy distribution in the ice film has the shape of a truncated cone whose side profile angle has important consequences for resist applications.

In the electron–matter interaction volume, ionization and chemical radical formation take place. The energetic molecules can then take part in several different chemical reaction pathways (figure 9.4). First, recombination can take place and the original molecule is reformed. Second, the chemical radical is formed on the ice thin-film surface, and it has very high VP and sublimates. As an example, for alkane ices, these high VP radicals are small molecules that contain hydrogen and a few carbon atoms, whose VP are in the range of methane (table 9.1). Third, the radical reacts with another molecule and forms a new network molecule with it, which has a very low VP such that it forms a stable deposit at room temperature. This is one of the main findings in the paper by Elsukova *et al* [19], which we encourage the reader to enjoy in full. In the same study the ultimate resolution of ice lithography is also presented and discussed.

If the reaction tips in favor of forming the network molecule, the deposit is stable at room temperature and it is most likely a cross-linked network composed of elements and molecule parts of the condensed ice. For instance, in the case of alkane

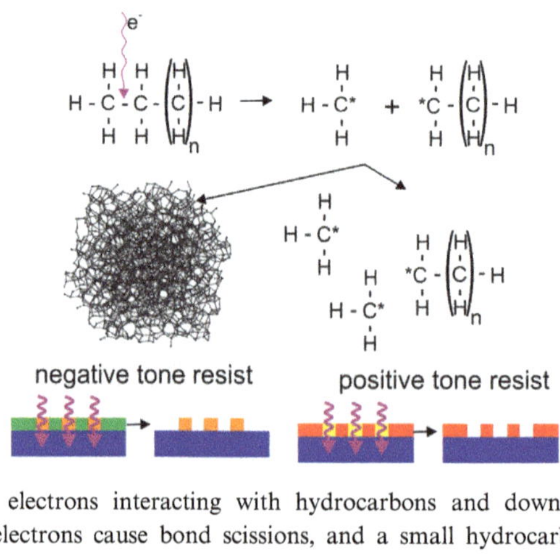

Figure 9.4. Chemistry of electrons interacting with hydrocarbons and down stream chemical reaction pathways. The energetic electrons cause bond scissions, and a small hydrocarbon radical containing one carbon and a large radical are created. The smaller radical may vaporize, recombine, or react with other molecules. The larger radical can recombine or react with other molecules. If the reaction with larger molecules dominates, large networks are formed. In terms of resist, the initial ice acts as a negative tone resist. Most organic ices published so far are considered as negative tone resists. If the evaporation dominates, no network will be formed, and the e-beam irradiated areas will be free from molecules. In terms of resist, the initial ice acts as a positive tone resist.

ices, the cross-linked molecule is expected to have many similarities to cross-linked polyethylenes. Meanwhile, for HMDSO ices, the cross-linked molecule is expected to behave as silicon rubbers such as PDMS (table 9.1). Relevant to lithography, these ices could be used as negative tone resists. In contrast, ices composed of elements that cannot form a stable network can only change chemistry by reaction path 2, which is the case for water and iodine. Here ices could be used as positive tone resists.

9.2.4 Summary

The ingredients of ice lithography are the energetic electrons and thin-films of condensed gases. The energy of the electrons are more than sufficient to facilitate chemical reactions. When the energetic electrons interact with the ice, a cascade of chemical reactions takes place, which leads to either cross-linking or chemical dissociation of the initial condensed gas. The product is either a polymer substance or smaller volatile molecules. There are thousands of different organic molecules that could be frozen into thin-films, and we have only studied a handful.

9.3 Lithography using electrons and ices

The basic principle of IL is simple; a vaporized material coats a cold surface and forms an ice resist layer, which is subsequently exposed to a focused electron beam provided by an electron microscope. This brief description shows IL has high similarity to both EBL and FEBID. In particular, IL is an indirect fabrication method that relies on resist masks like EBL; meanwhile, it employs a gas precursor that interacts with electrons like FEBID. However, there are also notable important differences among these methods, and IL provides value to the nanolithography process and toolbox. In this section, we focus on practical IL procedures, performance of ices, and developing custom IL instruments.

9.3.1 Ice lithography process

A detailed process flow of IL is depicted in figure 9.5 [22]. It reveals that IL skips spin-coating and chemical development required in EBL, resulting in a more streamlined process. It also shows that IL involves an electron–solid interaction, differing from FEBID where electrons interact with a gas that is condensed onto a surface [22]. During IL, the precursor gas firstly condenses onto the precooled sample and produces a uniform amorphous ice thin-film. Then the e-beam bombards the ice resist in specific regions for pattern generation. Depending on different types of electron–matter interaction (section 9.2.3), the e-beam patterning procedure in IL can be generally categorized into a subtractive process or an additive process.

For water ice, electrons remove the resist within the exposure areas (figure 9.5(b)) and the ice patterns subsequently are filled with required materials (e.g. metal) by deposition (figure 9.5(d)). Finally, the sample is immersed into alcohol at room temperature to remove all ice resist with residuals. Except for the last lift-off step or 'melt-off' step, the whole IL process for water ice is accomplished in vacuum.

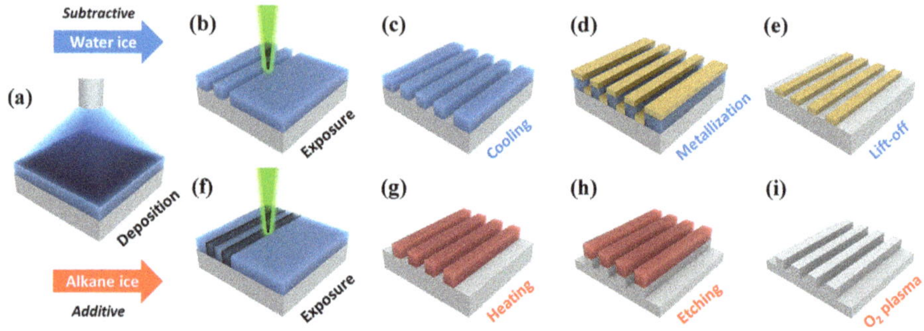

Figure 9.5. Ice lithography process flow. (a) Vaporized material is frozen onto the cold sample to form a layer of ice resist. (b and f) Focused e-beam exposes ice resist for patterning. For water ice, (c) the exposed ice vanishes and pattern transfer is realized after (d) metallization and (e) lift-off. For alkane ice, (g) only the exposed areas remain after heating and patterns are transferred by (h) plasma etching and (i) oxygen plasma for resist removal. Adapted with permission from [22]. CC BY 4.0.

For alkane ice, cross-linking occurs during electron–ice interaction. Heavily cross-linked molecules within the exposure areas form a solidified pattern, which remains in vacuum even after heating the sample to room temperature (figure 9.5(g)). This carbonaceous product is also nonvolatile and chemically inert at atmospheric pressure, thus suitable as an etching mask for transferring the resist pattern to the underlying material. Here, oxygen plasma cleaning can be used for resist removal.

9.3.2 Ice lithography resist

Like the resist classification in EBL, IL resists can be divided into positive-tone resists and negative-tone resists (section 9.2.3). We summarize and compare IL resists with EBL resists and a FEBID precursor in table 9.2. Although several chemicals have the potential to be IL resists (table 9.1), only water ice has been demonstrated as a positive-tone resist to date, and used to fabricate nanostructures with sub-10 nm features [18]. Organic ice from simple organic molecules, e.g. alkanes, alcohols, shows negative-resist-like capabilities. The linewidth of patterns defined in organic ice depends on both beam current of the instrument and hydrocarbon length of the ice resist molecules. In SEM based IL, the optimal linewidths in nonane (C_9H_{20}) and anisole (C_7H_8O) ice are usually 100 and 60 nm, respectively. However, the linewidth could be significantly reduced, e.g. <5 nm in octane (C_8H_{18}) ice [19], when using an e-beam provided by a transmission electron microscope (TEM). This value is comparable to the finest features achieved in EBL resists.

In addition to linewidth, contrast and sensitivity are also important resist properties. The plots in figure 9.6 show the contrast curves for two types of ice resists. For water ice, the thickness can be measured *in situ* through tilted SEM imaging. It is challenging to obtain the initial thickness of organic ice. Hence, we measure the thickness of exposed resist at atmospheric pressure by AFM, which is normalized to the thickness of completely exposed resist. Electron dose represents the number of electrons that are applied to a unit area of resists. Contrast is the slope of the curve.

Table 9.2. Linewidth, contrast and sensitivity of ice resists compared with EBL resists and a FEBID precursor. The total dose (equivalent to D_{100}) for FEBID depositing a single atomic layer denoted with asterisk.

Method	Resist/ Precursor	Tone	Linewidth (nm)	Contrast γ	Sensitivity (D_{50}, mC cm^{-2})	Reference
IL	H_2O	Positive	9	~2	450	[7, 18]
	C_7H_8O	Negative	60	1.0	15	[6, 20]
	C_8H_{18}	Negative	4.5	1.6	40	[19]
	C_9H_{20}	Negative	100	4.3	10	[6, 20]
	$C_{11}H_{24}$	Negative	5.5	1.4	20	[19]
	$C_{14}H_{30}$	Negative	8.8	0.2	20	[19]
EBL	PMMA	Positive	5	5.5–7	0.3–0.5	[23, 24]
	HSQ	Negative	4	5	0.15	[25, 26]
FEBID	$MeCpPtMe_3$	Negative	6	/	250*	[27, 28]

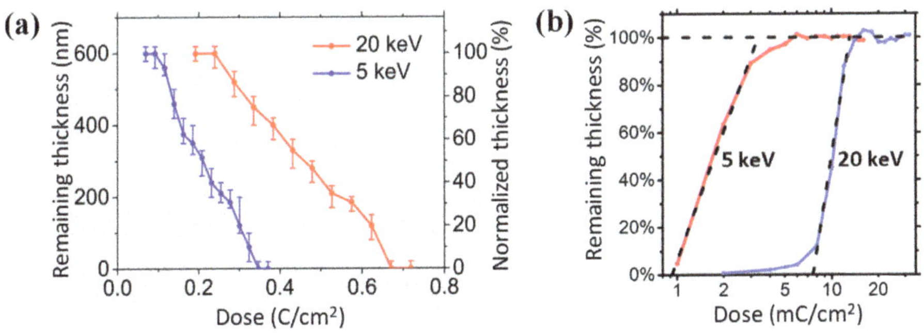

Figure 9.6. Contrast curves for (a) water ice and (b) nonane ice at 5 keV and 20 keV. Adapted with permission from [6, 7]. Copyright (2017, 2018) American Chemical Society.

Generally, an IL resist with high contrast enables close packing of patterned features, while that with low contrast allows flexible thickness control of partially exposed resists, which is beneficial for greyscale lithography. Sensitivity is defined as the dose at which the exposed features have half of the thickness of the initial positive-tone resist or of a completely exposed negative-tone resist. Sensitivity is related to the patterning speed. As shown in table 9.2, EBL resists need lower dose than IL resists, which results in a smaller number of electrons required during EBL, and a higher writing speed of patterning, thus a larger throughput. It is worth noting that FEBID requires 250 mC cm^{-2} to deposit a single atomic layer from an organometallic precursor. This dose stays on the same order of magnitude as IL resists. However, actual FEBID dose increases in proportion to the thickness of deposit and many more electrons are needed for thicker deposits. Compared to FEBID, the electron dose in IL is less dependent on resist thickness.

9.3.3 Ice lithography instrument

Ice lithography research needs the support of cutting-edge instrument development. Historically, the earliest experiments on nanopatterning ice by e-beam were conducted in an SEM with a cryosystem designed for biological applications, which was actually not suitable for IL. In 2010, the same research group demonstrated the successful IL process for fabricating electronic devices and reported the first dedicated IL instrument [5]. Inspired and guided by this system, another two instruments have been subsequently constructed at Technical University of Denmark (DTU) and Zhejiang University (ZJU), respectively. The DTU design (figure 9.7(a), from [22]) focuses on nanofabrication using organic ice resists. It is highly accessible and modular. Except for the load-lock, all subsystems can be assembled or disassembled quickly, enabling flexible switching between a conventional SEM imaging system and a specialized IL instrument. In China, a cryogenic metallization system is integrated to the IL instrument (figure 9.7(b)). Therefore, this instrument can also be used for the growth and *in situ* characterization of metal films at cryogenic temperature. To our best knowledge, the Chinese instrument is the most advanced today. However, the research field is

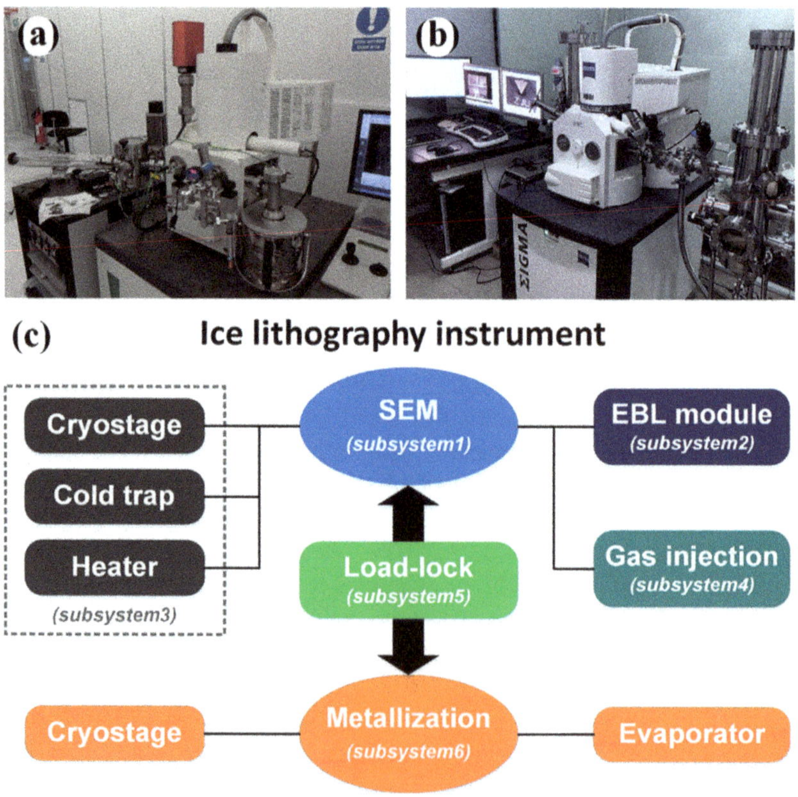

Figure 9.7. IL instruments at (a) Technical University of Denmark (DTU) and (b) at Zhejiang University (ZJU) in China. (c) Schematics of IL instrument consisting of six subsystems. Adapted with permission from [22]. CC BY 4.0.

growing, and at least four other instruments around the world are under construction and development.

As shown in figure 9.7(c), an IL instrument usually consists of five (e.g. DTU instrument) or six (e.g. ZJU instrument) subsystems, depending on the type of resists and processes involved. Here, *Subsystem 1* is a commercial SEM, aiming at providing a focused electron beam and a sample stage with high-precision displacement. *Subsystem 2* is an EBL module including a beam blanker, enabling rapid e-beam writing on ice resists. Both are off-the-shelf subsystems, while the others are highly customized as follows.

Subsystem 3: A cryogenic system includes a cryostage for cooling sample and a cold trap for shielding sample from vacuum contaminations. A heater is also required for thermoregulation or the sublimation of ice resists in vacuum. As in the case of DTU design (figure 9.8), the cryostage and cold trap are connected together with a copper braid, which has a large cross-section (~1 cm) to enhance thermal conductivity while still allowing mechanical flexibility during stage moving. The cold trap is mounted onto a custom flange at the right side of the SEM, and clamped to a long copper rod, whose external end is immersed into liquid nitrogen in a vacuum flask. Therefore, the cryostage is cooled through the copper braid with the same liquid nitrogen source as the cold trap. Figure 9.8 also shows the detailed design of the cryostage. Four polyether ether ketone (PEEK) pillars are used to

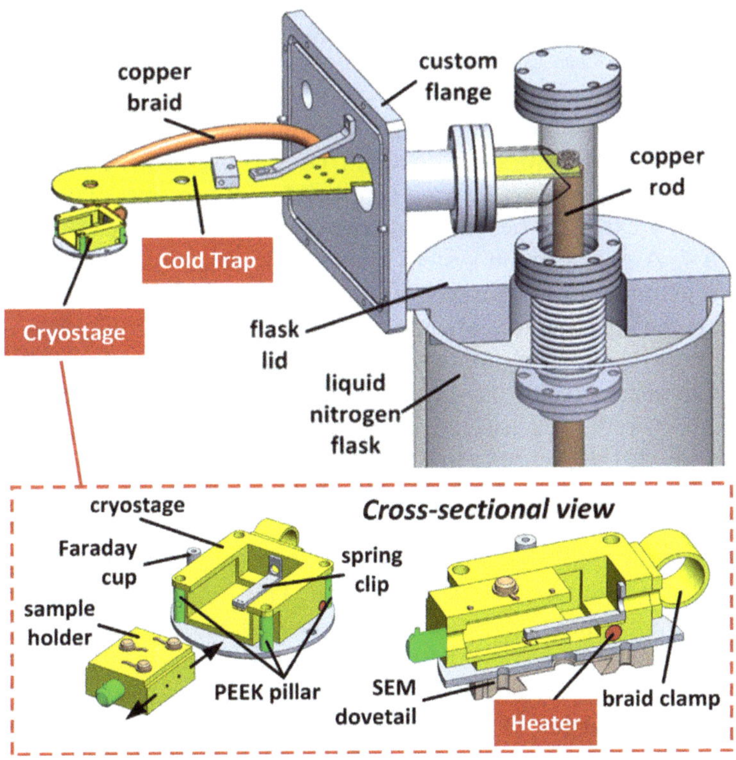

Figure 9.8. Schematics of the cryogenic system in DTU design. Inset shows assemblies of the cryostage and sample holder. Adapted with permission from [20]. Copyright (2018) Elsevier.

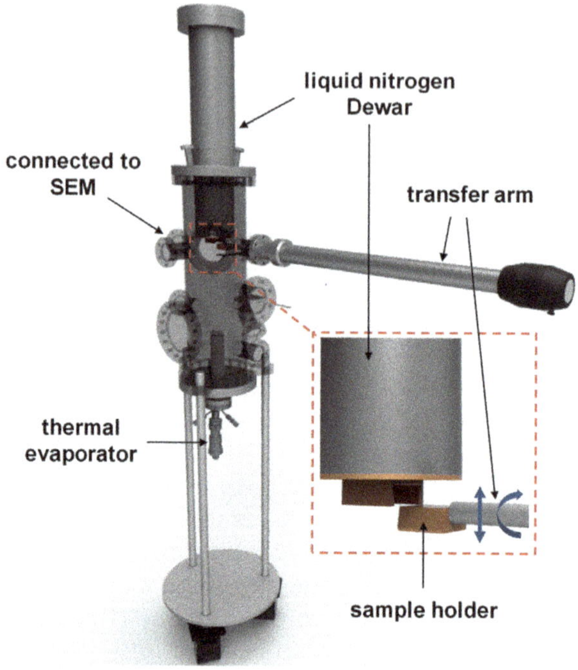

Figure 9.9. Schematics of the cryogenic metallization system in ZJU design. Adapted with permission from [7]. Copyright (2018) American Chemical Society.

thermally insulate the upper part of the cryostage from the lower part, which is attached to an adapted SEM stage dovetail. To improve heat transfer, a spring clip firmly holds the sample holder down when it slides into the cryostage. Temperature is adjusted with a cartridge heater inside the cryostage.

Subsystem 4: A gas injection system aims at introducing water or organic vapor into the SEM chamber through a stainless-steel nozzle. A leak valve connected to the nozzle restricts the injection rate of vapor, thus determining the thickness of ice resist. A vacuum gauge is used to monitor the gas pressure and its evolution during vapor deposition. Chemical precursors such as $MgSO_4{\cdot}7H_2O$ for water ice or C_9H_{20} for organic ice, are stored in a glass vial with vacuum connections. To avoid additional vibrations to SEM, the pumping line in this subsystem is typically removed during e-beam writing and attached only when loading a new precursor.

Subsystem 5: A load-lock allows fast exchange and transfer of samples without breaking the vacuum and cryogenic conditions in the SEM chamber. It can be also useful for conventional room-temperature EBL. This small side chamber has its own turbo and roughing pump, separated from those in SEM. The roughing pump runs a parallel line to *Subsystem 4*, but can be isolated and operated independently. Sample holders are manipulated by a rotatable transfer arm.

Subsystem 6: A cryogenic metallization system is required for processing water ice resist. A cryostage and a liquid nitrogen Dewar are included to maintain ice resist during thermal evaporation or magnetron sputtering of materials. As an example in ZJU design (figure 9.9), the cryostage is closely attached to the bottom of the Dewar for minimizing thermal loss along the cooling path. When transferred into this subsystem, the sample holder is rotated and then inserted into the cryostage with its face down to the thermal evaporator.

9.4 Applications

9.4.1 Nanofabrication on 3D structures

Nanofabrication on 3D structures has attracted growing attention in various research fields. For instance, a traditional aplanatic lens with aberration correction usually consists of two or more optical elements, while optical metasurfaces can manipulate light at the subwavelength scale and exhibit the expected function in a single element. However, the fabrication of an aplanatic metasurface is challenging as one necessary step is patterning on a 3D structure. Other examples are endoscopes and fibre-imaging systems. The realization of such miniature optical instruments also relies on advanced 3D nanofabrication techniques.

Since spin and spray coating can only attain uniform thin resist coatings on planar substrates, traditional EBL processing is limited to flat substrates. However, ice lithography appears highly promising for nanofabrication on 3D structures attributed to vapor-deposited ice resists. One example is patterning on AFM probes [18], as shown in figure 9.10. In IL, we can coat all accessible cold surfaces with water ice, even on a pyramidal tip of the AFM probe. Benefiting from the high dose for ice removal, a glance at the target structure in SEM will not affect the ice mask seriously but can help us position exposure areas without additional marks. Then we pattern a rectangle area centered on the tip by the focused e-beam to remove the water ice and reveal the underlying tip surface. After metallization, Au covers the probe including the exposed tip with an adhesion layer of Ti. Lift-off is implemented through immersing the sample in isopropanol and melting water ice

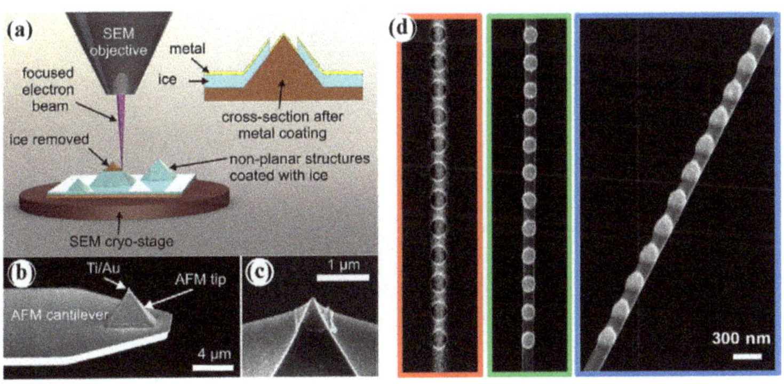

Figure 9.10. (a) Processes for IL to cap an AFM Si_3N_4 tip. (b) and (c) SEM images of the AFM probe whose tip is capped with 1-nm-thick Ti and 20-nm-thick Au. (d) Metallic nanoparticle arrays on a single Ag nanowire with a diameter of 160 nm. SEM images with red and green frames show the nanowire after e-beam exposure and metallization, respectively. Adapted with permission from [7, 18]. Copyright (2012, 2018) American Chemical Society.

at room temperature. Finally, a clean and residual-free AFM tip with a metal cap is obtained.

The next example is patterning on a single nanowire [7]. As just mentioned, we are able to observe the target structure under the ice resist through SEM imaging immediately. This advantage is clearly illustrated in figure 9.10(d). Here, we arrange the exposure areas upon the nanowire. After metallization and lift-off, we achieve a line of nanoparticles located on top of the nanowire. The diameter of the nanowire is 160 nm. It is clear that our alignment accuracy is better than 50 nm.

9.4.2 Nanofabrication on fragile structures

The second advantage of IL is patterning on fragile structures [18]. For instance, silicon nitride membranes are too fragile to withstand ultrasonication, and polymeric resists are usually required to assist lift-off in traditional EBL processing. In IL, we could easily pattern the surface of a fragile freestanding Si_3N_4 membrane, which is supported on a Si wafer frame. As shown in figure 9.11(a) and (b), metallic nanostructures with sub-10 nm features and bowtie structures with gaps less than 7 nm have been achieved on the 60-nm-thick Si_3N_4 membrane, respectively. These TEM images also show that the sample surface after lift-off is very clean without residuals or contaminations in the tiny gap. Figure 9.11(c) shows the back side of the Si_3N_4 membrane, which is located at the bottom of a 200-μm-deep pit and surrounded by Si. Traditional EBL processing would hardly apply a uniform resist through this pit. However, we still successfully perform IL and fabricate nanostructures on such a Si_3N_4 membrane with a thickness of 20 nm, even near the edge of the pit (figure 9.11(d)).

Another example is patterning on freely suspended single-walled carbon nanotubes. Here, the carbon nanotubes were located over a trench in a freestanding

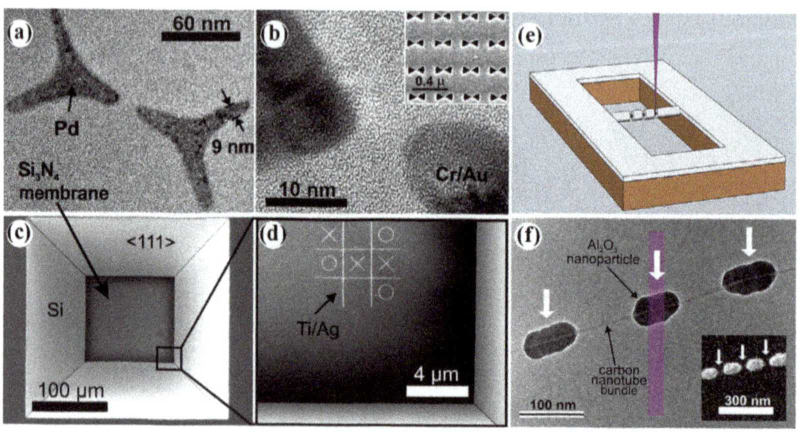

Figure 9.11. TEM images of (a) three-bladed 5-nm-thick Pd and (b) 5-nm-thick Au with 1-nm-thick Cr patterns on the planar top surface of a freestanding Si_3N_4 membrane. (c) and (d) SEM images of 10-nm-thick Ag with 1-nm-thick Ti patterns on the bottom-facing surface of a Si_3N_4 membrane. (e) and (f) Al_2O_3 nanoparticles grown by ALD coaxially on freely suspended nanotubes, where Ti seeds have been fabricated previously through IL. The white arrows indicate the areas where ice is removed. Adapted with permission from [18]. Copyright (2012) American Chemical Society.

silicon nitride membrane (figure 9.11(e)). As just mentioned, traditional EBL cannot deal with such a fragile structure either. However, after cooling down the sample, we can deposit water ice onto the tiny nanotubes. Then we control the e-beam to remove the ice selectively and deposit a Ti thin-film on the surface of the nanotube, which acts as a seed layer for growing Al_2O_3 by atomic layer deposition (ALD). As a result, Al_2O_3 particles are formed on the nanotubes whose locations and sizes are precisely determined by the IL patterning combined with ALD processing (figure 9.11(f)).

9.4.3 Rapid prototyping

Rapid prototyping, also known as additive manufacturing or 3D printing, refers to a group of manufacturing processes, which enables fast 3D fabrication of unprecedented engineering and production possibilities. Construction of a part is usually done by adding layers of one material on top of one another, and materials such as metals, ceramics and plastics are involved in the fabrication. Examples of rapid prototyping methods are inkjet printing, stereolithography, two-photon polymerization, and FEBID. In a sense, EBL can also be regarded as a rapid prototyping processing, since a stacking layer strategy is adopted for fabricating 3D structures. However, the fabrication of each layer needs a series of procedures including spin-coating, exposure, developing, deposition and lift-off, and additional markers should be fabricated previously for registration and alignment between layers. All these complex steps limit the 3D processing speed of EBL.

IL opens a new possibility for nanoscale rapid prototyping with e-beam [7]. For instance, we are able to perform a marker-less fabrication with water ice, which results in many fewer processing steps for 3D nanofabrication. Figure 9.12 shows detailed IL fabrication steps for a pyramidal nanostructure, just through repeating ice deposition, e-beam exposure and metallization three times. To realize a similar structure, at least 19 processing steps and 8 load–unload operations are involved in traditional EBL processing. In comparison, IL simplifies the processing, that only requires ten steps

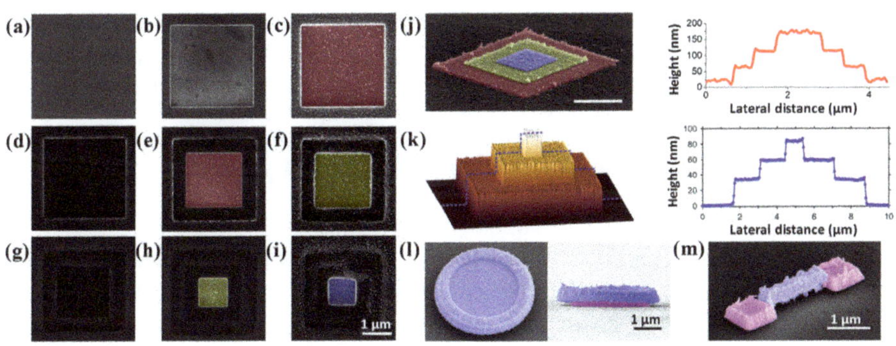

Figure 9.12. 3D layered and suspended nanostructures produced by IL. *In situ* SEM images after different processing steps including ice deposition (a, d, and g), e-beam exposure (b, e, and h) and metallization (c, f, and i). A pyramidal nanostructure obtained using water ice (j) or organic ice (k), with corresponding line scan by AFM. Similar to greyscale lithography, IL can also realize 3D mushroom-like (l) and bridge-like (m) nanostructures. Adapted with permission from [6, 7]. Copyright (2017, 2018) American Chemical Society.

and one load–unload operation. If using organic ice (figure 9.12(k)), the processing steps could be further compressed into the cycle of ice deposition and e-beam exposure followed by one-time lift-off, although alignment markers are needed. In IL, we also successfully fabricate partially suspended nanostructures like a mushroom (figure 9.12(l)) or a bridge (figure 9.12(m)), inspired by the strategy of greyscale lithography. Applying and combining these approaches will bring new and exciting opportunities in 3D nanofabrication.

9.4.4 High-resolution patterning

Resolution is a key criterion for the evaluation of a nanolithography method. Sub-10 nm features have been obtained using water ice. Meanwhile, the resolution of organic ice in SEM-based IL is usually above 50 nm. It is mainly due to the instrumental limitations of SEM, such as a broad beam spot and optics instabilities. To minimize these influences and achieve higher resolution, we perform IL in an environmental TEM equipped with a gas cell and a cryostage [19]. The TEM operates at 80 keV and the substrate is a 5-nm-thick Si_3N_4 membrane. Therefore, the incident electrons can be highly concentrated into a tiny spot (~1 nm) and pass through the substrate swiftly with a small cross section of scattered secondary electrons, finally resulting in expected high-resolution patterns.

As mentioned in section 9.2.3, two competing chemical reactions, chain scission and cross-linking typically take place in the resist during exposure. Here, cross-linking of the molecules dominates when exposing the organic ice layer to a focused e-beam, and forms new molecules with an increased molecular weight. The thickness of final solidified pattern is proportional to the number of molecules cross-linked enough to become nonvolatile at room temperature. As shown in figure 9.13, three organic alkane resists with different molecular lengths are tested. All these molecules have a simple linear structure with only C–C and C–H bonds. For each organic resist, the line patterns get wider with increasing beam current, due to the larger e-beam diameter. Furthermore, a positive correlation between the linewidth and the molecular length can be achieved under the same beam current. It demonstrates that the minimal feature size is fundamentally determined by the resist molecules. These evidences not only show IL has the capability for high-resolution patterning, but also set a clear direction for approaching the ultimate resolution of IL.

9.5 Future research, opportunities and challenges

9.5.1 Fundamental science research

A careful reader will quickly realize most of the fundamental scientific questions are wide open. Starting with the organic ice itself, there are thousands of organic molecules that can be frozen in vacuum. We have only reported eight chemicals, and we had poor measurements of freezing points of even fewer. Dr Schick's and Dr Ediger's teams used macroscopic AC nanocalorimetry to measure the transition temperature of alcohol ices [13, 29]. However, most physical properties of the organic ices are unknown. Technology is pushing ahead, and we must answer the most fundamental scientific questions:

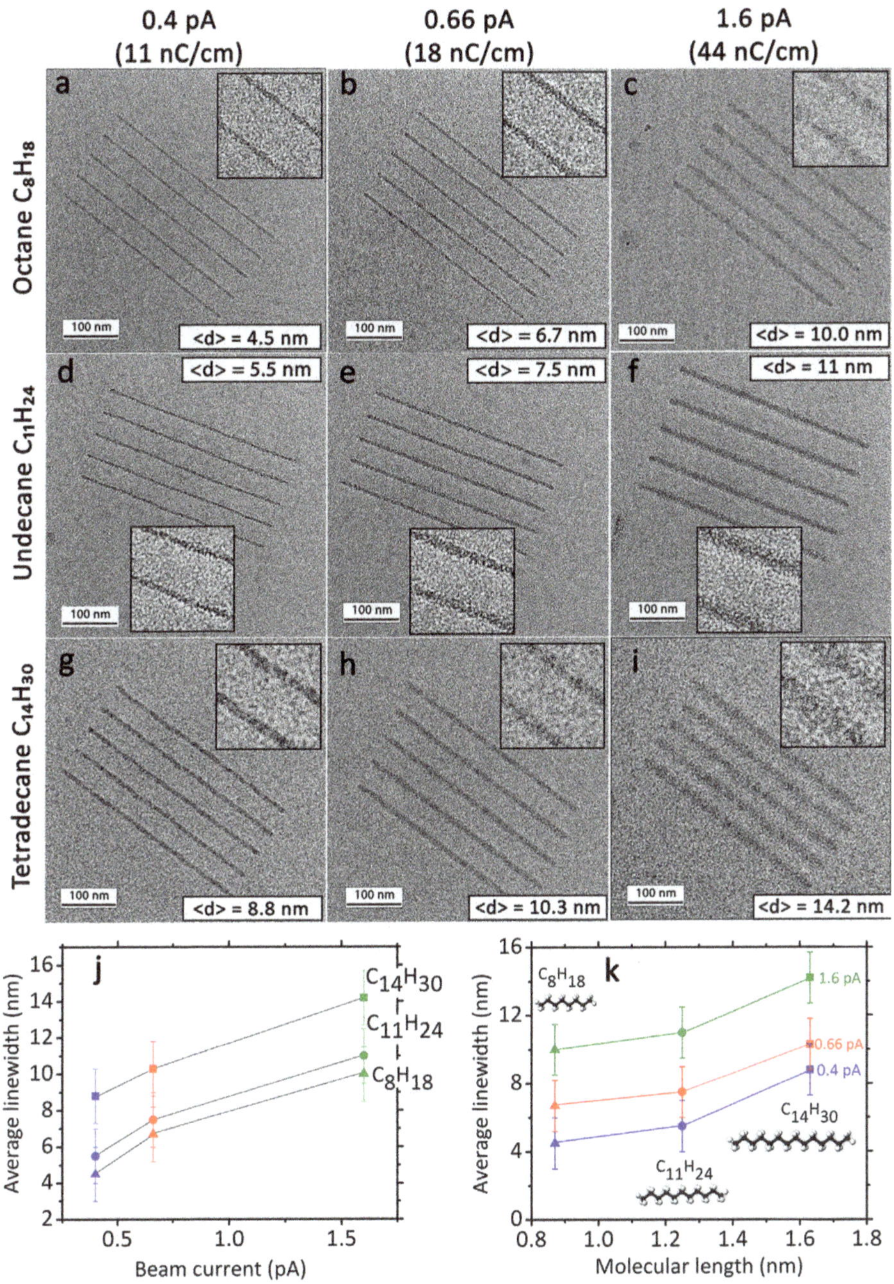

Figure 9.13. Line patterns on octane (a–c), undecane (d–f), and tetradecane (g–i) in TEM. Linewidth plotted as a function of (j) beam current and (k) resist molecular length. Adapted with permission from [19]. Copyright (2018) American Chemical Society.

1. What is the structure of organic ices? Are they crystalline or amorphous, or both?
2. What are the condensation, sublimation, glass transition temperatures?
3. What are the electrical properties such as conductivity, dielectric constant, and bandgap?
4. What are the mechanical properties such as thermal conductivity, density, and Young's modulus?
5. One big chemistry question is, what happens if we mix these ices together?

Critical readers will question if valuable efforts should be made to explore this empty space in science. If space travel is the future, then we should spend valuable effort in the research of organic ices. Space is the final frontier, and the environment is very different. It is cold and there is vacuum. Most organic chemicals, which are liquids or gases on Earth, will freeze to organic ices (solids) in space. Future space applications of organic ices might range from dynamic coatings of spaceships to lubricants at low temperatures.

9.5.2 Icetronics research

While the applications of organic ices themselves seem very distant and remote, the earthly applications of e-beam processed organic ices range from quantum technologies, telecommunications, to advanced production of computer chips. Lithography applications for e.g. MEMS, sensors, 2D materials, and photonics research are plentiful. The careful reader with a materials science background will quickly realize that it is possible to create new materials whose properties are completely unknown. Indeed, we envision nanoscale additive manufacturing using e-beam assisted engineering of electronics as an important research direction, i.e. to 3D print computer chips including functional materials. This technology can make patterns down to 10 nm, and enable acceptable production speed-up, which drives applications below 1 μm and in three dimensions. The main hypothesis is that by selecting and freezing organic compounds and exposing them with an optimized e-beam, it is possible to print functional materials such as metals, dielectrics and semiconductors. Because the two required ingredients are ice and electrons, and we aim to make electronic devices, we introduce the term '*Icetronics*'.

The three future research areas of Icetronics are:

1. To deposit metal conductors by e-beam patterning of organometallic ices.
2. To deposit organic dielectrics whose chemical compositions are similar to polymers.
3. To deposit organic semiconductors for field effect transistor devices.

For metal deposition, many metals could be deposited by using the full range of organometallics. We would like to focus on Au, because it has good electrical properties, and it is used in electronics as contact, electrode, and wiring material. Furthermore, Au is biocompatible, and it is widely used in medical implants. In 2016, Shawrav *et al* added water vapor to Au organometallic FEBID process to deposit almost pure Au, which have much better electrical properties than prior art [30]. Based on Shawrav *et al* studies, we hypothesize the formation of an almost pure and porous Au film. We showed that thin-film Au can be annealed and densified at 300 °C [31], which will increase the electrical performance of the deposit.

For dielectric deposition, we deposited alkanes and aromatic compounds [6] and silicone elastomer (PDMS) compounds [21]. Unpublished results show that the optical properties of the deposits are very different; this suggests that the original organic chemistry plays an important role in the deposit formation. This is fundamentally different from FEBID deposits, which is independent of the hydrocarbon precursor [32]. We believe this is caused by fundamental differences between electron–gas–surface (FEBID) and electron–solid–surface interactions (Icetronics). The e-beam cross-links the condensed organic ice film, and we hypothesize the resulting composition of the deposits and its physical properties can be tailored by selecting the appropriate chemistry. It is likely that the deposits properties are very similar to polymers, which are good barrier material for metal wires and common dielectric materials.

We can freeze and pattern, e.g. alkanes, siloxanes, and fluorocarbons such as C_4F_8. The deposited material is expected to have similar properties to polyethylene, silicone elastomers, and Teflon. Cyclic siloxanes can also do 'ring opening' polymerization reaction, and we hypothesize that this chain reaction could be initiated by direct injection of electrons using the e-beam. Furthermore, the reaction could be controlled by co-freezing inhibitors.

For semiconductor deposition, organic semiconductors (OS) are divided into small molecule OS (SMOS), and polymeric OS (POS) [33]. The fundamental building block of SMOS is the basic conjugated π-orbital units, which are benzene and thiophene. High mobility SMOS thin-films (e.g. pentacene) are polycrystalline, and the mobility is related to the size of the crystals. The correlation between thin-film structure and mobility is complicated, and studies show that larger crystals have higher mobility [34]. Apart from pentacene and its derivatives, other SMOS molecules are rubrene, and anthracine derivatives (9,10-diphenylanthracene, DAP) which is used for OLEDs [35, 36]. Anthracene, DAP and tetracene have low melting points and sufficient VPs for injection into the e-beam chamber and frozen to ices.

We could build field effect transistor (FET) devices using semiconducting organic ices, whose process is similar to our previous work on carbon nanotube FETs [17, 18]. For SMOS, the crystallinity is strongly correlated with the electrical performance. Hence, we anticipate that there is a strong co-relation among the number of electrons used for cross-linking of organic semiconducting ices, the crystallinity of the cross-linked product, and the electrical performance. The electrons needed for cross-linking will also damage and introduce defects into the ice crystal. The question is how robust these materials are? Finally, our goal is to 3D print a functional FET using *Icetronics* technology.

9.5.3 Applications

We have touched on applications throughout the chapter. It is important to point out that this is a scalable technology thanks to the development of multi-beam technologies, which are instruments that have 260 000 e-beams (www.ims.co.at). We classify applications into two areas. First, in IL applications, the ice serves as a masking or sacrificial material to pattern a functional material such as metals or semiconductors. IL is an additional tool in the powerful toolbox of nanolithography. Second, cross-linked organic ices could be used as functional materials, which opens up nanoscale 3D printing or rapid prototyping. The impact of this technology might be equivalent to the impact of 3D printing for mechanical and manufacturing engineering. Entirely new and unconventional applications might emerge.

9.6 End-of-chapter summary

IL has been used to deposit and pattern water ice and organic ice resists. Due to the high degree of conformity, IL has seen successful results in performing lithography on 3D structures, as well as nanofabrication on fragile structures. Nanoscale rapid prototyping can be done using the IL instrument. Furthermore, 3D structures are made by adopting layer-by-layer processing approach, which is similar to additive manufacturing. Finally, an important application is high resolution patterning, and we show line-width down to few nanometers by utilizing high energy e-beam exposure in a TEM instrument.

9.7 Student's corner

- What is the electron energy range in an SEM?
- Is it possible to condense CO_2 in an SEM using a liquid nitrogen based cryostage?
- What are the reaction products of electrons' interaction with CO_2 ice?
- What are the chemical reaction pathways for electrons–ice interactions?
- Which chemical would you want to work with and why?
- Can IL and EBL be used for lithography on 3D structures?
- List the factors that determine the patterning resolution of IL process.
- What device or structure would you like to build with IL?

References

[1] King G M, Schürmann G, Branton D and Golovchenko J A 2005 Nanometer patterning with ice *Nano Lett.* **5** 1157–60

[2] Steere R L 1957 Electron microscopy of structural detail in frozen biological specimens *J. Biophys. Biochem. Cytol.* **3** 45–60

[3] Singer S J and Nicolson G L 1972 The fluid mosaic model of the structure of cell membranes *Science (80-.)* **175** 720–31

[4] Deamer D, Akeson M and Branton D 2016 Three decades of nanopore sequencing *Nat. Biotechnol.* **34** 518–24

[5] Han A, Chervinsky J, Branton D and Golovchenko J A 2011 An ice lithography instrument *Rev. Sci. Instrum.* **82** 065110

[6] Tiddi W, Elsukova A, Le H T H T, Liu P, Beleggia M and Han A 2017 Organic ice resists *Nano Lett.* **17** 7886–91

[7] Hong Y *et al* 2018 Three-dimensional *in situ* electron-beam lithography using water ice *Nano Lett.* **18** 5036–41
[8] Williams D B and Carter C B 2009 *Transmission Electron Microscopy: A Textbook for Materials Science* (Berlin: Springer)
[9] Goldstein J I, Newbury D E, Michael J R, Ritchie N W M, Scott J H J and Joy D C 2017 *Scanning Electron Microscopy and X-ray Microanalysis* (Berlin: Springer)
[10] Landis S 2010 *Lithography* (New York: Wiley)
[11] Andreas E L 2007 New estimates for the sublimation rate for ice on the Moon *Icarus* **186** 24–30
[12] Swallen S F *et al* 2007 Organic glasses with exceptional thermodynamic and kinetic stability *Science (80-.)* **315** 353–6
[13] Tylinski M, Chua Y Z, Beasley M S, Schick C and Ediger M D 2016 Vapor-deposited alcohol glasses reveal a wide range of kinetic stability *J. Chem. Phys.* **145** 174506
[14] Singh S, Ediger M D and de Pablo J J 2013 Ultrastable glasses from in silico vapour deposition *Nat. Mater.* **12** 139–44
[15] Ediger M D 2017 Perspective: highly stable vapor-deposited glasses *J. Chem. Phys.* **147** 210901
[16] Ediger M D, de Pablo J and Yu L 2019 Anisotropic vapor-deposited glasses: hybrid organic solids *Acc. Chem. Res.* **52** 407–14
[17] Han A, Vlassarev D, Wang J, Golovchenko J A and Branton D 2010 Ice lithography for nanodevices *Nano Lett.* **10** 5056–9
[18] Han A, Kuan A, Golovchenko J and Branton D 2012 Nanopatterning on nonplanar and fragile substrates with ice resists *Nano Lett.* **12** 1018–21
[19] Elsukova A, Han A, Zhao D and Beleggia M 2018 Effect of molecular weight on the feature size in organic ice resists *Nano Lett.* **18** 7576–82
[20] Tiddi W, Elsukova A, Beleggia M and Han A 2018 Organic ice resists for 3D electron-beam processing: instrumentation and operation *Microelectron. Eng.* **192** 38–43
[21] Han A, Zhao D, Liu L and Skov A L 2019 Towards nanoscale 3d printing of PDMS-like polymers *2019 20th Int. Conf. on Solid-State Sensors, Actuators and Microsystems & Eurosensors XXXIII (TRANSDUCERS & EUROSENSORS XXXIII)* 1724–7
[22] Zhao D, Han A and Qiu M 2019 Ice lithography for 3D nanofabrication *Sci. Bull.* **64** 865–71
[23] Manfrinato V R *et al* 2017 Aberration-corrected electron beam lithography at the one nanometer length scale *Nano Lett.* **17** 4562–7
[24] Yasin S, Hasko D G and Ahmed H 2002 Comparison of MIBK/IPA and water/IPA as PMMA developers for electron beam nanolithography *Microelectron. Eng.* **61–62** 745–53
[25] Manfrinato V R *et al* 2013 Resolution limits of electron-beam lithography toward the atomic scale *Nano Lett.* **13** 1555–8
[26] Namatsu H, Yamaguchi T, Nagase M, Yamazaki K and Kurihara K 1998 Nano-patterning of a hydrogen silsesquioxane resist with reduced linewidth fluctuations *Microelectron. Eng.* **41–42** 331–4
[27] van Oven J C, Berwald F, Berggren K K, Kruit P and Hagen C W 2011 Electron-beam-induced deposition of 3-nm-half-pitch patterns on bulk Si *J. Vac. Sci. Technol.* B **29** 06F305
[28] Hari S, Hagen C W, Verduin T and Kruit P 2014 Size and shape control of sub-20 nm patterns fabricated using focused electron beam-induced processing *J. Micro/Nanolithogr., MEMS, MOEMS* **13** 033002

[29] Chua Y Z, Young-Gonzales A R, Richert R, Ediger M D and Schick C 2017 Dynamics of supercooled liquid and plastic crystalline ethanol: dielectric relaxation and AC nanocalorimetry distinguish structural α- and Debye relaxation processes *J. Chem. Phys.* **147** 014502

[30] Shawrav M M *et al* 2016 Highly conductive and pure gold nanostructures grown by electron beam induced deposition *Sci. Rep.* **6** 34003

[31] Todeschini M, Bastos Da Silva Fanta A, Jensen F, Wagner J B and Han A 2017 Influence of Ti and Cr adhesion layers on ultrathin Au films *ACS Appl. Mater. Interfaces* **9** 37374–85

[32] Bret T, Mauron S, Utke I and Hoffmann P 2005 Characterization of focused electron beam induced carbon deposits from organic precursors *Microelectron. Eng.* **78–79** 300–6

[33] Klauk H 2006 *Organic Electronics: Materials, Manufacturing and Applications* (New York: Wiley)

[34] Diao Y *et al* 2013 Solution coating of large-area organic semiconductor thin films with aligned single-crystalline domains *Nat. Mater.* **12** 665–71

[35] Zhu M and Yang C 2013 Blue fluorescent emitters: design tactics and applications in organic light-emitting diodes *Chem. Soc. Rev.* **42** 4963

[36] Aydemir M *et al* 2016 High efficiency OLEDs based on anthracene derivatives: the impact of electron donating and withdrawing group on the performance of OLED *Org. Electron* **30** 149–57

IOP Publishing

Nanofabrication
Nanolithography techniques and their applications
José María De Teresa

Chapter 10

Magnetic nanopatterning via thermal scanning probe lithography

Edoardo Albisetti, Daniela Petti, Riccardo Bertacco and Elisa Riedo

Finding new ways for controlling magnetism at the nanoscale is of great importance both for enabling fundamental studies on highly controlled systems and for developing novel device concepts. Recently, we introduced a technique called thermally assisted magnetic scanning probe lithography (tam-SPL), which allows one to craft the spin texture of magnetic multilayers with high flexibility and tunability. In this chapter, we review the principles and applications of tam-SPL. First, we introduce the main concepts and discuss the patterning of magnetic domains in exchange bias multilayer systems. Then, we describe different strategies for stabilizing complex spin textures with lower dimensionality and tailored properties, such as magnetic domain walls and solitons with deterministic topology. We show how to extend the technique to different magnetic systems, such as synthetic antiferromagnets. Finally, we discuss the application of tam-SPL to the field of magnonics, and realizing reconfigurable spin-wave circuits and optically-inspired nanomagnonic platforms.

10.1 Introduction

Searching novel tools for controlling the magnetic properties of thin films with nanometric resolution is of capital importance for studying fundamental phenomena and developing new device concepts. A notable example in this sense is the rise of engineered spin textures, such as magnetic domains, domain walls [1], and topological solitons (such as vortices [2, 3] and skyrmions [4]), as active elements in magnetic nanodevices for memory and computing.

The most widely used techniques for patterning magnetic materials in order to spatially control their magnetic properties include conventional micro- and nano-fabrication, such as optical and electron-beam lithography [5], and ion-beam irradiation [6–8]. In the first case, the functionality is determined by shaping the

doi:10.1088/978-0-7503-2608-7ch10

magnetic material via additive or subtractive manufacturing, i.e. by masking, deposition and subsequent lift-off, or by first depositing the material, then masking and subsequently etching away the unprotected material, e.g. via ion-milling. These methodologies represent the wide majority of the used approaches for fabricating e.g. domain-wall conduits, hall bar structures, spin-valves and magnetic tunnel junctions, spin-wave waveguides and magnonic crystals.

Ion-beam irradiation based-techniques, on the other hand, allow producing spatially controlled alterations of the magnetic properties of continuous films and multilayers. In the most common scenario, first the magnetic stack is masked via conventional lithography, then the whole sample is irradiated with a tunable dose of ions for either controllably damaging the magnetic material or multilayer interfaces, or producing localized chemical transformations which in turn change the local magnetic properties. Direct focused ion-beam writing can also be used, instead of masking, for irradiating nanoscale regions [9].

Finally, fabrication of three-dimensional magnetic nanostructures has been demonstrated by direct growth via focused electron-beam induced deposition (FEBID) [10], or using two-photon polymerization lithography in combination with electrochemical deposition [11].

In this chapter we will discuss a recently developed technique, called thermally assisted magnetic scanning probe lithography (tam-SPL) [12] which employs a heated nanometric scanning probe for reversibly controlling the magnetic properties of thin film heterostructures. In section 10.2 we introduce the working principle and main features of the technique, and show its application to the patterning of tailored spin textures in different systems. In particular, we discuss the stabilization of magnetic domains, domain walls and topological solitons in in-plane magnetized films and synthetic antiferromagnets. In section 10.3, we discuss the application of tam-SPL for manipulating spin waves in the emerging field of magnonics, focusing on the realization of reconfigurable spin-wave waveguides and nanoantennas based on tailored spin textures. Finally, we give an overview of future perspectives for the application of tam-SPL.

10.2 Thermally-assisted magnetic scanning probe lithography

Thermal scanning probe lithography (t-SPL) is based on scanning a nanoscopic heatable probe on the surface of the sample, e.g. in an atomic force microscope (AFM) system or dedicated thermal lithography tool. Such thermal probes were originally developed by IBM for data storage applications [13, 14], and consist of specially designed cantilevers which integrate a resistive micro-heater located above the tip. By flowing an electric current, the heater temperature can be precisely controlled from RT up to ~1100 °C. As the heat diffuses towards the apex of the nanometric tip, the lateral dimension of the hot spot becomes comparable to the tip size, and it can be used for producing a wide range of highly localized thermally-induced surface modifications. Throughout the years, several proof-of-principle demonstrations of applications of t-SPL to different systems were proposed [15]. For example, t-SPL can be used for locally removing or displacing material via thermally assisted nanoindentation or degradation, and therefore creating three-dimensional topographic patterns [16], or lithographic masks enabling the subsequent nanofabrication steps [17]. Another prolific field of application of t-SPL is the direct conversion of materials in the 'thermochemical nanolithography' approach, where the heated tip is used for promoting spatially confined thermochemical reactions in a wide range of systems [18]. Examples in this framework include the local deprotection of functional groups on the surface of polymeric films [19], which can then be used for immobilizing proteins and protein gradients [20], the direct fabrication of nanostructures from precursor materials [21], or the modification of the surface properties of 2D materials [22].

Tam-SPL is performed by sweeping the heated tip on the surface of a magnetic multilayer in the presence of an external magnetic field, for producing local, stable modifications to its magnetic properties.

10.2.1 Concept

One of the most widely studied and technologically relevant phenomena in magnetism is the *exchange bias* [23, 24]. It occurs at the interface between an antiferromagnetic material (AF) and a ferromagnetic material (FM) causing a shift in the hysteresis loop of the ferromagnet due to the exchange interaction between the two materials. In other words, the exchange bias introduces a unidirectional magnetic anisotropy in the ferromagnet, which contributes to determining its magnetization ground state. Exchange bias is widely used in magnetic recording for pinning the magnetization in magnetoresistive read-head devices such as spin valves or magnetic tunneling junctions. If, as in most cases, the Curie temperature of the ferromagnet is higher than the Néel temperature of the antiferromagnet, it is possible to set the exchange bias pinning direction by cooling the sample across its blocking temperature T_B (which is usually comparable to the Néel temperature of the antiferromagnet) while applying an external magnetic field, the so-called *field cooling*.

Tam-SPL consists of sweeping the heated probe on the surface of an exchange bias magnetic system in the presence of an external field, for performing a highly localized field cooling and locally re-setting the exchange bias anisotropy point-by-point with nanoscale resolution. In turn, this allows one to deterministically stabilize complex magnetization configurations. Figure 10.1 shows the working principle of the technique in the simplest case of an exchange bias FM/AF bilayer magnetized in the plane of the film. First (a), a global field cooling is performed for setting a uniform exchange bias and therefore a uniform magnetization orientation across the whole sample. The global pinning direction is determined by the direction of the 'initialization' field H_i. (b) Shows the corresponding hysteresis loop, which is shifted from the center of a quantity H_e called exchange bias field. The patterning (c) is performed by sweeping the tip, heated above the blocking temperature T_B, on the sample surface while applying an external 'writing' field H_w. As the tip is displaced across the sample, the previously heated region undergoes a highly localized field cooling, and the local exchange bias is re-set in the direction of the magnetization, set by H_w. As the external field and the heated tip are removed, the new magnetization ground state is determined by the unidirectional anisotropy landscape

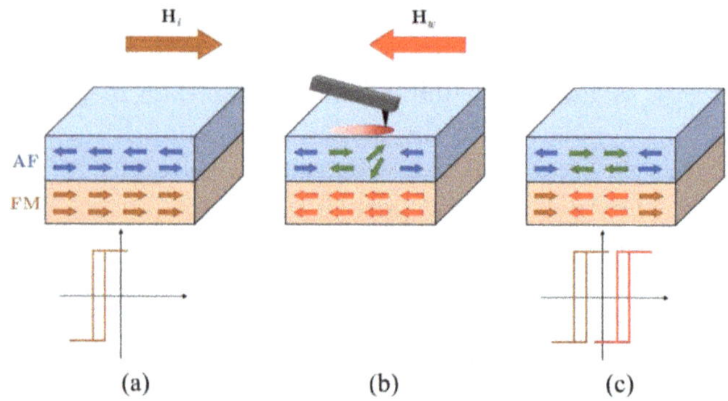

Figure 10.1. Sketch of tam-SPL performed on an exchange bias ferromagnet (FM)/antiferromagnet (AF) bilayer. (a) Initialization consists in performing a global field cooling in order to pin the magnetization of the FM in the direction of the external initialization field H_i. (b) tam-SPL is performed by sweeping a heated scanning probe on the sample in the presence of an external magnetic field H_w in order to locally reset the exchange bias. (c) Upon removal of the external field, the spin texture in the ferromagnet is stabilized by the local exchange bias field. The bottom panels show the hysteresis loops of the initialized sample (a) and of the patterned sample (c).

'stored' within the system via tam-SPL. This is visualized at the bottom of panel (c) by the opposite exchange bias shift, and therefore opposite unidirectional anisotropy, and in the local hysteresis loops of the non-patterned (orange loop) and patterned regions (red loop).

10.2.2 Patterning magnetic domains, domain walls and solitons

By controlling the patterning conditions and the geometry of the patterned area, it is possible to deterministically stabilize a wide range of spin textures with tailored properties and dimensionality. Here, we discuss the case of two-dimensional magnetic domains, one-dimensional domain walls and zero-dimensional topological solitons patterned in prototypical CoFeB/IrMn exchange bias bilayers magnetized in the plane of the film.

In general, magnetic domains of controlled size and magnetization orientation are patterned via tam-SPL by raster-scanning the heated tip on the sample, in a similar fashion as in atomic force microscopy imaging. The main technique used for characterizing the magnetic patterns at remanence was magnetic force microscopy (MFM), which shows the magnetic contrast associated with the stray field gradients arising in regions with non-uniform magnetization, such as the domain walls surrounding a uniform magnetic domain. Figure 10.2 shows the MFM images of arbitrary shaped domains with magnetization at remanence pointing in-plane in the opposite direction with respect to the magnetization outside the patterned regions. In particular, square, triangular and diamond shaped domains with size ranging from 1.5 to 2.5 μm are stabilized in $Ru(2)/Ir_{20}Mn_{80}(7)/Co_{40}Fe_{40}B_{20}(5)/SiO_2(1000)/Si$ (thickness in nm) multilayers grown via magnetron sputtering. The patterning was performed by raster-scanning the heated tip at a 3 $\mu m\ s^{-1}$ speed on the surface of the sample while applying an external writing field $H_w \sim 700$ Oe in the opposite direction

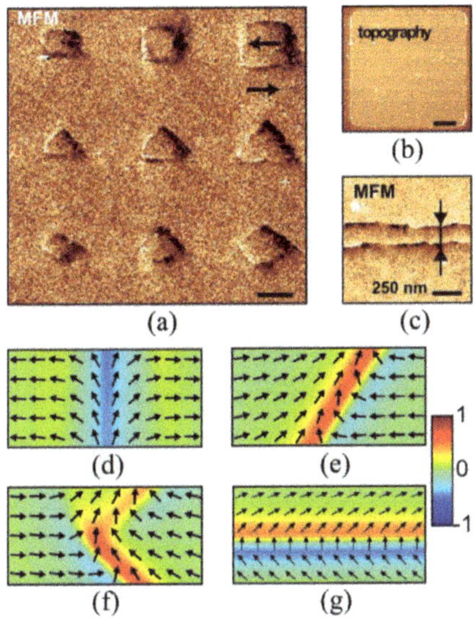

Figure 10.2. (a) Magnetic force microscopy image of tam-SPL written triangular, diamond and square shaped domains with different dimensions. Black arrows indicate the direction of the magnetization in the patterned and non-patterned regions. Scale bar: 2 μm. (b) Topography image of patterned region. Scale bar: 4 μm (c) High-resolution MFM image of a stripe domain written with a single pass of the heated tip. Scale bar: 500 nm. (d–g) Micromagnetic simulations showing the spin configuration of the domain walls of square, triangle, diamond and stripe domains, respectively. Adapted by permission from [12], copyright (2016) Springer.

with respect to the initialization field H_i. In this specific case, the heater temperature T_h was set to 560 °C, 675 °C and 765 °C for squares, triangles and diamonds, respectively. These values ensured that the surface temperature, which is lower than the heater temperature due to thermal dissipation, was kept above the blocking temperature of the exchange bias system $T_B \sim 160$ °C. Noteworthy, the topographic image associated with the MFM image (figure 10.2(b)) does not show any visible pattern, confirming the purely magnetic nature of tam-SPL patterning. For assessing the minimum domain size achievable in this system, stripe-shaped domains were patterned by sweeping the heated tip along a single 12 μm long line. Figure 10.2(c) shows the high-resolution MFM image of the resulting domain, which consists of two 180° Néel domain walls separated by 250 nm. This case provides a clear demonstration of the fact that the minimum domain size achievable via tam-SPL is related to the domain wall width of the magnetic system (which in turn depends on the micromagnetic properties of the system), rather than the spatial extension of the thermal profile generated by the tip, which could reach sub-10 nm resolution [25]

The possibility of vectorially orienting point-by-point the magnetization opens up a plethora of possibilities which go beyond the writing of two-dimensional domains. In fact, by controlling the magnetization within the domains, it is possible to tailor the properties and shape of the domain boundaries, as well. This capability is well clarified by the micromagnetic simulations shown in figures 10.2(d)–(g), which visualize the spin-configuration at the boundaries of the magnetic domains shown in figure 10.2(a), where the local spin-orientation is indicated by black arrows while the color code indicates the calculated MFM contrast, which allows one to directly compare the experiments with the simulations. Figures 10.2(d)–(g) show the stabilization of different types of domain walls, enabled by the control of the shape of the magnetic patterns with respect to the orientation of the writing field. Specifically, panel (d) shows the stabilization of a tail-to-tail 180° wall at the right boundary of the square domain, and the stabilization of side-to-side 180° walls at the top and bottom boundaries of square and stripe domains (panel (g)). In such domain walls, the spins at the center of the wall are oriented parallel (tail-to-tail) and perpendicular (side-to-side) with respect to the wall profile. Furthermore, domain walls characterized by a mixed character and complex shape profile develop at the boundaries of triangular and diamond shaped domains, and are shown in panels (e) and (f). This provides a compelling demonstration that tam-SPL allows the stabilization of complex tailored one-dimensional spin textures in continuous films e.g. by imposing non-uniform boundary conditions via domain writing.

By generalizing this concept, it is possible to develop a methodology [26] for patterning zero-dimensional magnetic structures such as vortices and Bloch lines within domain walls. Such topological spin textures are raising interest due to their potential in data storage and computing, so that finding novel ways for controlling their properties beyond the capabilities of conventional nanofabrication is highly appealing. For example, one of the most used strategies for stabilizing vortex configurations is by exploiting the flux closure in round or elliptical magnetic dots [27]. However, this does not allow one to straightforwardly control the circularity of the texture (clockwise or counter-clockwise), and its vorticity (vortex or antivortex). In the following, we show two strategies for stabilizing solitons with deterministically controlled properties in continuous exchange biased systems, by designing vectorially the unidirectional anisotropy landscapes via tam-SPL. Figure 10.3 shows the stabilization of vortex–antivortex (VA) pairs with tailored circularity. For doing so, as shown in panel (d), tam-SPL was used for patterning triangular shaped regions with counter-clockwise unidirectional anisotropy and a shared central apex. For obtaining a spatially rotating unidirectional anisotropy, each region was sequentially patterned with a different orientation of the writing field H_W. The arrows mark the direction of the exchange bias field in each region, which sets the direction of the magnetization at remanence, after patterning. This non-uniform magnetization configuration in turn generates four 90° domain walls, a potential minimum for the vortex core in the center of the patterned area (pink dot) and an antivortex located at the bottom left corner (orange dot). Panels (e) and (f) show the MFM image and micromagnetic simulation of the patterned structure, respectively, where the black arrows mark the local magnetization orientation, and the color contrast in the simulations indicates the divergence of the magnetization which is related to the experimental MFM contrast. It is worth noting that the circularity of the vortex and the position of the antivortex is determined by the sense of rotation of the unidirectional anisotropy pattern.

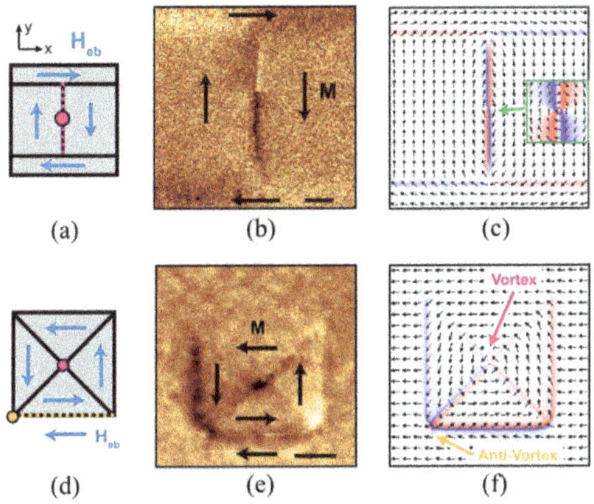

Figure 10.3. Patterning topological solitons. (a, d) Sketch of the direction of the exchange bias, blue arrows, for stabilizing a vortex Bloch line (pink dot) within a Néel wall (a) and a vortex (pink dot)-antivortex (orange dot) pair (d) in a continuous exchange bias system. (b, e) MFM image of the patterned regions at remanence. Black arrows indicate the direction of the magnetization. Scale bars: 3 μm (b) and 2 μm (e). (c, f) Micromagnetic simulation showing the spin configuration and the position of the vortex cores and antivortex. Adapted from [26], with the permission of AIP Publishing.

Panels (a, c) show a second strategy, which allows combining the patterning of domain walls and solitons. First, two magnetic domains with opposite magnetization orientation (e.g. pointing upwards and downwards, respectively) are patterned as shown in panel (a). This leads to the stabilization of a Néel side-to-side 180° wall at the domain boundary. Then, two domains with magnetization pointing horizontally (e.g. pointing towards the right and the left, respectively) are patterned at the two ends of the wall. As a result, the wall chirality (i.e. the direction of the magnetization within the wall) is determined, at each end, by the magnetization direction of the corresponding domain. A Bloch line is stabilized within the wall if the wall chirality is different at the two ends. Panel (a) shows the case of a vortex-type Bloch line with clockwise circularity (pink dot), stabilized within a straight Néel wall. Noteworthy, by controlling the relative direction of the magnetization of the four domains, it is possible to control the vorticity of the Bloch line (vortex or antivortex type), and its circularity (clockwise or counter-clockwise). Panels (b) and (c) show the MFM image and corresponding micromagnetic simulations, where dark arrows mark the local direction of the magnetization. The presence of the Bloch line is confirmed by the abrupt change in the wall chirality, which is visible in the MFM magnetic contrast. Interestingly, the domain wall is constituted by metastable points for the Bloch line, which could therefore be displaced along the wall, e.g. via external fields or currents.

10.2.3 Writing spin textures in synthetic antiferromagnets

So far, we focused on writing spin textures in exchange bias FM/AF bilayers, however, the very same tam-SPL concept can be generalized to more complex magnetic multilayers. In the following, we show the application of tam-SPL

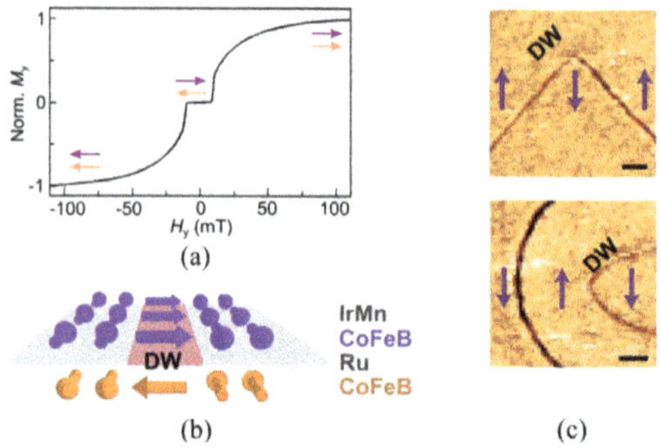

Figure 10.4. Patterning spin textures in synthetic antiferromagnets. (a) Hysteresis loop of a CoFeB (35)/Ru (0.6)/CoFeB (35)/IrMn (10) nm SAF system. The direction of the magnetization in the top (bottom) layer is indicated by the purple (orange) arrow. (b) Structure of a domain wall, in pink, in a synthetic antiferromagnet. (c) MFM images of straight and curved domain walls patterned via tam-SPL. The direction of the magnetization of the top FM layer is indicated by purple arrows. Scale bars 3 μm. Adapted from [36] John Wiley & Sons.

patterning to synthetic antiferromagnets (SAFs) [28], i.e. magnetic multilayers comprising two ferromagnetic layers exchange coupled antiferromagnetically via RKKY coupling [29, 30] across a thin nonmagnetic layer. Throughout the years, SAFs played a pivotal role in the development of spintronic devices, for example in spin valves [31] and magnetic tunneling junctions [32], and more recently they have been proposed for racetrack memories [33] and in the emerging field of magnonics [34].

In figure 10.4(a) we show the hysteresis loop and the magnetization direction of the top (purple arrow) and bottom (orange arrow) ferromagnetic layers of a Ru(2)/$Ir_{20}Mn_{80}$(10)/$Co_{40}Fe_{40}B_{20}$(45)/Ru(0.9)/$Co_{40}Fe_{40}B_{20}$(45) system measured by vibration sample magnetometry (VSM). The two lateral lobes mark the rotation of the ferromagnetic layers, while the plateau at low fields indicates robust antiferromagnetic coupling. In this region, the magnetization direction of the top CoFeB layer is set by exchange bias with the 10 nm IrMn layer, while the magnetization of the bottom layer is oriented antiferromagnetically. tam-SPL patterning is performed, as described above, by sweeping the probe heated above the blocking temperature of the system on the sample surface while applying a writing external field H_w strong enough to saturate the system [35]. This allows one to spatially control the exchange bias direction, and therefore the magnetization orientation of the top layer at remanence. The magnetization of the bottom layer is oriented point-by-point in the opposite direction by the antiferromagnetic coupling. Figure 10.4(b) shows a sketch of a 180° domain wall. Figure 10.4(c) shows two experimental examples of domain walls stabilized by writing oppositely oriented magnetic domains in SAF. In particular, MFM images show the magnetic contrast arising at the domain walls in the case of straight angled walls (top) and parabola-shaped domain walls with different curvature (bottom). In the following section, we show how tam-SPL written spin textures in SAF can be employed as tunable and highly controlled spin-wave emitters.

10.2.4 Features and capabilities of tam-SPL

Here, we review some of the features, capabilities and limitations of tam-SPL, taking as an example the simplest case of spin textures stabilized in exchange bias FM/AF bilayers.

As explained above, tam-SPL allows the vectorial point-by-point tuning of the unidirectional anisotropy. Below, we show that by varying the writing temperature during patterning it is possible to precisely tune the exchange bias strength in a continuous fashion. Figure 10.5(a) shows the MFM image at remanence of two squares patterned with writing field H_w in the opposite direction with respect to the initialization field H_i (top). The central and bottom panels show the corresponding (color coded) local hysteresis loops measured via magneto-optical Kerr effect microscopy (μ-MOKE) along the exchange bias direction. Each pattern was performed with a different heater temperature T_H = 240 °C (central square), T_H = 435 °C (left square). The non-patterned sample (top yellow square) does not show any magnetic contrast in the MFM image at remanence, as expected from its hysteresis loop, which is completely shifted from the center and featuring an exchange bias field H_e of −50 Oe set during initialization. By increasing the patterning temperature around the T_B of the system (central green square), the local loop shifts towards the center, and a magnetic contrast appears in the MFM image. By further increasing the patterning temperature above T_B (left red square), the loop completely shifts on the opposite direction up to H_{ep} = 50 Oe, and a well-defined magnetic domain is retrieved at remanence in the MFM image. This capability is enabled by the precise temperature control allowed by t-SPL, in combination with the statistical distribution of blocking temperatures in

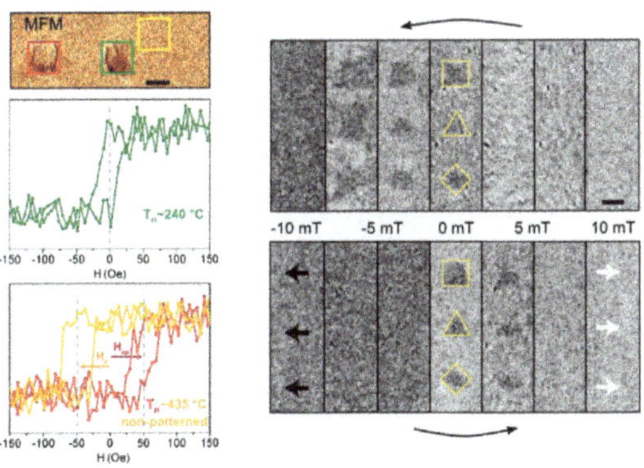

(a) (b)

Figure 10.5. Tunability of the unidirectional anisotropy and evolution of the domains as a function of the external field. (a) Top panel: MFM image of a non-patterned area (yellow), a square domain patterned with T_H = 240 °C (green) and a square domain patterned with T_H = 435 °C (red). Central and bottom panels: local μ-MOKE hysteresis loops color-coded as above. (b) μ-MOKE images of the domains shown in figure 10.2(a) acquired at different external fields, showing the evolution of the patterns as a function of the external field. In the bottom (top) panel the forward (reverse) 'loop' is shown. Scale bars 2 μm. Adapted by permission from [12], copyright (2016) Springer.

polycrystalline exchange bias systems, which arises from the distribution of grain sizes within the film [24].

Another appealing feature of tam-SPL is its reversibility. The process of setting the exchange bias via field cooling is initiated by temporary weakening (or destroying) the antiferromagnetic order with temperature. It is therefore a reversible process, provided that heating does not induce other irreversible transformations in the films. In this sense, heating polycrystalline sputtered multilayers at high temperature is known to cause interdiffusion and eventually damage the interfaces. In turn, this can cause changes in the magnetic properties of the film (coercivity, saturation magnetization) and weakening of the exchange bias. It is worth noting however, that the temperature resilience of sputtered magnetic materials has been extensively studied in the framework, e.g. of heat-assisted recording, MTJs (see e.g. [37, 38]), so that it is indeed possible to suitably design interfaces which are resilient to thermal treatment. In this framework, tuning the blocking temperature of the system by engineering the composition of the antiferromagnetic layer represents a viable additional route to reduce the patterning requirements in terms of peak temperature [39].

The reversibility of the patterning process leads to fully reconfigurable domains. In particular, as shown in [12], writing, erasing and rewriting patterns can be done simply by sweeping the heated tip on the same region while applying a different writing field $H_{w.}$ Noteworthy, as explained in figure 10.3, the reconfigurability of tam-SPL patterns is at the core of the ability to write complex spin textures, such as vortices and Bloch lines, and represents a distinctive feature with respect to the vast majority of nanopatterning methodologies.

While, so far, we mainly focused on the patterns at remanence, now we steer our attention to the evolution of the patterned domains as a function of the external field. For doing so, we acquired μ-MOKE images of the patterned domains as a function of the external field applied in the plane of the films. Figure 10.5(b) shows the image sequence and corresponding hysteresis loops of the domains shown in figure 10.2(a). In particular, the top (bottom) sequence was acquired by sweeping the field along to the exchange bias direction from +100 Oe to −100 Oe (from −100 Oe to +100 Oe). It is worth noting that two different regimes can be observed, one at low fields (around 0), one at high fields, comparable to the exchange bias field.

At low fields, the patterned domains, which appear as dark regions on bright background, are modulated by the application of the external field. In particular, they expand when the external field and their magnetization point in the same direction, and they shrink when they point in the opposite direction. This is due to the fact that the domain walls can be displaced upon application of relatively small fields. After removing the external field, the original domain configuration is retrieved. The possibility to reversibly and precisely tune the spin textures via external stimuli provides an additional flexible strategy to design real-time reconfigurable devices.

At higher external fields, i.e. when the field becomes comparable to the exchange bias field, the domain pattern disappears and a uniform magnetization orientation in the whole sample is obtained, as shown by the uniform dark (bright) contrast in the μ-MOKE images acquired at large negative (positive) fields. Importantly, after removing the external field the same magnetization configuration at remanence is recovered after 2000 loops. This stability of the pattern stands from the resilience of the patterned exchange bias against external magnetic fields.

10.3 Nanopatterned spin textures for magnonics

The ability to stabilize multidimensional spin textures in continuous films opens up a plethora of possibilities in the emerging field of magnonics. Magnonics seeks to control the excitation, propagation and transduction of spin waves, i.e. propagating perturbations in the spin texture of magnetic materials, for building highly-integrated and energy-efficient computing devices [34, 40, 41]. The main powerful enabling features of spin waves are the absence of Joule losses associated with spin-wave propagation, and the short spin-wave wavelength in the GHz–THz range, which can go down to a few nm, i.e. several orders of magnitude shorter than electromagnetic waves. These two features, together with the extremely-rich spin wave phenomenology and the improvement of micro-nanofabrication techniques makes magnonics a highly active research field. In this framework, the use of tam-SPL engineered spin textures for controlling the excitation and propagation of spin waves is an extremely promising avenue, ensuring reconfigurable functionality and high design flexibility towards both fundamental studies and fabrication of magnonic nanodevices. In this paragraph, we review our recent results on the use of 2D–1D and 0D engineered spin textures for controlling the emission, propagation and spatial confinement of spin waves in exchange bias single layer systems and synthetic antiferromagnets.

10.3.1 Controlling the spin-wave excitation and propagation with magnetic domains

Here, we show that the excitation efficiency and propagation of spin waves can be spatially controlled by patterning magnetic domains in a continuous film [12]. Figure 10.6(b) shows the MFM image of 2.5 μm wide vertical stripe-shaped domains with magnetization pointing towards the left and upwards direction (black arrows), patterned via tam-SPL in a CoFeB (5)/IrMn (7) nm exchange bias bilayer. The presence of sharp 90° Néel domain walls is confirmed by the dark/bright contrast. In order to excite spin waves, Ti (7)/Au (200) nm coplanar waveguides (CPW) were patterned via EBL on the top of the film after depositing an electrically insulating SiO_2 layer. An optical image of the sample is shown in figure 10.6(a). For exciting spin waves, an RF current is injected in the CPW, so that an RF Oersted magnetic field H_{rf} is generated in the vicinity of the CPW. In panel (a), the direction of H_{rf} is marked by red lines, and the domain walls separating alternating 90° domains are indicated by dashed orange lines.

For studying spin waves with high spatial resolution, we used micro-focused Brillouin light scattering (μ-BLS). In BLS, incident photons (generated by a laser focused on the sample surface) are scattered inelastically by the interaction with magnons. By analyzing the intensity, the wavevector and the frequency shift of the scattered photons, it is possible to retrieve information on the dispersion and

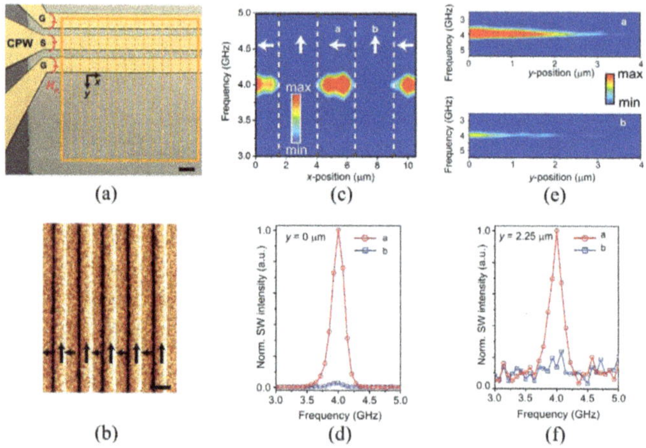

Figure 10.6. Controlling spin waves with magnetic domains. (a) Coplanar waveguide (CPW) for spin-wave excitation. The vertical dashed orange lines mark the domain walls. (b) MFM image showing vertical 2.5 μm stripe domains separated by 90° domain walls. The direction of the magnetization within the domains is indicated by black arrows. (c) Line-map of the spin-wave intensity, color-coded, along the x direction, acquired right outside the CPW. Dashed lines mark the domain walls and white arrows indicate the direction of the magnetization within the domains. (d) Spin-wave spectra acquired in the center of stripe a (Damon–Eshbach configuration) in red and stripe b (backward-volume) in blue. The spectra were acquired right outside the CPW. (e) Line-map of the spin-wave intensity, color-coded, acquired along the y direction in stripe a (top panel) and b (bottom panel). (f) Spin-wave spectra acquired at 2.25 μm from the CPW in stripe a (red) and b (blue). Adapted by permission from [12], copyright (2016) Springer.

intensity of the spin waves travelling within the magnetic film. Furthermore, by scanning the focused laser beam, it is possible to build one- and two-dimensional intensity maps with spatial resolution limited by the size of the laser spot, which in our case is around 250 nm. Figure 10.6(c) shows the one-dimensional spin-wave intensity map acquired along the x direction (parallel to the CPW) right outside the CPW. Spin waves were excited by running a 4 GHz current in the CPW. The spin-wave intensity is color-coded, the domain walls are indicated by dashed lines and the direction of the magnetization within each domain is indicated by the white arrows. Noteworthy, we observe well-defined maxima in the spin-wave intensity within the domains where the magnetization lies parallel to the CPW (Damon–Eshbach configuration, DE) and minima where the magnetization is perpendicular to the CPW (backward volume configuration, BA). Panel (d) shows the spin-wave spectra acquired in the center of regions a and b, showing a 36-fold higher excitation intensity in the DE stripes with respect to the BA stripes. This is due to the fact that the spin-wave excitation efficiency is maximum when the CPW Oersted field couples well with the dynamical magnetization of the spin-wave mode. Such coupling is more efficient in the DE configuration, because the Oersted field has a strong component along the y direction, i.e. along a major axis of the elliptical precession of the dynamic magnetization. In contrast, in the BA configuration, the coupling is less efficient, resulting in a much lower spin-wave intensity. In [42] we also showed that the spin-wave excitation efficiency within tam-SPL domains can be reversibly

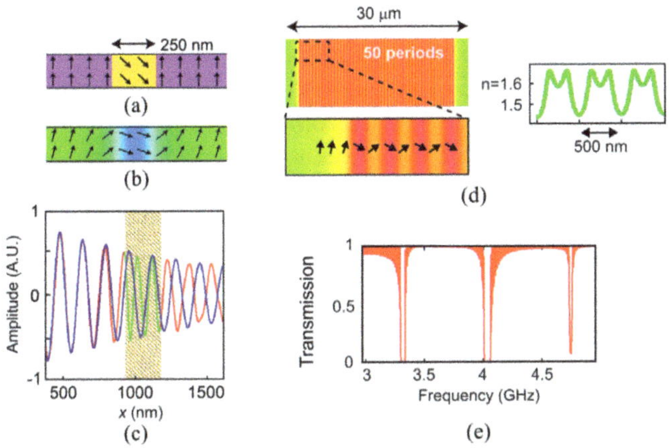

Figure 10.7. Micromagnetic simulations of phase-shifters and magnonic crystals. (a) Unidirectional anisotropy direction of a single 250 nm wide stripe. (b) Resulting micromagnetic configuration at remanence. (c) Amplitude of the spin waves propagating in presence of the stripe (green and red curve) and in the uniformly magnetized film (blue). (d) Sketch of the simulated magnonic crystal comprising 50 repetitions of the stripe shown in (a). On the right, profile of the effective refractive index. (e) Simulated transmission of the magnonic crystal. Adapted by permission from [12], copyright (2016) Springer.

modulated in real-time via an external static field, by temporarily reorienting their magnetization. Figure 10.6(e) shows a one-dimensional μ-BLS map acquired along the *y* direction (i.e. perpendicular to the CPW). The top panel, related to the DE configuration, shows the detection of propagating spin waves up to 4 μm from the CPW, while the bottom panel, related to the BA configuration, shows detection up to 2 μm from the CPW. Figure 10.6(f) shows the corresponding spin-wave spectra acquired at 2.25 μm from the CPW, demonstrating that while spin waves are still detectable in the DE stripe, no spin-waves are detected in the BA stripes. This demonstrates that, by locally controlling the direction of the magnetization, it is possible to straightforwardly spatially modulate in a reversible way the spin-wave excitation, propagation and confinement in a continuous film, without any physical patterning, and allowing for both off-line reconfigurability of the domain structure at remanence via tam-SPL and real-time reconfigurability via external stimuli.

Another appealing application of tam-SPL-written magnetic domains is the realization of reconfigurable magnonic metamaterials made by periodically modulating the magnetic properties (e.g. unidirectional anisotropy and/or magnetization orientation) with high spatial resolution. Figure 10.7 shows micromagnetic simulations presenting the design of a spin-wave phase-shifter based on a single 250 nm stripe domain. Panels (a, b) show the unidirectional anisotropy direction and resulting spin configuration of a single stripe, respectively, while panel (c) shows the effect of a single stripe on the spin-wave propagation, in contrast to the case of uniformly oriented magnetization. In particular, the presence of the stripe causes a 3π phase-shift in the propagating spin wave excited at 10 GHz (red and green line), with respect to the case of uniformly oriented magnetization (blue line). The phase-shifting mechanism is based on the strongly anisotropic spin-wave dispersion depending on the relative angle between the wavevector *k* and the magnetization *M*, which causes the spin waves to have a different wavelength when travelling within different magnetic domains. Such phase-shifting element can be used, e.g. as building blocks of Mach–Zehnder interferometers.

This change in the refractive index can be exploited for building magnonic crystals, i.e. periodic magnetic nanostructures for controlling the transmission of spin waves [43]. Panel (d) shows the micromagnetic configuration of a crystal (left), comprising 50 repetitions of the 250 nm wide stripes, and the periodic modulation of the spin-wave refractive index (right). Panel (e) shows the transmission of the crystal as a function of the frequency in the 3–5 GHz range, calculated with a one-dimensional transfer matrix method, displaying the characteristic magnonic band-gaps. The position and bandwidth of the gaps can be tuned by changing the width of the stripe, the number of repetitions and the separation between the stripes. Interestingly, as shown before, the magnonic crystal properties can be fully reconfigured by performing tam-SPL again, or in real time by applying an external field for temporarily re-orienting the magnetic domains.

10.3.2 Nanoscale spin-wave circuits based on spin textures

In the previous paragraph, we demonstrated that 2D micrometer-sized domains can be employed as waveguide for spin waves. Here, we show that tailored 1D spin textures can act as nanoconduits, for confining and steering spin waves and realizing spin-wave circuits [44]. Straight and curved domain walls were patterned via tam-SPL in a CoFeB (20)/IrMn (10)/Ru (2) nm exchange biased system, grown on Si/Si_3N_4 200 nm thick membranes via magnetron sputtering, as described previously. On top of the patterned samples, we fabricated 2 μm × 30 μm microstrip antennas via maskless optical lithography and lift-off, after depositing a 50 nm thick SiO_2 insulating layer via magnetron sputtering. The stripline consisted of a Cr 5 nm/Cu 200 nm bilayer deposited by thermal evaporation. We studied the confined spin-wave modes propagating along the spin textures via space and time-resolved scanning transmission x-ray microscopy (STXM), where monochromatic x-rays, tuned to the Co L3 absorption edge (photon energy of about 781 eV), are focused onto a spot on the sample, and the transmitted photons are recorded using an avalanche photodiode. The sample is scanned using a piezoelectric stage, and the transmitted x-ray intensity is recorded for each pixel to form the image. Magnetic contrast arises from the x-ray magnetic circular dichroism (XMCD) of the circularly polarized light employed. In the particular configuration we used to obtain the results reported in this paragraph, the black-white contrast is associated to the in-plane component of the magnetization (M_x).

In figure 10.8(a), we show the static STXM image of a curved 180° Néel domain wall, obtained by patterning two domains with antiparallel remanent magnetization. The orange arrows indicate the direction of the magnetization in the two domains. At the bottom of the figure, the edge of the antenna used for spin-wave excitation is visible. Panel (c) displays a sequence of STXM images showing the propagation of spin waves along the domain wall. The images are taken stroboscopically,

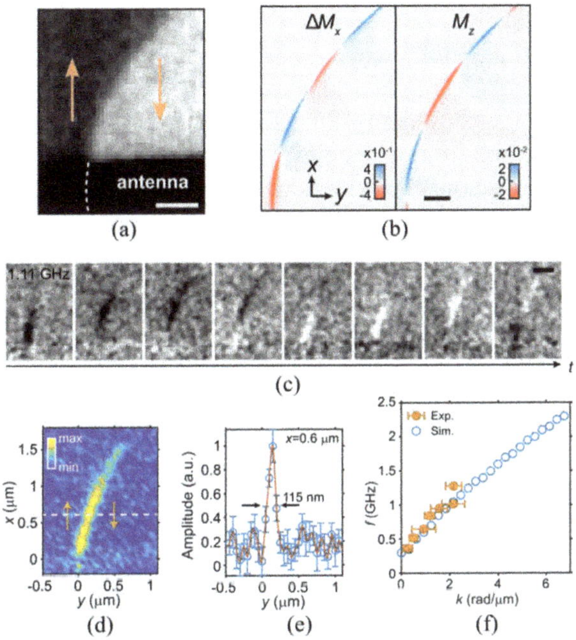

(a) (b) (c) (d) (e) (f)

Figure 10.8. Confining and steering spin waves with domain walls. (a) STXM image of a curved 180° Néel wall patterned in a CoFeB (20)/IrMn (10) nm system. Orange arrows indicate the direction of the magnetization within the domains. (b) Micromagnetic simulations showing spin waves propagating within the curved domain wall. (c) Time-resolved STXM images showing spin waves excited at 1.11 GHz, propagating along the curved 180° Néel wall. (d) Spatial map of the spin-wave excitation amplitude showing confinement at the domain wall. (e) Horizontal profile extracted from (d) at a distance of 0.6 μm from the antenna. (f) Experimental (filled orange) and simulated (empty blue) dispersion of spin waves propagating within a straight domain wall. Scale bars: 500 nm. Adapted from [44].

generating the spin waves with a sinusoidal excitation at 1.1 GHz and without any applied static external field. The normalized M_x signal reported in panel (c) is calculated as the magnetic deviation $\Delta M_x(t)$ from the time-averaged state $\langle M_x(t) \rangle$ and shows spin waves, confined at the domain wall, propagating away from the microstrip located at the bottom. In order to better understand the geometry of the propagating modes, in figure 10.8(b) are presented micromagnetic simulations of the spin-wave excitation in the presence of the curved domain wall. In particular, the left panel shows the magnetic deviation $\Delta M_x(t)$ from the time averaged value $\langle M_x(t) \rangle$, while in the right panel the M_z component is reported. The excited modes are called Winter magnons [45–47] and are characterized by a flexural motion of the domain wall, associated with an elliptical spin precession within the wall. Panel (d) reports the spatial map of the amplitude of the spin-wave excitation, which is obtained by plotting for each pixel the amplitude of the sinusoidal fit of the time-trace (raw data) acquired in it. The lateral extension of the spin-wave mode corresponds to 115 nm, as calculated from the full-width-at-half-maximum (FWHM) of the horizontal profile extracted from the spatial map at $x = 0.6$ μm from the stripline (panel (d)) and is reported in panel (e). This confirms that the mode is confined at the domain wall, and that no spin waves are propagating within the domains. In figure 10.8(f), the experimental (filled orange dots) and simulated (empty blue dots) dispersion relation is reported in the case of a straight 180° Néel domain wall. The experimental values were acquired from measurements at different excitation frequencies, while the simulations were obtained from a sinc-shaped field pulse excitation. The propagating character of the modes is confirmed by both simulations and experiments, displaying a positive dispersion.

A small bandgap is visible below 0.3 GHz and can be attributed to the presence of a residual effective field in correspondence of the domain wall, arising from crystalline anisotropy and the exchange bias. The spin-wave group velocity $v_g = \partial\omega/\partial k$, extracted at $k \to 0$, has been calculated both from the experimental data and from the simulations and is $v_{gExp} = 2.77 \pm 1.4$ km s^{-1} and $v_{gSim} = 2.30 \pm 0.34$ km s^{-1}, respectively. Noteworthy, the spin-wave modes confined at the domain wall are the only ones supported in this frequency range, which falls within the bandgap of the propagating modes in the uniform film. The confinement of the modes at the domain wall is therefore guaranteed. Noteworthy, propagating distances exceeding 4 μm have been measured in curved domain walls, with a confinement width < 200 nm. This fact, together with the positive dispersion confirms that shaped spin textures and the corresponding guided modes can be used for transporting information within integrated circuits.

In order to verify the possibility to fabricate nanocircuits, we patterned spin textures comprising two domain walls sharing a common apex. Figures 10.9(a) and (b) show the static STXM images of the two domain walls when a small static magnetic field of 2 mT and 1.68 mT was applied in the plane of the film, as indicated. It is possible to notice that the distance between the two domain walls (dashed white

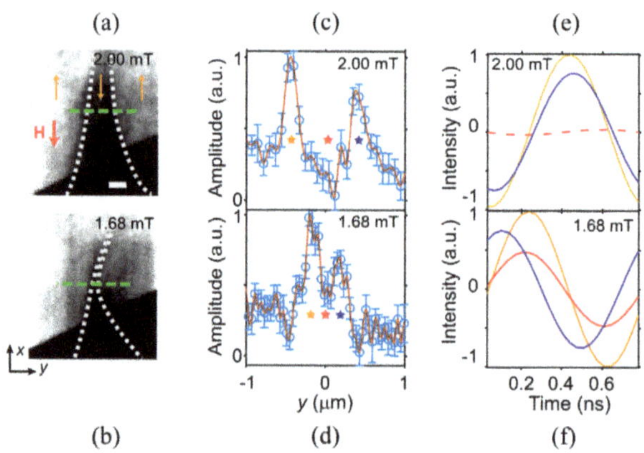

Figure 10.9. Reconfigurable spin-wave circuits based on domain walls. (a, b) STXM image of a spin texture comprising two 180° Néel walls (white dashed lines) sharing a common apex, acquired applying a static 2.00 mT (a) and 1.68 mT (b) external field H. The direction of the magnetization is indicated by the orange arrows. (c, d) Spin-wave excitation amplitude along the profiles indicated by the dashed green lines in panel (a) and (b), respectively. (e, f) Time traces extracted in correspondence of the colored stars in panels (c) and (d). When $H = 2.00$ mT the modes are spatially separate. By reducing the external field, the two modes can be tunably superimposed. Scale bars: 500 nm. Adapted from [44].

lines) can be tuned by applying a small external field. Indeed, by decreasing the field, the two domain walls, initially spatially separated (panel (a)), are brought closer and in panel (b) they share a common apex. As shown previously, the propagating spin waves are confined at the walls, so that by controlling the position of the walls it is possible to displace the spin-wave excitation with high accuracy.

The amplitude of the spin-wave modes along the horizontal profiles of panels (a-b) (green dashed lines) in the two configurations is shown in panels (c, d) and allows one to better appreciate the progressive overlapping of the two modes. The two guided excitations are visible as two peaks, with FWHM around 200 nm. In panel (c) the applied magnetic field corresponds to 2 mT and the two modes do not overlap, with a peak-to-peak distance of 810 nm. At a field of 1.68 mT, the two modes are separated by 340 nm and therefore partially overlap (panel (d)). In panels (e, f), by plotting the sinusoidal fit (time traces) extracted in correspondence of the stars (color coded) in panels (c, d), it is possible to better analyze the confined spin-wave modes as a function of the applied external field. Three curves are shown: those with higher amplitude (orange, blue) are related to the magnetization dynamics of the two guided modes, in correspondence of their maximum intensity. The phase difference of the two guided modes, as expected, depends on the applied field, which modifies the spin configuration and modulates the geometry of the nanowaveguides. The third, with lower amplitude (red), corresponds to the dynamics in the region where the two modes overlap. For an applied field of 2.00 mT, panel (e) shows that the two modes do not overlap, therefore no excitation is measured (red dashed line). Conversely, for lower fields, the two modes are spatially superimposed and the third lower-amplitude curve is visible in panel (f). This result shows that it is possible to achieve a sizeable modulation of the excitation amplitude and phase in the overlap region. The control of the superposition, phase difference and amplitude of the confined modes via external stimuli such as fields or currents, allows one to envision the implementation of logic devices based on nanoscale spin textures, such as Mach–Zehnder-type spin-wave interferometers.

10.3.3 Optically inspired nanomagnonics in synthetic antiferromagnets

In this paragraph, we discuss the use of patterned spin textures to generate spin waves for implementing an optically-inspired spin-wave platform [36]. We patterned domain walls and vortices in a Ru(2)/$Ir_{20}Mn_{80}$(10)/$Co_{40}Fe_{40}B_{20}$(45)/Ru(0.9)/$Co_{40}Fe_{40}B_{20}$(45) nm synthetic antiferromagnetic system (SAF) [48] (see section 10.2.3), grown on Si/Si_3N_4 200 nm thick membranes via magnetron sputtering. As described in the previous paragraph, we patterned 5 μm × 30 μm microstrips of Cr 5 nm/Cu 200 nm on a 50 nm thick SiO_2 deposited by thermal evaporation. We studied the spin-wave modes generated the spin textures via space and time-resolved STXM. In figure 10.10(a) a micromagnetic simulation sketches the concept of magnonic nanoantennas based on nanoscale patterned magnetic domain walls. The domain wall is set in oscillation by an external magnetic field H_{RF}, generated by a stripline located in close proximity to the wall, run by RF currents. In particular, the out-of-plane component of the oscillating H_{RF} field outside the microstrip couples

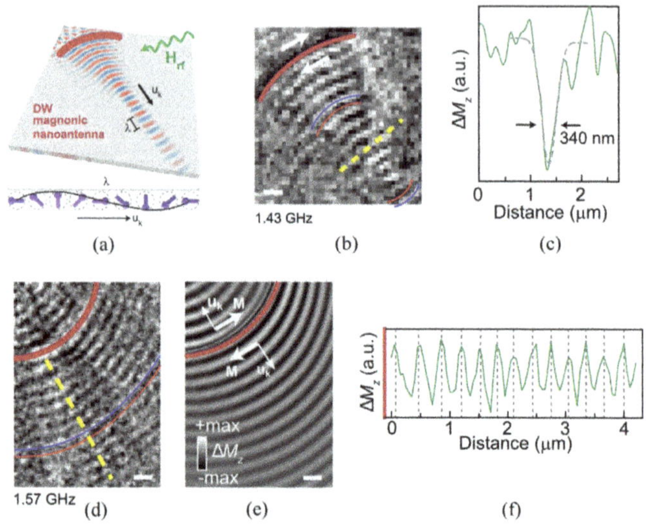

Figure 10.10. Domain wall based magnonic nanoantennas in synthetic antiferromagnets. (a) Simulations of the emission of a focused directional spin-wave beam by a domain wall based nanoantenna excited by an external oscillating field H_{rf}. u_k indicates the spin-wave propagation direction. On the bottom, the spin precession along in single oscillation is sketched. (b) STXM image showing the directional emission and focusing of convex spin-wave wavefronts by a curved wall. (c) Spatial profile of the spin-wave amplitude along the dashed line in panel (b). (d, e) STXM image and simulation of the emission and propagation of concave wavefronts. (f) Spatial profile extracted from the yellow dashed line in panel (d). Scale bars: 500 nm. Adapted from [36] John Wiley & Sons.

with the out-of-plane component of the magnetization of the SAF in correspondence to the wall. The coupling between the out-of-plane components has the advantage of being independent from the relative orientation of the stripline and the spin textures, so that curved, angled walls and vortices can be excited.

In figure 10.10(b), we show an experimental STXM image of the directional emission from a curved domain wall (red line) and focusing of spin-wave beams. The blue/red line corresponds to the emitted wavefronts and the white arrows represent the direction of the static magnetization of the upmost ferromagnetic layer. In this experimental configuration, the black/white contrast in the image corresponds to the oscillation of the out-of-plane component of the magnetization, normalized to the average value measured over a period of oscillation of the RF sinusoidal signal of the microstrip. The frequency of excitation is 1.43 GHz. The focus of the spin-wave beam is located at a distance of 2.5 μm from the spin texture, as can be evinced also from the change of concavity of the wavefronts. Panel (c) shows the spatial profile of the beams at the focal point, extracted along the yellow dashed line of panel (b). Consistently with a description analog to diffractive optics, the FWHM of the beam amplitude, which is $f = 340 \pm 50$ nm, is close to the value obtained considering the spin-wave wavelength and the numerical aperture of the source, i.e. the wall geometry: $f = \lambda/2\sin\,\alpha = 363$ nm. Panels (d, e) display the experimental STXM image and the corresponding micromagnetic simulation of spin waves emitted from

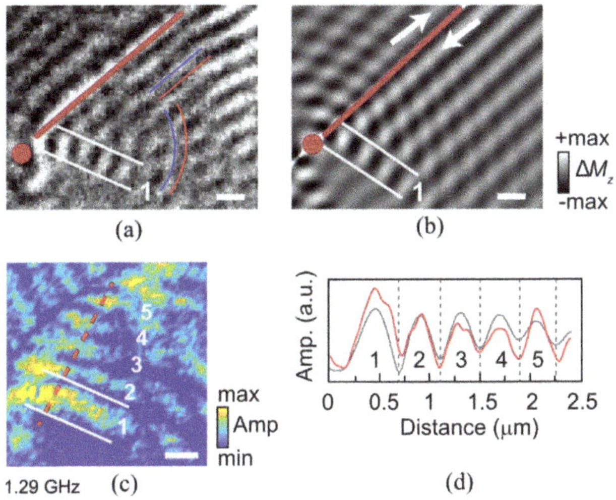

Figure 10.11. Generation of interference patterns using multiple nanoantennas. (a, b) STXM image and corresponding simulations of the interference from radial wavefronts emitted by a vortex (red dot) and a planar wavefront emitted by a straight wall (red line) excited at 1.29 GHz. The individual wavefronts are indicated by blue and red lines. (c) Spatial map of the spin-wave amplitude, showing the first five interference maxima (yellow) and minima (blue). (d) Amplitude profile extracted along the dashed line in panel (c) and corresponding simulations. Scale bars: 500 nm. Adapted from [36]. John Wiley & Sons.

an extended curved domain wall. In this case, the RF frequency was 1.57 GHz. Wavefronts retaining the same form of the domain wall are visible for several microns, limited only by the image acquisition window. Indeed, in figure 10.10(f) the experimental spatial profile extracted along the yellow dashed line of panel (d) shows a strong spin-wave signal visible for more than 15 periods, featuring a spin-wave wavelength of 300 nm. These experiments demonstrate how a straightforward wavefront engineering is possible by shaping the spin texture.

Figure 10.11 demonstrates that robust spin-wave interference patterns, spanning for several wavelengths, can be generated by patterning multiple spin-wave sources. In particular, panels (a-b) show the experimental STXM image (a) and micromagnetic simulations (b) of planar and radial wavefronts excited simultaneously by a straight domain wall (red line) and a vortex (red dot) located within the domain wall. The radial and planar wavefronts are highlighted by curved and straight red/blue lines, respectively, and the white arrows represent the direction of the magnetization of the top layer. The excitation frequency was 1.29 GHz.

As the black/white contrast represents the variation of the out-of-plane component of the magnetization at a specific time, constructive interference fringes are visualized as high intensity black/white regions. On the other hand, destructive interference is visible as low-intensity, grey areas and corresponds to regions where the spin waves sum in anti-phase. The region of the first interference maximum is indicated by white lines in both the panels. Figure 10.11(c) shows the spatial map of the spin-wave amplitude, obtained by fitting the time-trace of every acquired pixel of panel (a) with a sinusoidal function and plotting the corresponding amplitude.

The characteristic interference fringes, arising from the constructive (maxima) and destructive (minima) interference, are represented by yellow and blue regions, respectively, and are numbered in the panel. In panel (d), we show the experimental (red) and simulated (grey) spatial profile corresponding to the red dashed line of panel (c). The first five maxima, spaced by 450 nm are plotted, demonstrating a very good agreement between experimental and simulated data.

The spatial engineering of spin-wave wavefronts via tailored magnonic nano-antennas, the focusing of spin-wave beams down to the nanoscale, and the controlled generation of robust multibeam interference patterns are essential ingredients for analog processing. Future applications include reconfigurable micro-wave filters, isolators, and devices for pattern and speech recognition based on the emission, manipulation, and interference of coherent spin-wave wavefronts at the nanoscale.

10.4 End-of-chapter summary

In this chapter, we discussed our recent results on the development of the patterning technique tam-SPL, for reversibly controlling ferromagnetic spin textures with nanoscale precision. In particular, we presented the working principle and the main features of the technique. Figure 10.12 shows a summary of the capabilities of tam-SPL which were discussed in this chapter, and an overview of the magnetic systems where tam-SPL was applied. In particular, we demonstrated that tailored 2D–1D and 0D structures can be stabilized in continuous films (panel (a)): 2D magnetic domains with different shapes and extensions, 1D domain walls, such as Néel walls with tailored spin configuration, and 0D topological solitons, such as vortices and Bloch lines. We showed (panel (c)) that our technique can be applied to exchange biased systems with in-plane magnetization, but also to more complex systems such as synthetic antiferromagnets, where the magnetization orientation of the top layer at remanence can be patterned by controlling the exchange bias, while the magnetization of the bottom layer is set accordingly by the antiferromagnetic coupling. In perspective, tam-SPL can be applied also to exchange biased systems with perpendicular magnetic anisotropy. Interestingly, in these systems, Dzyaloshinskii–Moriya interaction stabilizes topological spin textures such as sky-rmions and chiral domain walls [49]. The nucleation, dynamics and annihilation of these topological spin configurations could be therefore controlled by patterning via tam-SPL.

Panel (b) presents some applications we demonstrated in the field of magnonics. In particular, we showed the spatial modulation of the spin-wave excitation and propagation in tailored magnetic domains, and the development of magnonic crystals based on the periodic modulation of the effective field. We then demon-strated that nanoscale 1D spin textures act as nanowaveguides, which effectively channel and steer spin waves, and can be the basic elements of tunable nanoscale circuits allowing for the control of the superposition and interference of confined spin waves. Finally, we presented a versatile, optically inspired platform using spin waves, in which magnonic nanoantennas based on tailored spin textures are used for

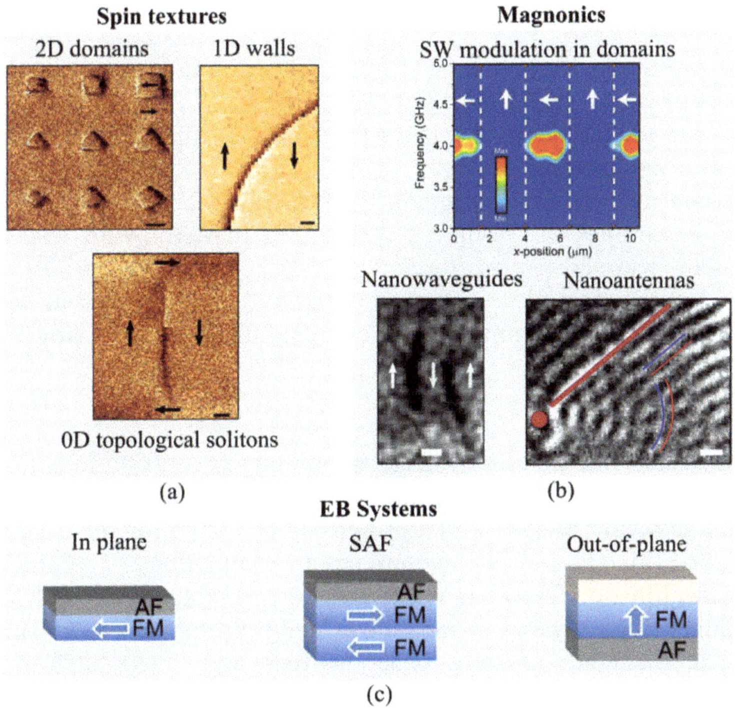

Figure 10.12. Current capabilities and applications of tam-SPL. (a) Multidimensional spin textures such as 2D magnetic domains [12, 42], 1D domain walls [44, 52] and tailored 0D topological solitons [26] are written reversibly in continuous films. (b) Applications of tam-SPL in the field of magnonics. Top panel: magnetic domains are used for spatially modulating the spin-wave excitation and propagation [12]. Bottom panels: nanoscale spin textures such as domain walls are used for building reconfigurable tunable circuits [44] and optically-inspired nanomagnonic platforms [36]. (c) Exchange bias systems where tam-SPL has been demonstrated so far: in-plane FM/AF bilayers, synthetic antiferromagnets and systems characterized by perpendicular magnetic anisotropy.

launching spatially-shaped coherent wavefronts, diffraction-limited spin-wave beams, and generating robust multi-beam interference patterns. In perspective, one of the most interesting applications in magnonics is represented by the study and control of spin waves in fully three-dimensional spin textures.

From a broader point of view, the full potential of engineered spin textures has still to be unveiled. For instance, the study and patterning of spin textures in different magnetic systems such as ferrimagnets and antiferromagnets represent extremely promising routes for exploiting the ultrafast velocity of domain walls [50], the spin dynamics up to the THz frequency range [51] and novel physical phenomena that can be eventually disclosed by tailoring the magnetic properties of these systems.

10.5 Student's corner

- What are the main factors determining the minimum magnetic domain size which can be patterned with tam-SPL?
- What are the factors determining the minimum tip temperature needed for tam-SPL patterning?
- What are the factors determining the maximum tam-SPL writing speed?
- What determines the resilience of tam-SPL patterns with respect to time, magnetic field and temperature?
- Would it be possible to pattern different spin textures in two stacked exchange bias antiferromagnet/ferromagnet bilayers? If so, how?

Acknowledgements

The authors would like to acknowledge the following contributions to the results reported in this chapter: Matteo Pancaldi and Paolo Vavassori for the micro-MOKE measurements; Marco Madami and Silvia Tacchi for the μ-BLS measurements; Raffaele Silvani for the micromagnetic simulations of propagating spin waves; Simone Finizio, Sebastian Wintz and Jörg Raabe for the STXM measurements; Adam Papp, Gyorgy Csaba, Wolfgang Porod for the micromagnetic simulations of magnonic crystals; Giacomo Sala and Giuseppe Scaramuzzi for sample preparation; William P King, Martin Spieser and Armin W Knoll for providing thermal nanolithography probes.

References

[1] Parkin S S P, Hayashi M and Thomas L 2008 Magnetic domain-wall racetrack memory *Science (80-.)* **320** 190–4

[2] Choe S-B 2004 Vortex core-driven magnetization dynamics *Science (80-.)* **304** 420–2

[3] Pribiag V S, Krivorotov I N, Fuchs G D, Braganca P M, Ozatay O, Sankey J C, Ralph D C and Buhrman R A 2007 Magnetic vortex oscillator driven by d.c. spin-polarized current *Nat. Phys.* **3** 498–503

[4] Fert A, Cros V and Sampaio J 2013 Skyrmions on the track *Nat. Nanotechnol.* **8** 152–6

[5] Ross C 2001 Patterned magnetic recording media *Annu. Rev. Mater. Res.* **31** 203–35

[6] Chappert C 1998 Planar patterned magnetic media obtained by ion irradiation *Science (80-.)* **280** 1919–22

[7] Holzinger D, Zingsem N, Koch I, Gaul A, Fohler M, Schmidt C and Ehresmann A 2013 Tailored domain wall charges by individually set in-plane magnetic domains for magnetic field landscape design *J. Appl. Phys.* **114** 013908

[8] Kim S, Lee S, Ko J, Son J, Kim M, Kang S and Hong J 2012 Nanoscale patterning of complex magnetic nanostructures by reduction with low-energy protons *Nat. Nanotechnol.* **7** 567–71

[9] Vieu C, Gierak J, Launois H, Aign T, Meyer P, Jamet J P, Ferré J, Chappert C, Mathet V and Bernas H 2000 High resolution magnetic patterning using focused ion beam irradiation *Microelectron. Eng.* **53** 191–4

[10] De Teresa J M, Fernández-Pacheco A, Córdoba R, Serrano-Ramón L, Sangiao S and Ibarra M R 2016 Review of magnetic nanostructures grown by focused electron beam induced deposition (FEBID) *J. Phys. D Appl. Phys.* **49** 243003
[11] Williams *et al* 2018 Two-photon lithography for 3D magnetic nanostructure fabrication *Nano Res.* **11** 845–54
[12] Albisetti *et al* 2016 Nanopatterning reconfigurable magnetic landscapes via thermally assisted scanning probe lithography *Nat. Nanotechnol.* **11** 545–51
[13] Vettiger *et al* 2002 The 'millipede'-nanotechnology entering data storage *IEEE Trans. Nanotechnol.* **1** 39–54
[14] Vettiger *et al* 1999 Ultrahigh density, high-data-rate NEMS-based AFM data storage system *Microelectron. Eng.* **46** 11–7
[15] Garcia R, Knoll A W and Riedo E 2014 Advanced scanning probe lithography *Nat. Nanotechnol.* **9** 577–87
[16] Pires D, Hedrick J L, De Silva A, Frommer J, Gotsmann B, Wolf H, Despont M, Duerig U and Knoll A W 2010 Nanoscale three-dimensional patterning of molecular resists by scanning probes *Science (80-.)* **328** 732–5
[17] Zheng *et al* 2019 Patterning metal contacts on monolayer MoS_2 with vanishing Schottky barriers using thermal nanolithography *Nat. Electron.* **2** 17–25
[18] Szoszkiewicz R, Okada T, Jones S C, Li T D, King W P, Marder S R and Riedo E 2007 High-speed, sub-15 nm feature size thermochemical nanolithography *Nano Lett.* **7** 1064–9
[19] Carroll K M, Giordano A J, Wang D, Kodali V K, Scrimgeour J, King W P, Marder S R, Riedo E and Curtis J E 2013 Fabricating nanoscale chemical gradients with thermochemical nanolithography *Langmuir* **29** 8675–82
[20] Albisetti E, Carroll K M, Lu X, Curtis J E, Petti D, Bertacco R and Riedo E 2016 Thermochemical scanning probe lithography of protein gradients at the nanoscale *Nanotechnology* **27** 315302
[21] Kim S, Bastani Y, Lu H, King W P, Marder S, Sandhage K H, Gruverman A, Riedo E and Bassiri-Gharb N 2011 Direct fabrication of arbitrary-shaped ferroelectric nanostructures on plastic, glass, and silicon substrates *Adv. Mater.* **23** 3786–90
[22] Wei *et al* 2010 Nanoscale tunable reduction of graphene oxide for graphene electronics *Science (80-.)* **328** 1373–6
[23] Nogués J and Schuller I K 1999 Exchange bias *J. Magn. Magn. Mater.* **192** 203–32
[24] O'Grady K, Fernandez-Outon L E E, Vallejo-Fernandez G, O'Grady K, Fernandez-Outon L E E, Vallejo-Fernandez G, O'Grady K, Fernandez-Outon L E E and Vallejo-Fernandez G 2010 A new paradigm for exchange bias in polycrystalline thin films *J. Magn. Magn. Mater.* **322** 883–99
[25] Ryu Cho Y K, Rawlings C D, Wolf H, Spieser M, Bisig S, Reidt S, Sousa M, Khanal S R, Jacobs T D B B and Knoll A W 2017 Sub-10 nanometer feature size in silicon using thermal scanning probe lithography *ACS Nano* **11** 11890–7
[26] Albisetti E, Calò A, Spieser M, Knoll A W, Riedo E and Petti D 2018 Stabilization and control of topological magnetic solitons via magnetic nanopatterning of exchange bias systems *Appl. Phys. Lett.* **113** 162401
[27] Shinjo T, Okuno T, Hassdorf R, Shigeto K and Ono T 2000 Magnetic vortex core observation in circular dots of permalloy *Science (80-.)* **289** 930–2

[28] Grünberg P, Schreiber R, Pang Y, Walz U, Brodsky M B and Sowers H 1987 Layered magnetic structures: evidence for antiferromagnetic coupling of Fe layers across Cr interlayers *J. Appl. Phys.* **61** 3750–2
[29] Bruno P and Chappert C 1991 Oscillatory coupling between ferromagnetic layers separated by a nonmagnetic metal spacer *Phys. Rev. Lett.* **67** 1602–5
[30] Parkin S S P, More N and Roche K P 1990 Oscillations in exchange coupling and magnetoresistance in metallic superlattice structures: Co/Ru, Co/Cr, and Fe/Cr *Phys. Rev. Lett.* **64** 2304–7
[31] Dieny B, Speriosu V S, Gurney B A, Parkin S S P, Wilhoit D R, Roche K P, Metin S, Peterson D T and Nadimi S 1991 Spin-valve effect in soft ferromagnetic sandwiches *J. Magn. Magn. Mater.* **93** 101–4
[32] Yuasa S and Djayaprawira D D 2007 Giant tunnel magnetoresistance in magnetic tunnel junctions with a crystalline MgO(0 0 1) barrier *J. Phys. D Appl. Phys.* **40** R337–54
[33] Yang S H, Ryu K S and Parkin S 2015 Domain-wall velocities of up to 750 m s-1 driven by exchange-coupling torque in synthetic antiferromagnets *Nat. Nanotechnol.* **10** 221–6
[34] Chumak A V, Vasyuchka V I, Serga A A and Hillebrands B 2015 Magnon spintronics *Nat. Phys.* **11** 453–61
[35] Albisetti E, Scaramuzzi G, Rinaldi C, Cantoni M, Bertacco R and Petti D 2020 Temperature dependence of the magnetic properties of IrMn/CoFeB/Ru/CoFeB exchange biased synthetic antiferromagnets *Materials (Basel)* **13** 387
[36] Albisetti *et al* 2020 Optically inspired nanomagnonics with nonreciprocal spin waves in synthetic antiferromagnets *Adv. Mater.* **32** 1906439
[37] Hayakawa J, Ikeda S, Matsukura F, Takahashi H and Ohno H 2005 Dependence of giant tunnel magnetoresistance of sputtered CoFeB/MgO/CoFeB magnetic tunnel junctions on MgO barrier thickness and annealing temperature *Jpn. J. Appl. Phys. Lett.* **44** 587–9
[38] Lee Y M, Hayakawa J, Ikeda S, Matsukura F and Ohno H 2006 Giant tunnel magnetoresistance and high annealing stability in CoFeB/MgO/CoFeB magnetic tunnel junctions with synthetic pinned layer *Appl. Phys. Lett.* **89** 87–90
[39] Sharma P, Albisetti E, Monticelli M, Bertacco R and Petti D 2016 Exchange bias tuning for magnetoresistive sensors by inclusion of non-magnetic impurities *Sensors* **16** 1030
[40] Stancil D D and Prabhakar A 2009 *Spin Waves: Theory and Applications* (Berlin: Springer)
[41] Csaba G, Papp Á and Porod W 2017 Perspectives of using spin waves for computing and signal processing *Phys. Lett.* A **381** 1471–6
[42] Albisetti E, Petti D, Madami M, Tacchi S, Vavassori P, Riedo E and Bertacco R 2017 Nanopatterning spin-textures: A route to reconfigurable magnonics *AIP Adv.* **7** 055601
[43] Krawczyk M and Grundler D 2014 Review and prospects of magnonic crystals and devices with reprogrammable band structure *J. Phys. Condens. Matter* **26** 123202
[44] Albisetti *et al* 2018 Nanoscale spin-wave circuits based on engineered reconfigurable spin-textures *Commun. Phys.* **1** 56
[45] Winter J M B 1961 Wall excitation. application to nuclear resonance in a bloch wall *Phys. Rev.* **124** 452–9
[46] Garcia-Sanchez F, Borys P, Soucaille R, Adam J-P, Stamps R L and Kim J-V 2015 Narrow magnonic waveguides based on domain walls *Phys. Rev. Lett.* **114** 247206
[47] Wagner K, Kákay A, Schultheiss K, Henschke A, Sebastian T and Schultheiss H 2016 Magnetic domain walls as reconfigurable spin-wave nanochannels *Nat. Nanotechnol.* **11** 432–6

[48] Wintz S, Tiberkevich V, Weigand M, Raabe J, Lindner J, Erbe A, Slavin A and Fassbender J 2016 Magnetic vortex cores as tunable spin-wave emitters *Nat. Nanotechnol.* **11** 948–53
[49] Yu *et al* 2018 Room-temperature skyrmions in an antiferromagnet-based heterostructure *Nano Lett.* **18** 980–6
[50] Siddiqui S A, Han J, Finley J T, Ross C A and Liu L 2018 Current-induced domain wall motion in a compensated ferrimagnet *Phys. Rev. Lett.* **121** 057701
[51] Li *et al* 2020 Spin current from sub-terahertz-generated antiferromagnetic magnons *Nature* **578** 70–4
[52] Albisetti E and Petti D 2016 Domain wall engineering through exchange bias *J. Magn. Magn. Mater.* **400** 230–5

Chapter 11

Nanofabrication of three-dimensional magnetic structures

Dédalo Sanz-Hernández, Claire Donnelly, Lucas Pérez and Amalio Fernández-Pacheco

The extension of nanoscale magnetism to three dimensions provides exciting opportunities to explore new physical phenomena, and at the same time, opens fascinating prospects to create 3D magnetic devices which could be exploited in future sensing and data storage technologies. The move of nanomagnetism to three dimensions faces, however, similar challenges to other nanotechnology fields: the fabrication of 3D nanoscale materials is extremely difficult and requires new patterning methods better suited for this purpose. In this chapter, we review some of the most promising emerging fabrication techniques for 3D nanopatterning of magnetic systems. We focus first on direct-write methods such as focused electron beam induced deposition (FEBID) and two-photon lithography, which have the ability to 3D print nano- and micro-structures with great precision, and which can either directly write magnetic materials, or create 3D scaffolds for subsequent deposition of magnetic films. We complement the chapter with a section dedicated to electrodeposition of magnetic materials, a technique with the ability to conformally deposit magnetic metals and lead to a wide variety of magnetic systems, including nanowires, multilayers and complex objects. Our text introduces first the fundamentals of these techniques, followed by an in-depth discussion of recent key examples where these techniques have been used to explore new 3D magnetic effects. An extensive list of references and materials is included, to help those interested in the 3D nanofabrication of magnetic systems.

11.1 Introduction

11.1.1 3D magnetic nanostructures for the future

Magnetic materials play a large role in our lives in a large variety of ways, due to their tailorable properties [1]. On the macro scale, their interaction with external

doi:10.1088/978-0-7503-2608-7ch11

phenomena is key: permanent magnets, that display high coercivities and produce high stray magnetic fields, are used widely in energy production and harvesting, particularly for renewable energy, making them a key component of the fight against climate change. Inductive magnetic materials, on the other hand, that are highly susceptible to external fields, are widely used as sensors.

On the micro- and nanoscales, the impact of miniaturised magnets is no less extensive. Multilayered magnetic films are used as sensors in the automotive industry, whilst grainy and patterned magnetic thin films are most widely used in perhaps the most well-known application of magnets: data storage. Following the discovery of effects such as giant magnetoresistance (GMR), the possibility for the electrical detection of the magnetic state has revolutionised data storage capabilities [2]. With the advent of big data, and the Internet of Things, data and its storage is of increasing importance in our lives, and magnetic storage devices—both tapes and hard drives—remain one of the most reliable long-term options.

Over the years, progress has relied on the continuous development of magnetic materials with improved properties—however, for further progress, including higher functionalities and more efficient and 'green' devices, a paradigm shift is required. One example is the case of permanent magnets: since the discovery of NdFeB in the 1980s, slight progress has been made via the doping of grain boundaries with exotic metals (Dy). However, progress on the level of the game-changing improvements provided by the discovery of NdFeB have been elusive. Recently, exciting developments in the 3D printing of bonded permanent magnets [3] have demonstrated that higher functionality can be obtained by patterning the magnetic material in specific 3D geometries: therefore providing significant improvements in device properties via the geometry, and not the material itself.

There is growing excitement/consensus in the field that such a *change in dimensionality* may provide the much-needed paradigm shift for nanomagnetism [4, 5]. Indeed, up until now, micro- and nanomagnetism have been limited to planar systems that are essentially two-dimensional, and much progress has relied on the highly controlled deposition of high quality magnetic thin films. As with permanent magnets, with the extension to three dimensions, there is the potential for control over the geometry to provide completely new functionalities. In curved magnetic films, for example, there have been predictions of curvature-induced magnetic anisotropy, as well as curvature-induced Dzyalozinskii–Moriya interactions (DMI) that induce chirality into the system [6]. As a result, the introduction of curvature to a film provides an alternative route to tailoring the magnetic material properties, with the potential for high levels of control, and access to properties not achievable in two dimensions. When one moves to three-dimensional nanostructures, such as magnetic nanotubes, or cylindrical magnetic nanowires, theoretical predictions indicate that the new geometries available will lead to domain wall velocities significantly higher than those observed in two-dimensional systems, as well as the introduction of new degrees of freedom, such as magnetochirality, non-reciprocal spin wave propagation, amongst others [7]. As a result, three-dimensional nanomagnetism could offer the paradigm shift that the magnetism community has been waiting for.

11.2 Overview of nanofabrication techniques for magnetism

Despite the huge potential envisaged for 3D nanomagnetism provided by theoretical and computational studies, the experimental realization of 3D magnetic nanostructures has so far been scarce, due to the huge difficulty in patterning nanoscale materials in three dimensions [4]. This fundamental bottleneck faced by nanomagnetism is common to other branches of nanotechnology, which would also benefit from moving to 3D. As an introduction to new types of 3D nanofabrication techniques, and specifically to those covered in this book chapter, let us first briefly outline an example of how the fabrication of a standard magnetic nanostructure device takes place, consisting of the combination of thin film physical vapour deposition (PVD) and standard lithography (optical/electron-beam lithography) processes. In a PVD step, a single or multilayered thin film with nanometric thickness is deposited on a substrate at high or ultra-high vacuum conditions via techniques such as evaporation or sputtering. This results in a high-quality *planar* magnetic material, where stoichiometry, crystallographic structure and defect density can be very finely controlled. Importantly, materials deposited by PVD also benefit from high-quality interfaces, where sophisticated magnetic interfacial effects [8] such as surface perpendicular magnetic anisotropy, indirect exchange coupling or antisymmetric exchange coupling can be engineered. Complementing PVD processes, lateral patterning of thin films at the nanometric scale is achieved by a series of lithography steps, where an organic resist is first spun on the substrate, followed by its exposure to a well-controlled photon/electron dose that modifies its solubility and enables it to be selectively removed by a chemical developer. The remaining resist is then used as a mask for subsequent steps, where the film areas not protected by this mask are selectively removed, a process that involves, e.g. physical etching (via ion irradiation) or chemical etching processes. This type of patterning approach to obtain a magnetic nano-device may comprise tens of steps involving the deposition of metallic and insulating layers, metallization steps, lift-off procedures… all of them critical for the final realization of the device. Despite this, 2D patterning is now a mature technology used in industry and research institutions around the world.

These denominated top-down methods for nanofabrication, where material is removed appropriately to form a magnetic nanostructure (see figure 11.1), excel in the fabrication of planar systems, and can readily reach nanoscale resolutions if the correct method is used. However, despite some remarkable advances in realizing 3D nanostructures [9], these are not designed, and in general are not suitable for 3D nanopatterning. The field of 3D nanomagnetism thus requires a different set of nanofabrication methods following a bottom-up approach, where small building constituents are put together to form a nanostructure. There exists a great variety of bottom-up techniques currently under development for nanotechnology, such as block co-polymer technology [10], DNA self-assembly [11], vapor–liquid solid methods [12], to cite a few. In this book chapter, we focus on two particular methodologies that are starting to provide exciting new experimental results in 3D nanomagnetism, namely (i) direct-write techniques for 3D nanofabrication, which include direct building of

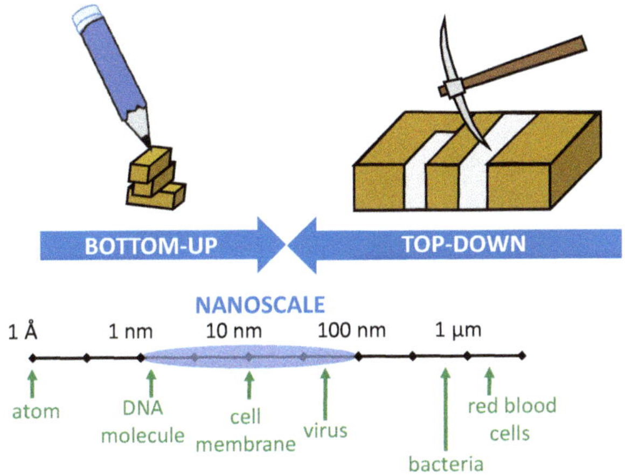

Figure 11.1. Schematic representation of bottom-up versus top-down lithography techniques for nanotechnology, involving systems on the scale of 1–100 nm. Some natural systems are included in the scale for comparison.

3D-shaped magnetic materials, as well as the possibility of creating 3D scaffolds that are subsequently covered by magnetic thin films, and (ii) deposition of advanced 3D magnetic structures using electrodeposition techniques. We note that this chapter content does not constitute a comprehensive overview of the 3D nanofabrication of magnetic materials, but provides a good introduction to the fundamentals of direct-writing and electrodeposition of 3D magnetic nanostructures.

11.3 Direct-write techniques for 3D nanofabrication of magnetic materials

In recent years, significant advances in 3D printer technologies have enabled the fabrication of 3D geometries that are not possible with standard 'subtractive' techniques [13, 14], making possible the prototyping of small-scale devices in an accessible and low-cost fashion. This ability to design and fabricate 3D patterns and geometries at the micro- and nanoscale has led to creative developments and advances in several fields, including optics [15–17], electronics [18–20] and biomedical technologies [21, 22].

For magnetic devices, which are widely used in sensing, the automation industry, and for energy harvesting and production, 3D fabrication techniques are also of significant interest. Recent developments in the 3D printing of magnetic materials have led to lightweight polymer bonded magnets that can be patterned in complex shapes. As a result, enhancements of the properties of permanent magnets [23, 24] and the production of specific magnetic fields [3], ideally suited for specific scientific infrastructure applications [25], has been realised. In addition to permanent magnets, patterned 3D magnetic systems have been incorporated in shapeable magnetoelectronics [26] and rolled up sensor microfluidics [27]. An important advantage of magnetic structures is the possibility that they provide for remote

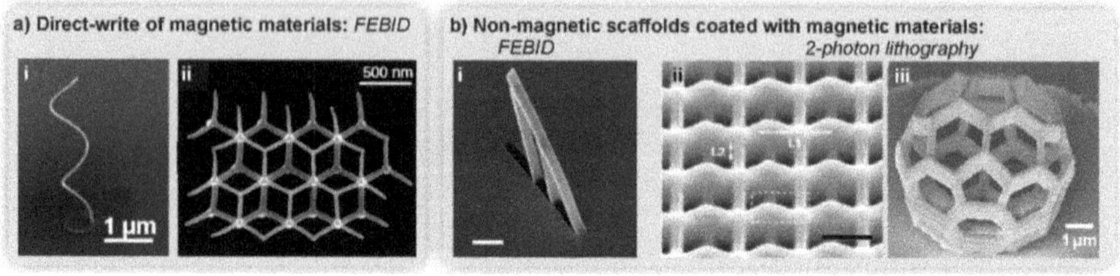

Figure 11.2. Direct-write of 3D magnetic nanostructures using two main approaches. (a) The direct write of magnetic materials in 3D, with FEBID, results in isolated structures such as (i) the helix, as well as (ii) extended lattices. (b) The patterning of non-magnetic scaffolds, which are then coated with magnetic materials to form shell structures, resulting in (i) a domain wall conduit, (ii) an artificial spin ice lattice, (iii) an artificial buckyball structure. (a-i) Reproduced from [51]. (a-ii) Reproduced from [52]. (b-i) Reproduced from [53]. (b-ii) Reproduced from [54]. (b-iii) Reproduced from [55].

control, which has important implications for *in vivo* applications such as drug delivery, and indeed has been applied in microrobots [28] for the transport of targeted cells in mice and zebrafish [29], the orientation and alignment of electronics in 3D with floating magnetic micro-tweezer robots [30], as well as for athermal shape-memory [31] with the possibility of 4D bioprinting [32] in the future.

The 3D structuring of materials becomes particularly interesting when fundamental lengthscales are approached which, for magnetism, are the magnetic exchange length and domain wall width (typically ~tens to hundreds of nanometres) [33]. On this scale, the geometry starts to influence the magnetic structure via effects such as shape anisotropy, configurational anisotropy and curvature-induced phenomena. In particular, 3D confined geometries have been predicted to lead to a variety of new magnetic properties [4, 6], including new types of domain walls with ultra-high velocities [7, 34, 35], effective anisotropies and chiral interactions in curved magnetic shells [36–38] and torsion effects in magnetic helices [39]. First experimental results have confirmed the presence of Bloch-point domain walls [40] and significant curvature-induced chirality [41]. However, domain wall dynamics have turned out to be significantly more complex than predicted [42, 43], highlighting the need for extensive experimental characterisation, including sophisticated nanofabrication.

There have been several demonstrations of the fabrication of 3D systems via, for example, self-assembly [10, 44], rolled-up technologies [27, 45], grazing incidence physical vapour deposition [46–48] and anodised alumina/polymer templates [49, 50], where specific geometries or shapes can be fabricated. For the realisation of a wide variety of geometries, however, a more flexible, tunable technique is required. In this section, we focus on recent advances in the fabrication of arbitrary 3D magnetic nanostructures via nanoscale 3D printing direct write techniques, which provide the capability to design and fabricate arbitrary geometries.

For the fabrication of arbitrary 3D magnetic nanostructures using direct-write techniques, there are two main routes available (see figure 11.2): the direct growth of magnetic materials, and the growth of scaffolds, which can then be mapped into magnetic structures via a variety of deposition methods. We will give an overview of both routes, including the different techniques available and the state-of-the-art, along with long-standing challenges and limitations.

11.4 Focused electron beam induced deposition

11.4.1 FEBID for magnetism

As described in section 1.2, in recent years, magnetic nanostructures have mainly been fabricated with top-down techniques that combine lithography with the thin film deposition of ferromagnetic materials. For the direct printing of 3D magnetic materials, however, a bottom-up fabrication process is required. Such a process is currently provided by FEBID, which enables the direct-write fabrication of materials with nanoscale resolution (figure 11.2(a)). FEBID is a powerful patterning technique in which a material is locally deposited via the cracking of a precursor gas by a focused electron beam [56] (see also chapter 4 in the present book). Due to the flexibility in the precursor used, as well as the small focal points of electron beams in electron microscopes, it is possible to deposit a variety of materials with very high spatial resolutions, on the order of tens of nanometres, thus opening up exciting perspectives in condensed matter physics [57].

When one considers the fabrication of 3D *magnetic* materials, one of the main concerns is the quality of the deposited material, which has critical implications for magnetic properties of the nanoscale system. In this regard, FEBID could be slightly concerning: generally, the organic nature of the deposition process leads to a high carbon content of the deposited structure [56]. However, in recent years the direct deposition of functional—and in many cases, high purity—cobalt and iron structures, with good magnetic properties has been demonstrated [58, 59]. This key progress is possible thanks to the usage of carbonyl cobalt and iron precursor gases, whose decomposition is particularly efficient under a FEBID process [60]. Furthermore, recent advances in precursor synthesis have led to the deposition of CoFe alloys [61]. This, together with the co-injection of several gases [62] and the growth of core–shell geometries via sequential growth [63] paves the way for the deposition of tailored magnetic materials. With the development of advanced modelling and simulations of FEBID, a deeper understanding of the underlying mechanisms has been achieved, which in turn has led to the fabrication of tuneable materials with a range of different properties [64–66].

Of course, perhaps the most relevant aspect of FEBID for nanomagnetism is not simply the deposition of magnetic materials, but the capability it provides for the structuring of magnetic materials on lengthscales of the order of tens of nanometres. At these scales, the geometry and curvature of patterned systems can be used to dictate the magnetic properties, and thus result in magnetic nanodevices. In two dimensions, 2D nanowire devices have been applied successfully, e.g. to domain wall logic [67–69], while in three dimensions, freestanding pillars for nanomagnetic logic [70] and magnetic actuation [71] have been realised. Furthermore, the ability of FEBID for direct-writing on arbitrary substrates has been exploited to realise a range of customised scanning-probe microscopy and spectroscopy sensors with different 3D geometries [72–76].

From first demonstrations of the fabrication of simple cylindrical nanowires with a static electron beam, advances in the design of stream files—that are used to direct the electron beam—have led to enhanced control in the fabrication of high-purity 3D magnetic structures. This control has led to greater control over the growth and magnetic properties of the more simple geometry ferromagnetic cylindrical nanowires [77, 78], to the ability to pattern more complex 3D structures (see figure 11.2(a)) such as ferromagnetic nano-helices [51] and 3D artificial lattices [52]. Yet more complex 3D structures have been achieved with recent advances in algorithms for the nanofabrication of 3D structures [52, 79, 80] ranging from 3D nanowire architectures to the implementation of CAD-based design for truly 3D nanostructures, from the topologically-non-trivial Möbius strip to a variety of curved 3D surfaces. The new feasibility of the fabrication of these more complex structures opens up exciting perspectives for additive nano-manufacturing, thus advancing FEBID towards a 3D printer at the nanoscale.

Whilst the direct patterning of 3D magnetic nanostructures that FEBID provides is unparalleled by other techniques, one of the main challenges it faces is the material properties of the deposited nanostructures. For the realisation of the predicted potential of 3D nanomagnetism, greater control over the global magnetic properties, such as the exchange lengths, saturation magnetisation and coercive fields are required. Moreover, despite some attempts to control interfacial effects in FEBID materials [81], the technique is currently not sufficient to study and exploit this type of phenomena with the level of control and reproducibility required for spintronics. Aside from interfacial effects, higher magnetic functionality can also be obtained via the combination of 3D nanomagnetism with alternative *active* materials including, for example, magnetostrictive materials. Both routes will require the development of further expertise in the development of new precursor molecules, as well as in the co-deposition of materials, whilst post-processing techniques such as annealing, which has proven effective in obtaining high purity crystalline cobalt [82], and electron radiation in a hydrogen atmosphere [83] promise accessibility to higher purity materials.

Although our first thought may be to focus on developing alternative treatments and post-processing techniques to avoid low-purity materials, it is important to consider the advantages that 'dirty' carbon-transition metal alloys might provide for magnetic devices. In fact, it has been shown that by working in the low transition metal-content regime, effects such as intergranular tunnelling [84] an enhanced extraordinary Hall effect [85] can be achieved. Such effects have been successfully exploited for ultra-high sensitivity sensing in other non-magnetic precursors for a variety of effects such as metal–insulator transitions [86], strain [87] and gas sensors [88].

11.4.2 FEBID fundamentals

FEBID employs the beam of electrons of a scanning electron microscope (SEM), to controllably induce the decomposition of a precursor gas into a solid deposit and volatile fragments [79, 80, 89, 90]. The tightly-focused beams of electrons produced by SEMs enable nanoprinting structures as small as a few tens of nm and can be computer-controlled with nanometric precision. Figure 11.3(a) schematically describes this process.

The FEBID process, however, is not simple and involves a large set of interactions and interlinked physical effects [56, 64, 91, 92] and newcomers diving into the scientific literature may find themselves overwhelmed by the large number of variables, parameters and experimental conditions that play a role in the process (figure 11.3(b)). In this section, we introduce the tools and models developed by the FEBID research community for the understanding of the FEBID process through the combination of experiments, continuum models and advanced simulation tools. Furthermore, user-friendly empirical calibration procedures and beam-pattering algorithms that enable non-expert users to design and fabricate structures will be discussed to provide a starting point to those considering using FEBID for 3D nanopatterning. Four independent subsections follow, covering resolution, basic theory and modelling, patterning algorithms and hybrid techniques. These may be read in series or individually.

11.4.3 FEBID resolution

The *spatial resolution* of the technique (defined as the thinnest printable feature) is determined by how tightly the electrons are compressed by the electromagnetic lenses in the SEM (beam width) and how much does the beam spread when the electrons collide with the atoms in the substrate and the growing nanostructure (beam-substrate interaction volume). It can vary between 200 and 15 nm depending on the conditions employed [90].

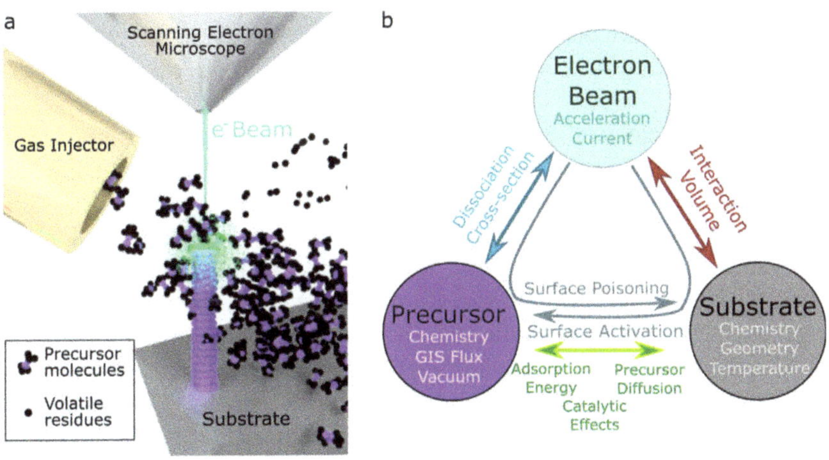

Figure 11.3. The FEBID process. (a) Schematic representation of the decomposition of precursor molecules being performed by the electron beam of an SEM. (b) Cross-interaction between parameters during FEBID. Three circles represent the three elements that interact during FEBID (electron beam, precursor and substrate) and key parameters related to each element are included inside each circle. Key interactions between elements are represented by arrows.

Beyond being able to produce such small 3D volumetric features, FEBID also excels at controlling the position of such structures, the *positioning resolution* of FEBID is determined by the precision of SEM beam-scanning hardware, which can reach sub-nm performance at high magnification. Positioning resolution is seldomly discussed, as volume resolution is the standard benchmark across all techniques, but it is, however, of great interest for the nanofabrication of 3D magnetic nanostructures as magnetic coupling is highly dependent on nanostructure separation and overlap [93].

Additionally, one may define a *layer resolution* as the thinnest possible layer that the technique can deposit. In theory, the layer resolution in FEBID is limited by the shortest pulse that the SEM can produce and sub-monolayer deposition of material could be possible using a sufficiently low precursor flux and a sufficiently short pulse of electrons. In practice, however, this process is limited by vacuum quality in the SEM chamber. Depending on the microscope's base pressure, monolayers of residual gas will adhere to the surface at different rates [94]. This residual gas is always decomposed by the electron beam and limits the smallest layer thickness that can be deposited without suffering from excessive contamination. Layer thickness in the order of a few nanometers is generally employed during 3D nanofabrication [80].

11.4.4 FEBID theory, modelling and practical recommendations

The FEBID process involves three key physical entities: a *precursor* gas which is decomposed into a solid deposit and volatile fragments, a focused *beam* of electrons, and a *substrate* where the precursor molecules adsorb to generate nanostructure growth. Once the nanoprinting process has started, the nanostructure being printed becomes the substrate. These three entities are represented as solid circles in figure 11.3(b), and key experimental parameters related to each are included inside the circles. This section describes this relationship diagram in detail, discussing different tools available to gain a deeper understanding of some of the interactions represented.

Starting with the electron beam, its *acceleration potential* determines the kinetic energy with which electrons arrive at the substrate and is critical for the scattering that takes place once electrons interact with the substrate [95]. This beam–substrate scattering is indicated as a double arrow in figure 11.3(b) labelled *interaction volume* and results in a cloud of electrons that forms inside and around the substrate after beam impact. The shape of this interaction volume depends both on the energy of the beam and the substrate material and density. It also strongly affects the resolution of the process and can generate the appearance of parasitic co-deposition below or around the exposed regions [79, 89, 90]. Figure 11.4(a) shows as an example, the interaction volume for an electron beam incident onto a Pt tip with an energy of 3 keV (left) and 30 keV (right), calculated using Monte Carlo simulations. At 3 keV most electrons are stopped or backscattered by the tip leading to a high interaction volume, whereas at 30 keV most of the beam passes through with little scattering.

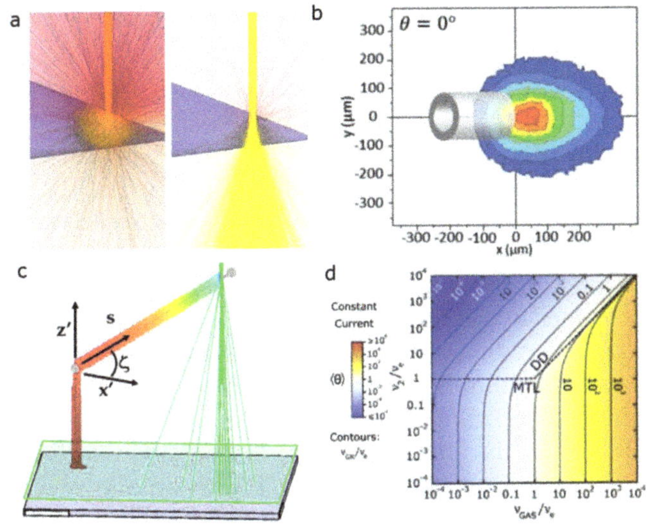

Figure 11.4. Simulation and modelling tools. (a) Electron trajectories at 3 keV (left) and 30 keV (right), simulated using CASINO [96]. Reproduced from [97]. (b) Simulated gas flow from a gas injector system. Reproduced from [64]. (c) Simulated nanostructure growth using a finite difference model including temperature, precursor diffusion and electron trajectories. Reproduced from [98]. (d) Analytical solution of precursor concentration and growth rate as a function of fundamental FEBID frequencies represented in a 2D map. Reproduced from [65].

The amount of scattering also impacts the energy deposited on the substrate and consequent heating. If electrons are stopped by the substrate, the kinetic energy of the beam will directly be converted into heat, whereas if the electrons are mostly subject to elastic scattering events, negligible energy is transferred. A versatile and freely accessible tool to model and understand electron–substrate interactions is CASINO [96].

The second critical parameter related to the incident beam is the *beam current*, which determines the amount of electrons that impact the substrate every second. A high beam current density will likely cause a fast depletion of the precursor in the regions where the electron concentration is higher. It will also result in a higher release of energy into the substrate (heating). High current beams are associated with larger beam spot sizes (due to electrostatic repulsion inside the beam), and will result in higher growth rates as more molecules are dissociated per second [64, 65, 89, 99].

We continue this discussion with the precursor. In FEBID, the precursor is injected into the chamber of an electron microscope using a small nozzle or *gas injector system* (GIS), with a typical diameter of a few hundreds of microns. A GIS needs to be employed in order to inject the gas locally to the growth region only, keeping a high vacuum in the rest of the chamber. This enables electrons to travel with negligible scattering, at the same time making sure that no parasitic decomposition of the precursor gas takes place at other parts of the chamber, which could result in damage if a metallic (or insulating) film was accidentally deposited on them. The use of a GIS means that the concentration of precursor gas is not uniform across the whole sample, but is space dependent [100]. Figure 11.4(b) shows a simulated gas distribution, where a central maximum corresponding to the GIS axis is observed, with a progressive decrease in gas concentration as one deviates from this axis. Additionally, the arrival of the gas is also directional, and shadowing effects can occur during the nanofabrication process. Due to this spatial dependence of the gas flow, a critical experimental parameter when performing FEBID is the GIS alignment. Due to the small size of the nozzles employed (50–300 μm diameters) an incorrect alignment of the GIS can easily lead to very little gas arriving at the growth region and unsuccessful growth. In general, to achieve maximum gas flow, one should make the center of the GIS axis coincide with the point where growth will take place. It is important for inexperienced users to keep in mind that this alignment occurs in 3D space and cannot be assessed solely from the 2D image provided by the SEM. The height encoder of the sample stage should be employed to assess the height of the GIS during the alignment process, to ensure that the axis of the GIS truly aligns to the growth region and not below or above it. The sketch of figure 11.3(a) may serve as a rough illustration of this type of GIS misalignment; in this figure, the GIS is pointed to the top of the nanowire, but the gas lands on the substrate very far to the right, where the electron beam cannot reach. In a real experiment the nanowire of the figure would never have started growing in the first place!

The *chemistry and composition* of the precursor is a second variable critical to the process, as it determines the composition of the deposit. Due to their availability, the first FEBID precursors were borrowed from the field of chemical vapour deposition (CVD). CVD chemistries [56], however, are optimized for thermal instead of electron decomposition, and the search for more suitable FEBID precursors is an active field of research [101]. For the direct nanofabrication of magnetic nanostructures, $Co_2(CO)_8$ and $Fe(CO)_5$ are two commonly employed precursors due to the high purity of magnetic deposits that they produce [59, 102]. Although not directly related to the precursor itself, we include *vacuum quality* as a precursor variable. Inevitably, the vacuum in an SEM chamber is not perfect and there always exists a residual level of gases in the chamber that compete with the precursor for the adsorption sites available on the substrate. The number of monolayers of residual carbon impinging on the substrate per second decreases with increasing vacuum quality [94] and therefore low chamber pressures and the use of cryogenic traps inside the chamber is recommended to achieve high-purity growth of magnetic materials.

Finally, the substrate plays also a key role. Its composition and microstructure not only determine the amount of electron scattering during a FEBID process, but it may also play a critical role in the growth mechanisms via its interaction with the precursor. When a precursor molecule falls on the substrate, it could potentially form a bond that will depend on the chemical state of the precursor molecule and the substrate's surface. One or more *adsorption energies* may be associated with the formation of this bond, and complex adsorption potentials, with barriers and wells may be present [103]. If the bond between the substrate and the precursor results from a strong chemical interaction, *catalytic effects* may occur, and the strength of such bonding will also affect *precursor diffusion*, which is the movement of precursor molecules over the substrate due to concentration gradients.

The precursor dynamics related to all the above energy barriers is critically influenced by the *substrate temperature*. A high substrate temperature, for example, can enable the formation of chemical bonds that are otherwise prevented by an energy barrier [103], or may cause a strong desorption of the precursor when the thermal energy of the molecules increases above the adsorption energy [65]. Cryogenically cooled substrates, for example, are employed in order to allow the absorption of a large number of precursor molecules on the substrate and generate accelerated FEBID growth [104].

In addition to the elementwise interactions discussed above, two second-order interactions are possible between the electron beam, the precursor, and the substrate. These are *surface activation* and *surface poisoning*. Surface activation consists of the creation of chemically active species by the electron beam on the surface of the substrate that reduce the energy of adsorption of the precursor or cause autocatalytic effects. On the other hand, surface poisoning refers to the effect in which the decomposition of precursor by the electron beam results in chemically inactive species covering the substrate, inhibiting potential autocatalytic activity [105].

To understand the consequences of the large number of interactions described above, two complementary approaches are employed: continuum models [64, 65] in which the fundamental precursor-substrate-beam variables are functions of continuous variables (space and time), and finite-difference simulations (figure 11.4(c)) in which each variable is modelled within small 3D-voxels and nanostructure growth is calculated with fixed time steps [66].

Continuum models are most suitable for the study of fundamental interactions and establishing correlations between the different FEBID parameters. Typically finding analytical solutions, predictions from continuum models can be computed at low cost and can be efficiently fitted to experimental data. To help in the simplification of model construction and data interpretation, the framework of fundamental FEBID frequencies has been recently introduced [65]. These frequencies correspond to the coefficients found in the differential equations that drive precursor dynamics on the surface of growing substrates and their unit is [s^{-1}] or [Hz]. When these frequencies are compared to each other (figure 11.4(d)), different growth regimes can be directly inferred and allow researchers to simplify the modelling, interpretation, and planification of a FEBID experiment.

To model more complex 3D dynamics for FEBID growth, only finite-difference simulations can currently incorporate the evolving nature of the growth surfaces albeit with a significant computational cost. The emergence of GPU (graphical processing unit) software acceleration successfully employed in, e.g. finite-element micromagnetic modelling [106], and the rapidly increasing ratio of GPU computational power/GPU price (driven by the strong role that GPUs play in the efficiency of artificial intelligence algorithms) can provide significant acceleration to finite-difference FEBID modelling and even open the path to more complex modelling or even iterative optimization of 3D-nanoprinting pattern generation, discussed in the next section.

11.4.5 FEBID: 3D nanopatterning algorithms

For many users of FEBID techniques, the study of fundamental interactions and the development of growth optimization algorithms will be beyond the scope of their application. For this reason, the FEBID community has recently made an important effort to develop plug-and-play nanopatterning softwares that require minimum user expertise and therefore open the use of the technique to a user base as wide as possible. This approach was pioneered by Fowlkes *et al* [66] and is based on the creation of interconnected nanowire networks. This algorithm starts with a definition of a segment array such as the one displayed in figure 11.5(a), using a customized software package. After defining the nanowire array, a calibration protocol is executed in the microscope (figure 11.5(b)) to measure the dependence between segment angle and exposure time [79]. The measured dependence is then fed into the same software and an exposure file is generated automatically. Execution of the exposure then leads to the nanoscale printing of the nanowire array (figure 11.5(c)). Further correction parameters to, e.g. heating during growth, can be added during the nanostructure design process based on extra calibration experiments or finite-element modelling [107]. A similar patterning strategy was developed by Keller and Huth [108] and efforts are being made to integrate both solutions into a single software package [90].

Following on the successful implementation of Fowlkes *et al*, a layer-by-layer method for 3D nanopatterning was developed by Skoric *et al* [80] to generalise the

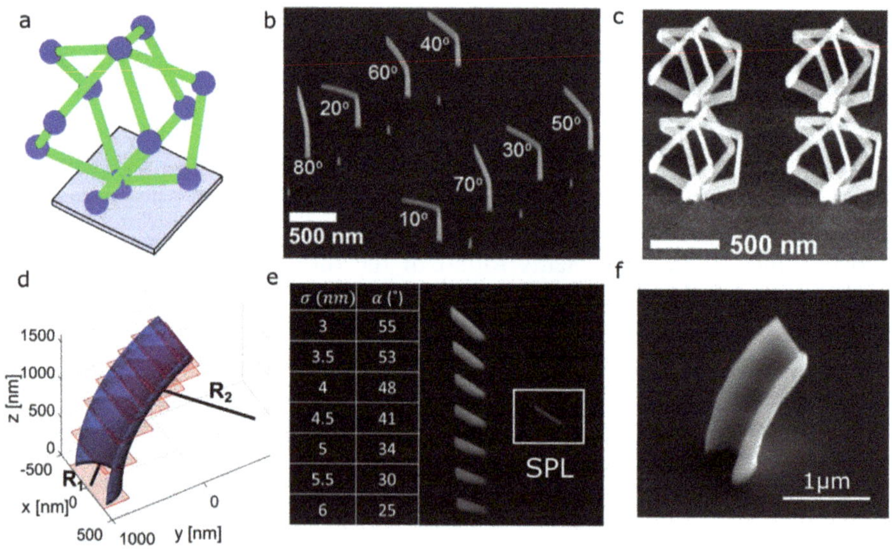

Figure 11.5. (a) Nanowire network defined as a set of vertices. (b) Segment angle calibration example. (c) 3D nanoprinted structures generated by the combination of (a) and (b) into dedicated nanopatterning software. (d) 3D nanostructure defined using a standard stereolithography (STL) file and segmented by dedicated software. (e) Calibration procedure to fit the geometrical properties of the electron beam interaction volume with the substrate. (f) 3D nanoprinted structure defined in the STL file of (d). (a–c) Reproduced from [79]. (d–f) Reproduced from [80].

technique beyond nanowire networks. The input method for this algorithm is a standardised .stl file that can be generated with common computer aided design (CAD) software. In this case the .stl file is adaptively segmented (layer thickness depends on local curvature) by a customized software package (figure 11.5(d)) that in addition calculates the heat resistance of the nanostructure below each layer to correct for heating effects. A two-step calibration protocol designed to measure the space-dependent dose distribution resulting from a single dwell, and the heat conductivity of the printed material, are implemented [80]. Step 1 of this calibration is displayed in figure 11.5(e). Using the calibrated parameters, a dwell time solver optimizes each layer exposure employing a per-layer proximity matrix and generates a beam scanning file that is then executed by the microscope (figure 11.5(f)). Further corrections such as beam defocusing and angular dependence of gas flow can be incorporated at the expense of extra calibration experiments [80].

11.4.6 FEBID and thin film deposition: hybrid approach for 3D nanopatterning

In addition to the use of FEBID for direct-writing of ferromagnetic materials, it is also possible to fabricate 3D nanostructures using a 'hybrid' scaffold-based approach, where a non-magnetic 3D architecture is patterned, and then coated or filled with a magnetic material, making possible the fabrication of both 'shell'-like and 'bulk'-like 3D magnetic nanostructures. With this segmented approach, the stringent requirements on the material properties of the 3D printed material are lifted, and the geometry and dimensions of the scaffold alone can be optimised [80, 109] while processes for the deposition of thin-films on 3D surfaces can be developed independently. As we will see in section 11.5, this approach can be implemented with other 3D nanofabrication techniques different from FEBID.

Regarding FEBID, a key advantage is using well-established chemistries, e.g. the archetypical carbon–platinum compound via the $MeCpPtMe_3$ precursor. FEBID of such a material is a more established and well-understood process [89] than FEBID of ferromagnetic materials, where less reproducible properties [51], electron-mediated autocatalytic processes [110] and sudden changes in growth regimes [77] are observed. Furthermore, using the scaffold approach, growth conditions can be tuned to operate at or close to the reaction-rate-limited regime, which lead to higher spatial resolutions and greater 3D growth fidelity [56]. This is not generally possible when directly depositing ferromagnetic materials, where conditions at the mass-transport-limited regime are preferable to achieve high metallic purity [58, 111]. This hybrid approach, combining FEBID with thin film deposition, is a promising route for the fabrication of core–shell nanowire architectures, as well as curved magnetic thin films [80].

In a first demonstration, this hybrid nanofabrication approach has been employed to demonstrate the injection of magnetic information from a 2D substrate into a 3D nanostructure for the first time [53]. For this, structures such as the one represented in figure 11.6 have been used. Additionally, the versatility of FEBID enables 3D shadowing strategies, such as the one shown in figure 11.6(d), where a 3D shade is employed to prevent the deposition of magnetic material at the base of the 3D

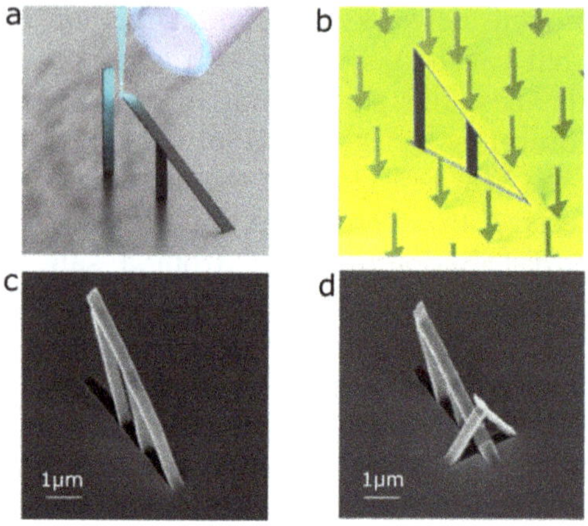

Figure 11.6. (a) 3D nanofabrication of a non-magnetic scaffold. (b) A layer of magnetic material is deposited over the scaffold and the substrate. (c) Resulting hybrid 3D nanostructure. (d) A 3D nano-shade may be employed to prevent magnetic material from being deposited at the base of the 3D nanowire and create a magnetically-disconnected reference sample. (a–d) Reproduced from [53].

nanowire, which generates a magnetically-disconnected reference structure. Extensive details regarding this fabrication strategy are available in [97].

11.5 Two-photon lithography

11.5.1 Two-photon lithography for 3D nanomagnetism

An alternative method for the fabrication of non-magnetic 3D scaffolds introduced in section 11.4.6, is the use of two-photon lithography [112], see figure 11.7 (see also chapter 2 in the present book). In two-photon lithography, an optical laser is focused with an objective lens into a photoresist that is sensitive to UV light, and unaffected by longer wavelength light. As a result, the chemical structure of the resist is only modified where there is sufficient photon density for two-photon absorption to take place, which is only within the focal point of the laser beam. In this way, high spatial resolutions (sub-diffraction-limit) can be reached and, as the resist is only modified in the focal point of the laser, truly 3D structures can be written by moving this focal voxel relative to the substrate.

Compared to FEBID, the main advantages of two-photon lithography are the speed and accessibility—which currently come at the cost of spatial resolution. With the combination of sophisticated lasers, piezo stages and galvo-mirrors, two-photon lithography offers a significant increase in the writing speed of 3D nanostructures—reducing the typical writing time of a 1 μm structure from minutes to seconds—thus opening up the possibility for high throughput processing/fabrication. In addition, the technique makes use of a laser setup, instead of the need for an electron microscope, and recent commercialisations have led to an increase in accessibility for non-expert users. However, the spatial resolutions achievable with an optical technique are still limited: although sub-wavelength spatial resolutions can be reached, with standard two-photon lithography techniques spatial resolutions are

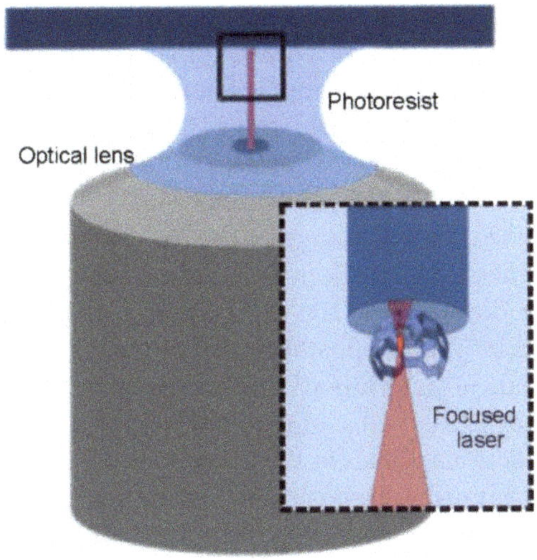

Figure 11.7. Two-photon lithography, where a photoresist is exposed in the focal point of a focused laser, creating a non-magnetic scaffold. Reproduced from [113].

still on the order of hundreds of nanometres—at the limit of the relevant magnetic lengthscales—and with an elongation of the focal voxel in the vertical direction.

In recent years, two-photon lithography combined with a negative tone resist—where the exposed region remains after development—has been applied to 3D nanomagnetism, enabling the creation of structures such as an isolated artificial Buckyball [114, 115], and extended artificial 3D spin lattices [116, 117] for the study of magnetic frustration in 3D (see figure 11.2(b)(ii–iii)). It is also possible to combine two-photon lithography with positive tone resists, where the exposed regions are removed after development. These 3D molds can then be filled using electroplating (see section 11.6), as has been demonstrated for gold metamaterials [118], as well as for magnetic materials [117] which provide a route for the fabrication of nano- and micro-wire and bulk magnetic architectures, albeit on larger lengthscales.

With the main challenge for two-photon lithography being the achievable spatial resolution, it should come as no surprise that there have been significant efforts in recent years to reduce these to more relevant lengthscales, as described in detail in [116]. There has been work on the development of new systems, new photoresists, and post-processing techniques, which all show significant promise for the fabrication of structures with sub-100 nm features. In particular, post-processing techniques such as pyrolysis, in which the crosslinked polymer scaffold can be reduced to a glassy carbon structure by heating at high temperatures in a nitrogen atmosphere [119, 120], leading to the shrinking of an architecture by up to 80% and spatial resolutions below 100 nm for 3D structures. This new technique has been recently combined with homogeneous deposition of magnetic material to form an artificial buckyball shell structure [114, 115]. In addition, two-photon lithography has been mostly performed with a laser of 780 nm wavelength, which results in a minimum lateral feature size of approximately 200 nm. Recently, higher spatial resolutions have been achieved by the move to lower wavelength incident light, resulting in sub-100 nm spatial resolutions with a wavelength of 405 nm [121]. Comparative increases in spatial resolution have been achieved through chemical changes that lower the density of active radicals in the resist [122] as well as the use of overlapping incident laser beams [123], thus demonstrating the wide range of routes to higher spatial resolutions that are currently being explored.

As previously discussed, an advantage of the 3D scaffold + PVD approach to the fabrication of 3D magnetic nanostructures involves the tuneable magnetic material properties of thin-film depositions: for the mapping of the scaffold into magnetic materials, standard thin film deposition techniques already optimised for the deposition of high-quality materials can often be used, thus providing access to significantly higher purities, as well as the possibility for the integration of interfacial or magnetostrictive functionalities. On the other hand, a main challenge comes with the fact that many thin film deposition techniques are optimised for planar substrates. In 3D geometries, shadowing effects typically interfere with the homogeneous coating [114], and typically result in strongly-textured materials [124].

Such effects can be avoided with a number of chemical deposition techniques, which are better suited for the deposition of thin films on 3D geometries. Perhaps the most well-known of these is atomic layer deposition (ALD), a process consisting of multiple 'self-limiting' reactions that results in the deposition of homogeneous thin films with precise thicknesses, well-suited to high aspect ratio, or 3D surfaces. While well-established for dielectric materials such as Al_2O_3, ALD of metallic materials generally requires high temperature processes, except for particular metals including tungsten, platinum and iridium. The direct deposition of magnetic materials via ALD is also possible, and involves the deposition of a transition metal oxide, and its subsequent reduction under a hydrogen atmosphere in high temperatures [125–127], thus requiring scaffold structures with high thermal stability.

Electrochemical techniques (section 11.6) also offer a route to the homogeneous deposition of magnetic materials on a 3D surface. With electrodeposition, a magnetic layer can be grown on a 3D surface as long as a surface is conductive. Indeed, recently the lower deposition temperatures possible with non-magnetic metals have been exploited to first deposit a conductive layer on a 3D polymer scaffold, onto which a Ni layer was subsequently electroplated [113, 115]. An alternative electrochemical technique that does not require a conductive layer is electro*less* deposition, where a solution is used to activate a surface uniformly, therefore making possible the deposition of a metallic layer. Magnetic materials, both hard [128] and soft [129] have recently been deposited in this way, including on polycarbonate membranes to produce magnetic nanotubes [129, 130].

11.5.2 Examples of scaffold lithography using two-photon lithography

As mentioned in the previous section, two-photon lithography provides a flexible route to fabricating 3D magnetic nanostructure via the fabrication of a non-magnetic scaffold. As an example, we will now describe a case study of this technique in detail, that was the first demonstration of the use of two-photon lithography for 3D nanomagnetism: the fabrication of an artificial buckyball structure [114]. The 6 μm diameter 3D artificial buckyball structure consisted of a polymer scaffold that was coated with a magnetic material to form a 3D interlinked tubular structure (a scanning electron microscope image is given in figure 11.2). Along with providing an opportunity to investigate the role of vertices in a linked magnetic nanotube structure, the buckyball geometry is also an extremely stable structure which has been found to survive radiation damage well.

In the study, in order to provide feedback on the effectiveness of the fabrication method, the structural and chemical properties of the magnetic nanostructure were characterised by resonant phase tomography. To ensure a clear view of the sample about 360° for the tomography experiments, the structure needed to be mounted on tomography pins offering both high stability and an unobstructed view. An overview of the sample fabrication is given in figure 11.8.

To ensure a 360° view for the tomography measurements, the 3D magnetic nanostructure was fabricated on pillars manufactured on a silicon substrate, that were created via a combination of electron beam lithography to define a hard mask, and rcactive ion etching, resulting in high aspect ratio silicon pillars of height 36 μm and diameter 10 μm (for more details we refer you to [114, 115]).

The buckyball was then fabricated directly on top of the pillars using two-photon absorption lithography with a Nanoscribe 3D direct laser write system. Specifically, in two-photon lithography a near infra-red laser (λ = 780 nm) is focused with an objective lens into a photoresist that is sensitive to UV light, and unaffected by longer wavelength light. As a result, the chemical structure of the resist is only modified where there is sufficient photon density for two-photon absorption (with an effective wavelength of 390 nm) to take place, which is only within the focal point of the laser beam. In this way, a spatial resolution of around 150 nm can be reached and even surpassed by tailoring the laser power and exposure time. In addition, as

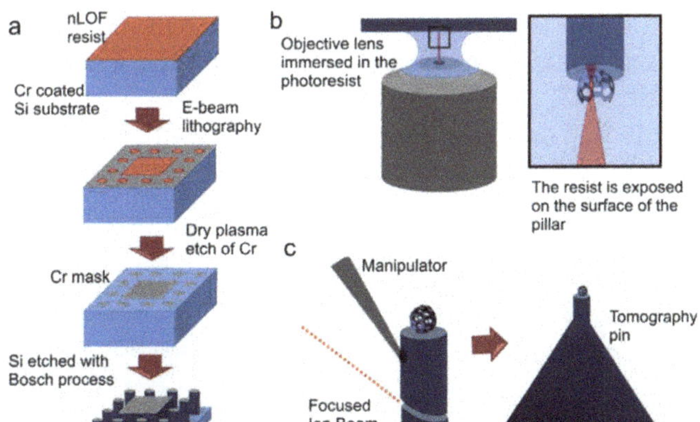

Figure 11.8. Fabrication of the artificial magnetic buckyball. (a) High aspect ratio Si pillars are prepared. (b) The 3D structure of interest is exposed on the Si pillar using two-photon absorption lithography. (c) The pillar is transferred to the tomography pin using the focused ion beam. Reproduced from [113].

the resist is only modified in the focal point of the laser, truly 3D structures can be written by moving this focal spot within the resist relative to the substrate. With the 'dip-in' mode, the objective lens of the system is immersed in the photoresist, allowing for the 3D structure to be exposed directly on the pillar, as seen in figure 11.8(b), which is located and brought into focus manually using the microscope of the system. After exposure, the sample is developed by immersing in PGMEA for 1 min, after which a final cleaning process involving immersion in isopropanol is performed.

Once the non-magnetic scaffold for the buckyball structure was successfully fabricated on the pillars, the cross-linked polymer structure was then coated with 30 nm of Co capped with 6 nm of gold using a UHV sputtering system of base pressure 5×10^{-9} mbar. To optimise the homogeneity of the coating on the 3D structure, the sample was mounted at an angle of approximately 45° with respect to the source and during the deposition the specimen was rotated continuously.

As a final step, the buckyball was transferred to the tomography pin using a focused ion beam (FIB). In particular, a micromanipulator was attached to the pillar approximately 10 μm below the top, where the buckyball was located, using focused ion beam deposition of carbon. The pillar was subsequently cut from the Si substrate as shown in figure 11.8(c) and transferred to the tomography pin. SEM images of the sample mounted on the tomography pin, and the buckyball itself, are given in figure 11.9.

The quality of the 3D magnetic buckyball structure was subsequently determined using element-specific x-ray phase tomography, a high-resolution imaging technique. In addition to providing a map of the structure in three dimensions, by tuning the energy of the incident x-rays, this technique provides element specificity, revealing the location of a particular element, in this case cobalt, within the sample [114]. In the case of the buckyball, the 3D structure of the sample, and the location of the cobalt, was determined to 25 nm spatial resolution in three dimensions. An element specific rendering of the artificial buckyball structure is shown in figure 11.9(c), with the resist shown in blue and the Co in orange. The resist visible on the inside and underside of the buckyball illustrates the shadowing effects of the Co sputter deposition [55].

As well as information on the location of the cobalt within the structure, element-specific phase tomography provides a quantitative map of the electron and atomic

Figure 11.9. Characterisation of the artificial magnetic buckyball (a, b) Scanning electron micrographs of the Co-coated artificial buckyball mounted on the tomography pin. (c) A 3D rendering of the artificial buckyball with elemental contrast, with the cobalt layer shown in orange, and the polymer in blue. Reproduced from [55].

density within the structure, which was used to determine information on the oxidation state of the cobalt. In this case, it was determined that the cobalt layer on the polymer scaffold was in fact oxidized, which was confirmed with spectroscopic measurements [114].

From the element specific phase tomography of the buckyball, two main problems with the fabrication processes were identified. Firstly, the cobalt was oxidised. As the oxidation of cobalt destroys its ferromagnetic properties, in order to obtain a good quality ferromagnetic 3D structure for investigations of the effect of the 3D structure on the magnetic properties, further optimisation of the sample fabrication is required. As the Co was capped on the outside with a thin layer of gold, it is likely that the oxidation and disorder in the film is due to diffusion of oxygen or other impurities from the polymer scaffold to the magnetic film. To avoid this, a protective layer could be introduced between the polymer and magnetic film. Secondly, the magnetic film deposited on top of the polymer scaffold is not homogeneous: there are rather severe shadowing effects and in some areas the structure was not coated at all.

For thc optimal production of magnetic structures, it was therefore concluded that the deposition of magnetic thin films onto 3D scaffolds is the most challenging area. Although the shadowing effects of more traditional deposition techniques can be exploited in the fabrication of 3D magnetic nanostructures based on curved surfaces, such as in the case of the 3D artificial spin ice lattices [116, 117], the homogeneous deposition of magnetic thin films on a 3D surface is particularly challenging.

11.6 Electrodeposition of 3D materials

11.6.1 Electrodeposition for 3D nanomagnetism

Complementing our discussion on direct-write nanofabrication techniques, we focus this last section on the synthesis of nanomaterials and 3D materials via electrodeposition. Electrodeposition has appeared in recent years as a non-expensive and versatile alternative to 3D nanofabrication, in particular for the growth of cylindrical nanowires and arrays of nanowires. Electrodeposition is a chemical method for the synthesis of materials based on a cathodic process, where ions in an electrolyte are incorporated on a conductive substrate forming a film via a reduction reaction [131]. As with most chemical techniques, electrodeposition is a conformal growth technique, that allows the deposition of a wide variety of materials, mainly metals and metallic alloys but also semiconductors [132] and polymers [133], onto areas of complex shape. No vacuum or high temperature is needed, which reduces the cost of electrodeposition compared with other growth techniques and makes the process highly suitable for industrial use, mainly for the fabrication of protective coatings but also in technological areas including micro-electronics and magnetic recording [134]. However, there is an important draw-back when comparing with other physical or chemical techniques normally used for the growth of thin films and coatings: the need for conductive substrates. Electrodeposition is based on an electron-transfer reaction [135] and, therefore, it can only take places on metals, metallized surfaces—as we will show later—or highly doped semiconductors [136, 137].

However, this drawback turns into an advantage for the growth of patterned thin films and 3D structures. The key concept is that electrodeposited material can only grow on the conductive parts of the substrates. We can illustrate this with an example. Let us take a silicon wafer in which a Ti/Au bilayer has been sputtered to make it conductive. Then, we cover the wafer by a resist and, by means of e-beam lithography and resist developing, we drill an ordered pattern of nanoholes. When we use this structure as substrate in an electrodeposition process, the reduction reaction can only take place on the bottom part of the holes, where the Au is in contact with the electrolyte and the material growth from bottom to top, filling the pores. Figure 11.10(a) shows a collection of ordered electrodeposited Co, whose shape is exactly that of the pores—slightly conical. When reaching the top part of the resist, the growth is not confined by the geometry and their material overgrows the resist, growing from the top part of the columns, forming 'nano-mushrooms' (figure 11.10(b)). This example clearly shows the two main attributes that make electrodeposition so well suited for the growth of 3D materials: the need for conductive substrates and conformal growth.

11.6.2 Electrodeposition of nanowires

In the above-described combination of electrodeposition and lithography, there are no limitations in the aspect ratio of the structures [138] and, in fact, the LIGA (lithographie, galvanoformung and abformung) technique uses electrodeposition through molds prepared by x-ray lithography to fabricate 3D-structures with a very large aspect ratio [139]. However, the most extended used of template-assisted electrodeposition has been the synthesis of nanowires using nanoporous templates as masks [140]. Among the different possible options that can be used as masks, two stand out: etched ion-track polymeric nanoporous membranes [141] and anodized aluminium oxide self-organized templates [142].

The etched ion-track ones are normally made of polymers (polymide, polycarbonate...) and their fabrication involves the irradiation of the polymers with energetic heavy ions, forming damage areas called ion tracks, and chemical etching

Figure 11.10. Electrodeposition for the synthesis of patterned and 3D samples: an introductory example (a) Co nanowires grown through nanoholes prepared in lithographic resist by means of e-beam lithography. The material grows from the conductive substrate filling the holes. (b) When the material reaches the top part of the resist, the growth is not limited by the mask, starting 3D growth from the top of the column.

to selectively dissolve the ion-track to form channels. The resulting channels are oriented parallel in the direction perpendicular to the surface [143]. The density of pores is quite low, and they are randomly distributed (see figure 11.11(a)). The control over the irradiation and etching conditions enables the production of channels with different geometries and sizes. Although these templates are very versatile, their fabrication process requires the use of large accelerator facilities, like GSI in Germany, Brookhaven in the USA or Lanzhou in China, which are normally not available for most researchers. Therefore, most examples in the literature using polymeric templates used commercial ones, that can be purchased with pores ranging from 10 to 100 nm diameter.

Alumina nanoporous templates can also be purchased from several companies, but this kind of template can be easily prepared by electrochemical anodization of Al, which provides a low cost and highly scalable technique to produce nanoporous templates for the synthesis of nanowires [144]. In this case, it is possible to obtain a highly ordered hexagonal array of parallel pores, as shown in the top view and cross section images of a membrane in figure 11.11(b). The Fourier transformation of the image (inset of the figure) clearly reveals the high order of the template. The pore diameter and the interpore distance strongly depends on the anodization conditions (applied voltage, electrolyte and temperature). Therefore, by controlling these parameters, it is possible to tailor the templates, obtaining nanoporous with

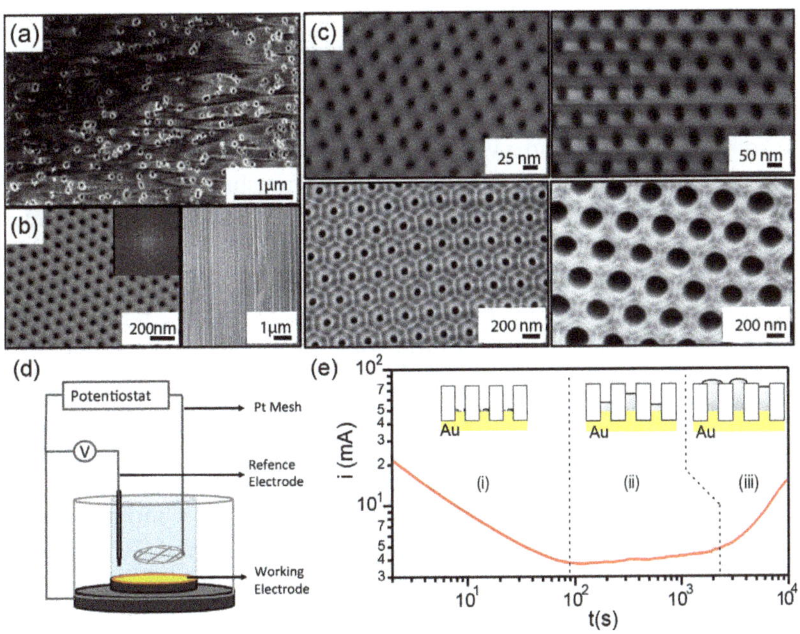

Figure 11.11. Electrodeposition of nanowires, (a) polycarbonate template and (b) alumina template. The difference between randomly oriented pores in polycarbonate templates and ordered in alumina templates is clearly observed. (c) Alumina templates with different diameter. (d) Schematics of the electrochemical cell for the growth of nanowires. (e) Evolution of current with time during the growth of nanowires. Dashed lines divide the curve in three different regimes: (i) nucleation, (ii) steady sate growth and (iii) saturation.

diameters ranging from hundreds, down to a few nanometers (figure 11.11(c)) [145]. Atomic force microscopy nanoindentations in the Al before anodization [146], as well as lithography [147] can be used to define the symmetry and the lattice parameter of the ordered arrays.

Once the templates are prepared, the procedure for the synthesis of nanowires is the same in all cases. Before electrodeposition, a conductive layer, typically a gold layer, should be evaporated, via thermal evaporation or sputtering, on one side of the membrane. This conductive layer will act as cathode or working electrode in the electrodeposition of nanowires, and its thickness should be thick enough to guarantee good electrical conductivity in the whole template area. The nanowires are normally grown under potentiostatic conditions, i.e. applying a constant potential between the working electrode and the reference one, using an electrochemical cell similar to the one shown in figure 11.11(c). This control is normally preferred when growing nanostructures, because it is almost independent of the substrate conductive area, which is difficult to know and control in template-assisted deposition using a nanoporous template.

The evolution of the electrical current flowing through the electrochemical cell can be used to monitor the growth. Figure 11.11(e) shows a standard current-versus-time transient state, measured during the growth of nanowires under potentiostatic deposition. In this case, the curve corresponds to the growth of Bi in a polycarbonate template. Three different regions can be distinguished in the curve, corresponding to three different growth regimes. Firstly, there is a drop of the current due to nucleation of the material onto the conductive layer. Immediately after the nucleation process, there is a steady regime due to the constant growth rate while the channels are being filled, since the active surface is constant. Finally, when the channels are completely filled, the material begins to grow over the surface of the template, connecting the different pores, similarly to the example of the nano-mushrooms in figure 11.10. When targeting the fabrication of nanowires, the growth is stopped in the second regime.

Therefore, by template-assisted deposition, it is possible to grow metallic nanowires with independent control over diameter (fixed by the pore diameter of the template), length (controlled by the electrodeposition time) and composition and crystal quality (controlled by the composition of the electrolyte and the growth potential). It is important to have in mind that electrochemical deposition is not a stoichiometric technique, there is therefore not a general relationship between the ratio of electroactive ions in the electrolyte and the one of the different elements in the growth material. In particular, in the case of magnetic alloys formed by metals from the iron-group (Fe, Co, Ni), alloys that are particularly important in magnetism, the so-called *anomalous co-deposition* should be considered. In this effect, taking place when plating an alloy of two or more metals belonging to this group, the less noble of the metals deposits preferentially to the more noble one [148]. This depends on the electrochemical parameters, as well as on the geometry of the template [149], making it essential to calibrate the composition under similar electrochemical conditions, before each growth experiment.

Electrodeposited nanowires can be studied and/or used inside the template, but also released from the template by etching the conductive layer and dissolving the template using the appropriate organic (for the polycarbonate) or inorganic (for the alumina) chemical. When released from the substrate, their individual magnetic properties can be studied by means of a range of microcopy techniques, such as magnetic force microscopy [150], transmission electron microscopy [151] and synchrotron radiation-based techniques [40, 42]. They can also be contacted by means of focused-ion-beam induced deposition [152], e-beam lithography [153, 154], or laser lithography [132], to study their magnetotransport properties or other experiments involving the application of electrical currents. It is also possible to prepare a suspension of nanowires in different solvents, functionalized them and used in biomedical applications [155].

In recent years, there have been some proposals to create a freestanding tridimensional structure formed by vertical nanowires released from the template. For this purpose, the conductive layer used as working electrode should be thickened either before or after the electrodeposition of the nanowires, but before dissolving the template to release them. Although the efforts are focused, up to now, in energy [156] and neurobiology [157] applications, there are emerging applications in the field of magnetism that will be developed in the near future.

There is a nice video description of the full procedure to fabricate nanowires, including the formation of the porous alumina membrane that is used as the template, the electrodeposition of the nanowires inside the template and the release of the nanowires into a solution in a recent publication of Jürgen Kosel's group [158].

11.6.3 Nanostructured multilayered nanowires

In the 1980s, different groups developed a method for the electrodeposition of metallic multilayers from a single electrolyte, with control of the layer thickness down to tens of nanometers. To illustrate how this method works, let us consider as an example the electrodeposition of a FM/Cu multilayer, where FM is a ferromagnetic material with reduction potential more negative that Cu. Therefore, there is a potential A at which only Cu is deposited and a potential B at which both, Cu and FM are deposited, forming an alloy. Pulsing the potential between A and B (pulse plating deposition), it is possible to grow a multilayer film consisting of alternating layers of pure Cu and a FM–Cu alloy. By using a very low concentration of Cu^{2+} ions in the electrolyte, the amount of Cu in the alloy will be small, with the magnetic properties of the FM–Cu being very similar to those for the pure FM. Clearly, this method presents some limitations regarding the composition of the different layers and, indeed, in the case of a magnetic material is mostly limited to the electrodeposition of FM/Cu multilayers were FM is an alloy of Fe, Ni and Co. Nevertheless, it allowed one to prepare structures with high interfacial quality, showing interfacial effects normally limited to multilayers grown by physical vapour deposition techniques, such as GMR [159], enabling, e.g. the creation of spin-valves [160].

The combination of template-assisted growth using nanoporous templates and pulse plating deposition allows the preparation of multilayered nanowires, consisting in a stack of disks or cylinders of alternating composition along the wire axis. This strategy allows, therefore, synthesizing materials with submicrometric structure in the three dimensions: laterally through the use of the nanoporous templates—where the size is controlled by pore diameter—and along the growth direction by pulse-plating electrodeposition—where the composition and length of the different segments or layers is controlled by the electrochemical parameters and time [161]. This approach allows for a huge control over the magnetic properties (coercivity, remanence, anisotropy) of the grown structures [162–164], even obtaining super-paramagnetic nanowires [165] or nanowires showing magnetization reversal ratchet effects [166].

The above-described pulse plating procedure is limited to the fabrication of alternating ferromagnetic and non-magnetic layers along the nanowire, due to limitation imposed by anomalous co-deposition, which makes it impossible to grow pure Co, Ni or Fe from an electrolyte containing different ions. A different approach to prepare all-magnetic multilayers consists of using a dual-bath technique, based on changing the electrolyte after the growth of each layer [150, 167] This technique opens more possibilities for lateral nanostructuration, but is slow and complex as the number of layers in the stack increases. Recently, a novel approach has been reported that takes into account the dependence of anomalous co-deposition on growth potential, to produce multilayers formed by all-ferromagnetic layers using single-bath pulse-plating electrodeposition [168]. We hereafter describe this procedure in detail as an example of the preparation of multisegmented nanowires, in particular $Fe_{20}Ni_{80}/Fe_{80}Ni_{20}$ nanowires.

Figure 11.12(a) shows the large variation of the Fe content in a Ni–Fe alloy when varying the growth potential for electrolytes with different a Fe^{2+}/Ni^{2+}. This variation can be tuned by changing the Fe^{2+}/Ni^{2+} ratio in the electrolyte. In particular, for a Fe^{2+}/Ni^{2+} ratio of 2%, it is possible to have a variation from $Fe_{20}Ni_{80}$ (for a growth potential of −1.5 V) to $Fe_{80}Ni_{20}$ (for a growth potential of −1.0 V), in the case of nanowires of 200 nm of diameter. Now, we can apply a pulsed potential between these two values to grow multisegmented nanowires. Figure 11.12(c) shows the temporal evolution of the cathodic current during the electrodeposition. Varying the temporal duration of each pulse, it is possible to control the length of each segment. In this example, equispaced local changes in composition were introduced, as shown in figure 11.12(b). The accurate control that can be achieved for nanostructuring of nanowires using this method is revealed by transmission electron microscopy images (figure 11.12(c)).

Interestingly, the same idea of pulse plating deposition can be also applied for nanostructuring alumina templates. It was mentioned before that the diameter of porous in alumina templates depends, among other parameters, on the anodization voltage. Therefore, pulsing the voltage between two different values makes it possible to obtain a controlled modulation, in terms of pore diameter and length of the different segments of the pore [169]. These modulated templates can be used as templates for electrodeposition to fabricate nanowires with modulation in the diameter along their axis [170]. Furthermore, the different steps of anodization can be also accompanied by wet etching and atomic layer deposition processes, in order to widen or shrink the pores, respectively. This provides a larger versatility in terms of control of diameter and length of the different segments [128].

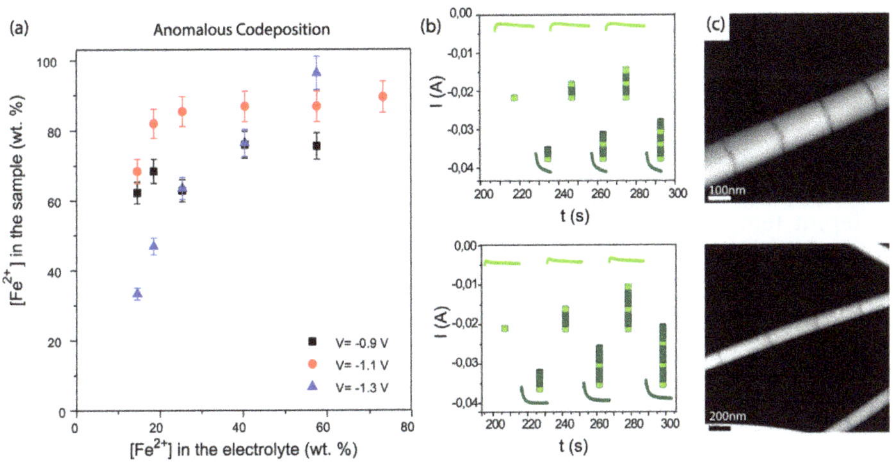

Figure 11.12. Electrodeposition of multisegmented nanowires. (a) Variation of the Fe content as a function of the concentration of Fe^{2+} ions in the electrolyte in the electrodeposition of Fe–Ni alloys. (b) Evolution of the current with time during nanowire growth and (c) transmission electron microscopy images of the grown nanowires in which the different periodic segments can be observed.

11.6.4 Electrodeposition of complex structures

In the last few years, exciting perspectives for the study of 3D nanomagnetism [4] and new opportunities to apply magnetic nanostructures in different areas [146, 171–173] are triggering the search for novel methods for the synthesis of more complex magnetic structures. In this context, electrodeposition has been also employed for the advanced synthesis of 3D structures using different approaches, all of them based on the two main properties mentioned before: the need for conductive substrates and conformal growth characteristics that enable the coating of complex 3D conductive structures [174].

For most biomedical and environmental applications, it is important to protect the magnetic material with a functionalizable coating. For magnetic applications, a coating can provide the material with additional contributions to magnetic anisotropy. Therefore, in addition to nanostructuration along the axis of the nanowire, it is important to tune the composition along the radial direction, making possible the growth of 3D magnetic core–shell structures, using electrodeposition in combination with other techniques. The easiest approaches consist of the growth of metallic nanowires, which are partially oxidized to form core–shell structures after releasing the nanowires from the template [175]. This technique is easy to implement but it is difficult to control the composition and thickness of the different layers. The second approach is the so-called widening method [176]. In this method, a material is first electrodeposited into the porous of a template following the standard form. Then, the template is immersed in a solution that partially removes the nanoporous walls, leaving an annular gap around the nanowires. A second material is subsequently electrodeposited into the gap forming a core–shell nanostructured array. Finally, the third reported method to prepare core–shell nanostructures is again a modification of the template-assisted method for the synthesis of nanowires: once the template is prepared, the inner part of the pore is covered by an insulating material using sol–gel [177, 178], proceeding with the growth afterwards, filling the pores by electrochemical deposition. For applications involving the release of the nanowires, it is important to select a material for the coating that is not etched by the chemicals and solvent used to remove the template.

Different templates with more complex shapes can be prepared and filled with magnetic materials using electrochemical deposition. Recently, it has also been proposed that cylindrical nanoporous alumina membranes can be formed by anodization of Al wires, a geometry that can be used for the synthesis of radially distributed nanowire arrays [179, 180]. Different approaches have been also proposed to produce nanostructured templates with a tridimensional structure of nanopores that can be filled by electrochemical deposition. Two-photon lithography can be used to define complex tridimensional nanopatterns in polymers [112]. Interconnected nanowire networks have been fabricated by ion-track-etching of a polycarbonate template by using a double ion irradiation with different incidence angles [181]. Finally, the anodization process can be also modified to produce tridimensional networks of perpendicularly arranged nanoporous in templates [182], templates that can be filled by magnetic materials, allowing the tailoring of the magnetic anisotropy [168].

To sum up, electrodeposition is a very suitable technique for the growth of tridimensional materials providing that an appropriate template is available. The development of new nanofabrication tools in the present and future will allow, in combination with electrodeposition, the manufacture of new magnetic materials, with new opportunities for the study of magnetism in 3D systems, opening the door for new applications.

11.7 End-of-chapter summary

- FEBID is a versatile nanopatterning technique that can achieve cutting-edge volumetric resolutions as well as very high nanostructure positioning accuracy.
- A large number of interactions take place between the electron beam, the substrate and the precursor during the FEBID process.
- Numerous modelling and experimental tools enable an in-depth understanding of these interactions.
- Automatic patterning algorithms with standardised calibration protocols have been developed to enable a wide range of users to use FEBID for nanoscale 3D printing.
- FEBID may be combined with other material deposition techniques in order to access materials for which a precursor gas is not available.
- Two-photon lithography can be used, in combination with RF magnetron sputtering, to fabricate a 3D magnetic nanostructure such as an artificial Buckyball.

- Element-specific phase tomography is a powerful technique to characterize 3D nanostructures with a spatial resolution of 25 nm. In the case of the buckyball, this method revealed that the Co layer was not homogeneously distributed over the 3D surface, and that it was in fact oxidized, thus highlighting the challenges of fabricating 3D magnetic nanostructures using 3D fabrication hybrid methods.
- Electrodeposition is a growth technique for the synthesis of materials, mostly metallic. Electrodeposited material can only grow on the conductive parts of the substrates. This, together with its ability for conformal growth, makes electrodeposition very suitable for the growth of patterned thin films and 3D structures.
- Electrodeposition has been extensively used in combination with nanoporous template to produce nanowires with accurate control of diameter (down to a few nanometers), length (up to hundreds of micron) and composition. These nanowires can be also nanostructured in diameter and composition along the axial or radial directions, to produce more complex tridimensional structures.
- The templates can be totally removed to study individual nanowires. It is also possible to prepare suspensions of nanowires in different solvents, and also create a freestanding tridimensional structure of vertical nanowires released from the template.
- The use of more complex templates prepared by different techniques (anodization of 3D networks, ion-track etching with double irradiation, FEBID, two-photon lithography…) leads to the synthesis of more complex 3D magnetic structures.

11.8 Student's corner

Further reading:

- *Reviews*: 3D nanomagnetism [4–6], magnetic nanostructures grown by FEBID [59, 102], 3D FEBID nanoprinting [183], additive manufacturing of metallic micro- and nanostructures [13], two-photon lithography methods for 3D nanomagnetism [112].
- *Modelling*: CASINO [96], GIS [100], finite difference simulation [66], continuum models [64], FEBID frequency framework and maps [65].
- *FEBID patterning algorithms*: nanowire networks [79, 89], layer-by-layer [80].
- *3D magnetic hybrid systems*: scaffold-based nanopatterning [97].

Exercises and further work:

- Use CASINO [96] to model the electron distribution occurring in different materials and substrate geometries at different beam energies.
- Estimate the angle that the GIS forms with respect to the electron column in your FEBID microscope using a ruler and basic geometry and consult the corresponding gas distribution in [100].

- Measure the vertical distance between your substrate and the GIS by slowly rising the substrate until it touches the GIS while imaging with the SEM. (Do this **very** carefully with an experienced user if it is the first time you try!!)
- Calculate using basic geometry if the axis of the GIS is really pointing to where the beam hits the substrate for your standard growth conditions. (Remember this is a 3D problem)
- Familiarize yourself with the table of experimental parameters that should always be recorded in an experiment in order to use the fundamental frequency framework ([65] and supplementary material therein).
- Find out where the above parameters are displayed in your electron microscope and familiarize yourself with their continuous monitoring during experimental sessions and the use of the calculator provided in [65] supplementary material.
- Understand why it is challenging to grow freestanding horizontal elements, and familiarize yourself with the use of substrate tilting described in [97].
- Search bibliography dedicated to the advanced characterisation of 3D magnetic nanostructures using electron and x-ray microscopy techniques. Identify the main challenges related to a 3D geometry when using these methods.
- Regarding the fabrication of 3D magnetic systems using 3D scaffolds and deposition of magnetic thin films, can you think of why a non-homogeneous coating might be useful, and why it might not?
- For this method, can you think of how one might improve the fabrication method and what techniques could be suitable?
- What are the two main characteristics of the electrodeposition technique that make it so versatile for the synthesis of 3D structures?
- How can you combine two-photon lithography or FEBID with electrodeposition to prepare 3D structures?
- What kind of 3D magnetic nanostructure would you like to make, and how would you make it?

References

[1] Coey J M D 2010 *Magnetism and Magnetic Materials* (Cambridge: Cambridge University Press)

[2] McCray W P 2009 How spintronics went from the lab to the iPod *Nat. Nanotechnol.* **4** 2–4

[3] Huber C *et al* 2016 3D print of polymer bonded rare-earth magnets, and 3D magnetic field scanning with an end-user 3D printer *Appl. Phys. Lett.* **109** 162401

[4] Fernández-Pacheco A, Streubel R, Fruchart O, Hertel R, Fischer P and Cowburn R P 2017 Three dimensional nanomagnetism *Nat. Commun.* **8** 15756

[5] Fischer P, Sanz-Hernández D, Streubel R and Fernández-Pacheco A 2020 Launching a new dimension with 3D magnetic nanostructures *APL Mater.* **8** 010701

[6] Streubel R *et al* 2016 Magnetism in curved geometries *J. Phys. D Appl. Phys.* **49** 363001

[7] Hertel R 2016 Ultrafast domain wall dynamics in magnetic nanotubes and nanowires *J. Phys. Condens. Matter.* **28** 483002

[8] Hellman F *et al* 2017 Interface-induced phenomena in magnetism *Rev. Mod. Phys.* **89** 025006

[9] Testing memory downsizing limits *Nat. Nanotechnol.* **14** 1 2019
[10] Ross C A, Berggren K K, Cheng J Y, Jung Y S and Chang J-B 2014 Three-dimensional nanofabrication by block copolymer self-assembly *Adv. Mater.* **26** 4386–96
[11] Li H, Carter J D and LaBean T H 2009 Nanofabrication by DNA self-assembly *Mater. Today* **12** 24–32
[12] Morber J R *et al* 2006 PLD-assisted VLS growth of aligned ferrite nanorods, nanowires, and nanobelts-synthesis, and properties *J. Phys. Chem.* B **110** 21672–9
[13] Hirt L, Reiser A, Spolenak R and Zambelli T 2017 Additive manufacturing of metal structures at the micrometer scale *Adv. Mater.* **29** 1604211
[14] Vaezi M, Seitz H and Yang S 2013 A review on 3D micro-additive manufacturing technologies *Int. J. Adv. Manuf. Technol.* **67** 1721–54
[15] Juodkazis S 2016 Manufacturing: 3D printed micro-optics *Nat. Photonics* **10** 499–501
[16] Moore D G, Barbera L, Masania K and Studart A R 2020 Three-dimensional printing of multicomponent glasses using phase-separating resins *Nat. Mater.* **19** 212–7
[17] Gissibl T, Thiele S, Herkommer A and Giessen H 2016 Sub-micrometre accurate free-form optics by three-dimensional printing on single-mode fibres *Nat. Commun.* **7** 11763
[18] Kong Y L *et al* 2014 3D printed quantum dot light-emitting diodes *Nano Lett.* **14** 7017–23
[19] Lewis J A and Ahn B Y 2015 Three-dimensional printed electronics *Nature* **8** 42–3
[20] Valentine A D *et al* 2017 Hybrid 3D printing of soft electronics *Adv. Mater.* **29** 1703817
[21] Murphy S V and Atala A 2014 3D bioprinting of tissues and organs *Nat. Biotechnol.* **32** 773–85
[22] Tian B and Lieber C M 2019 Nanowired bioelectric interfaces *Chem. Rev.* **119** 9135–52
[23] Li L *et al* 2016 Big area additive manufacturing of high performance bonded NdFeB magnets *Sci. Rep.* **6** 36212
[24] Paranthaman M P *et al* 2016 Binder jetting: a novel NdFeB bonded magnet fabrication process *JOM* **68** 1978–82
[25] Kersten W *et al* 2019 Additive-manufactured and topology-optimized permanent-magnet spin rotator for neutron interferometry *Phys. Rev. Appl.* **12** 014023
[26] Makarov D, Melzer M, Karnaushenko D and Schmidt O G 2016 Shapeable magnetoelectronics *Appl. Phys. Rev.* **3** 011101
[27] Mönch I *et al* 2011 Rolled-up magnetic sensor: nanomembrane architecture for in-flow detection of magnetic objects *ACS Nano* **5** 7436–42
[28] Medina-Sánchez M, Magdanz V, Guix M, Fomin V M and Schmidt O G 2018 Swimming microrobots: soft, reconfigurable, and smart *Adv. Funct. Mater.* **28** 1707228
[29] Li J *et al* 2018 Development of a magnetic microrobot for carrying and delivering targeted cells *Sci. Robot.* **3** eaat8829
[30] Barbot A, Tan H, Power M, Seichepine F and Yang G-Z 2019 Floating magnetic microrobots for fiber functionalization *Sci. Robot.* **4** eaax8336
[31] Testa P *et al* 2019 Magnetically addressable shape-memory and stiffening in a composite elastomer *Adv. Mater.* **31** 1900561
[32] Gao B, Yang Q, Zhao X, Jin G, Ma Y and Xu F 2016 4D bioprinting for biomedical applications *Trends Biotechnol.* **34** P746–56
[33] Hubert A and Schäfer R 2008 Introduction *Magnetic Domains* (Berlin: Springer) pp 1–10
[34] Yan M, Andreas C, Kákay A, García-Sánchez F and Hertel R 2011 Fast domain wall dynamics in magnetic nanotubes: suppression of Walker breakdown and Cherenkov-like spin wave emission *Appl. Phys. Lett.* **99** 122505

[35] Yan M, Kákay A, Andreas C and Hertel R 2013 Spin-Cherenkov effect and magnonic Mach cones *Phys. Rev.* B **88** 220412(R)

[36] Sheka D D *et al* 2015 Curvature effects in statics and dynamics of low dimensional magnets *J. Phys.* A **48** 125202

[37] Gaididei Y, Kravchuk V P and Sheka D D 2014 Curvature effects in thin magnetic shells *Phys. Rev. Lett.* **112** 257203

[38] Kravchuk V P *et al* 2016 Topologically stable magnetization states on a spherical shell: curvature-stabilized skyrmions *Phys. Rev.* B **94** 144402

[39] Sheka D D, Kravchuk V P, Yershov K V and Gaididei Y 2015 Torsion-induced effects in magnetic nanowires *Phys. Rev.* B **92** 054417

[40] Da S C *et al* 2014 Observation of Bloch-point domain walls in cylindrical magnetic nanowires *Phys. Rev.* B **89** 180405

[41] Volkov O M *et al* 2019 Experimental observation of exchange-driven chiral effects in curvilinear magnetism *Phys. Rev. Lett.* **123** 077201

[42] Wartelle A *et al* 2019 Bloch-point-mediated topological transformations of magnetic domain walls in cylindrical nanowires *Phys. Rev.* B **99** 024433

[43] Schöbitz M *et al* 2019 Fast domain wall motion governed by topology and Œrsted fields in cylindrical magnetic nanowires *Phys. Rev. Lett.* **123** 217201

[44] Hsueh H-Y, Huang Y-C, Ho R-M, Lai C-H, Makida T and Hasegawa H 2011 Nanoporous gyroid nickel from block copolymer templates via electroless plating *Adv. Mater.* **23** 3041–6

[45] Streubel R, Kronast F, Fischer P, Parkinson D, Schmidt O G and Makarov D 2015 Retrieving spin textures on curved magnetic thin films with full-field soft X-ray microscopies *Nat. Commun.* **6** 7612

[46] Phatak C, Liu Y, Gulsoy E B, Schmidt D, Franke-Schubert E and Petford-Long A 2014 Visualization of the magnetic structure of sculpted three-dimensional cobalt nanospirals *Nano Lett.* **14** 759–64

[47] Eslami S *et al* 2014 Chiral nanomagnets *ACS Photonics* **1** 1231–6

[48] Gibbs J G, Mark A G, Lee T-C, Eslami S, Schamel D and Fischer P 2014 Nanohelices by shadow growth *Nanoscale* **6** 9457–66

[49] da Câmara Santa Clara Gomes T, De La Torre Medina J, Lemaitre M and Piraux L 2016 Magnetic and magnetoresistive properties of 3D interconnected NiCo nanowire networks *Nanoscale Res. Lett.* **11** 466

[50] Liakakos N *et al* 2014 Solution epitaxial growth of cobalt nanowires on crystalline substrates for data storage densities beyond 1 Tbit/in2 *Nano Lett.* **14** 3481–6

[51] Fernández-Pacheco A *et al* 2013 Three dimensional magnetic nanowires grown by focused electron-beam induced deposition *Sci. Rep.* **3** 1492

[52] Keller L *et al* 2018 Direct-write of free-form building blocks for artificial magnetic 3D lattices *Sci. Rep.* **8** 6160

[53] Sanz-Hernández D, Hamans R F, Liao J W, Welbourne A, Lavrijsen R and Fernández-Pacheco A 2017 Fabrication, detection, and operation of a three-dimensional nanomagnetic conduit *ACS Nano* **11** 11066–73

[54] May A, Hunt M, Van Den Berg A, Hejazi A and Ladak S 2018 Realisation of a frustrated 3D magnetic nanowire lattice *arXiv:*1812.02442

[55] Donnelly C *et al* 2015 Element-specific X-ray phase tomography of 3d structures at the nanoscale *Phys. Rev. Lett.* **114** 115501

[56] Utke I, Hoffmann P and Melngailis J 2008 Gas-assisted focused electron beam and ion beam processing and fabrication *J. Vac. Sci. Technol.* **B 26** 1197
[57] Huth M, Porrati F and Dobrovolskiy O V 2018 Focused electron beam induced deposition meets materials science *Microelectron. Eng.* **185–186** 9–28
[58] Fernández-Pacheco A, De Teresa J M, Córdoba R and Ibarra M R 2009 Magnetotransport properties of high-quality cobalt nanowires grown by focused-electron-beam-induced deposition *J. Phys. D Appl. Phys.* **42** 055005
[59] De Teresa J M, Fernández-Pacheco A, Córdoba R, Serrano-Ramón L, Sangiao S and Ibarra M R 2016 Review of magnetic nanostructures grown by focused electron beam induced deposition (FEBID) *J. Phys. D Appl. Phys.* **49** 243003
[60] Utke I *et al* 2005 Cross section investigations of compositions and sub-structures of tips obtained by focused electron beam induced deposition *Adv. Eng. Mater.* **7** 323–31
[61] Porrati F *et al* 2015 Direct writing of CoFe alloy nanostructures by focused electron beam induced deposition from a heteronuclear precursor *Nanotechnology* **26** 475701
[62] Porrati F, Kämpken B, Terfort A and Huth M 2013 Fabrication and electrical transport properties of binary Co–Si nanostructures prepared by focused electron beam-induced deposition *J. Appl. Phys.* **113** 053707
[63] Pablo-Navarro J, Magén C and De Teresa J M 2016 Three-dimensional core-shell ferromagnetic nanowires grown by focused electron beam induced deposition *Nanotechnology* **27** 285302
[64] Toth M, Lobo C, Friedli V, Szkudlarek A and Utke I 2015 Continuum models of focused electron beam induced processing *Beilstein J. Nanotechnol.* **6** 1518–40
[65] Sanz-Hernández D and Fernández-Pacheco A 2017 Modelling focused electron beam induced deposition beyond Langmuir adsorption *Beilstein J. Nanotechnol.* **8** 2151–61
[66] Fowlkes J D, Winkler R, Lewis B B, Stanford M G, Plank H and Rack P D 2016 Simulation-guided 3D nanomanufacturing via focused electron beam induced deposition *ACS Nano* **10** 6163–72
[67] Fernández-Pacheco A *et al* 2009 Domain wall conduit behavior in cobalt nanowires grown by focused electron beam induced deposition *Appl. Phys. Lett.* **94** 192509
[68] Fernandez-Pacheco A *et al* 2009 Magnetization reversal in individual cobalt micro-and nanowires grown by focused-electron-beam-induced-deposition *Nanotechnology* **20** 475704
[69] Gavagnin M, Wanzenboeck H D, Belić D and Bertagnolli E 2013 Synthesis of individually tuned nanomagnets for nanomagnet logic by direct write focused electron beam induced deposition *ACS Nano* **7** 777–84
[70] Gavagnin M *et al* 2014 Free-standing magnetic nanopillars for 3D nanomagnet logic *ACS Appl. Mater. Interfaces* **6** 20254–60
[71] Vavassori P, Pancaldi M, Perez-Roldan M J, Chuvilin A and Berger A 2016 Remote magnetomechanical nanoactuation *Small* **12** 1013–23
[72] Stiller M *et al* 2017 Functionalized Akiyama tips for magnetic force microscopy measurements *Meas. Sci. Technol.* **28** 125401
[73] Mattiat H *et al* 2020 Nanowire magnetic force sensors fabricated by focused-electron-beam-induced deposition *Phys. Rev. Appl.* **13** 044043
[74] Guo F, Belova L M and McMichael R D 2013 Spectroscopy and imaging of edge modes in permalloy nanodisks *Phys. Rev. Lett.* **110** 017601
[75] Jaafar M *et al* 2020 Customized MFM probes based on magnetic nanorods *Nanoscale* **12** 10090–7

[76] Plank H *et al* 2019 Focused electron beam-based 3D nanoprinting for scanning probe microscopy: a review *Micromachines* **11** 48
[77] Pablo-Navarro J, Sanz-Hernández D, Magén C, Fernández-Pacheco A and De Teresa J M 2017 Tuning shape, composition and magnetization of 3D cobalt nanowires grown by focused electron beam induced deposition (FEBID) *J. Phys. D Appl. Phys.* **50** 18LT01
[78] Córdoba R, Sharma N, Kölling S, Koenraad P M and Koopmans B 2016 High-purity 3D nano-objects grown by focused-electron-beam induced deposition *Nanotechnology* **27** 355301
[79] Fowlkes J D *et al* 2018 High-fidelity 3D-nanoprinting via focused electron beams: computer-aided design (3BID) *ACS Appl. Nano Mater.* **1** 1028–41
[80] Skoric L, Sanz-Hernández D, Meng F, Donnelly C, Merino-Aceituno S and Fernández-Pacheco A 2019 Layer-by-layer growth of complex-shaped three-dimensional nanostructures with focused electron beams *Nano Lett.* **20** 184–91
[81] Costanzi B N, Riazanova A V, Dan Dahlberg E and Belova L M 2014 In situ manufacture of magnetic tunnel junctions by a direct-write process *Appl. Phys. Lett.* **104** 222401
[82] Pablo-Navarro J, Magén C and De Teresa J M 2018 Purified and crystalline three-dimensional electron-beam-induced deposits: the successful case of cobalt for high-performance magnetic nanowires *ACS Appl. Nano Mater.* **1** 38–46
[83] Begun E *et al* 2015 Post-growth purification of Co nanostructures prepared by focused electron beam induced deposition *Nanotechnology* **26** 075301
[84] Lavrijsen R *et al* 2011 Fe:O:C grown by focused-electron-beam-induced deposition: magnetic and electric properties *Nanotechnology* **22** 025302
[85] Gabureac M, Bernau L, Utke I and Boero G 2010 Granular Co-C nano-Hall sensors by focused-beam-induced deposition *Nanotechnology* **21** 115503
[86] Fernández-Pacheco A, De Teresa J, Córdoba R and Ibarra M 2009 Metal-insulator transition in Pt-C nanowires grown by focused-ion-beam-induced deposition *Phys. Rev.* B **79** 174204
[87] Dukic M *et al* 2016 Direct-write nanoscale printing of nanogranular tunnelling strain sensors for sub-micrometre cantilevers *Nat. Commun.* **7** 12487
[88] Arnold G *et al* 2018 Tunable 3D nanoresonators for gas-sensing applications *Adv. Funct. Mater.* **28** 1707387
[89] Winkler R, Lewis B B, Fowlkes J D, Rack P D and Plank H 2018 High-fidelity 3D-nanoprinting via focused electron beams: growth fundamentals *ACS Appl. Nano Mater.* **1** 1014–27
[90] Winkler R, Fowlkes J D, Rack P D and Plank H 2019 3D nanoprinting via focused electron beams *J. Appl. Phys.* **125** 210901
[91] Huth M *et al* 2012 Focused electron beam induced deposition: a perspective *Beilstein J. Nanotechnol.* **3** 597–619
[92] van Dorp W F and Hagen C W 2008 A critical literature review of focused electron beam induced deposition *J. Appl. Phys.* **104** 081301
[93] Sanz-Hernández D *et al* 2020 Artificial double-helix for geometrical control of magnetic chirality *ACS Nano* **14** 8084–92
[94] Roth A 1990 *Vacuum Technology* (Amsterdam: Elesevier)
[95] Joy D C and Joy C S 1996 Low voltage scanning electron microscopy *Micron* **27** 247–63
[96] Drouin D, Couture A R, Joly D, Tastet X, Aimez V and Gauvin R 2007 CASINO V2.42 a fast and easy to use modeling tool for scanning electron microscopy and microanalysis Users *Scanning* **29** 92–101

[97] Sanz-Hernández D *et al* 2018 Fabrication of scaffold-based 3D magnetic nanowires for domain wall applications *Nanomaterials* **8** 483
[98] Fowlkes J D, Winkler R, Mutunga E, Rack P D and Plank H 2019 Simulation informed CAD for 3D nanoprinting *Micromachines* **11** 8
[99] Utke I 2004 Thermal effects during focused electron beam induced deposition of nanocomposite magnetic-cobalt-containing tips *Microelectron. Eng.* **73–74** 553–8
[100] Friedli V and Utke I 2009 Optimized molecule supply from nozzle-based gas injection systems for focused electron- and ion-beam induced deposition and etching: simulation and experiment *J. Phys. D Appl. Phys.* **42** 125305
[101] Carden W G, Lu H, Spencer J A, Fairbrother D H and McElwee-White L 2018 Mechanism-based design of precursors for focused electron beam-induced deposition *MRS Commun.* **8** 343–57
[102] De Teresa J M and Fernández-Pacheco A 2014 Present and future applications of magnetic nanostructures grown by FEBID *Appl. Phys.* A **117** 1645–58
[103] Bishop J, Lobo C J, Martin A, Ford M, Phillips M and Toth M 2012 Role of activated chemisorption in gas-mediated electron beam induced deposition *Phys. Rev. Lett.* **109** 146103
[104] Bresin M, Toth M and Dunn K A 2013 Direct-write 3D nanolithography at cryogenic temperatures *Nanotechnology* **24** 035301
[105] Oudar J and Wise H 1985 Deactivation and poisoning of catalysts (New York: Marcel Dekker)
[106] Vansteenkiste A, Leliaert J, Dvornik M, Helsen M, Garcia-Sanchez F and Van Waeyenberge B 2014 The design and verification of MuMax3 *AIP Adv.* **4** 107133
[107] Mutunga E, Winkler R, Sattelkow J, Rack P D, Plank H and Fowlkes J D 2019 Impact of electron-beam heating during 3D nanoprinting *ACS Nano* **13** 5198–213
[108] Keller L and Huth M 2018 Pattern generation for direct-write three-dimensional nanoscale structures via focused electron beam induced deposition *Beilstein J. Nanotechnol.* **9** 2581–98
[109] Mutunga E, Winkler R, Sattelkow J, Rack P D, Plank H and Fowlkes J D 2019 The impact of electron beam heating during 3D nanoprinting *ACS Nano* **13** 5198–213
[110] Tu F *et al* 2016 On the magnetic properties of iron nanostructures fabricated via focused electron beam induced deposition and autocatalytic growth processes *Nanotechnology* **27** 355302
[111] Serrano-Ramón L *et al* 2011 Ultrasmall functional ferromagnetic nanostructures grown by focused electron-beam-induced deposition *ACS Nano* **5** 7781–7
[112] Hunt M *et al* 2020 Harnessing multi-photon absorption to produce three-dimensional magnetic structures at the nanoscale *Materials* **13** 761
[113] Donnelly C 2018 Hard X-ray tomography of three dimensional magnetic structures *Microsc. Microanal* **24** 80–1
[114] Donnelly C *et al* 2015 Element-specific x-ray phase tomography of 3D structures at the nanoscale *Phys. Rev. Lett.* **114** 115501
[115] Gliga S, Seniutinas G, Weber A and David C 2019 Architectural structures open new dimensions in magnetism: magnetic buckyballs *Mater. Today* **26** 100–1
[116] May A, Hunt M, Van Den Berg A, Hejazi A and Ladak S 2019 Realisation of a frustrated 3D magnetic nanowire lattice *Commun. Phys.* **2** 13
[117] Williams G *et al* 2018 Two-photon lithography for 3D magnetic nanostructure fabrication *Nano Res.* **11** 845–54

[118] Gansel J K *et al* 2009 Gold helix photonic metamaterial as broadband circular polarizer *Science (80-.)* **325** 1513–5

[119] Seniutinas G, Weber A, Padeste C, Sakellari I, Farsari M and David C 2018 Beyond 100 nm resolution in 3D laser lithography—post processing solutions *Microelectron. Eng.* **191** 25–31

[120] Bauer J, Schroer A, Schwaiger R and Kraft O 2016 Approaching theoretical strength in glassy carbon nanolattices *Nat. Mater.* **15** 438–43

[121] Mueller P, Thiel M and Wegener M 2014 3D direct laser writing using a 405 nm diode laser *Opt. Lett.* **39** 6847–50

[122] Sang H P, Tae W L, Yang D Y, Ran H K and Lee K S 2006 Improvement of spatial resolution in nano-stereolithography using radical quencher *Macromol. Res.* **14** 559–64

[123] Wollhofen R, Katzmann J, Hrelescu C, Jacak J and Klar T A 2013 120 nm resolution and 55 nm structure size in STED-lithography *Opt. Express* **21** 10831–40

[124] Smith D O, Cohen M S and Weiss G P 1960 Oblique-incidence anisotropy in evaporated permalloy films *J. Appl. Phys.* **31** 1755–62

[125] Daub M, Knez M, Goesele U and Nielsch K 2007 Ferromagnetic nanotubes by atomic layer deposition in anodic alumina membranes *J. Appl. Phys.* **101** 09J111

[126] George S M 2010 Atomic layer deposition: an overview *Chem. Rev.* **110** 111–31

[127] Johnson R W, Hultqvist A and Bent S F 2014 A brief review of atomic layer deposition: from fundamentals to applications *Mater. Today* **17** 236–46

[128] Bochmann S, Döhler D, Trapp B, Staňo M, Fruchart O and Bachmann J 2018 Preparation and physical properties of soft magnetic nickel-cobalt three-segmented nanowires *J. Appl. Phys.* **124** 163907

[129] Staňo M *et al* 2018 Flux-closure domains in high aspect ratio electroless-deposited CoNiB nanotubes *SciPost Phys.* **5** 038

[130] Richardson D and Rhen F M F 2015 The diverse nanostructure of electroless plated CoNiFeB alloy: thin film, nanotubes and nanoparticles *Phys. Procedia* **75** 1158–66

[131] Schwarzacher W Electrodeposition: a technology for the future *Electrochem. Soc. Interface*

[132] Lincot D 2005 Electrodeposition of semiconductors *Thin Solid Films* **487** 40–8

[133] Beck F 1988 Electrodeposition of polymer coatings *Electrochim. Acta* **33** 839–50

[134] Datta M 1998 Applications of electrochemical microfabrication: an introduction *IBM J. Res. Dev.* **42** 563–6

[135] Pletcher D, Greff R, Peat R, Peter L M and Robinson J 2001 *Instrumental Methods in Electrochemistry* (Cambridge: Woodhead Publishing)

[136] Magnetic anisotropy phenomena in electrodeposited heterostructure devices - KU Leuven. [Online]. Available: https://limo.libis.be/primo-explore/fulldisplay?docid=LIRIAS1752282&context=L&vid=Lirias&search_scope=Lirias&tab=default_tab&lang=en_US&fromSitemap=1 [Accessed: 27-Mar-2020].

[137] Plaza M, Abuín M, Mascaraque A, González-Barrio M A and Pérez L 2012 Epitaxial growth of Bi ultra-thin films on GaAs by electrodeposition *Mater. Chem. Phys.* **134** 523–30

[138] Romankiw L T 1997 A path: From electroplating through lithographic masks in electronics to LIGA in MEMS *Electrochim. Acta* **42** 2985–3005

[139] Genolet G and Lorenz H 2014 UV-LIGA: from development to commercialization *Micromachines* **5** 486–95

[140] Prida V M, Vega V, García J, Iglesias L, Hernando B and Minguez-Bacho I 2015 Electrochemical methods for template-assisted synthesis of nanostructured materials

Magnetic Nano- and Microwires: Design, Synthesis, Properties and Applications (Amsterdam: Elsevier) 3–39

[141] Whitney T M, Jiang J S, Searson P C and Chien C L 1993 Fabrication and magnetic properties of arrays of metallic nanowires *Science (80-.).* **261** 1316–9

[142] Martin C R 1994 Nanomaterials: a membrane-based synthetic approach *Science (80-.).* **266** 1961–6

[143] Toimil-Molares M E 2012 Characterization and properties of micro- and nanowires of controlled size, composition, and geometry fabricated by electrodeposition and ion-track technology *Beilstein J. Nanotechnol.* **3** 860–83

[144] Sousa C T, Leitao D C, Proenca M P, Ventura J, Pereira A M and Araujo J P 2014 Nanoporous alumina as templates for multifunctional applications *Appl. Phys. Rev.* **1** 031102

[145] Lee W and Park S J 2014 Porous anodic aluminum oxide: anodization and templated synthesis of functional nanostructures *Chem. Rev.* **114** 7487–556

[146] Jiang W, Rutherford D, Vuong T and Liu H 2017 Nanomaterials for treating cardiovascular diseases: a review *Bioactive Mater.* **2** 185–98

[147] Vega V *et al* 2016 Long-range hexagonal arrangement of TiO_2 nanotubes by soft lithography-guided anodization *Electrochim. Acta.* **203** 51–8

[148] Brenner A 1963 *Electrodeposition of Alloys* (New York: Academic)

[149] Llavona Á, Pérez L, Sánchez M C and De Manuel V 2013 Enhancement of anomalous codeposition in the synthesis of Fe-Ni alloys in nanopores *Electrochim. Acta* **106** 392–7

[150] Berganza E *et al* 2017 Multisegmented nanowires: a step towards the control of the domain wall configuration *Sci. Rep.* **7** 11576

[151] Andersen I M *et al* 2020 Exotic transverse-vortex magnetic configurations in CoNi nanowires *ACS Nano* **14** 1399–405

[152] Marcano N *et al* 2010 Weak-antilocalization signatures in the magnetotransport properties of individual electrodeposited Bi Nanowires *Appl. Phys. Lett.* **96** 3328101

[153] Costas A *et al* 2018 Magnetism and magnetoresistance of single Ni–Cu alloy nanowires *Beilstein J. Nanotechnol.* **9** 2345–55

[154] Proenca M P *et al* 2019 Deterministic and time resolved thermo-magnetic switching in a nickel nanowire *Sci. Rep.* **9** 17339

[155] Martínez-Banderas A I *et al* 2019 Iron-based core-shell nanowires for combinatorial drug delivery and photothermal and magnetic therapy *ACS Appl. Mater. Interfaces* **11** 43976–88

[156] Moreno J A *et al* 2019 Growth of ordered iron oxide nanowires for photo-electrochemical water oxidation *ACS Appl. Energy Mater.* **2** 8473–80

[157] Home | ByAxon. [Online]. Available: http://byaxon-project.eu/ [Accessed: 27-Mar-2020].

[158] Patel N S, Lago-Cachón D, Mohammed H, Moreno J A and Kosel J 2019 Iron nanowire fabrication by nano-porous anodized aluminum and its characterization *J. Vis. Exp.* **152** e60111

[159] Alper M *et al* 1993 Giant magnetoresistance in electrodeposited superlattices *Appl. Phys. Lett.* **63** 2144–6

[160] Attenborough K, Boeve H, De Boeck J, Borghs G and Celis J P 1999 Electrodeposited spin valves on n-type GaAs *Appl. Phys. Lett.* **74** 2206

[161] Prida V M, García J, Hernando B, Bran C, Vivas L G and Vázquez M 2015 Electrochemical synthesis of magnetic nanowires with controlled geometry and magnetic

anisotropy *Magnetic Nano- and Microwires: Design, Synthesis, Properties and Applications* (Amsterdam: Elsevier) 41–104

[162] Sun L, Hao Y, Chien C L, Searson P C and Searson P C 2005 Tuning the properties of magnetic nanowires *IBM J. Res. Dev.* **49** 79–102

[163] Núñez A, Pérez L, Abuín M, Araujo J P and Proenca M P 2017 Magnetic behaviour of multisegmented FeCoCu/Cu electrodeposited nanowires *J. Phys. D Appl. Phys.* **50** 155003

[164] Staňo M and Fruchart O 2018 Magnetic nanowires and nanotubes *Handbook of Magnetic Materials* vol 27 (amsterdam: Elsevier) pp 155–267

[165] Tang X T, Wang G C and Shima M 2006 Superparamagnetic behavior in ultrathin CoNi layers of electrodeposited CoNi/Cu multilayer nanowires *J. Appl. Phys.* **99** 123910

[166] Bran C *et al* 2018 Magnetization ratchet in cylindrical nanowires *ACS Nano* **12** 5932–9

[167] Fernández J G, Martínez V V, Thomas A, de la Prida Pidal V M and Nielsch K 2018 Two-step magnetization reversal FORC fingerprint of coupled bi-segmented Ni/Co magnetic nanowire arrays *Nanomaterials* **8** 548

[168] Ruiz-Clavijo A, Ruiz-Gomez S, Caballero-Calero O, Perez L and Martin-Gonzalez M 2019 Tailoring magnetic anisotropy at will in 3D interconnected nanowire networks *Phys. Status Solidi Rapid Res. Lett.* **13** 1900263

[169] Lee W, Ji R, Gösele U and Nielsch K 2006 Fast fabrication of long-range ordered porous alumina membranes by hard anodization *Nat. Mater.* **5** 741–7

[170] Rodríguez L A *et al* 2016 Quantitative nanoscale magnetic study of isolated diameter-modulated FeCoCu nanowires *ACS Nano* **10** 9669–78

[171] Parkin S S P, Hayashi M and Thomas L 2008 Magnetic domain-wall racetrack memory *Science* **320** 190–4

[172] Ivanov Y P *et al* 2016 Tunable magnetic nanowires for biomedical and harsh environment applications *Sci. Rep.* **6** 24189

[173] Zhang Q, Yang X and Guan J 2019 Applications of magnetic nanomaterials in heterogeneous catalysis *ACS Appl. Nano Mater.* **2** 4681–97

[174] Ruiz-Gómez S *et al* 2015 Graphene foam functionalized with electrodeposited nickel hydroxide for energy applications *Diam. Relat. Mater.* **57** 63–7

[175] Cortés-Llanos B *et al* 2017 Thermal route for the synthesis of maghemite/hematite core/shell nanowires *J. Phys. Chem.* C **121** 23158–65

[176] Loh P Y, Liu C, Sow C H and Chin W S 2014 Coaxial hetero-nanostructures with controllable shell thickness: a 'pore widening' method *RSC Adv.* **4** 8735–40

[177] Khan U *et al* 2016 Magnetic response of hybrid ferromagnetic and antiferromagnetic core-shell nanostructures *Nanoscale* **8** 6064–70

[178] Martín-García L, Ruiz-Gómez S, Abuín M, Montaña Y, Carmona N and Pérez L 2015 Multifunctional core-shell Co-SiO_2 nanowires via electrodeposition and sol-gel techniques *RSC Adv.* **5** 97503–7

[179] Sanz R, Hernández-Vélez M, Pirota K R, Baldonedo J L and Vázquez M 2007 Fabrication and magnetic functionalization of cylindrical porous anodic alumina *Small* **3** 434–7

[180] Garcia C *et al* 2018 Magnetization reversal in radially distributed nanowire arrays *J. Phys. Chem.* C **122** 5124–30

[181] Araujo E *et al* 2015 Artificially modified magnetic anisotropy in interconnected nanowire networks *Nanoscale* **7** 1485–90

[182] Martín J, Martín-González M, Francisco Fernández J and Caballero-Calero O 2014 Ordered three-dimensional interconnected nanoarchitectures in anodic porous alumina *Nat. Commun.* **5** 5130
[183] Plank H, Winkler R, Sattelkow J, Fowlkes J D and Rack P D 2018 3D Nanoprinting via focused electron beams *Microsc. Microanal.* **24** 346–7

Chapter 12

FEBIP for functional nanolithography of 2D nanomaterials

Songkil Kim and Andrei G Fedorov

Lithographic patterning on emerging nanomaterials is one of the essential ingredients for the development of new functional nanomaterials and device systems. Focused electron beam induced processing (FEBIP) in combination with an on-demand reconfigurable precursor injection system enables the highest resolution and controllability of nanolithography on two-dimensional (2D) nanomaterials. FEBIP allows for localized atomic-to-nanoscale engineering of nanomaterials and devices in a multi-mode format including geometrical patterning, functionalization/defect formation/doping, metal/metal oxide deposition, interface fusion and activation, active channel cleaning. This chapter discusses the scientific bases and showcases the state-of-the-art FEBIP technologies with a specific focus on lithography of 2D nanomaterials and provides future prospects of FEBIP technologies to address real industrial needs for the development of high-performance, next-generation functional devices.

12.1 Introduction

Localized manipulation of nanomaterials can provide unprecedented opportunities for design of new functionalities as well as breakthrough improvements of existing properties of nanomaterials. Having a high degree of control with atomic-to-nano-scale resolution, a focused electron beam with a broad energy range in combination with molecular precursors delivered locally in an electron microscope environment is a promising imaging probe and a powerful 'direct write' tool enabling local excitation and modification of nanomaterials. Focused electron beam induced processing (FEBIP) is a next-generation technology for a *nanofactory* where deposition, etching, fusion, doping and other manipulations of nanomaterials occur in a multi-mode format and is simultaneously combined with ultra-high-resolution *in situ* imaging to provide unique opportunities in new materials design and analysis.

doi:10.1088/978-0-7503-2608-7ch12

Two-dimensional (2D) nanomaterials have been extensively explored owing to their unique mechanical, optical, thermal and electrical properties [1]. Atomically thin 2D layered materials have been a subject of much attention after the successful mechanical exfoliation of graphene from graphite by Novoselov and Geim [2]. Graphene is a single atom-thick hexagonal carbon structure, which is optically transparent and has ultra-high carrier mobility, high thermal conductivity and Young's modulus at room temperature [1]. The discovery of graphene with its extraordinary properties motivated exploration of other 2D nanomaterials, including an insulating 2D hexagonal boron-nitride (h-BN) and especially 2D inorganic semiconducting nanomaterials with non-zero electronic bandgaps [1]. Transition metal dichalcogenides (TMDs) are layered materials with a stoichiometry of MX_2 where M is transition metals, including Mo, Ti, Ta, and W, and X is chalcogens, such as S, Se and Te. In particular, a single-layer MoS_2, a direct band gap semiconductor, has been under intense attention since it can be a component for flexible high-performance 2D electronics and optoelectronics and can also be stacked with semi-metallic graphene and insulating h-BN to form a van der Waals heterostructure for novel functional devices [3]. Owing to the tunable nature of 2D nanomaterials' properties, they present an interesting platform as a model multi-layer nanomaterial for application of FEBIP capabilities for site-specific, high-resolution modification of the composite interface formation and modulation of interfacial and transport properties. This chapter provides an overview of FEBIP as a direct-write lithography toolbox to atomically engineer the topological, structural and compositional features of 2D composite nanomaterials with an aim to introduce new functional properties and devices. It presents the background on atomic manipulation of 2D nanomaterials, with in depth focus on electron-beam-assisted surface doping, defect formation, etching, interface fusion and examples of FEBIP applications to 2D nanomaterial-based electronic devices.

12.2 Atomic manipulation of 2D nanomaterials

12.2.1 Defect engineering of 2D nanomaterials

Directed atomic manipulation of 2D nanomaterials can be achieved using a focused electron beam in (scanning) transmission electron microscopy. A large number of *in situ* experiments have been conducted to investigate electron beam irradiation effects on atomic structures of 2D nanomaterials along with atomic-resolution imaging [4, 5]. High energy focused electrons transfer energy to 2D nanomaterial systems so that controllable defect formation can be achieved by tuning beam parameters such as energy, current, irradiation time/area and the corresponding electron beam dose. Defects can cause the mechanical and electronic properties of pristine nanomaterials in general to deteriorate, but they can also enhance chemical and magnetic properties or impose new functional properties to nanomaterials. Thus, controlling the formation of specific defects using an electron beam at atomic resolution will provide a unique opportunity for structure-induced tuning of materials' properties.

Typical point defects in graphene include sp^3-type defects, Stone–Wales defects and vacancy formation [6]. They can be formed by electron irradiation at varying levels of beam accelerating energy. An sp^3-type defect is defined as sp^3-hybridization appearing locally so that new covalent bonds can be formed between adatoms and the underlying carbon atoms in the graphene basal plane, which can be induced by mild electron irradiation at a relatively low energy in a range of a few tens of keVs. The energetically favored adsorption configuration of adatoms is the bridge site over the top of the carbon–carbon bond of graphene having a binding energy of 1.5–2 eV [6]. Other metastable configurations are feasible and the energy difference between different configuration states is as small as ~0.3 eV, which indicates that the adatoms are very mobile over the graphene plane and can be stimulated to migrate by the external energy input such as electron irradiation or do so spontaneously due to thermal fluctuations, even at room temperature [6]. The adatoms can be carbon atoms from e-beam dissociated hydrocarbons adsorbed on the surface (e.g., from surface contamination) or other foreign atoms introduced onto a surface, e.g., via FEBID. The adatom can change properties of graphene depending on whether it strongly interacts with graphene or not. Electron irradiation can activate the chemical interactions of foreign adatoms and graphene, which establish new functionality and/or modify the intrinsic properties of graphene.

Stone–Waals (SW) defects result from the rotation of one carbon–carbon bond at an in-plane 90° with no removal or displacements of carbon atoms [6–8]. Figure 12.1(a) shows the calculated energy barrier for bond rotation. The kinetic barrier for the formation of the SW defect is about 9–10 eV and the energy barrier for the reverse bond rotational transformation is 5 eV, indicating that the SW defect is quite stable at room temperature [7]. Due to the high kinetic barrier, it can be rarely found in an equilibrium state after graphene synthesis that typically occurs at temperatures below 1000 °C. However, electron irradiation can lead to the non-equilibrium change of local states of graphene and an electron beam at an accelerating energy less than 80 keV can form SW defects with no vacancy

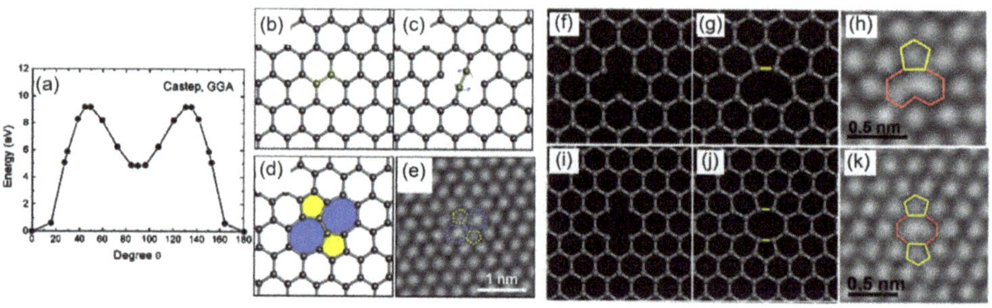

Figure 12.1. (a) DFT calculations of formation energies for C–C bond rotations depending on angles showing that an SW defect can be generated by overcoming an energy barrier of ~10 eV. (Reprinted with permission from [7], Copyright (2005) by the American Physical Society.) (b)–(d) Atomic models and (e) the corresponding aberration-corrected (AC) TEM image to describe the bond rotations to form an SW defect. (Reproduced from [8] with permission of The Royal Society of Chemistry.) (f) and (g) Atomic models of a mono vacancy formation describing a single carbon atom sputtered and the structural relaxation to a final form of a monovacancy (5–9), which is identical to the (h) AC TEM image. (i) and (j) Atomic models of the divacancy (5–8–5) formation with (k) the corresponding AC TEM image. Reproduced from [4] with permission.

formation [6]. Figures 12.1(b)–(d) show the atomic model demonstrating the C–C bond rotation to form the SW defect, and the corresponding aberration-corrected TEM image of the SW defect is shown in figure 12.1(e).

Vacancy formation occurs by direct sputtering of carbon atoms from the lattice by high energy electron beam irradiation. The threshold energy for knock-on displacements of carbon atoms in graphene is 18–22 eV, which can be transferred by electron impact at an accelerating energy higher than 90 keV, based on the theoretical estimations [4, 6]. Figures 12.1(f)–(k) show atomic models of single vacancy (SV) and divacancy (DV) defects and the corresponding TEM images. Vacancies are unstable due to the dangling bonds which can lead to the reconstruction of defected sites and attract carbon adatoms or other foreign elements to passivate the edges of graphene or to form a perfect graphene lattice under electron irradiation. Prolonged electron beam irradiation can result in sequential removal of carbon atoms, but also facilitate the transformation of defected graphene into more stable structures. Vacancy formation can also occur with the intense beam current density (corresponding to the increased electron beam dose) at an electron beam energy of 80 keV, lower than a threshold electron beam energy theoretically estimated [4]. Electron accelerating energy of 80 keV is not sufficient to result in direct knock-on sputtering of carbon atoms by electron impingement, but any adatom under electron beam irradiation can interact with graphene disrupting the sp^2 hexagonal network and weakening the binding of C–C bonds [4]. In addition, the prolonged electron beam irradiation transits any perturbation to pristine graphene, including SW defects and interstitials, to vacancy formation since the C–C bond strength of graphene becomes weakened and thus, the threshold displacement energy of carbon atoms can be lowered [4–6]. Defect formation on graphene is a very complex process, and many different types of defects such as point defects, line defects (dislocations) and phase transformation to 2D amorphous structures can occur sequentially and concurrently. Annealing and quenching of existing defects, while new defects are being formed under electron beam irradiation, make it difficult to clearly define the contributions of electron beam conditions to defect evolution and properties in graphene, unless the electron irradiation conditions are set with the sufficient beam energy and dose to exceed the threshold limits for generating a certain type of defects.

Defects in graphene can be reactive sites to attract functional elements and build up a composite material system whose activation energy can be easily overcome under electron beam irradiations. For example, surface Fe residual atoms driven by electron beam irradiations are structured to a freestanding Fe monolayer in suspended graphene nanopores for the enhanced magnetic moment [9]. The low-voltage spherical aberration-corrected TEM was utilized as a platform to image and irradiate graphene with Fe residual atoms with the electron beam energy of 80 keV. The fast Fourier transformation of the TEM image verified the Fe crystal lattice structure formed within the graphene nanopore along with the atomic model by the density functional theory (DFT) calculation.

Beyond the graphene, many different types of 2D nanomaterials including hexagonal boron nitride (h-BN) and transition metal dichalcogenides (TMDs) have been explored for local modification by focused electron beam irradiation.

Bi-component 2D nanomaterials under electron beam irradiation provide many intriguing opportunities for new functionalities and designs, which include atomic fabrication of heterogeneous structures and directed phase patterning. h-BN is an insulating 2D nanomaterial which has the closest structural similarity to graphene [4, 10]. The difference in the threshold energy for knock-on displacements between B atoms (74 keV) and N atoms (84 keV) enables the selective sputtering or displacements of B atoms under the controlled electron beam irradiation conditions [11]. High energy electron beam can perforate a h-BN monolayer to replace sputtered B or N atoms with carbon atoms formed by e-beam induced dissociation of surface adsorbed hydrocarbons, which creates a new composite BCN 2D nanomaterial [10].

Many 2D transition metal dichalcogenides nanomaterials were extensively explored to study electron beam induced production of defects by calculating the displacement threshold energy for sputtering an atom using the density functional theory (DFT) calculations, which concluded that chalcogenides atoms are more easily sputtered than transition metals [12]. The displacement threshold energies of chalcogenides atoms were calculated ranging between ~3.5 and ~7.2 eV for $TaTe_2$ and TaS_2, respectively. One of the representative transition metal dichalcogenides nanomaterials is molybdenum disulfide (MoS_2). It is a three-atom-thick semiconducting nanomaterial at the energetically stable semiconducting phase, which can, however, have multiple metastable phases with different electronic properties [13]. Electron beam irradiation can lead to the selective sputtering of sulfur atoms due to the lower threshold energy for the atomic displacement than that for the molybdenum atoms. Sulfur atoms can be knocked out with an electron energy at 80 keV or higher [5, 12]. Vacancy formation and bond rotation result in diverse defect types, and six representative point defects can be induced in a monolayer MoS_2 by electron beam irradiation, including a monosulfur vacancy, a disulfur vacancy and an anti-site defect in which a Mo atom substitutes a S_2 column [5]. Line defects are the extended defects as the agglomeration of vacancies, which can be controllably patterned under electron beam irradiation [5, 14, 15]. Grain boundary is also one of the line defects in MoS_2, which can be mirror and tilt twin boundaries as can be revealed by photoluminescence (PL) images, and the increased conductance was measured in the current flow direction parallel to the grain boundary [14]. Line defects can provide local conductive paths in a semiconducting matrix and some line defects were theoretically proven to have a magnetic property. Importantly, the electronic band gap reduces with the increased number of line defects, as shown in figure 12.2, which can be 'inserted' by the highly controllable electron beam irradiation to develop novel functional devices [15].

Another interesting characteristic of MoS_2 is that it has two primary phases: 2H semiconducting and 1T metallic phases. High-energy electron beam irradiation can induce the transformation of its phase from the stable 2H phase to the metastable 1T phase [13]. Sufficient charge accumulation in 2H semiconducting MoS_2 under electron beam irradiation in the STEM environment led to the transformation of the phase by gliding the lattice-plane of sulfur. The local phase transformation and of MoS_2 in electronic device applications enables low-resistant MoS_2–metal contacts, and it can realize the development of high-performance, low-powered MoS_2-based electronic devices. In addition, nanosculpturing of 2D MoS_2 using the focused electron beam can fabricate a conductive nanowire [16]. Figure 12.3(a) shows

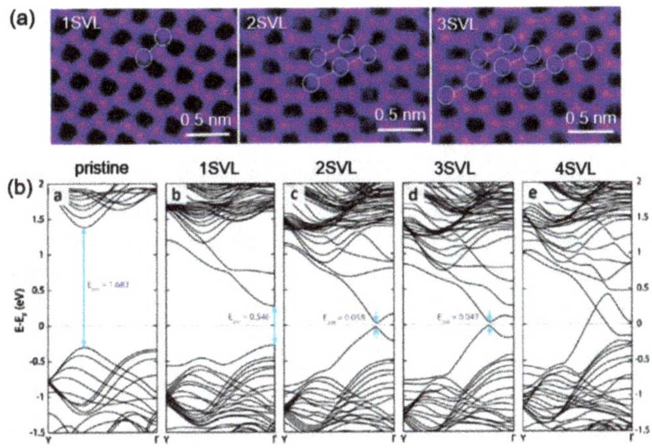

Figure 12.2. (a) STEM images of short single-vacancy line defects (1SVL, 2SVL and 3SVL) generated by electron beam irradiation (adapted from [14], copyright (2013) with permission of Springer), and (b) the corresponding electronic band structures highlighting the reduction of the band gap energy with the increase in the number of line defects indicative of the transition from semiconducting to metallic phases. Reprinted with permission from [15]. Copyright (2016) American Chemical Society.

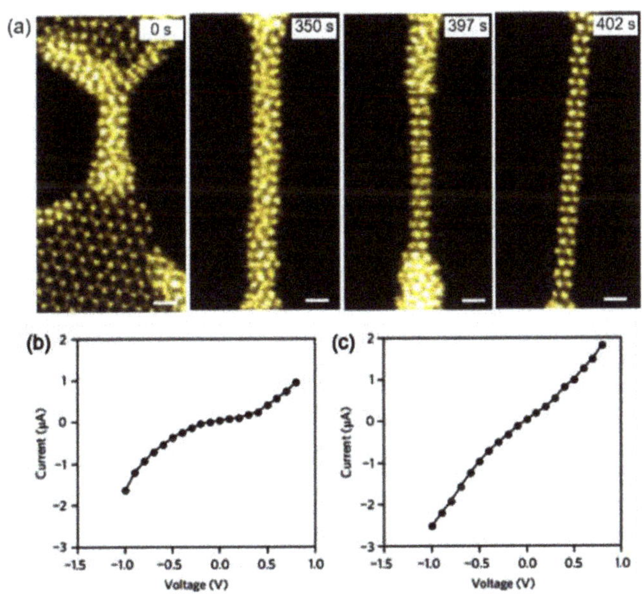

Figure 12.3. Fabrication of a nanowire from a single layer 2D MoS_2 using 60 keV electron beam irradiation in the STEM environment. (a) False-colored STEM images showing the reduction of the nanowire widths under electron beam irradiation, and the electrical measurement (*I–V* curves) results for (b) a relatively thick nanowire and (c) a further thinned nanowire, demonstrating the increase of conductivity. Reprinted by permission from Springer [16], copyright (2014).

the fabrication of a MoS nanowire from a single layer MoS_2 by directed electron beam irradiation. Two nanoholes were first fabricated on a suspended MoS_2, and scanning the electron beam over the bridge between two holes continuously thinned the nanowire at the atomic precision. The thinner nanowire was measured to have higher electrical conductivity, as shown in figures 12.3(b) and (c).

12.2.2 Directing matter: atomic forging with *in situ* imaging and material manipulations

The advances in electron microscopy and computing power have brought new opportunities in materials design and imaging at atomic-to-nanoscales, bringing Feynman's vision of forming the functional micro/nano-systems via atom-by-atom addition/removal of material into a reality with some recently demonstrated remarkable examples. Aberration-corrected scanning transmission electron microscopy (STEM) enables atomically-resolved imaging of materials with its focused electron beam as a probe whose size is ~0.1 nm. Energetic focused electron beam with an atomic diameter can locally excite atoms of materials to manipulate into programmed locations and to build molecular structures. Local interaction of electrons with solid state matters has been explored to verify feasibility of material manipulations in atomic scales, enabling a true bottom-up synthesis of nanomaterials. A vision of fully automated, atomic scale fabrication of nanomaterials, called 'atomic forge,' was recently suggested to tune material properties in a programmed manner [17]. The 'atomic forge' requires the advancement of electron beam control, real-time monitoring and image data analytics, and swift feedback to control the beam. Recently, much effort has been on the development and demonstration of real atomic fabrication and image data processing [18–21].

Susi *et al* first demonstrated that 60 keV electron irradiation can cause structural rearrangements at silicon dopant sites in the graphene lattice which undergo a random walk due to structural relaxation with the mechanism described in figure 12.4(a) [18]. Molecular dynamic simulation of electron impact delivering 15 eV to a carbon atom neighboring Si–C_3. The ejected C atom travels slightly away and then towards the Si atom. While the Si atom falls under relaxation into the vacated carbon site, the ejected carbon atom is drawn back to the position of the starting point of the Si atom [18]. After the first demonstration of a possibility for atomic manipulation under electron beam irradiation, they recently conducted the very detailed study about single-atom dynamics of phosphorus dopants on graphene under selective electron beam irradiation using the experiments and the first-principles calculations [20]. An underlying theory was developed to estimate the probabilities of atomic configurational transformations. The researchers in the Oak Ridge National Laboratory first demonstrated the directed displacements of a Si atom in graphene lattice [19]. Parking a focused electron beam in the next lattice site of the target dopant (silicon atom) drives the movement of the silicon atom to the site under electron irradiation at the beam energy of 60 keV. They demonstrated the linear motion of a silicon atom (figure 12.4(b)) as well as formation of dimer

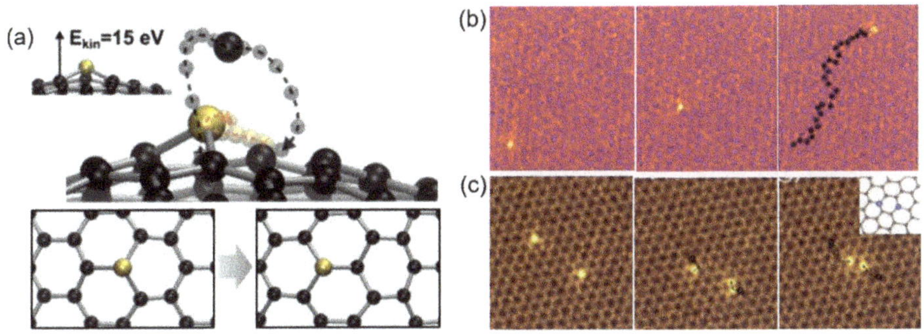

Figure 12.4. (a) Molecular dynamic simulation result demonstrating the mechanism for electron beam induced displacement of a Si dopant from one atomic position to the next neighboring position in graphene lattice structure. (Reprinted with permission from [18], copyright (2014) by the American Physical Society.) (b) STEM images showing the electron beam-induced atomic linear motion of a Si dopant on graphene along a path length of ~4 nm and (c) the formation of a Si dimer by manipulating two Si atoms using the electron beam. ([19] John Wiley & Sons. Copyright 2018 WILEY-VCH Verlag GmbH & Co. KGaA, Weinheim.)

(figure 12.4(c)) and trimer and their rotational motion, along with silicon atom manipulation at the edges of a graphene hole for passivation and de-passivation. To provide better understanding of the dynamics of dopant atoms on graphene, a deep learning network was applied to analyze low-resolution STEM movies and to extract atomic configurations and coordinates of carbon and silicon atoms enabling the study of electron beam-induced chemical transformations at the edge and in the bulk of graphene [21]. Many technological challenges still remain to be solved before the atomic forging of nanomaterials can be applied to realize operational functional devices on a large scale, but continuous development of instrumentation, theory and data science such as artificial intelligence and machine learning will significantly accelerate the realization of true atomic manufacturing of new functional nanomaterials, structures and devices.

12.3 Directed surface and interface modification of graphene-based nanomaterials using focused electron beam and precursor molecules

The scanning electron microscopy (SEM) environment enables nanoscale-to-microscale manipulation of 2D materials with a large electron beam size (few nanometers) and lower accelerating voltages from few hundreds of electron volts (eV) to 30 keV. Single atomic manipulation of 2D nanomaterials cannot be realized at such relatively low voltages, but advantageously a larger area patterning and fabrication of operational devices can be achieved using the SEM beams in combination with directed introduction of molecular precursors to the beam irradiation site. Nanoscale resolution of 2D nanomaterial patterning and the corresponding functional mapping of nanoelectronic devices can be accomplished, where, otherwise, multiple steps are required and uncontrollable damage or undesired contamination of graphene substrate can occur when using the conventional lithographic techniques.

12.3.1 Graphene oxide

An intriguing characteristic of graphene is its 2D structure which can be chemically modified, so as to endow new functional properties to a baseline material or its derivatives. One representative form of graphene-based nanomaterials is graphene oxide, whose 2D graphitic structure has surface functional groups such as epoxy, hydroxyl and carboxyl [22]. Depending on the content of surface functional groups, graphene oxide can be an insulator or a semiconductor. The conventional bulk synthesis method of graphene oxide makes ~70% of surface functional groups to the number of carbon atoms in graphene which is an insulator, and then graphene oxide is reduced to have conductivity and controllable electronic bandgap [22]. Conventional methodologies for reduction of graphene oxide are to utilize toxic chemicals and a high temperature thermal process, which are not environment-friendly and can damage electronic device substrates [23]. Moreover, patterned reduction of graphene oxide requires multiple steps of lithography procedures, which are time-consuming and often result in additional defects or contaminations. Kim *et al* developed a direct-write method using a focused electron beam in the SEM environment for nanoscale patterned reduction of graphene oxide with no need for toxic chemicals and multiple lithographic steps [24]. As shown in figures 12.5(a) and (b), they fabricated a graphene oxide device on a SiO_2/Si substrate, whose channel was simply irradiated by focused electron beam at the energy of 25 keV and the dose in the order of 10^{17}–10^{18} electrons/cm^2. No current was measured through the insulating graphene oxide before irradiation, but after electron beam irradiation, the current started to flow through graphene oxide and it increased with the increase of the electron beam dose for irradiation, as shown in figure 12.5(c). The density function theory (DFT) calculations demonstrated the likely mechanism underlying

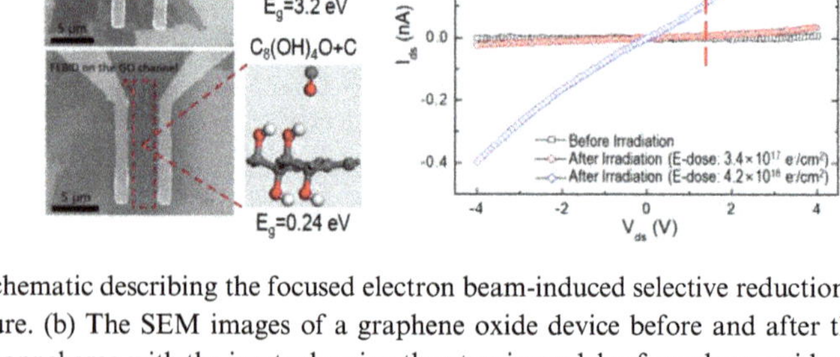

Figure 12.5. (a) Schematic describing the focused electron beam-induced selective reduction of graphene oxide in a device structure. (b) The SEM images of a graphene oxide device before and after the e-beam induced reduction at the channel area with the insets showing the atomic models of graphene oxide and locally reduced graphene oxide. (c) Electrical measurements of the graphene oxide device as a function of the electron dose, as compared to the as-fabricated device without electron irradiation, indicating the increase of the channel conductivity with electron irradiation. Reprinted from [24], with the permission of AIP Publishing.

the experimental observation of the electron beam induced nanoscale reduction of graphene oxide. Hydrocarbon contamination molecules omnipresent on graphene oxide were dissociated by secondary electrons generated by inelastic collisions of primary electrons with the graphene oxide/supporting substrate. The dissociated hydrocarbon or carbon species interacted with graphene oxide to take epoxy or hydroxyl groups from graphene oxide, which finally reduced graphene oxide.

12.3.2 Other members of graphene-based nanomaterials

In addition to graphene oxide, graphene functionalization was extensively studied with elements such as hydrogen, fluorine, and carbon for tuning properties of graphene [25–27]. The focused electron beam enables selective, nanoscale surface functionalization or de-functionalization of graphene-based nanomaterials. Surface functionalization of graphene under electron beam irradiation requires the introduction of precursor molecules containing a target element to be chemically attached to a graphene basal plane [25]. Interaction of electron-graphene-precursor molecules entails the excitation of graphene structure with a high energy electron beam and the dissociation of precursor gas molecules adsorbed on graphene with the secondary electrons generated as the results of interaction between the primary focused electron beam and the graphene/substrate. Fully hydrogenated graphene is called 'graphane', which has a wide bandgap tunable depending on the extent of hydrogenation. At some extents of hydrogenation, hydrogenated graphene can exhibit ferromagnetism, and the magnetic properties of hydrogenated graphene are especially useful in applications such as information technology and superconducting quantum devices [28, 29]. To controllably hydrogenate graphene using electron beam, NH_3 and water molecules were tested in ultra-high vacuum conditions and micro Raman spectroscopic analysis were implemented to evaluate the formation of hydrogenated graphene under electron beam irradiation with different energies (100 eV–10 keV) and electron doses (2.8–210 mC cm^{-2}) [25]. Being supported by the control experiments with other gases (O_2, Ar, N_2, H_2, and CO_2), it was concluded that the electron beam-induced dissociation of the adsorbed NH_3 and H_2O molecules on graphene played a significant role in producing H^+ ions and H radicals for the formation of fully or partially hydrogenated graphene [25]. Figure 12.6(a) shows the Raman spectra of graphene irradiated by an electron beam at an energy of 5 keV and a dose of 1.2 mC cm^{-2} after exposing graphene to various gases. The appearance of a D-band peak in the Raman spectra of graphene with NH_3 and H_2O gases confirms the hydrogenation of graphene compared to the results of no D-band peak appearance in the spectra with other gases. Hydrogen gas molecules are difficult to adsorb on graphene and thus, are not suitable as a direct route to hydrogenation.

Other studies used a hydrogen silsesquioxane (HSQ) film, which is a negative photoresist for e-beam lithography, as a source of hydrogen [28, 30]. Breaking a bond of Si–H of the HSQ film produced hydrogen radicals to be chemically bound to a graphene basal plane forming a hydrogenated graphene. Hydrogenated graphene under electron beam irradiation can be reversibly dehydrogenated by

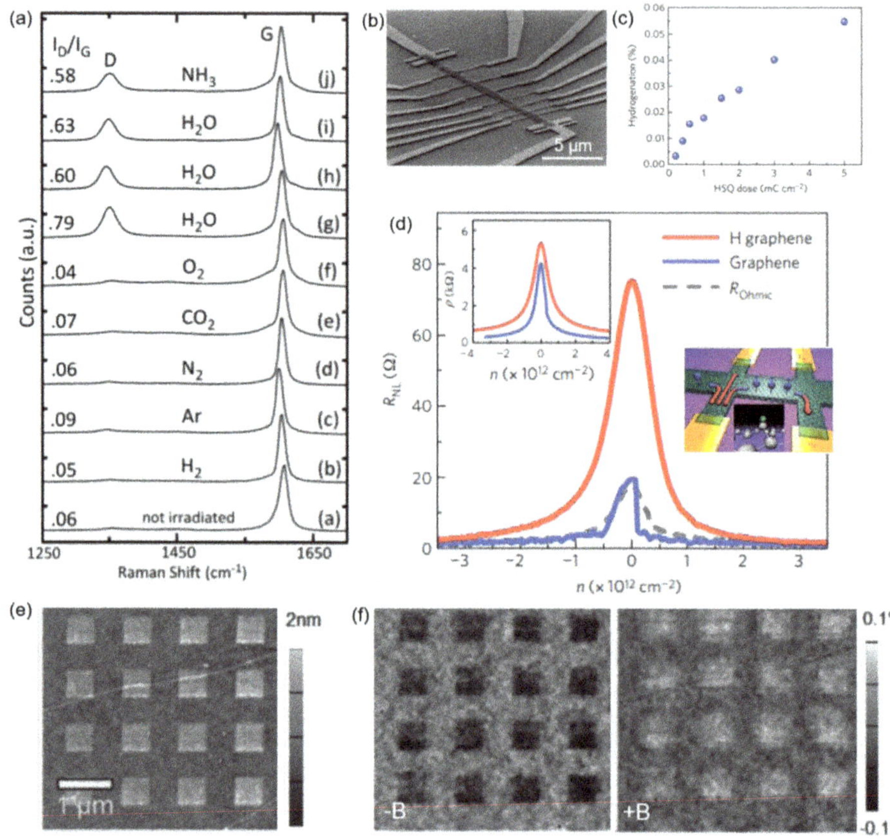

Figure 12.6. (a) Raman spectra of graphene under electron beam irradiation with a variety of precursor gases, which demonstrate that water and NH_3 gases can be a precursor source for hydrogenation of graphene. (Reprinted from [25], copyright (2010), with permission from Elsevier.) (b) SEM image of a partially hydrogenated graphene device showing multiple Hall bar junctions and (c) evolution of the hydrogenation on graphene by increasing the electron dose from 0.4 to 5 mC cm^{-2} for HSQ dissociation. (d) Room temperature measurements of non-local signal (R_{NL}) depending on charge densities (n) for partially hydrogenated (~0.02%) graphene (red) and a pristine graphene (blue) highlighting that even weak hydrogenation of graphene enhanced the spin–orbit coupling by ~400% compared to the pristine graphene device. The insets show the resistance measurements on charge densities and the measurement schematics for non-local spin Hall measurement with the atomic structure of a partially hydrogenated graphene. (Reprinted by permission from Springer [28], copyright (2013).) (e) Surface topography image of the partially hydrogenated square pattern (500 nm × 500 nm) array which was generated on a fully hydrogenated graphene surface by means of dehydrogenation using an electron beam patterning, and (f) the corresponding MFM phase images with a tip magnetized with the south pole (−B) and with the north pole (+B), indicative of magnetism on the square pattern array of partially hydrogenated graphene. (Reprinted from [29], copyright (2015) John Wiley & Sons.)

thermal annealing at the relatively low temperature of ~200 °C [30]. A weakly hydrogenated graphene was proven to have the enhanced spin–orbit coupling which can overcome the limitation in graphene's applications due to extremely small spin–orbit interactions [28]. HSQ was used to hydrogenate graphene under electron beam irradiation. The localized patterning of graphene hydrogenation using electron beam is beneficial to precisely control the process without introducing undesired vacancies, compared to the bulk, destructive radio frequency (RF) hydrogen plasma process. Figures 12.6(b)–(d) show the results of the enhanced spin–orbit coupling of the hydrogenated graphene device selectively patterned by electron beam irradiation and the hydrogenation revealed by Raman measurements depending on electron beam energies and doses [28]. The focused electron beam was also used to pattern magnetic regions in hydrogenated graphene by selectively removing hydrogen atoms and quenching the magnetism to generate non-magnetic graphene domains [29]. Figures 12.6(e) and (f) show the patterned magnetic array of partially hydrogenated graphene which was imaged using magnetic force microscopy (MFM).

Fluorinated graphene is a stable, wide-bandgap nanosheet, whose carbon atoms are covalently bonded to F atoms [31]. It is an emerging member of the graphene family owing to its unusual properties, including a large negative magnetoresistance, a wide optical bandgap (3.8 eV) and an ultra-high room-temperature electrical resistance, which allows for fabrication of an atomically-thin 2D insulator [26, 31, 32]. Since the properties of fluorinated graphene strongly depend on the ratio of F to C atoms, it is important to control with fine degree the level of fluorination for achieving the desired behavior. Focused electron beam irradiation tunes the properties of fluorinated graphene by local de-fluorination on exposure to electron doses (0–100 mC cm^{-2}) at the accelerating voltage of 10 keV and the beam current of 0.13 nA, as shown in figure 12.7(a) [32]. The temperature dependence of the square resistance of fluorinated graphene was investigated following irradiation with different electron doses, as shown in figure 12.7(b). Increasing the electron dose reduced the coverage of F atoms on graphene from its initial composition $CF_{0.28}$ which was an insulating with a square resistance of ~1 TΩ. De-functionalization of fluorinated graphene under electron beam irradiation resulted in the decrease of electronic bandgap, converting the fluorinated graphene from insulating (no electron irradiation) to metallic (exposed to electron dose ~1 C cm^{-2}) materials. After the

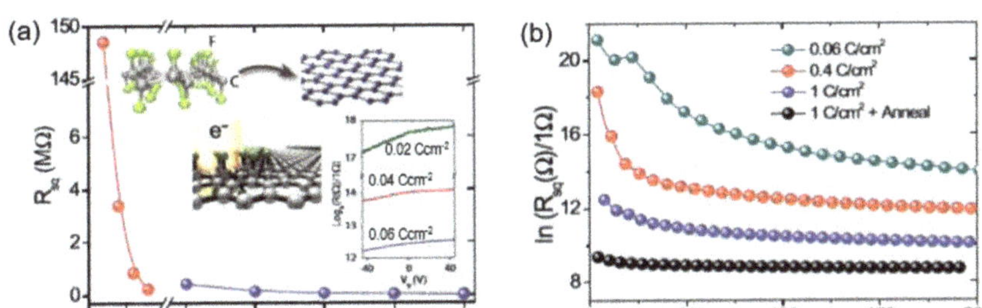

Figure 12.7. Controlled fluorination on graphene by electron beam irradiation. (a) The reduction of fluorinated graphene device electrical resistance upon the increase of electron beam irradiation dose as a result of defluorination (dissociating F atoms from graphene surface) and (b) the temperature dependence of the sheet resistance of fluorinated graphene subject to the electron beam irradiation with different doses, demonstrating a transition from an insulating state of $CF_{0.28}$ to a metallic state after the complete defluorination at 1 C cm^{-2} followed by the mild thermal annealing to remove contaminants. Reproduced from [32] with permission. Copyright IOP Publishing. All rights reserved.

irradiation at the dose of 60 mC cm^{-2}, the fluorinated graphene showed the characteristics of a lightly doped semiconductor. The insulating fluorinated graphene flake was used in a device configuration by depositing source and drain metal contacts and patterning a conductive channel to various widths by selectively exposing fluorinated graphene to the electron beam. This led to de-fluorination by selective fragmentation of C–F bonds, as revealed with AFM topography imaging and *I–V* measurements [26]. The electron beam induced de-fluorination ultimately transforms the insulating fluorinated graphene to pristine graphene reversibly and demonstrates the capability of controlled electron beam patterning to fabricate local, conductive channels in an insulating fluorinated graphene.

Other interesting members of graphene-based nanomaterials include carbon-adsorbed graphene. DFT calculations using the Vienna *ab initio* simulation package (VASP) were performed to model carbon-adsorbed graphene with various carbon coverages ranging from 3.1% to 50% and calculate adsorption energies, electronic structures and magnetic properties, as shown in figure 12.8(a) [27]. At the bridge site (on top of C–C bond of graphene) which is the most stable adsorption site of a carbon adatom, graphene was found to have a magnetic moment of ~0.36 μ_B per adsorbed carbon adatom at coverages of 3.1% and 12.5%, while pristine graphene is non-magnetic. The further increase of the carbon adatom coverage up to 50% changes the adsorption site to the top of carbon atoms on graphene, having no magnetic moment, but opening a band gap of 2.5 eV. Experimental implementation of carbon adsorption on graphene can be realized using focused electron beam induced deposition (FEBID) [33]. Hydrocarbon molecules physisorbed on graphene surface can be dissociated by the secondary electrons, and chemisorption of dissociated hydrocarbon species simultaneously occurs on the sp^3-type defect sites on graphene which are induced by being exposed to high energy (25 keV) primary electrons [33]. Simplified descriptions of the chemisorption phenomena were demonstrated with a model hydrocarbon molecule of CH_4, using the DFT calculations, showing the transitions of adsorption configurations when hydrogen atoms are sequentially dissociated by the secondary electrons, as shown in figure 12.8(b). The DFT calculations were confirmed with experimental measurements of two different adsorption states, physisorption versus chemisorption, of hydrocarbon species on graphene. The primary electrons played a role in inducing chemisorption, while the graphene areas only exposed to the secondary electrons have physisorbed carbon deposits, as shown in the Raman D-band peak mapping of figure 12.8(c). After FEBID, a molecularly thin layer of chemisorbed carbon deposits was achieved by Raman laser-assisted postdeposition thermal ablation, whose thickness was measured as ~0.8 nm by atomic force microscope (AFM), shown in figure 12.8(d), and the Raman I(D)/I(G) ratios upon thermal ablation in figure 12.8(e) confirm the chemisorption at the region where the primary electrons irradiated in contrast to the physisorption at the region where only secondary electrons contributed to deposit carbons. This study exhibits the dual capability of FEBID to make chemical functionalization of graphene as well as establish weak electronic coupling (doping) with carbon atoms.

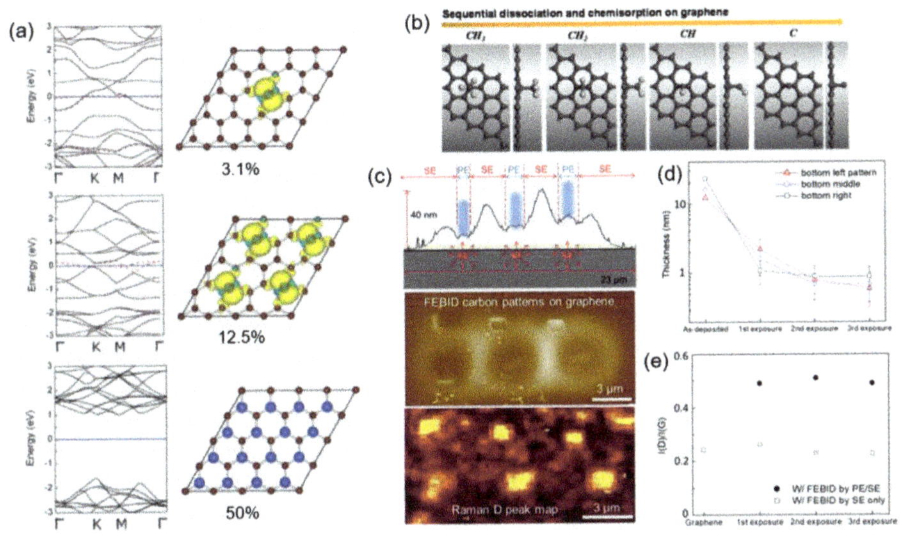

Figure 12.8. (a) DFT calculations of electronic band structures and adsorption configurations of carbon adatoms on graphene depending on the carbon coverage (3.1%–50%), showing that carbon adsorbed graphene at a low surface coverage can have a magnetic property while a high surface coverage about 50% can open up the electronic band gap in graphene. (Reproduced from reference [27] with permission from the PCCP Owner Societies.) (b) DFT calculations for describing an atomic mechanism of the FEBID process of a simple model of a CH_4 molecule on graphene, demonstrating the sequential dissociation and chemisorption of the dissociated species on graphene induced by focused electron beam. (c)–(e) Experimental demonstration of chemisorption and physisorption of carbon adatoms deposited by the FEBID on graphene. (c) FEBID of carbon on graphene surface showing two regions where the primary electron (25 keV) directly irradiated for patterning squares and the secondary electron only affected depositing carbon (surrounding the squares), whose surface topography and adsorption states were measured by the AFM imaging and the corresponding Raman D-band peak mapping. (d) The change of thickness after Raman laser-induced thermal ablation of the deposited carbon square structures indicating the fabrication of the molecularly thin carbon deposit film (~0.8 nm), and (e) the Raman I(D)/I(G) ratios upon laser-induced thermal ablation on two different regions confirming two different adsorption states of carbon on graphene. Reprinted with permission from [33]. Copyright (2014) American Chemical Society.

12.4 Focused electron beam induced etching (FEBIE) of 2D nanomaterials

Focused electron beam induced etching (FEBIE) is another mode of 'direct-write' nanolithographic techniques [34]. The FEBIE process can be applied to pattern inorganic and organic materials, for example, Ti, Si, SiO_2, Si_3N_4, GaAs, carbon nanomaterials and diamond, at high resolutions down to few nanometers with reactive gases such as H_2O, N_2, XeF_2, O_2, Cl_2 and NH_3 which are generally flooded into the SEM chamber through a gas injection nozzle with an inner diameter in an order of few hundreds of micrometers [34, 35]. The gas molecule adsorbates are dissociated by secondary or backscattered electrons from the interaction of incident primary electrons with the target materials (substrates), and they turn into reactive radicals or ions and etch the target materials creating volatile compounds.

Carbon-based nanomaterials, e.g., carbon nanotube, graphene and diamond, can be etched mostly with water vapors and oxygen gas under electron beam irradiation [34, 36–40]. Residual water vapors in the SEM chamber at the total pressure of 2×10^{-6} Torr were sufficient to cut multi-walled nanotubes under electron irradiation [36]. Figure 12.9(a) shows the TEM image showing a fine cutting (~40 nm) of the nanotube with a nominal beam spot size of ~3 nm. Water molecules were dissociated into highly reactive OH·, H· and HO_2· to react with carbon atoms forming volatile CO, CO_2 or hydrocarbon species. High partial pressure (2×10^{-5} Torr) of water led the etching process in the chemical reaction limited regime, where dissociation of water molecules becames a dominant factor in etching carbon atoms, and thus, increasing the beam current at the fixed beam energy of 1 keV significantly reduced the cutting time of the nanotubes, as shown in figure 12.9(b). In contrast, increasing the beam energy slowed down the cutting speed since the ionization cross section of water molecules is inversely proportional to the electron energy, as shown in figure 12.10(c). These results conclude that the etching efficiency can be improved with a higher beam current and at a lower beam energy; however, the beam is broadened which can lower the etching resolution.

Suspended graphene can be also etched using FEBIE by directly injecting water vapors into the SEM chamber [37]. High resolution line etching with the cut width of ~7 nm was achieved with the high electron beam accelerating energy at 30 keV (nominal beam spot size of ~0.9 nm) as shown in figure 12.10(a). High resolution etching is also attributed to graphene's atomically thin layer whose interaction volume with the electron beam is limited leading to a very confined area of secondary/backscattered electron generations with the reduced yields. Using the FEBIE, suspended graphene nanoribbon was fabricated with a width of ~20 nm directly on the pre-fabricated graphene field-effect transistor, and the graphene nanoribbon was electrically measured showing a band gap energy of 60 meV. The supported graphene on an SiO_2/Si substrate was patterned using the 3 keV electron beam with oxygen gas molecules injected

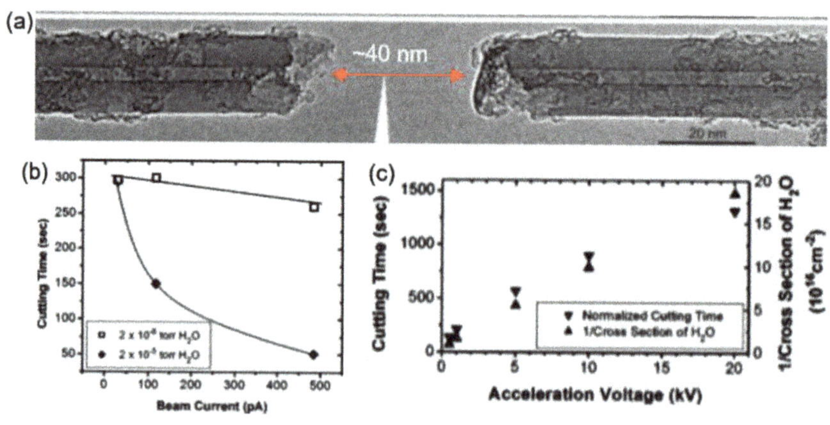

Figure 12.9. (a) TEM image of a multi-walled carbon nanotube etched by focused electron beam with residual water vapor, whose etching rate can be affected (b) proportionally by the electron beam current, but (c) inverse proportionally by the beam energy. Reprinted from [36], with permission of AIP Publishing.

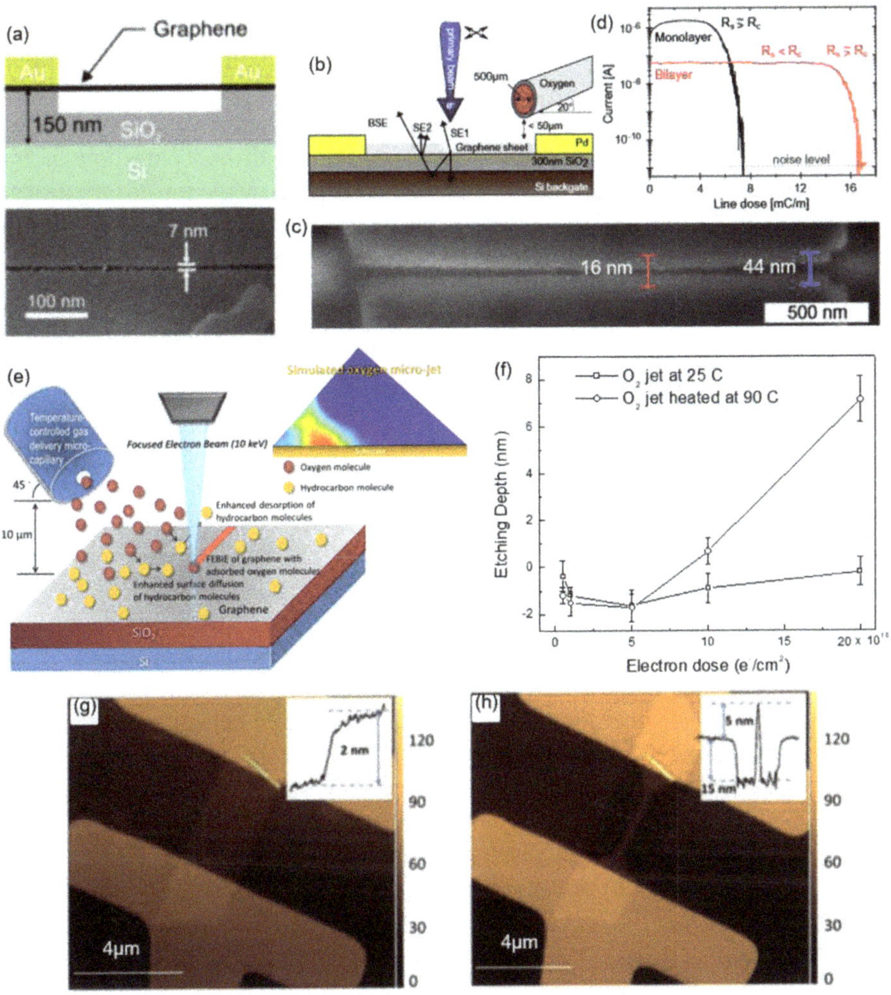

Figure 12.10. (a) FEBIE of suspended graphene showing 7 nm etching resolution with water vapor. (Reprinted by permission from Springer [37], copyright (2015).) (b) Schematic showing the FEBIE set-up with a 500 μm inner diameter injection nozzle for flowing oxygen gas. (c) SEM images of line etch showing the tens of nanometer resolution on supported monolayer graphene and (d) the required electron doses for the complete etch of mono- and bi-layer graphene. (Reprinted from [38], copyright (2013), with permission from Elsevier.) (e) Schematic describing the FEBIE process with a micro nozzle (~10 μm inner diameter) to flow an oxygen supersonic micro-jet and (f) the etching outcomes of supported graphene with a room-temperature oxygen micro-jet and a heated micro-jet at 90 °C by increasing the electron doses. (Reprinted from [39], with permission of AIP Publishing.) AFM images of a few-layer MoS_2 FET device (g) before and (h) after the channel was patterned. The insets show the heights of the channel. Reproduced from [41] with permission from The Royal Society of Chemistry.

through a ~500 μm inner diameter nozzle (figure 12.10(b)), achieving tens of nanometer resolutions in a line etching mode as shown in figure 12.10(c) [38]. *In situ* measurements of current during line cutting found the electron doses for the complete etch were ~8 and ~16 mC m^{-1} for a single-layer graphene and a bi-layer graphene, respectively, as shown in figure 12.10(d). In addition, the detailed study provided two underlying mechanisms of etching the supported graphene [38]. Imaging the intermediate state of graphene etching revealed that the etching started at the edges of graphene and continued to remove carbon atoms towards the center of the line. Also, the electron dose study of etching on the graphene basal plane revealed that, at a low electron dose (<3 C cm^{-2}), hydrocarbon deposition occurred first due to initially high concentration of hydrocarbon surface contamination, and then the localized depletion of hydrocarbon contamination molecules allowed more adsorption sites of oxygen molecules to be available on the surface leading to etch hydrocarbon deposits and graphene when increasing the electron dose. These experimental findings highlight that leveraging surface contamination is of significance to improve the etching rate and resolution of the etched structures in the supported graphene. To improve the etching resolution, an inert N_2 gas was used: the ionized nitrogen interacting with the electron beam diffused onto the electron irradiation spot and etched carbon atoms on graphene. Since the ionized nitrogen is less reactive compared to water vapor or oxygen, finer control of etching was accomplished, which enabled the fabrication of sub-5 nm nanopores on suspended few-layer graphene with the focused electron beam at 5 keV [40].

Most of the FEBIE studies used a relatively large diameter injection nozzle, which leads to the flooding of a reactive gas in a free-molecular flow regime. Flooding of a reactive gas might contaminate the SEM chamber, increase the chamber resulting in greater e-beam scattering, and the non-interacting molecular flow of the gas does not enhance the outcomes of the FEBIE. To control the flow of a reactive gas, a micro-capillary nozzle with an inner diameter of 10 μm was used to deliver an oxygen gas in a continuum flow regime at a supersonic speed of the expanding gas jet, as shown in the schematic of the FEBIE process using a temperature-controlled oxygen micro-jet in figure 12.10(e) [39]. Modification of the oxygen gas flow provides an additional knob to control the FEBIE process. A localized oxygen supersonic micro-jet with the minimal jet spreading impinged on graphene surface at a high kinetic energy as described in the inset (direct simulation Monte Carlo result) of figure 12.10(e). Increasing oxygen jet temperature by heating the nozzle can additionally tune the FEBIE process *via* two physical mechanisms: (i) impact-stimulated activation of oxygen chemisorption on graphene, (ii) enhanced surface diffusion versus desorption of hydrocarbon contamination molecules depending on the jet kinetic energy. Graphene was etched free of parasitic halo deposits (common to FEBIP) using a supersonic oxygen micro-jet at ~30 times lower electron beam dose and even at the higher beam energy (10 keV) than using a conventional large diameter nozzle (500 μm) for delivery of oxygen. The heated micro-jet at 90 °C enabled faster etch of graphene as shown in figure 12.10(f), but enhanced the surface diffusion of hydrocarbon molecules which resulted in the faster growth of parasitic hydrocarbon deposits after the complete etch of graphene. These experiments suggest that the further improvement of graphene etching can be accomplished by high-temperature oxygen micro-jets whose impact kinetic energy is above the threshold for hydrocarbon molecule desorption.

Recently, a mechanically exfoliated few-layer MoS_2 channel in a field-effect transistor was successfully patterned using a pulsed focused electron beam induced etching with XeF_2 gas injected through a microneedle, as shown in figures 12.10(g) and (h) [41]. The etching process altered the doping status of a MoS_2 FET device which was evidenced by the blue-shifted threshold voltage indicating the change from n-type doping to more intrinsic or p-type doping. In addition, the field-effect mobility of the device was reduced from 1.96 to 0.39 $cm^2 V^{-1} s^{-1}$. The change of the electrical characteristics of the device was attributed to the formation of sulfur vacancies on the channel area, not directly exposed to the electron beam during the etch, and the atomic defects was also supported by micro-Raman and micro-photoluminescence spectroscopic measurements.

12.5 Applications of the FEBIP techniques to 2D nanomaterial-based electronic devices

12.5.1 Nanoscale welding

Focused electron beam induced deposition (FEBID) capability for site-specific, high-resolution patterning of nanomaterials was broadly utilized to enable a number of unique applications, including interfacial engineering of graphene–metal contacts, electromechanical welding of nanotubes, novel nanoscale electro-resistive or magnetic sensors, photonic and plasmonic devices [42]. One of the principal FEBIP appeals is in the local modifications of heterogeneous electrical contacts in low-dimensional nanomaterial-based electronic device structures. Weak couplings between low-dimensional nanomaterials and metal contacts is inevitable when developing electronic devices using the conventional lithography techniques such as metal evaporation or sputtering. Improving contact properties is a critical prerequisite to developing reliable, high-performance nanodevices by incorporation of many different types of low-dimensional nanomaterials into a composite opto-electro-magnetically active structure [43, 44]. Multi-walled carbon nanotubes (MWCNTs) were aligned on drain-source metal electrodes for electric interconnect applications, and both ends of the MWCNTs were electromechanically connected to the metal electrodes through graphitic contacts deposited by FEBID of carbon and annealed at 350 °C for ~30 min [43]. The successful, electrical connection of multiple inner shells of the MWCNTs to the metals reduced the device resistance down to the order of ~100 Ω from initially 1 GΩ. Similarly, thermo-electro-mechanical properties of graphene–metal interfaces were significantly enhanced after the nanoscale welding using the FEBID technique [44]. Figure 12.11(a) shows the nanoscale welding process of a graphene field-effect transistor using FEBID. A field-effect transistor was fabricated with the graphene channel underneath the metal source–drain contacts. The graphene–metal contacts of the as-fabricated device have poor interfacial properties due to their intrinsically weak interaction as well as PMMA lithographic polymer residues. The buried interfaces were engineered by irradiating a focused electron beam at graphene–metal contact areas, and the high energy electrons traveled through the metal contacts to be accessible for modifying the interfacial properties in two mechanisms; generation of sp^3-type defects on graphene and chemical interconnection of graphene and metal with the electron beam dissociated carbon/graphitic layers. Transport characteristics of the graphene FETs before and after the interfacial engineering verified that the FEBID nanoscale welding process improved thermo-mechanical properties of the interfaces with the reduction of device resistances by ~50% at a Dirac point and by

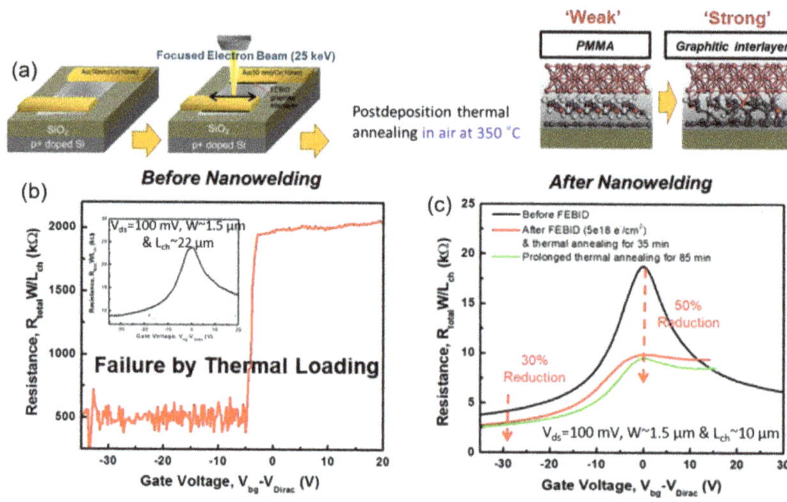

Figure 12.11. (a) Schematic describing the nanoscale welding process of graphene–metal interfaces using the FEBID. Two-terminal electrical resistance measurements of graphene devices by changing the gate voltage demonstrating that (b) the as-fabricated device with no nanoscale welding was failed by thermal loading, but (c) the significant improvement of interfacial properties of graphene–metal contacts was achieved with the nanoscale welding process. Adapted with permission from [44]. Copyright (2016) American Chemical Society.

~30% at the gate voltage far from the Dirac point, as shown in figures 12.11(b) and (c), and the mechanism of the improvements was revealed experimentally by Raman spectroscopic and AFM analysis.

12.5.2 Reconfigurable graphene doping

Modulation of doping levels in graphene nanoelectronic devices is useful to control device functionalities and performance. P-type doping of graphene can be easily realized by water molecule adsorption and polymer resist residual during wet-transfer of graphene and lithography procedures for device fabrication [45]. However, n-type doping requires multiple steps of treatments such as plasma treatment, electro-thermal reaction, chemical treatment for surface adsorption of n-type dopants or electron beam irradiation [46]. Controlled, localized n-type doping was demonstrated during the FEBID [46]. Figure 12.12(a) shows the strategy for the controlled n-type doping of the channel on a graphene field-effect transistor. The controlled electron dose of a primary beam on graphene–metal contact areas varied the number of secondary electrons generated from the long-traveling primary electrons (backscattered electrons escaping from the graphene channel area), and resultantly, changed the coverage of carbon deposition (doping) on the channel area which was deposited from the source of hydrocarbon contamination molecules. Increasing the electron dose led to the change of the doping level from p-type doping (pristine) to n-p-n doping at a low dose (10^{18} electron/cm^2) and n-type doping at a high dose (10^{19} electron/cm^2), as shown in figure 12.12(b). More interestingly, diffusion and redistribution of weakly interacting FEBID carbon atoms on graphene towards the low-concentrated area on the

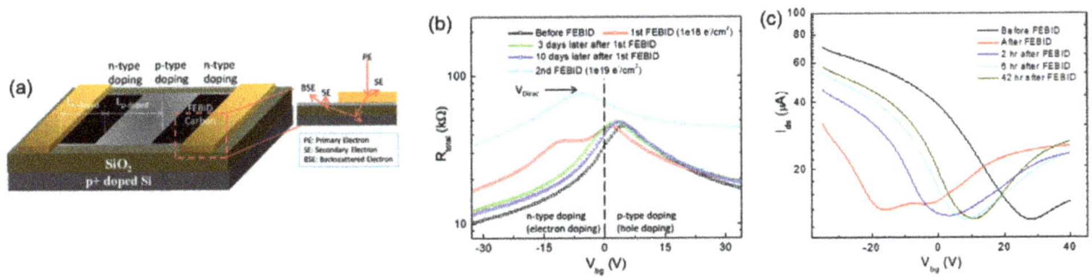

Figure 12.12. (a) Schematic showing the carbon doping process on a graphene device using the FEBID technique. (b) Electrical measurement results showing the controllability of graphene doping states from n-p-n junction formation to fully n-type doping by controlling the electron beam dose and (c) the reconfigurable doping state which is transiently changed from the n-p-n doping state to the p-type doping by the diffusion and redistribution of carbon adatoms on graphene surface. Reproduced from [46] with permission from The Royal Society of Chemistry.

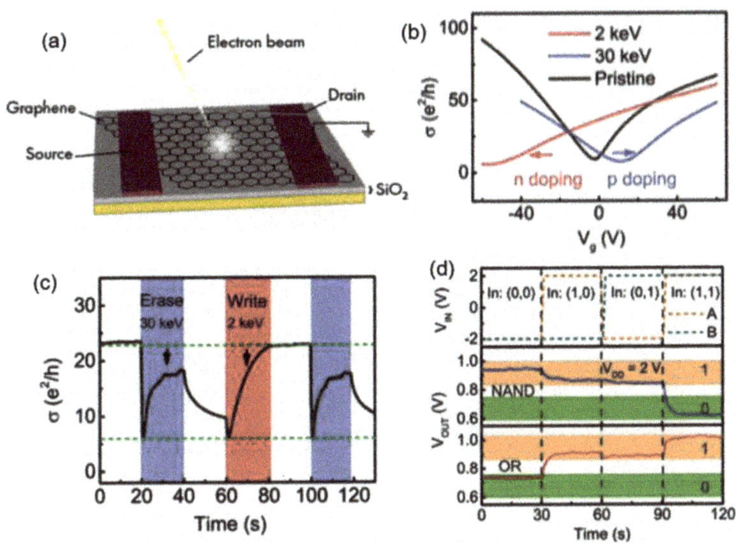

Figure 12.13. (a) Schematic describing the direct irradiation of a graphene channel using a primary electron beam for the controlled modulation of graphene doping states. (b) Conductivity measurement results by modulating gate voltages demonstrating the capability of the site-specific bipolar graphene doping using the direct primary electron irradiation with different beam energies which enables (c) the realization of 'writing' and 'erasing' modes of the graphene device and (d) the development of graphene-based logic devices such as NAND and OR gates. Reproduced from reference [47] with permission from The Royal Society of Chemistry.

graphene channel reconfigured the doping state to the p-type doping, as shown in figure 12.12(c). This work highlights the capability of the FEBID technique for development of dynamically reconfigurable graphene devices.

Direct irradiation of a graphene channel in a FET device by high-energy primary electrons can also control doping states as a result of the electron beam-induced modulation of the charge distribution on the SiO_2 substrate [47]. The site-specific reversible bipolar graphene doping was achieved by changing an accelerating energy of a primary beam, as shown in figure 12.13(b). Irradiation at a low beam energy induced n-type doping and increasing the beam energy led to a transition from n-type (Dirac point at −55 V after 2 keV electron irradiations) to p-type doping (11 V after 30 keV electron irradiations). *In situ* electrical measurement set-up in the SEM allowed for observing the swift change of the doping state from n-type to p-type doping (and vice versa) by altering the beam energies. 'Write-erase' cycles (figure 12.13(c)) were programmable up to 10^4 cycles owing to the relatively low electron dose required for changing doping levels, and finally, an electron beam-driven programmable graphene logic device was demonstrated for NAND and OR gates, as shown in figure 12.13(d). The developed technique was also applied to a MoS_2 FET device, and the similar effect of electron beam irradiation on a MoS_2 channel was observed at 2 keV. Clearly distinguishable 'ON' and 'OFF' states were demonstrated by being exposed to the electron beam at 2 and 30 keV, respectively.

12.5.3 Tuning MoS_2-based nanoelectronic devices

The representative inorganic 2D nanomaterial is a single- or few-layered MoS_2 which is a semiconductor and whose semiconducting phase (2H-MoS_2) can be

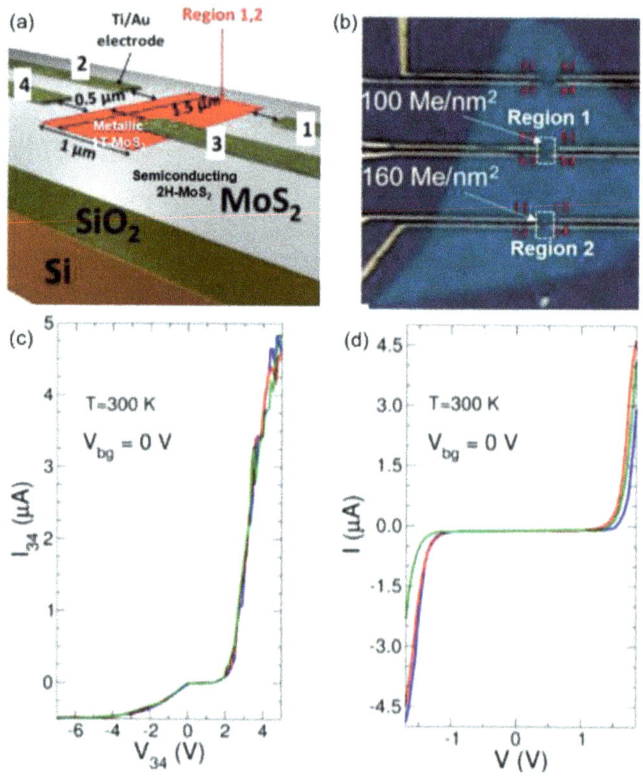

Figure 12.14. (a) Schematic view of few-layer MoS_2 devices with electron beam-induced Schottky junction formation and (b) the optical microscope image of the devices. *I–V* measurements of the devices irradiated with the electron dose of (c) 1.6×10^{20} electrons/cm^2 (Schottky junction formation) and (d) 10^{20} electrons/cm^2 (no Schottky junction formation). Reprinted with permission from [49]. Copyright (2016) American Chemical Society.

transformed to a metallic phase (1T-MoS_2) by external energy deposition [13, 48]. Electron beam irradiation can controllably convert the phase of MoS_2 site-specifically and allows for the fabrication of atomically thin lateral 2H/1T Schottky junctions and SB-FETs [49, 50]. Figures 12.14(a) and (b) show the schematic of 1T/2H Schottky junction fabrication on few layer n-type MoS_2 flakes by electron beam irradiation and the optical microscope image of the devices along with the schematic of a final device structure [49]. The electron dose of 10^{20} electrons/cm^2 was not enough for phase transition, but the increase of the dose to 1.6×10^{20} electrons/cm^2 provoked the formation of a 1T metallic phase at room temperature. The electrical measurements of the fabricated in-plane SB-FETs in figure 12.14(c) demonstrated the formation of a Schottky junction which showed the clear asymmetric behavior (rectification property), while the symmetric *I–V* curve was measured for the device with no Schottky junction, as shown in figure 12.14(d).

A CVD monolayer MoS_2 (figure 12.15(a)) was also investigated to study the effect of electron beam irradiation on its transport properties and a photoelectric performance of a MoS_2 device shown in figure 12.15(b) [51]. The elemental analysis of the monolayer MoS_2 was conducted using x-ray photoelectron spectroscopy (XPS) before and after electron irradiation at 10 keV for 10 min The results demonstrated the change of stoichiometry from 1.84:1 to 1:1.65 for S:Mo, which is indicative of S vacancy formation. Electron beam irradiation reduced the source–drain current and the field-effect mobility of the devices with an on/off current ratio unchanged, which were measured in air. However, the photoelectric performance of the device was improved after irradiation. The light response and release time were significantly reduced from 11 s and 105 s to 6 s and 5 s, respectively, as shown in figures 12.15(c) and (d). The light dark current ratio also increased from 7.6 s to 34.6

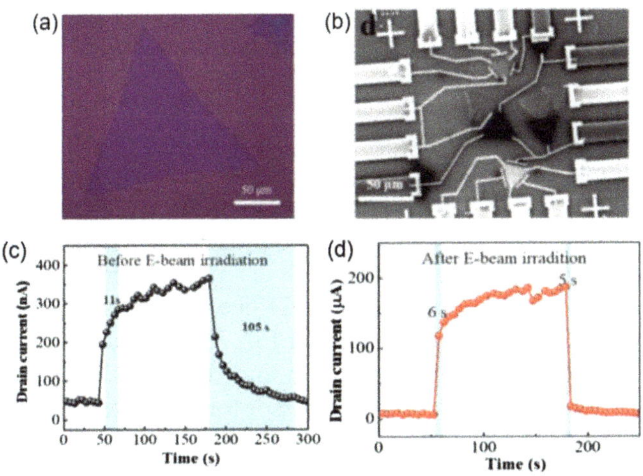

Figure 12.15. (a) Optical microscope image of a CVD-grown single-layer MoS_2 on the SiO_2/Si substrate, and the SEM image of the MoS_2 device with the Ti/Au contacts. The drain current measurements with light on and off for the device (c) before and (d) after electron beam irradiation, which shows the enhanced photoelectron property with electron beam induced modification of MoS_2. (Reprinted from [51], copyright (2018) John Wiley & Sons.)

s. The improvement of the photoelectric performance was due to the defect-assisted electron–hole recombination, which was revealed by Kelvin probe force microscopy (KPFM) measurements.

12.6 End-of-chapter summary

This chapter discusses the scientific bases and showcases the multi-modal nanofabrication capabilities of FEBIP as a next-generation lithography technique, with specific focus on processing of 2D organic and inorganic nanomaterials. From atomic scale opto-electro-mechanical and chemical property modification to enabling communication and quantum information applications, FEBIP holds promise in local physicochemical engineering of nanomaterials and fabrication of unique device architectures combining 1D, 2D and 3D topologies with unprecedented resolution and functional feature control at the heterogeneous material interfaces. Importantly, in the realm of 2D atomically thin materials, which are highly sensitive to contamination and material handling in between processing steps, FEBIP exceeds the capabilities of conventional lithography methods by enabling the multi-step fabrication of composite nanodevices within the same processing environment. Using the standard toolbox of an electron microscope but in combination with an on-demand reconfigurable chemical precursor injection system allows for all the steps required for fabrication of a fully functional nano-opto/electro/mechanical device accomplished in the same chamber without ever breaking vacuum. These include structural element functionalization/defect formation/doping; definition and geometrical patterning (via etch process by FEBIE); metal/metal oxide coating (via deposition process by FEBID); interface fusion and activation, active channel cleaning from parasitic adsorbates and changing the 2D layer redox state with atomic resolution.

12.7 Student's corner

FEBIP capabilities are particularly compelling to enrich the material processing capabilities and to create new nano-system architectures in the emerging field of quantum computing and information processing. The next frontier in FEBIP as a nanolithography tool of 2D materials is exploration for the use of liquid phase precursors to enable new (electro)chemical modes of material processing using solvated electrons and highly reactive radiolytic species, which enable an extension of patterning topologies from conventional 2D line and dot structures on the surface to 3D bridges and overpasses over the surface [52, 53]. The use of a cryogenic cooling provides another promising route for expanding the FEBIP capabilities by taking advantage of the condensed state of the precursor forming a thin low-mobility 'frozen' layer on a substrate for improved control of processing outcomes [54].

Recent advances in technological capabilities of FEBIP, including development of CAD tools [55, 56] for *in silico* process design and on-demand switchable "additive/subtractive" patterning capability of 2D materials [57], enable both device architects to develop new concepts of nanomaterials or functional systems and basic science researchers to conduct fundamental research on 2D material behavior with the *in situ* imaging capability, exploring new physical phenomena that are triggered by local perturbation of the material state on a single atom scale. From the manufacturing perspective, FEBIP is currently most suitable for laboratory scale 'prototyping' the new concepts of 2D nanomaterials and devices. The 'direct-write' nature of FEBIP is cost (and processing time) prohibitive for large-scale production, which is possible with the conventional batch semiconductor processing. Potentially, the FEBIP scale-up to wafer-level can be achieved with continuous development of instrumentation and reduction of the process time, such as massively parallel multi-beam systems as well as process automation, which can be realized with the faster control hardware for electron beam and gas injection coupled to the real-time monitoring and image data analytics through an instant feedback system.

Acknowledgements

This book chapter was supported by the U.S. Department of Energy (DOE), Office of Science, Basic Energy Sciences (BES), under Award #DE-SC0010729 and by the National Research Foundation of Korea (NRF) grant funded by the Korea government (MSIT) (No. 2019R1C1C1010556).

References

[1] Zhang H 2015 *ACS Nano* **9** 9451–69
[2] Novoselov K S, Geim A K, Morozov S V, Jiang D, Zhang Y, Dubonos S V, Grigorieva I V and Firsov A A 2004 *Science* **306** 666–9
[3] Geim A K and Grigorieva I V 2013 *Nature* **499** 419–25
[4] Sun L T, Banhart F and Warner 2015 *J MRS Bull* **40** 29–37
[5] Zhao X, Kotakoski J, Meyer J C, Sutter E, Sutter P, Krasheninnikov A V, Kaiser U and Zhou W 2017 *MRS Bull.* **42** 667–76
[6] Banhart F, Kotakoski J and Krasheninnikov A V 2011 *ACS Nano* **5** 26–41
[7] Li L, Reich S and Robertson J 2005 *Phys. Rev.* B **72** 184109
[8] Robertson A W and Warner J H 2013 *Nanoscale* **5** 4079–93
[9] Zhao J, Deng Q, Bachmatiuk A, Sandeep G, Popov A, Eckert J and Rummeli M H 2014 *Science* **343** 1228–32
[10] Wei X L, Wang M-S, Bando Y and Goldberg D 2011 *ACS Nano* **5** 2916–22
[11] Zobelli A, Gloter A, Ewels C P, Seifert G and Colliex C 2007 *Phys. Rev.* B **75** 245402
[12] Komsa H P, Kotakoski J, Kurasch S, Lehtinen O, Kaiser U and Krasheninnikov A V 2012 *Phy. Rev. Lett.* **109** 035503
[13] Lin Y C, Dumcenco D O, Huang Y S and Suenaga K 2014 *Nat. Nanotechnol.* **9** 391–6
[14] van der Zande A M, Huang P Y, Chenet D A, Berkelbach T C, You Y M, Lee G H, Heinz T F, Reichman D R, Muller D A and Hone J C 2013 *Nat. Mater.* **12** 554–61
[15] Wang S S, Lee G D, Lee S, Yoon E and Warner J H 2016 *ACS Nano* **10** 5419–30
[16] Lin J H, Cretu O, Zhou W and Suenaga K *et al* 2014 *Nat. Nanotechnol.* **9** 436–42
[17] Kalinin S V, Borisevich A and Jesse S 2016 *Nature* **539** 485–7
[18] Susi T, Kepaptsoglou D, Mangler C, Lovejoy T C and Krivanek O L *et al* 2014 *Phys. Rev. Lett.* **113** 115501
[19] Dyck O, Kim S, Jimenez-Izal E, Alexandrova A N, Kalinin S V and Jesse S 2018 *Small* **14** 1801771

[20] Su C, Tripathi M, Yan Q B and Wang Z *et al* 2019 *Sci. Adv.* **5** eaav2252
[21] Ziatdinov M, Dyck O, Jesse S and Kalinin S V 2019 *Adv. Funct. Mater.* **29** 1904480
[22] Kulkarni D D, Kim S, Chyasnavichyus M, Hu K, Fedorov A G and Tsukruk V V 2014 *J. Am. Chem. Soc.* **136** 6546–9
[23] Dreyer D R, Park S, Bielawski C W and Ruoff R S 2010 *Chem. Soc. Rev.* **39** 228–40
[24] Kim S, Kulkarni D D, Henry M, Zackowski P, Jang S S, Tsukruk V V and Fedorov A G 2015 *Appl. Phys. Lett.* **106** 133109
[25] Jones J D, Mahajan K K, Williams W H, Ecton P A, Mo Y and Perez J M 2010 *Carbon* **48** 2335–40
[26] Withers F, Bointon T H, Dubois M, Russo S and Craciun M F 2011 *Nano Lett.* **11** 3912–6
[27] Zhou Y G, Wang Z G, Yang P, Zu X T, Xiao H Y, Sun X, Khaleel M A and Gao F 2011 *Phys. Chem. Chem. Phys.* **13** 16574–8
[28] Balakrishnan J, Koon G K W, Jaiswall M, Castro Neto A H and Ozyilmaz B 2013 *Nat. Phys.* **9** 284–7
[29] Lee W K, Whitener K E, Robinson J T and Sheehan P E 2015 *Adv. Mater.* **27** 1774–8
[30] Ryu S, Han M Y, Maultzsch J, Heinz T F, Kim P, Steigerwald M L and Brus L E 2008 *Nano Lett.* **8** 4597–602
[31] Feng W, Long P, Feng Y and Li Y 2016 *Adv. Sci.* **3** 1500413
[32] Martins S E, Withers F, Dubois M, Craciun M F and Russo S 2013 *New J. Phys.* **15** 033024
[33] Kim S, Kulkarni D D, Davis R and Kim S S *et al* 2014 *ACS Nano* **8** 6805–13
[34] Toth M 2014 *Appl. Phys.* A **117** 1623–9
[35] Schoenaker F J, Cordoba R, Fernandez-Pacheco R, Magen C, Stephan O, Zuriaga-Monroy C, Ibarra M R and De Teresa J M 2011 *Nanotechnology* **22** 265304
[36] Yuzvinsky T D, Fennimore A M, Mickelson W, Esquivias C and Zettl A 2005 *Appl. Phys. Lett.* **86** 053109
[37] Sommer B, Sonntag J, Ganczarczyk A, Braam D, Prinz G, Lorke A and Geller M 2015 *Sci. Rep.* **5** 7781
[38] Thiele C, Felten A, Echtermeyer T J, Ferrari A C, Casiraghi C, Lohneysen H and Krupke R 2013 *Carbon* **64** 84–91
[39] Kim S, Henry M and Fedorov A G 2015 *Appl. Phys. Lett.* **107** 233102
[40] Fox D S, Maguire P, Zhou Y, Rodenburg C, O'Neill A, Coleman J N and Hongzhou Z 2016 *Nanotechnology* **27** 195302
[41] Urbanos F J, Black A, Rbernardo-Gavito R, Vazquez de Parga A L, Miranda R and Granados D 2019 *Nanoscale* **11** 11152–8
[42] Huth M, Porrati F and Dobrovolskiy O V 2018 *Microelectron. Eng.* **185** 9–28
[43] Kim S, Kulkarni D D, Rykaczewski K, Henry M, Tsukruk V V and Fedorov A G 2012 *IEEE Trans. Nanotechnol.* **11** 1223–30
[44] Kim S, Russell M, Kulkarni D D and Henry M *et al* 2016 *ACS Nano* **10** 1042–9
[45] Shin D W, Lee H M, Yu S M and Lim K S *et al* 2012 *ACS Nano* **6** 7781–8
[46] Kim S, Russell M, Henry M and Kim S S *et al* 2015 *Nanoscale* **7** 14946–52
[47] Zhou Y B, Jadwiszczak J, Keane D, Chen Y, Yu D and Zhang H 2017 *Nanoscale* **9** 8657–64
[48] Kappera R, Voiry D, Yalcin S E, Branch B, Gupta G, Mohite A D and Chhowalla M 2014 *Nat. Mater.* **13** 1128
[49] Katagiri Y, Nakamura T, Ishii A and Ohata C *et al* 2016 *Nano Lett.* **16** 3788–94
[50] Katagiri Y, Nakamura T, Ohata C, Katsumoto S and Haruyama J 2017 *Appl. Phys. Lett.* **110** 143109

[51] Li F, Gao F, Xu M, Liu X, Zhang X, Wu H and Qi J 2018 *Adv. Mater. Interfaces* **5** 1800348

[52] Fisher J, Kottke P A, Kim S and Fedorov A G 2015 *Nano Lett.* **15** 8385–91

[53] Fisher J, Kottke P A and Fedorov A G 2018 *Mater. Today Phys.* **5** 87–92

[54] Cordoba R, Orus P, Strohauer S, Torres T E and De Teresa J M 2019 *Sci. Rep.* **9** 14076

[55] Fowlkes J D, Winkler R, Lewis B B and Fernández-Pacheco A *et al* 2018 *ACS Appl. Nano Mater.* **1** 1028–41

[56] Henry M R and Fedorov A G 2021 *Comput. Mater. Sci.* **186** 109993

[57] Kim S, Jung S, Lee J, Kim S and Fedorov A G 2020 *ACS Appl. Mater. Interfaces* **12** 39595–601

IOP Publishing

Nanofabrication
Nanolithography techniques and their applications
José María De Teresa

Appendix A

Responses and comments to the section 'Student's corner' in each chapter

Chapter 1 Introduction to nanolithography techniques and their applications

Draw similar diagrams to those shown in figure 1.11 for the following lithography techniques: FIB, FEBID, Stencil lithography

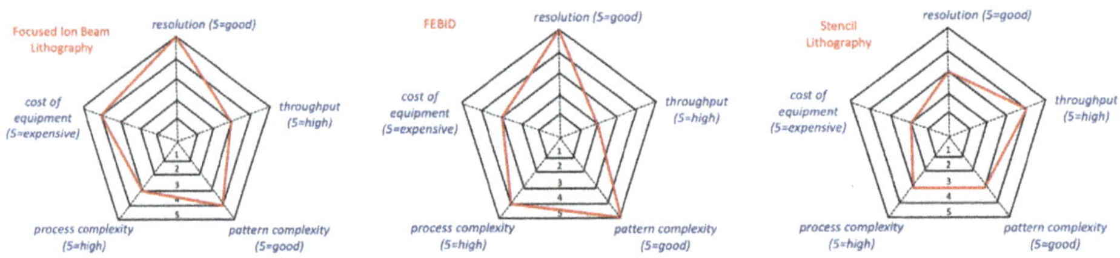

Chapter 2 Optical lithography

- **In a good approximation, we can assume that the value of the technological constant k_1 has been decreasing linearly over the years, with a constant decrease rate of 0.017/year. If the value of k_1 was 0.8 in 1980, indicate which information can be obtained about the rest of parameters that define the resolution in projection optical for the six cases indicated in figure 2.4.**
  The resolution and wavelength in projection optical lithography extracted from the data of figure 2.4, are the following:

doi:10.1088/978-0-7503-2608-7ch13

Year	Resolution (nm)	λ (nm)
1982	2000	436
1990	800	365
1997	250	248
2003	90	193
2009	32	193
2019	7	13.5

Assuming a yearly decrease of k_1 equal to 0.017/year, the value of k_1 for the same years can be approximated to:

Year	k_1
1982	0.766
1990	0.63
1997	0.511
2003	0.409
2009	0.307
2019	0.137

The obtained value of k_1 for the year 2019 was below the fundamental limit of 0.25, and in consequence, it cannot be considered. The resolution value for 2019 corresponds to EUV lithography, which implies a drastic decrease in wavelength, then a larger value of k_1 provides the required resolution.

Resolution in projection optical lithography is given by equation (2.3)

$$R = k_1 \frac{\lambda}{NA}$$

The only unknown parameter is the numerical aperture, NA. Then, the values that can be assumed for NA are:

Year	NA
1982	0.2
1990	0.3
1997	0.5
2003	0.9
2009	1.9

The numerical aperture in 2009 is larger than 1, as in 2009 immersion lithography was already in place, with a value of k_1 close to the theoretical limit of 0.25. Further advances in resolution during the following days were achieved thanks to the use of multiple patterning.

Regarding EUV lithography, high *NA* systems with NA = 0.55 have been reported recently (Zahlten C *et al* 2019 High-NA EUV lithography: pushing the limits *Proc. SPIE* **11177** 43). Assuming a resolution of 7 nm, the corresponding value of k_1 would be:

$$k_1 = R\frac{NA}{\lambda} = 7\frac{0.55}{13.5} = 0.29$$

- **The parameter γ determines the contrast of a photoresist after exposure and development, and, in consequence, it is a parameter that influences the resolution. Evaluate the interplay between photoresist contrast and optical exposure. For this, indicate the extension of the dimensions of a pattern for three photoresist with γ = 1, 5 and 10, and assuming that the profile of the optical exposure at the edges of the exposed area follows a Gaussian like distribution with σ = 1 µm (case 1) and σ = 3 µm (case 2). Make the analysis for a positive photoresist.**

  The relation between remaining photoresist and exposure dose for a positive photoresist is given by equation (2.5):

$$\frac{d'}{d_o} = \gamma \log_{10}\left[\frac{D_c}{D}\right]$$

The Gaussian profile of the optical radiation at the edges of the exposed area will be given by:

$$I = Ioe^{-\frac{x^2}{2\sigma^2}}$$

where x is the distance from the edge of the exposed area. If we consider *Io* to be the optical intensity to provide the clearance dose D_C, we can obtain a relation of the remaining dose as a function of the distance from the edge of the exposed area:

$$\frac{d'(x)}{d_o} = \gamma \log_{10} e^{\frac{x^2}{2\sigma^2}} = \gamma\left(\frac{x^2}{2\sigma^2}\right)\log_{10} e$$

Then, the value of x for which the thickness of the resist remains constant (i.e. $d'(x)/\ d_o = 1$) is given by:

$$x = \sigma\sqrt{\frac{2}{\gamma \log_{10} e}}$$

Using this relation with the indicated values of γ and σ, the extension of the dimensions of the pattern for each case will be:

	$\gamma = 1$	$\gamma = 5$	$\gamma = 10$
Case 1: σ = 1 µm	2.1 µm	1.0 µm	0.7 µm
Case 2: σ = 3 µm	13.8 µm	2.8 µm	1.4 µm

- **In an EUV lithography system the light is reflected in 12 multilayered mirrors from the intermediate focus of the EUV source to the exposed wafer. The reflectance of each mirror is 0.68, the exposure dose of the used photoresist is 40 mJ cm^{-2}, each die is 50 mm × 30 mm in size, and the exposition time is 1.5 s. If the energy conversion efficiency at the light source is 4%, estimate the power consumption at the EUV source.**

  The energy consumption per die is:

  $$40\ \text{mJ/cm}^2 \times 15\ \text{cm}^2 = 600\ \text{mJ}$$

  If the exposition time is 1.5 s, then the EUV power at the wafer is:

  $$600\ \text{mJ}/1.5\ \text{s} = 400\ \text{mW}$$

  This power must be the result after 12 reflections in the mirrors, so that the power at the intermediate focus should be, at least

  $$400\ \text{mW}/(0.68)^{12} = 40.9\ \text{W}$$

  Then, considering the conversion efficiency, the power consumption at the EUV source would be:

  $$\text{Power}_{\text{source}} = 40.9\ \text{W}/0.04 = 1.02\ \text{kW}$$

- **In the case that you had to fabricate a prototype of a planar circuit including superconducting and/or magnetic materials, with patterning resolution of 0.8 μm, would you choose NUV optical lithography, EUV optical lithography, laser direct write lithography or two-photon-absorption lithography? Why?**

  I would choose laser direct write lithography. As the circuit is planar, we do not need two-photon-absorption lithography, which is only chosen for 3D patterning. EUV optical lithography is very expensive and only used when very high resolution (a few nm) is needed. NUV optical lithography could be used too in this application, given that the 0.8 μm resolution can be obtained, but we would need to fabricate a physical mask, which takes money and time. Only if many of these planar circuits need to be fabricated, would we use NUV optical lithography. However, in this case, we only want to fabricate a prototype. If the prototype is not suitable, we would likely have to change the mask design and fabricate another mask, taking money and time. By using direct write lithography, the change of design would be straightforward, through software modifications.

- **If you were working in the MEMS domain and needing to fabricate structures with aspect ratio (height/width) of 20 and widths down to 2 μm resolution, propose a lithography fabrication process based on optical lithography, including the used technique, the photoresist tone (positive or negative) and thickness, the post-exposure steps, etc. What if you need to fabricate an overhanging cantilever structure of similar dimensions?**

The process will be different if the cantilever must be fabricated using a particular material or it can be of any material like the resin itself. Thus, we propose two different approaches:

(a) If the structure's material can be the resist itself, I would use a photoresist that is appropriate for mechanical devices, such as the negative-tone SU8 resist. This photoresist can be patterned with thicknesses as large as a few hundreds of microns, which would be appropriate if we need widths down to 2 μm resolution and heights larger than 40 μm. I would use NUV optical lithography or laser direct write lithography to expose the resist. The spin coating parameters should be optimized to achieve photoresist thickness as large as the required height in the structures. After exposure with optimized parameters, a post-exposure bake and development would be enough to reveal the targeted micro-structures and finish the processing.

(b) If the structure's material is different, after development, the required material should be grown. One popular growth technique for large thicknesses is electroplating. However, not all resists behave appropriately during the electroplating process. In the reference Koukharenko *et al* 2005 *J. Mater. Sci. Mater. Electron.* **16** 741–47, several photoresists were tested and SPR 220-7 photoresist was found to behave during the electroplating process in sharp contrast with SU8.

In the case of an overhanging cantilever structure built with the resist itself, one process that could work would be based on two-photon-absorption lithography, which is appropriate for 3D patterning including cantilevers, as shown in figure 2.15(c).

- **The following images correspond to SEM images of self-assembled patterns of PS-b-PMMA block copolymer, where the darker areas correspond to the PMMA phase and brighter areas to the PS phase. The block copolymer film has been deposited on a silicon oxide surface and the thickness is around 40 nm. In the three cases, a neutralization surface layer has been formed on the surface before the block copolymer deposition to create the optimal neutralization. Indicate:**

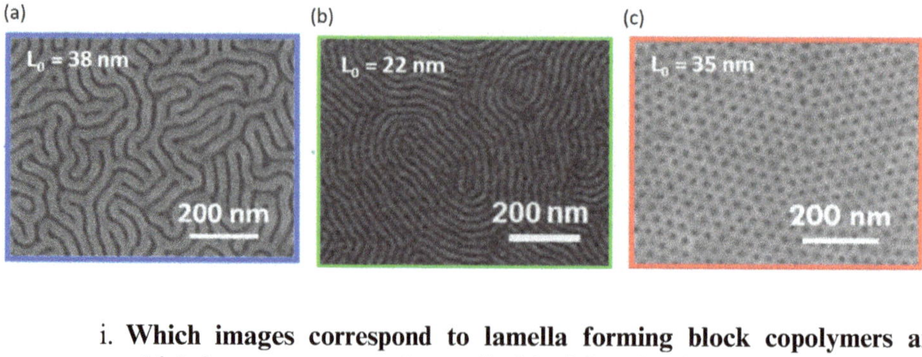

i. **Which images correspond to lamella forming block copolymers and which images correspond to cylindrical forming block copolymer.**

ii. **Are the block copolymer molecules oriented parallel or perpendicular with respect to the surface plane?**
iii. **What can be said about the presence of a neutralization layer on the surface of the substrate.**
iv. **Which is the position of each block copolymer in the phase diagram of block copolymers (figure 2.20(a)).**

i. Images (a) and (b) correspond to lamella forming block copolymer and image (c) to cylindrical forming block copolymer. In consequence the relative weight fraction of PMMA with respect to the PS-b-PMMA molecule is around 50% for (a) and (b). For the case (c), PMMA weight fraction is lower, typically around 30%.
ii. In the three cases, the block copolymer molecules are oriented parallel to the surface. As a consequence, for cases (a) and (b) they form vertical lamellae, and for case (c) they form vertical cylinders of PMMA surrounded by a PS matrix.
iii. In the three cases a neutralization layer on top of the surface is required to ensure that the molecules position horizontally to the surface and the observed patterns are formed. The silicon oxide surface presents larger affinity to PMMA than PS. In the case of (a) and (b), the neutralization layer has to have equal affinity to PS and PMMA. For case (c), the neutralization layer has slightly more affinity to PS.
iv.

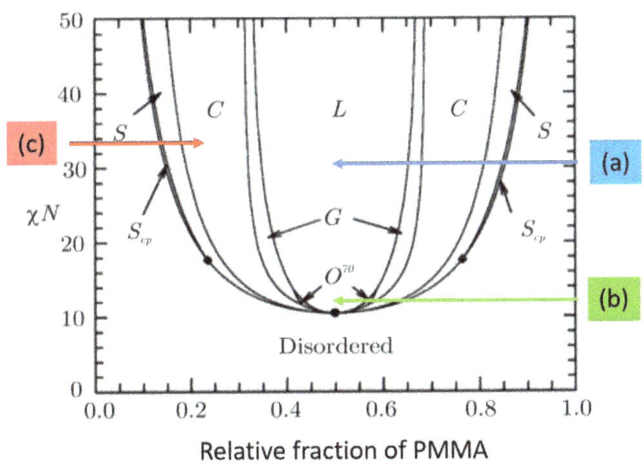

Chapter 3 Electron beam lithography

Correct answers (in bold) to the multiple-choice questions:

- A FE-SEM allows working at low potentials which is crucial when dealing with …
    A. Metallic films.
    B. Semiconducting films.
    C. **Dielectric films.**

- If you need to take an SEM image for topographic information at a rather high tilting angle (>45°), what would you choose?
  - A. It is not possible to take tilted SEM images as the SEM configuration is fixed (beam, detectors, etc.)
  - B. In-lens detector is the best choice.
  - C. **ET detector is the best choice.**
- For an EBL process, which of the following statements are true?
  - A. **After the exposure and development processes, the positive tone resist is removed in the exposed areas.**
  - B. After the exposure and development processes, the negative tone resist is removed in the exposed areas.
  - C. Development determines the tone of the resist.
- Choose the WF with the nominal higher resolution.
  - A. **100 μm.**
  - B. 500 μm.
  - C. 1 mm.
- What does the concentration of PMMA in solvent affect?
  - A. The homogeneity of the coating.
  - B. **The thickness of the resist.**
  - C. **The optical contrast of the resist.**
- If you need to expose an array (larger than 1 mm) of the same nanostructured pattern, which of the following approaches are the most suited for a correct lithography?
  - A. Expose the whole WA in a unique (U,V) center.
  - B. **A smaller area (according the used WF) CAD file is prepared and the whole WA is exposed thanks to the creation of a matrix of CAD elements with shifted (U,V) centers.**
  - C. **The whole CAD file is prepared, but we translate the WA exposure at different (U, V) centers (according the used WF).**
- How can we overcome the stitching problem?
  - A. **Through the use of an interferometric position stage.**
  - B. Through the use of a micrometric position stage.
  - C. **Through the use of markers.**
- Suppose that during an alignment process with markers their centers are not visible in the windows that are opened by the EBL program. How would you proceed?
  - A. I would repeat the three-point alignment and probably I would need to repeat also the global origin and angle correction.
  - B. I would change the dimensions of the markers square in the CAD file (I would make them bigger so that I can see their center in the next four markers alignment).
  - C. **I would repeat the alignment process on the four markers as many times as necessary. In the first alignment, I would center each window in the visible part of the marker so that in the subsequent alignments I will finally identify the center of the markers.**

- What are the advantages of the HSQ resist with respect to the PMMA?
  - A. Versatility.
  - B. **Resolution.**
  - C. Lower exposition time due to the lower used doses.
- How would you reduce the PE problem through fragmentation?
  - A. **In an isolated feature by reducing the dose at the center and increasing the dose at the edges.**
  - B. In an isolated feature by increasing the dose at the center and decreasing the dose at the edges.
  - C. In packed structures by increasing the dose between neighbouring features.
- How would you solve a lift-off problem due to the crosslinking of the resist?
  - A. Using thinner resists.
  - B. **Using multilayer resists.**
  - C. **Understanding if crosslinking is related first to certain positions of the position of the sample in the evaporator.**

Chapter 4 Focused electron beam induced deposition

- **Which growth variables have an influence on the lateral resolution of FEBID, and why?**

  In general, higher lateral resolutions are obtained by playing with numerous growth variables:
  - Higher acceleration voltage provides smaller probe size, thus potentially higher spatial resolution.
  - Lower beam current is associated to smaller spot size, thus higher spatial resolution.
  - Short growth times minimize the effect of drift and instabilities, contributing to higher spatial resolution.
  - Conducting substrates minimize unwanted charging effects, which degrade the resolution.
  - High density substrates induce a more localized emission of secondary electrons, thus a smaller deposition area.
  - If very thin substrates (such as membranes) are used, the interaction volume will be smaller and the FEBID resolution will be better.
- **Which measures would you take to reduce the halo around FEBID deposits?**

  There are two main strategies that minimize the halo produced around the FEBID deposits, which are related to using the smallest-possible electron beam spot and reducing the interaction volume. The electron beam size can be reduced by increasing the acceleration voltage (minimizing the objective lens aberrations) and reducing the beam current. The interaction volume can be reduced independently of the beam properties by using high density (Z number) substrates or very thin substrates.
- **Are insulating substrates suitable for FEBID growth?**

In general, insulating substrates are not suitable for FEBID due to the appearance of detrimental charging effects around the growth area. However, there are different approaches that can be taken to enable a satisfactory FEBID growth on insulating substrates. For instance, the fabrication of metal pads around the growth area for charge evacuation, or the use of (voltage biased) metallic masks such as the ARAGON-Chip discussed in section 4.5 can be used to dissipate charge. In this case, a metallic mask allows the evacuation of the charge generated during the FEBID growth (Pablo-Navarro J *et al* 2019 Diameter modulation of 3D nanostructures in focused electron beam induced deposition using local electric fields and beam defocus *Nanotechnology* **30** 505302). The use of lower voltage and beam currents could also be beneficial for insulating or poorly-conducting substrates.

- **Growth rate is one of the key issues of FEBID. *For a given electron dose*, which growth parameters could you modify to increase the growth rate?**
The growth rate is the result of numerous factors influencing the decomposition and growth processes in FEBID, and not all of them are independent. *For a given electron dose*, the growth rate will increase by:

  - Decreasing the acceleration voltage, given that the amount of secondary electrons reaching the surface of the sample will be higher, thus being more efficient to decompose the precursor molecules (De Teresa J M *et al* 2009 Origin of the difference in the resistivity of as-grown focused-ion- and focused-electron-beam-induced Pt nanodeposits *J. Nanomater.* **2009** 936863).
  - Decreasing the beam current, as there will be less heating on the growth area, avoiding the decrease of the precursor residence time (Mutanga E *et al* 2019 Impact of electron-beam heating during 3D nanoprinting *ACS Nano.* **13** 5198–213).
  - Increasing the precursor gas flux, in case we are in the precursor-limited regime in which the deposit does not grow faster for a lack of available precursor molecules (Pablo-Navarro J *et al* 2017 Tuning shape, composition and magnetization of 3D cobalt nanowires grown by focused electron beam induced deposition (FEBID) *J. Phys. D: Appl. Phys.* **50** 18LT01).
  - Increasing the refresh time, so the depleted precursor molecules have longer time to be replenished.
  - Decreasing the dwell time to values lower than the effective residence time of the precursor gas molecules, so the molecule depletion is avoided (Serrano-Ramón L *et al* 2011 Ultra-small ferromagnetic nanostructures grown by focused electron beam-induced deposition *ACS Nano* **5** 7781).
  - Decreasing the distance from the GIS nozzle to the area of interest on the substrate. This should result in a higher growth rate if we are in the precursor-limited regime because it would increase the number of available precursor molecules.

- ○ Increasing the density of the material (higher average Z), so a higher number of secondary and backscattered electrons reach the surface of the sample and can decompose the precursor.
- ○ Decreasing the temperature of the substrate to increase the precursor residence time (Huth M *et al* 2020 Temperature-dependent growth characteristics of Nb- and CoFe-based nanostructures by direct-write using focused electron beam-induced deposition *Micromachines* **11** 28).

Chapter 5 Focused ion beam induced processing

- **Compare the advantages and disadvantages of FEBID and FIBID in these cases: growth of electrical contacts to a nano-object, growth of a 3D magnetic nanostructure, and growth of a protective layer on a given material.**
  Both FIBID and FEBID are grounded on the same fundamental principle—a focused beam of charged particles, either ions or electrons, induces the decomposition of a gaseous precursor adsorbed on the surface of a substrate, mostly via the emission of SEs by the latter. Due to the mass and charge difference between the two types of charged particles, the processes involved in the ion–matter interaction are different. Generally speaking, FIBID is faster than FEBID (due to a greater SE yield induced by the impinging ions when compared to the electrons), and also yields a higher metallic content than FEBID. However, when using FIBID, ion implantation occurs, which can change the properties of the substrate. Secondly, ion bombardment induces substrate amorphization and defect creation, which can be detrimental in certain cases. Moreover, in FIBID a competition between growth and milling is always present (which can be to a certain extent controlled by adjusting the dose and precursor flux), while in FEBID, milling is not relevant.

  Hence:
  - ○ In the case of electrical contacts, FIBID should be first considered, for it will yield metallic contacts when used with the appropriate precursor and in short processing time. However, the sensitivity of the substrate and of the object that is being contacted to ion irradiation must be considered—will local amorphization of the sample (below the point of contact placement and on its vicinity) significantly alter the electrical properties of the object? If so, (for example, in the case of thin samples) then FEBID should also be considered, although the contacts may be less metallic and therefore less conductive (Hernández-Ramírez F *et al* 2006 Characterization of metal-oxide nanosensors fabricated with focused ion beam (FIB) *Sens. Actuators B* **118** 198–203).
  - ○ Similar arguments hold for the growth of a 3D magnetic nanostructure. Since a significant amount of the focused beam ions will get implanted, they may play a role on the overall magnetic properties of the sample. Advantages of using FEBID for this application are the high resolution and, in general, the high purity and magnetization obtained (Fernández-

Pacheco A *et al* 2009 Magnetotransport properties of high-quality cobalt nanowires grown by focused-electron-beam-induced deposition *J. Phys. D Appl. Phys.* **42** 055005). On the other hand, FIBID will be a faster growth technique but milling effects can be detrimental (Sanz-Martín C *et al* 2019 High volume-per-dose and low resistivity of cobalt nanowires grown by Ga^+ focused ion beam induced deposition *Nanomaterials* **9** 1715).

- In order to grow a protective layer over a material, FEBID is preferred to FIBID if amorphization, implantation and defect creation are to be avoided. However, growing large, thick layers using FEBID can take unreasonably long processing times. A common procedure in protective deposits for TEM sample preparation and for cross-sectional SEM imaging is to make use of *both* techniques to grow such layers (FEBID and FIBID can be easily performed in FIB/SEM instruments)—first, a relatively thin protective layer is deposited via FEBID on top of the sample, and then a second thicker layer is grown by FIBID on top of the first, hence only damaging the (irrelevant for further experiment purposes) FEBID-grown layer.

- **Compare the advantages and disadvantages of FIB and FEBIE for optical mask repair.**
  In optical mask repair, it is important not to modify the optical properties of the underlying substrate of the mask, which is better accomplished using FEBIE instead of Ga^+-FIB given that Ga^+ implantation produces optical changes. However, the use of light-ions in gas field FIB sources could change this in future (Yasaka A *et al* 2016 Nanoscale imaging, material removal and deposition for fabrication of cutting-edge semiconductor devices ion-beam-based photomask defect repair technology *Hitachi Rev.* **65** 71–75).
- **Compare the advantages and disadvantages for compositional analysis of EDX using an SEM and SIMS using an HIM.**
  While both SIMS and EDX are compositional characterization techniques, their working principles, scopes and purpose are vastly different. EDX is performed by measuring the x-rays that the sample emits when it is irradiated by an SEM. Different elements yield x-rays of different energies, which manifest themselves in the form of peaks in a spectrum, whose extent in terms of energy is capped by the energy of the beam (no x-rays with higher energy than that provided with the beam can be emitted). Applying higher voltages to extend the spectrum results in an enhanced electron penetration (bigger interaction volume). On the other hand, SIMS is based on the physical sputtering of the surface of a sample, with the ejected atoms being mass filtered to separate the different constituents.

  Therefore:
  - EDX is *simpler*. It is usually readily available in most characterization facilities, typically implemented as an additional module in instruments that naturally use an SEM, and it is relatively easy and fast to perform. SIMS, on the other hand, in most cases requires a dedicated instrument

(although it can also be implemented in HIMs), and the procedure is more complex.

- EDX should, generally speaking, be thought of as a *bulk* technique. While low SEM energies allow only for limited spectra, high acceleration voltages result in more area being probed. In addition, neighboring peaks arising from different elements may overlap, blurring the conclusions. If EDX is performed in a TEM instrument, however, both, its resolution and fidelity, are enhanced.
- SIMS is, on the other hand, a *surface* technique, which enables for high-resolution discerning of chemical elements constituting the surface of the sample. It can also be performed in dynamic mode, tracing a depth profile of the sample.
- EDX is reasonably *non-invasive* (although it does yield a relatively high electron dose on the sample), while SIMS is *destructive*, since it physically sputters the surface that is being analyzed.
- SIMS is, generally, more *precise*. The spatial resolution of an HIM FIB is greater than that of an SEM, and the mass separation procedure allows for more fidelity in element separation than that based on x-ray spectra analysis. It is even possible to obtain grain-boundary compositions with good accuracy.

Chapter 6 Scanning probe lithography

- **Which of the following SPL methods do pattern a resist in the first nanolithographic step?**
  A. Oxidation SPL.
  B. **Thermal SPL.**
  C. Bias SPL.
  D. Dip-pen nanolithography.
- **Select from the list the factors that influence the feature size in oxidation SPL**
  A. **The tip's radius.**
  B. The temperature of the tip.
  C. **The applied voltage.**
  D. The thickness of the resist.
- **What are the functionalities of the oxide structures fabricated by o-SPL ?**
  The oxide nanostructures might be used as dielectric barriers or masks for an etching process or templates.
- **(a) Describe thermal scanning probe lithography. (b) Mention the dominant factors that control the feature size and (c) its order of magnitude (1, 10, 100 or 1000 nm ?).**
  (a) Thermal SPL uses a hot tip to pattern a resist. The tip induces the local melting and removal of the resist. The pattern is used as a mask in an etching process.
  (b) Tip's size, tip's temperature and resist properties.

(c) 10 nm.

- **Dip-pen nanolithography is based on the deposition of diffusive inks on a surface. Describe the main steps of the transport process of diffusive inks.**
The process might be divided in three main steps: dissolution of the ink molecules in the water, motion of the ink molecules in the water meniscus formed between the tip and the surface, and assembly of the ink molecules on the surface.
- **Provide an estimation of the electric field generated by applying 10 V for a tip that is placed 1 nm above a gold surface. Is this field larger or smaller than the average electric fields applied in electron beam lithography?**
The mean electric field without considering the tip's finite geometry is of 10 GV m^{-1}. That value is extremely high. In EBL, very high voltages are applied to emit and accelerate the electrons. Typical voltages are in the 1–30 kV range. Those voltages are applied over ~1–30 cm distances. The equivalent electric field (10 kV applied over 1 cm and without including geometry factors) is of 1 MV m^{-1}, this is, two-to-three orders of magnitude smaller than the fields that might be generated in SPL. This example illustrates how a small voltage might generate an extremely high electric field.
- **Design a micro-optical tag made of multilayered MoS_2 that alternates blue, green and yellow regions. The color of MoS_2 flakes is related to the number of monolayers (see scheme below). Each optical region of the tag is a square of 800 nm × 800 nm. A triangular MoS_2 flake of 10 μm (side length) and 30 monolayers in thickness is deposited on a Si substrate (see below). It is the raw material for the tag. Hint: The number of MoS_2 monolayers oxidized per unit of time in o-SPL is known. The MoS_2 oxides generated in o-SPL are soluble in water (see figure 6.2(c)).**

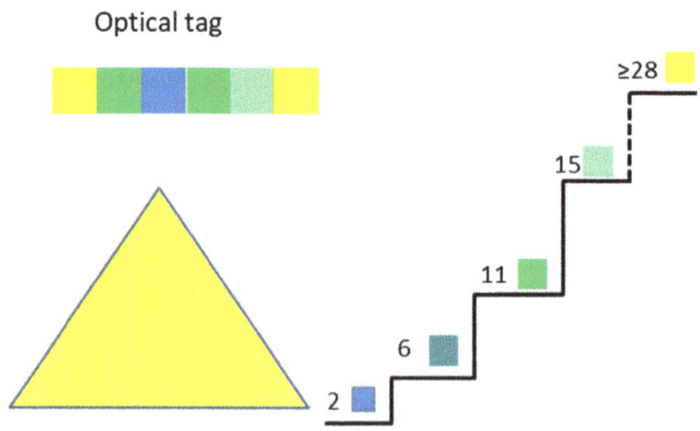

The process involves three steps. First, the initial flake is patterned by o-SPL according to the figure. The o-SPL parameters are chosen to oxidize completely the flake (30 mL). Second, the optical regions are patterned according to the figure. The numbers in parenthesis indicate the thickness of the patterned oxides in terms of monolayers. Third, the sample is immersed in water to remove the oxidized regions.

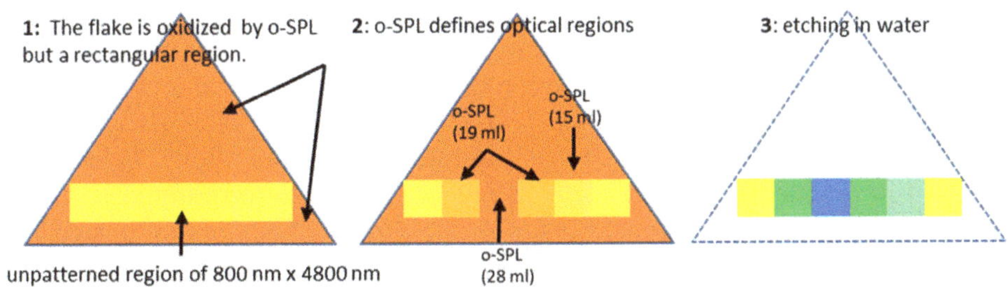

Chapter 7 Soft thermal nanoimprint and hybrid processes to produce complex structures

- **How is it possible to use PDMS as mold for T-NIL of nanostructures on PC films if the elastic modulus of the PC is three orders of magnitude higher?**
  The thermal NIL process is performed at temperatures well above the polymer's Tg, in practice this temperature is set about 50 °C–90 °C above Tg so that the polymer reaches the viscous flow state. At this high temperature, the elastic modulus and yield strength decrease several orders of magnitude and as a result, the compressive stress on the mold will be much reduced avoiding mold deformation while the heated polymer chains flow and fill the cavities of the mold. In the particular case of PC, the glass transition temperature is about 150 °C and the elastic modulus has a value of 2.2 GPa at 25 °C. By raising the temperature up to 170 °C, the elastic modulus already decreases three orders of magnitude, down to 2 MPa, which is comparable to that of the PDMS.
- **During T-NIL of HAR nanostructures on PMMA it is observed that upon demolding a large number of nanopillars are affected by ripping. What could be done to improve the replica quality in this case?**
  Ripping is a consequence of stress concentration at the base of high aspect ratio nanostructures during the demolding step. The demolding defects can be minimized by decreasing the surface energy of the mold and by adjusting the demolding temperature. Surface energy can be tuned within the range of 10–20 mJ m^{-2} by using silane-based anti-sticking layers. In the case of PMMA, a demolding temperature of 70 °C was observed to be optimum to avoid ripping by inducing the appropriate levels of softness and elasticity, while reducing friction and adhesion forces.
- **During the fabrication of nanopillars by T-NIL on a PS film, it is observed that a large number of pillars are taller than expected according to the mold dimensions, what could be the reason? How can you select the demolding temperature?**

The most plausible reason to cause the stretching of nanopillars is demolding at excessively high temperature. At temperatures close to the glass transition of the polymer (110 °C for PS), the elastic modulus and mechanical strength is still reduced which allows for stretching of soft material due to the strong demolding forces that upon cooling, results in permanent plastic deformation of the nanostructures. The optimum demolding temperature is achieved when the adhesive force is decreased and the friction force is at its lowest. The optimum demolding temperature should be somewhere slightly below the Tg and in practice, it is experimentally found for each imprinting task.

- **How does the aspect ratio influence the detachment force during demolding of nanoimprinted structures? What is the advantage of using a flexible mold during the demolding step of HAR nanostructures?**
  The demolding process consists in the separation of the mold from the imprinted structures. The required demolding force depends, among other factors, on the adhesion force between the mold and the polymer and the friction force of the structures on the side walls roughness as they are being released. These forces are determined by the contact area which depends directly on the geometrical parameters of the mold, the aspect ratio and density of the structures. A flexible mold allows performing a sequential action of peeling of the mold during the demolding step, resulting in a smaller area, thus a lower demolding force compared to a rigid mold. As a result, the number of defects are reduced improving overall the replica quality.
- **Which are the main factors related to the pattern collapse of HAR?**
  Principally, the limitation to the aspect ratio of a nanostructure depends on its mechanical properties (Young's modulus and Poisson's ratio), which determine the effective stiffness at the nanoscopic dimensions. In addition, there are two main factors which affect the nanofeature's stability: surface energy and pattern density. HAR nanostructures may suffer collapse by adhesion forces between pillars (lateral collapse) or between pillars and substrate (ground collapse).
- **Given a polymer with viscosity = 10^6 [Pa s], surface tension ~20 mN m^{-1} and wetting contact angle of 80°**
  a. **What is the time needed to fill 0.5 × 2 μm pore size cavities by capillary action?**
  b. **What viscosity should the polymer have to reduce that time to about 7 min?**
  c. **If mold silanization is accomplished in order to facilitate the demolding process, how is the polymer wetting contact angle expected to be affected?**

Use the relations derived from the Lucas-Washburn equation:

$$h^2 = \frac{r\gamma \cos\theta}{2\eta}t, \quad t = \frac{2\eta h^2}{r\gamma \cos\theta}$$

(a) t = 77 min

(b) viscosity = 10^5 Pa s
(c) In order to facilitate mold release the silanization process would reduce the mold surface energy and, as such, the wetting contact angle would increase.

- **What provides the gecko with its amazing climbing capability?**
The gecko's climbing ability relies on the fibrillar structures on its toe pads. The pads are covered by densely packed, high aspect ratio hierarchical pillars called *setae*, with diameters between 5–10 μm. Each of these setae presents fibrillar branches at the tip with diameters ranging from 200 up to 500 nm and aspect ratios between 10 and 25. This hierarchy induces a progressive lowering of the effective stiffness and increases dramatically the available contact area. The hierarchical setae can then easily conform and come close to a surface where van der Waals forces develop and give rise to strong frictional adhesive forces.

Chapter 8 Stencil lithography

- **If you want to deposit 100 nm structures, and you can afford ±10 nm, would you use sputtering, e-beam evaporation, or thermal evaporation? Why?**
In this case we would use a stencil membrane with 100 nm apertures. In order to stay below 10 nm variation from the nominal value, we would have to calculate our geometric blurring from equation (8.1)

$$\Delta w_{\text{geom}} \approx G \cdot \frac{d}{D} \leqslant 10\ \text{nm}.$$

This is a rather small value for blurring. As we mentioned in the chapter, the typical gap is within the tens of microns, let's assume it is 10 μm. Therefore, we need a ratio $\frac{d}{D} > 10^{-3}$. Sputtering is discarded because D is short and d is large (larger than 1 cm). In addition, the relatively high pressures at which the process would take place will make the atoms from the source collide with each other and the deposition will not be directional. Evaporation is preferred. Between thermal and e-beam evaporation, the latter is preferred because it offers smaller source dimensions d.

- **Considering a distance source–substrate of 75 cm and a typical source size of 2 mm, how much should the gap be to ensure the requirements above?**
Just following from the equation above:

$$\Delta w_{\text{geom}} \approx G \cdot \frac{d}{D} \leqslant 10\ \text{nm} \rightarrow G \leqslant 10\ \text{nm} \cdot \frac{75\ \text{cm}}{2\ \text{mm}} \leqslant 3.75\ \mu\text{m}$$

- **If these 100 nm structures are supposed to be 200 nm thick, what are the main issues that we would face, and why?**
The main issue that would be encountered is the fact that lateral deposition would happen on the lateral walls of the stencil membrane, thus progressively

reducing the aperture dimensions and ending up with the deposition of a pyramidal structure (triangular cross section) likely thinner than expected. This is the phenomenon we call *clogging*.

- **How would you pattern with SL a ring of 10 μm in diameter and 200 nm in width, plus 20 nm thick?**
  A direct approach would be to do this via dynamic stencil lithography. Patterning on the membrane a hole of 200 nm in diameter and then drawing a circle. In the same way as is shown in the support video #3, available at https://iopscience.iop.org/book/978-0-7503-2608-7. A second option would be using quasi-dynamic stencil lithography, by fabricating a membrane with two complementary patterns that together would form a full ring. These complementary patterns could be, for example, two half-rings. Finally, with standard SL, ring-like patterns are difficult to perform because you cannot fabricate a ring-shaped aperture. A solution would be to fabricate a ring-shaped apertured split into smaller pieces, as the figure below shows. These narrow supports for the inner circle should be narrow-enough so that blurring would eventually join each of the apertures. Typically, they should be below 50 nm to have a conductive layer. With this method, we use one disadvantage of SL in order to obtain the desired pattern.

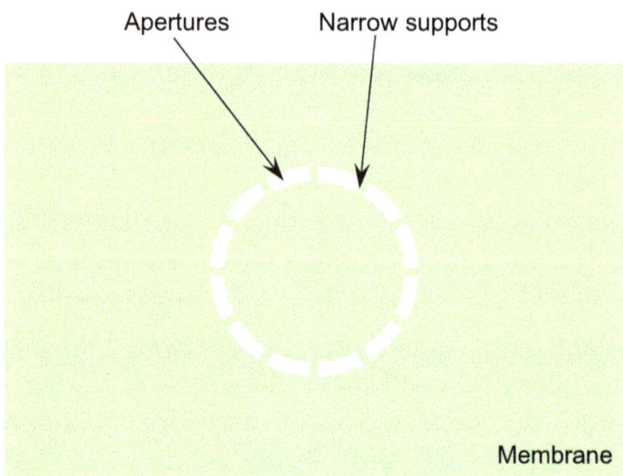

Chapter 9 Ice lithography

- **What is the electron energy range in an SEM?**
  Primary electron energy range between 100 eV to 30 keV. Secondary electron energy range between 1 eV and 30 keV.
- **Is it possible to condense CO_2 in an SEM using a liquid nitrogen based cryostage?**
  No, because the vapor pressure of CO_2 at 150 K is 6 Torr, which is much higher than SEM vacuum.

- **What are the reaction products of electrons' interaction with CO_2 ice?**
  CO (g), CO solid network, oxygen.
- **What are the chemical reaction pathways for electrons -ice interactions?**
  Secondary electrons cause ionization or radicalization of the organic ice. The reactive chemicals can recombine, cross-link into networks or larger molecules, or the smaller molecule fragments can vaporize.
- **Which chemical would you want to work with and why?**
  Please contact the authors, and we can discuss your ideas.
- **Can IL and EBL be used for lithography on 3D structures?**
  IL produces uniform and conformal thin-films on the sample surface with complex topography. Therefore, it is suitable for performing lithography on 3D structure. On the other hand, because EBL resists are coated by spin-coating or spray coating, the resist uniformity is not guaranteed on a 3D structure, and therefore, EBL is less suited for the application in question.
- **List the factors that determine the patterning resolution of IL process.**
  Material chemistry, electron beam parameter (energy, beam spot, etc), mechanical limitation (stability, positioning, etc), and ice property (thickness, crystallite, etc).
- **What device or structure would you like to build with IL?**
  All answers are good answers, and some answers can be great answers.

Chapter 10 Magnetic nanopatterning via thermal scanning probe lithography

- **What are the main factors determining the minimum magnetic domain size which can be patterned with tam-SPL?**
  There are two main factors to be considered: the lateral extension of the thermal profile produced by the tip on the sample, and the magnetic properties of the sample. In the first case, the thermal profile produced on the sample depends both on the probe size (down to a few nm), and on the thermal conductivity of the substrate. In general, the higher the thermal conductivity of the substrate, the more confined is the thermal profile in the lateral direction. Concerning the magnetic properties: the minimum size of the domains that can be stabilized in a ferromagnetic film is comparable to the domain wall width, which in turn is determined by the micromagnetic properties of the system. Therefore, a small minimum domain size is obtained by combining highly confined thermal profile and narrow domain walls.
- **What are the factors determining the minimum tip temperature needed for tam-SPL patterning?**
  The minimum temperature on the sample needed for patterning should be higher than the blocking temperature of the exchange bias system. The temperature set on the heater (on the cantilever) is typically higher, due to the need to compensate for the thermal dissipation across the sample.
- **What are the factors determining the maximum tam-SPL writing speed?**

The main limiting factor is the thermal constant of the heated tip, which is typically in the order of tens of microseconds. This poses a limit to how fast the heat can by switched on and off controllably while sweeping the tip on the sample. Maximum writing speeds around 1 mm s^{-1} can be reached in optimized systems.

- **What determines the resilience of tam-SPL patterns with respect to time, magnetic field and temperature?**
  The resilience of tam-SPL patterns arises from the stability of the exchange bias system. Provided that the blocking temperature is well above room temperature, the unidirectional anisotropy of the system is not affected by the external magnetic field, and is stable over time. The stability of the system is lost when approaching the blocking temperature. This allows the system to be reconfigured either by resetting a uniform exchange bias on the whole film, or by performing tam-SPL again.
- **Would it be possible to pattern different spin textures in two stacked exchange bias antiferromagnet / ferromagnet bilayers? If so, how?**
  In order to pattern effectively two stacked layers via tam-SPL, one should be able to address each layer independently. In order to do so, either the blocking temperature of the systems, or the effective temperature reached by each layer during patterning should be different. Once this condition is reached, different magnetic domain structures could be stabilized within the films independently.

Chapter 11 Nanofabrication of three-dimensional magnetic structures

- **Search bibliography dedicated to the advanced characterisation of 3D magnetic nanostructures using electron and x-ray microscopy techniques. Identify the main challenges related to a 3D geometry when using these methods.**
  For the characterization of 3D magnetic structures, single 2D projections are not always sufficient to determine the 3D configuration. 3D imaging techniques such as tomography and laminography provide the opportunity to map the 3D magnetization (x-rays), or magnetic induction (electrons) vector fields. These advanced techniques require sophisticated setups to enable to measurement of the appropriate data. In addition, advanced data analysis using sophisticated reconstruction algorithms is required to recover the 3D vector field, which will require significant computing power.
- **Regarding the fabrication of 3D magnetic systems using 3D scaffolds and deposition of magnetic thin films, can you think of why a non-homogeneous coating might be useful, and why it might not?**
  There are interesting uses for both homogeneous and non-homogeneous coatings, reflecting the wide variety of interesting 3D magnetic systems. Homogeneous coatings are required for the realization of structures such as magnetic nanotubes, which are predicted to exhibit ultra high domain wall velocities and magnetochiral effects, whilst inhomogeneous coatings can be used to fabricate curved surfaces and frustrated magnetic systems. The key is

the control over the fabrication process, and the ability to fabricate the designed structure.

- **For this method, can you think of how one might improve the fabrication method and what techniques could be suitable?**
  This method can be improved by using techniques allowing nanostructuring (PVD techniques, pulse plating electrodeposition, …). In this way, non-homogeneous coating can be produced in a more controlled way, expanding the possible 3D magnetic structures to explore.
- **What are the two main characteristics of the electrodeposition technique that make it so versatile for the synthesis of 3D structures?**
  As with most chemical techniques, electrodeposition is a conformal growth technique. In addition, electrodeposition can only take place on metals or metallized surfaces. Therefore, using a 3D non-conductive template with an opening to a conductive contact, electrodeposition is going to proceed from the contact, fully fill in the template without shadows.
- **How can you combine two-photon lithography or FEBID with electrodeposition to prepare 3D structures?**
  We could use either of the two 3D nanofabrication techniques to create a template for a subsequent electrodeposition process. For this, you can think about two possible strategies. In the first one, 3D insulating nanostructures with the inverse design as the final structure would be created. These would have openings to a conductive substrate, enabling the electrodeposited material to fill these holes. In the second strategy, you could create FEBID metallic nanostructures which would be afterwards covered by a metallic shell via electrodeposition.
- **What kind of three-dimensional magnetic nanostructure would you like to make, and how would you make it?**
  Depending on the 3D magnetic structure you want to make, there are different approaches shown in this chapter. If your desired structure is not very demanding in terms of the quality of the magnetic materials, or if you want a pure Co or Fe structure, you can directly write the structure using FEBID. However, if you need a better control of the magnetic material, you should combine different techniques. First, you need to create a non-magnetic scaffold using FIBID of two-photon lithography and, afterwards, coat it or fill in it with a magnetic material using the electrodeposition or one of the techniques mentioned throughout the chapter.

www.ingramcontent.com/pod-product-compliance
Ingram Content Group UK Ltd.
Pitfield, Milton Keynes, MK11 3LW, UK
UKHW051908100926
8556IPUK00005B/808

9 780750 326063